III	IV	V	VI	VII	0
					2·2 **He** 4.00
2·3 5 **B** 10.81	2·4 6 **C** 12.01	2·5 7 **N** 14.01	2·6 8 **O** 16.00	2·7 9 **F** 19.00	2·8 10 **Ne** 20.18
2·8·3 13 **Al** 26.98	2·8·4 14 **Si** 28.09	2·8·5 15 **P** 30.97	2·8·6 16 **S** 32.06	2·8·7 17 **Cl** 35.45	2·8·8 18 **Ar** 39.95

2·8·15·2 27 **Co** 58.93	2·8·16·2 28 **Ni** 58.71	2·8·18·1 29 **Cu** 63.55	2·8·18·2 30 **Zn** 65.37	2·8·18·3 31 **Ga** 69.72	2·8·18·4 32 **Ge** 72.59	2·8·18·5 33 **As** 74.92	2·8·18·6 34 **Se** 78.96	2·8·18·7 35 **Br** 79.90	2·8·18·8 36 **Kr** 83.80
2·8·18·16·1 45 **Rh** 102.91	2·8·18·18 46 **Pd** 106.4	2·8·18·18·1 47 **Ag** 107.87	2·8·18·18·2 48 **Cd** 112.40	2·8·18·18·3 49 **In** 114.82	2·8·18·18·4 50 **Sn** 118.69	2·8·18·18·5 51 **Sb** 121.75	2·8·18·18·6 52 **Te** 127.60	2·8·18·18·7 53 **I** 126.90	2·8·18·18·8 54 **Xe** 131.30
2·8·18·32·15·2 77 **Ir** 192.2	2·8·18·32·17·1 78 **Pt** 195.09	2·8·18·32·18·1 79 **Au** 196.97	2·8·18·32·18·2 80 **Hg** 200.59	2·8·18·32·18·3 81 **Tl** 204.37	2·8·18·32·18·4 82 **Pb** 207.19	2·8·18·32·18·5 83 **Bi** 208.98	2·8·18·32·18·6 84 **Po** (210)	2·8·18·32·18·7 85 **At** (210)	2·8·18·32·18·8 86 **Rn** (222)

Atomic weights are based on carbon-12;
values in parentheses are for the most stable or the most familiar isotope.
† Symbol is unofficial

2·8·18·24·8·2 62 **Sm** 150.35	2·8·18·25·8·2 63 **Eu** 151.96	2·8·18·25·9·2 64 **Gd** 157.25	2·8·18·27·8·2 65 **Tb** 158.92	2·8·18·28·8·2 66 **Dy** 162.50	2·8·18·29·8·2 67 **Ho** 164.93	2·8·18·30·8·2 68 **Er** 167.26	2·8·18·31·8·2 69 **Tm** 168.93	2·8·18·32·8·2 70 **Yb** 173.04	2·8·18·32·9·2 71 **Lu** 174.97

2·8·18·32·23·9·2 94 **Pu** (244)	2·8·18·32·24·9·2 95 **Am** (243)	2·8·18·32·25·9·2 96 **Cm** (247)	2·8·18·32·26·9·2 97 **Bk** (247)	2·8·18·32·27·9·2 98 **Cf** (251)	2·8·18·32·28·9·2 99 **Es** (254)	2·8·18·32·29·9·2 100 **Fm** (257)	2·8·18·32·30·9·2 101 **Md** (258)	2·8·18·32·31·9·2 102 **No** (259)	2·8·18·32·32·9·2 103 **Lr** (260)

CHEMISTRY AND
THE LIVING ORGANISM

CHEMISTRY AND THE LIVING ORGANISM

SECOND EDITION

Molly M. Bloomfield

John Wiley & Sons

New York • Chichester • Brisbane • Toronto

This book was set in Helvetica Light by
Ruttle, Shaw & Wetherill, Inc.
It was printed and bound by Halliday Lithograph.
The designer was Angie Lee. The drawings were
designed and executed by John Balbalis with the
assistance of the Wiley Illustration Department.
The copyeditor was Deborah Herbert. Picture
research was done by Kathy Bendo. Lilly Kaufman
supervised production.
Cover photo by Alfred T. Lammé, FBPA

Library of Congress Cataloging in Publication Data

Bloomfield, Molly M 1944-
Chemistry and the living organism.

Includes bibliographies and index.
1. Chemistry. 2. Biological chemistry.
I. Title.
QD33.B672 1980 540 79-20753
ISBN 0-471-04754-6

Printed in the United States of America

10 9 8 7 6 5 4 3 2 1

To Stefan

Preface to the Second Edition

This second edition of *Chemistry and the Living Organism* provides a highly motivating and student-oriented approach to the study of chemistry. It is intended for survey courses for students seeking a basic understanding of chemical principles, and especially for students interested in the biological applications of those principles. This book, therefore, is particularly appropriate for students in the allied health sciences and related disciplines, such as physical education and home economics. All too often such students dread chemistry, imagining it as a subject taught in a highly technical language, with little relevance to their personal needs. This textbook addresses such concerns by including the student as an integral part of the subject matter. Because the principles of chemistry are presented in the context of their clinical and biological applications, the relevance of these concepts to the student's personal and professional life is constantly emphasized. Throughout, my purpose is to provide a complete and accurate introduction to the basic principles of chemistry in a style that is easy to understand and enjoyable to read.

Revising this textbook, however, has been a difficult exercise in self-restraint. New, fascinating, and important developments have continually been reported since the first edition appeared. Moreover, I have received suggestions to include additional technical topics which were deliberately omitted from the first edition. But the easy solution—that of including all of these worthwhile topics in the book—would have inevitably led to a text that contained much more material than could be adequately covered in many courses at this level. Thus, this revision is the result of a selective process of topic addition and modification. In choosing the concepts to be added I have tried to remain faithful to the spirit of the first edition, selecting only those recent developments and additional topics best illustrating the basic principles of chemistry as reflected in the life processes of the human body. In addition, the material remaining from the first edition has also been carefully revised to correct any weaknesses or deficiencies pointed out by colleagues, students, and reviewers, and to further build on those instructional techniques that have been well received.

Special Features of This Edition

Student-Oriented Approach

This book is dedicated to making chemistry as interesting as possible to the student. The principal strength of the text is its ability to capture the student's attention by emphasizing the direct application of fundamental chemical principles to the student's personal life. In this edition I have further stressed the usefulness of this subject through additional explanations of the basic chemistry underlying the tasks and standard procedures students will likely encounter in their fields of endeavor.

Conversational Writing Style

A major strength of the first edition was the comfortable, conversational style of its discussion. For the second edition I have worked to improve further the readability of the text, and have also made a conscientious effort to eliminate any sexual bias that might have been present in the first edition.

Scientific Terminology and Glossary

I have increased the scientific terminology in this edition to include more of the vocabulary commonly encountered in the classroom and medical laboratory. As in the first edition, careful attention has been given to defining all scientific terms the first time that they appear in the text, and the Index remains extremely detailed to permit easy reference to all scientific names and terms. In addition, a large Glossary has been included at the end of the textbook to further aid the learning of these terms.

Mathematical Concepts and Manipulations

The students using this textbook vary considerably in their mathematical skills. To provide as much help as possible I have included many more solutions to example problems within the chapters, and have considerably expanded the end-of-chapter problems to allow for more drill exercises. As a further learning aid, an Appendix has been provided that gives answers to selected end-of-chapter problems. And finally, a detailed review of basic mathematical concepts has been added to the *Student Study Guide* for those students who need such fundamental review. Some of the mathematical concepts added to this edition of the textbook are Graham's law, the universal gas constant, equivalents, and osmolarity. SI units are introduced and defined, but metric units (as well as other units commonly used in various medical fields) are also presented.

Case Histories

The chapter-opening case histories in the first edition proved extremely effective in capturing the student's interest and motivating the study of the fundamental concepts being presented. The second edition contains several new case histories. Moreover, some of the original cases have been modified to reflect changing medical developments and practices.

Learning Objectives

As in the first edition, each chapter is preceded by a list of learning objectives. In many instances I have rewritten these objectives to better help the student identify the important topics covered in the chapter, and to serve as a study guide for later review.

Chapter Summaries

A new feature of this second edition is the inclusion of summaries at the end of each chapter. These summaries highlight and review for the student the major concepts that have been discussed.

Topic Coverage

This second edition, like the first, is divided into four sections. The first section, "Introduction," uses a discussion of the disease phenylketonuria to motivate the study of chemistry. The second section, "A Chemical Background," then introduces the basic vocabulary of chemistry. To improve topic delineation in lectures and to facilitate review and testing of the material, I have divided several chapters of the first edition into shorter chapters, each covering fewer topics. Thus, the material formerly found in Chapter 2 ("Matter and Energy") now appears in three chapters entitled "Matter," "Energy," and "The Three States of Matter." Similarly, the topics formerly discussed in Chapter 6 ("Combinations of Atoms") are now found in the two chapters "Combinations of Atoms" and "Chemical Equations and the Mole." And the material formerly found in Chapter 8 ("Water and Solution Chemistry") is now found in two chapters "Water and Solutions" and "Acids and Bases." The third section of the book, "The Elements Necessary for Life," examines the functions of the elements critical to living systems. This section also has been restructured, with the material formerly found in Chapter 9 ("Carbon and Hydrogen") now appearing in the two chapters "Introduction to Organic Chemistry" and "Carbon and Hydrogen." The fourth section, "The Compounds of Life," discusses the large molecules important to living organisms and the interactions among these molecules. A major change in this section is the revision and expansion of the material on vitamins (Chapter 21).

Although this second edition contains no specific chapter headings covering body fluids or concepts of nutrition, these important topics have been integrated into the discussion of associated chemical concepts. Thus, they are fully treated in a manner that emphasizes their fundamental chemical nature.

I have planned this book to accommodate many different approaches to teaching the course. In particular, I have tried to make each section within a chapter as self-contained as possible to allow maximum flexibility in selecting those sections that are most pertinent to the students' needs. Although most courses taught at this level are a semester or more in length, this textbook may also be used for courses lasting only one quarter. The following is a suggested list of sections that may be most easily omitted when time is limited:

Chapter 3, Sections 3.7 to 3.8.
Chapter 4, Sections 4.5 to 4.11.
Chapter 5, Sections 5.11 to 5.14.
Chapter 6, Sections 6.6 to 6.10.
Chapter 10, Section 10.9.
Chapter 11, Sections 11.16 to 11.19, and 11.21.
Chapter 12, Section 12.10.
Chapter 14, Sections 14.5 and 14.17.
Chapter 16, Sections 16.10, 16.11, and 16.13.
Chapter 17, Sections 17.7 to 17.14.
Chapter 18, Sections 18.3, 18.6, 18.8, and 18.17, and 18.29 to 18.31.
Chapter 19, Sections 19.5 to 19.7, 19.12 to 19.14, and 19.20.
Chapter 20, Section 20.3.
Chapter 21, Sections 21.15 to 21.17.
Chapter 22, Section 22.10.

Supplemental Materials

As with the first edition, this textbook is part of a complete and carefully integrated learning package that includes the following materials.

A **student's study guide** is available for this second edition. The study guide contains a brief summary and a list of important terms for each section in a chapter, as well as many worked out examples of mathematical problems and an extensive set of self-test questions and answers for each chapter. An important new feature of the study guide is a basic review of mathematical concepts and manipulations.

The **laboratory manual** written by Joseph Bauer of William Rainey Harper College has been completely revised for this second edition. Four new experiments have been included, and all experiments have been very carefully designed to avoid the use of hazardous reagents. Another feature of this laboratory manual is the addition of a new section that describes fundamental laboratory operations.

The **teacher's manual** for the second edition contains the answers to all exercises and problems in the textbook, answers to the laboratory exercises, and a list of chemicals and equipment (as well as some helpful hints) for the laboratory experiments. A set of **overhead transparency masters** selected from figures in the textbook, and including other more advanced illustrations, will be supplied to adopters on request.

Acknowledgments

The revision of this textbook profited considerably from the help and advice of many people—students, colleagues, reviewers, and friends in the medical profession. Among the many individuals who helped in this task, I especially thank

James Stewart, Cypress College, California
Miriam Smith, Pasadena City College, California
Mahesh Sharma, Columbus College, Georgia
Carol Swezey, Purdue University, Indiana
Joseph Bauer, William Rainey Harper College, Illinois
C. R. Winkel, Ricks College, Idaho
Leslie Loew, SUNY Binghamton, New York
Robert M. Hawthorne, Jr., Purdue University-North Central Campus, Indiana
Henry Benz, Normandale Community College, Minnesota
David Shaw, Madison Area Technical College, Wisconsin
W. J. Wasserman, Seattle Central Community College, Washington
Margaret Goodrich, Seattle Central Community College

for their detailed comments and suggestions during various stages of this revision. I am particularly grateful to Lawrence Stephens, Elmira College, William L. Leoschke, Valparaiso University, Thomas V. Rowland, University of Puget Sound, and Kenneth J. Wright, North Idaho College, who read the entire manuscript and offered their valuable suggestions. Dave Macaulay of William Rainey Harper College must also be thanked for his meticulous verification of the answers to exercises and problems in the textbook and student's study guide.

Many instructors have commented on the important role played by the chapter-opening case histories in sparking and maintaining the student's interest throughout the course. To this end I gratefully acknowledge the substantial help I have received from the doctors and staff of Good Samaritan Hospital in Corvallis, Oregon, who carefully checked the accuracy of each case history and provided many valuable insights into the relationship between chemistry and medical practice. I especially thank Dr. James Riley, Dr. William Lloyd, and Bob Vanderford for the time and effort they so generously provided.

Much of the final revised manuscript was prepared during a year at the University of Louvain in Louvain-la-Neuve, Belgium. During this period I profited greatly from the suggestions and helpful assistance of Yves Eeckhout of the International Institute of Cellular and Molecular Pathology. Thanks also to Mady Leroy for her fine job of typing in this very foreign language. As always, I am indebted to Gary Carlson and the staff at Wiley for their continuing encouragement and editorial support. But most of all I must thank my husband Stefan, whose editing and meticulous regard for details made my ideas a reality, and without whose support and understanding this second edition could not have been completed.

Louvain-la-Neuve, Belgium — Molly M. Bloomfield

CHEMISTRY AND
THE LIVING ORGANISM

section I
introduction

chapter 1

PKU—A Case for Understanding Chemistry

1.1 Why Study Chemistry?

Chemistry is the study of the composition and interaction of substances.

Now, this may sound like a pretty general definition for such a specialized field of study. However, the broad scope of this definition is one way of indicating just how thoroughly chemistry is involved in each of our lives. For example, you drink water from your tap at home without a second's hesitation, for someone has added chemicals to the water to insure its safety. You seldom need to use an iron, thanks to the development of chemicals that give your clothes a permanent press. Just picture your daily routine: You wake up in the morning under sheets made of synthetic fibers that were chemically produced in a factory, or sheets made of cotton fibers—which were created through chemical reactions in the blossom of the cotton plant. You put on clothes made largely of synthetic materials, brush your teeth with a toothpaste containing fluoride, and eat a breakfast fortified with minerals and synthetic vitamins. You may drive to school in a car powered by the energy released through chemical reactions in the engine, or perhaps you pedal a bicycle—powered by the energy released through chemical reactions in your muscles. And now you're reading this textbook, whose paper was created through a chemical process and whose ink is a blend of chemicals. Literally every part of your life is closely related to the field of chemistry, whether it be the synthetic chemistry of the test tube and the modern laboratory, or the natural chemistry that makes up all of nature.

Chemistry also affects each of our lives in very personal ways. For example, your physical appearance is governed by chemicals. Chemical substances called hormones help determine your height, your weight, your build, and your sexual characteristics. Your good health depends upon chemicals that preserve the food you eat, chemicals that protect you from disease, and chemicals (in the form of food) that supply your body with the nutrients it needs to function properly. Chemicals influence your behavior and your emotional feelings. Much of your memory may be chemical; your thoughts and experiences may be stored in your brain in the form of chemical compounds. What we are trying to say is that your entire life is chemical, so a basic knowledge of chemistry will allow you to be more aware of your total self and the way in which you interact with your environment.

1.2 A Case for Understanding Chemistry

To illustrate how a knowledge of chemistry might be helpful in understanding the events surrounding us, let's consider a story about a family—it could have been your family or the family of a friend. Don't concern yourself too much with the exact chemistry in this story now. We will return to this case later in the book when we have developed the basic vocabulary and background necessary to understand the chemical processes we will now be describing.

Billy was brought home from the hospital as a happy, healthy baby. People who came to visit the family commented on how fair his skin and hair looked compared with the rest of the family. As Billy entered his fourth month, his mother started noticing that he no longer watched his mobile as it turned above the crib, and he rarely returned her smiles. Billy was slow in learning to sit up by himself. But then again, his brother had also been late in doing such things and he had turned out to be a very active young child. As Billy grew older, however, his parents became increasingly concerned about his development and behavior. He had become irritable, and would have temper tantrums for no reason at all. Although his parents worked very hard to teach Billy to talk, he was able to learn only a few words. Furthermore, his mother noticed a strange musty odor about him when she changed his diapers, and his skin was often inflamed and flakey. As Billy neared the age of three, his parents began to admit that he was retarded. He still didn't walk or talk, and he was becoming uncontrollable.

Finally, a friend convinced them that the best thing for Billy and for themselves would be to take him to a clinic for diagnosis. At the clinic Billy was given a set of tests, which indicated that he had a disorder called phenylketonuria—PKU for short. Billy's parents were quite upset, and asked if there was any hope for Billy. The doctor replied that nothing could be done to reverse the retardation. When untreated, PKU causes irreversible brain damage, and Billy's IQ was found to be 40. (The average person's IQ is 100, and individuals with an IQ below 70 are considered retarded.) However, the doctor told them that Billy could be placed on a special diet which would improve his behavior and his skin condition. She further explained that PKU was a recessive inherited disease, which meant that each parent must be a carrier of the defective gene for that disease. Therefore, there was a 1 out of 4 chance that any other child that they might have would inherit the disease (Figure 1.1). Nevertheless, the doctor did not discourage Billy's parents from having more children. She explained that the disorder was now understood and could be treated, and that if PKU were diagnosed soon after birth, children having the disease could lead normal lives. Recent laws in most states have required the testing of newborn infants for PKU, so most new cases of PKU were being diagnosed soon after birth.

Billy was taken home, and his parents placed him on a special diet. Within a few weeks, his parents were pleased to find that he no longer had the musty odor, that his skin and hair color darkened, that his behavior improved, and that he even began to smile. To her relief, Billy's mother found that she was now able to take care of him.

Billy's sister Susan was born two years later. Before she was brought home from the hospital, a sample of blood was taken from her heel and placed on a piece of filter paper for testing. Three weeks later Susan was brought back to the clinic to have the test repeated. The doctor then told Susan's parents that the tests had revealed high levels of a substance called phenylalanine in Susan's blood. This meant that Susan, like Billy, had PKU. However, unlike Billy, the outlook for Susan was very hopeful. In order to reduce the abnormally high level of phenylalanine in her blood, Susan was immediately put on a low phenylalanine diet.

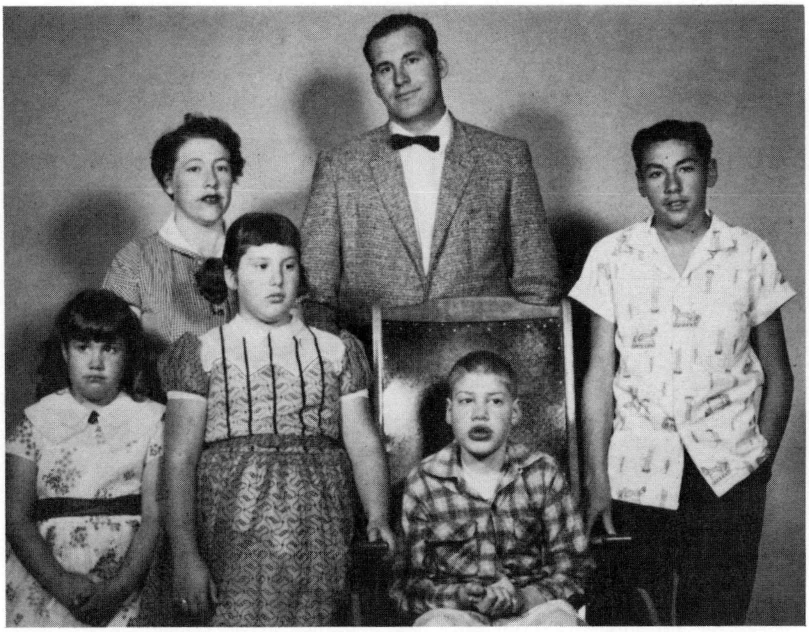

Figure 1.1 This family illustrates the characteristic genetic distribution of phenylketonuria. The parents are both carriers of the trait. By the laws of probability, each child has a 25% chance of being normal, a 50% chance of being a carrier of the trait, and a 25% chance of having PKU. The son at the right is normal, the two daughters are carriers, and the son in the wheelchair has PKU. (Courtesy Willard R. Centerwall, M.D., from *Phenylketonuria,* Frank L. Lyman, ed., Charles C. Thomas, 1963.)

To her parents' delight, Susan is growing into a healthy, active child with an above average IQ. As she grows into maturity and her brain completes its development, Susan's diet can become more varied. The contrast between Billy and Susan is remarkable, and it resulted entirely from high levels of one chemical compound in the blood (Figure 1.2).

The physical symptoms of PKU result from a chemical imbalance within the human body. You are probably aware of instances in which the chemical balance that exists within your own body has been disrupted. Hangovers or muscle cramps are the unpleasant results of some common minor disruptions in the body's chemistry.

The cause of the chemical disruption occurring in PKU is an error in the body's process for converting phenylalanine into another substance called tyrosine. Both of these chemicals, in proper amounts, are required for the normal functioning of the body—especially for the proper formation of nerve cells in the rapidly growing brain of a young child. It is the abnormally high level of phenylalanine, and the accompanying low level of tyrosine, that cause the various symptoms observed in Billy. If PKU is

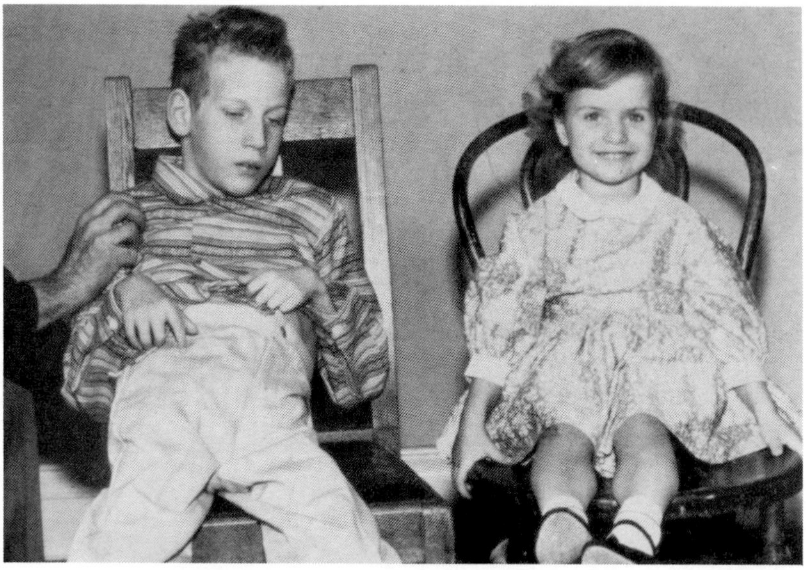

Figure 1.2 Treated and untreated siblings with phenylketonuria. The eleven-year-old boy is severely retarded, while his $2\frac{1}{2}$-year-old sister is normal (Photo by Willard R. Centerwall, M.D. From *Phenylketonuria,* Frank L. Lyman, ed., Charles C. Thomas, 1963.)

diagnosed soon after birth, a child can be placed on a special diet that will allow his brain to develop under normal chemical conditions. Otherwise the brain will be growing in an abnormal chemical environment, leading to severe and irreversible mental damage and retardation. Even then, however, some of the other clinical symptoms such as skin disease, musty odor, abnormally light skin and hair color, seizures, and destructive behavior are reversible, and will improve once the child is placed on the proper diet. This is possible because the body systems responsible for these abnormalities can begin to function normally when they are placed in the correct chemical environment.

There are other examples of similar diseases which will be discussed in later chapters, but the point of this story is to emphasize the extreme importance of the proper chemical balance in the body, from conception throughout life. As researchers become more knowledgable about the chemistry of living organisms, they will be able to control many more of the diseases that result from chemical irregularities.

1.3 This Textbook

If you had a PKU child in your family, you would certainly want to learn as much as you could about the disease, its cause, symptoms, and treatment. To understand the material written about PKU you would first need to learn the vocabulary used in such discussions, and perhaps you would want to read some general books on chemistry, biology, and anatomy to gain a good background in this subject.

Actually, this is an approach with which you should be quite familiar. You might want to learn how to change the spark plug on a car, adjust the gears on a bicycle, or cook a Chinese dinner. In each case you would have to be familiar with the vocabulary used in the instruction manual, and would need to have at least some general knowledge about cars, bicycles, or kitchens. In the same way, before we can completely understand the many ways in which chemistry affects our lives we must first become familiar with the vocabulary used in chemical discussions, and with some of the basic principles and laws that govern the chemical reactions in living organisms. In Section II of this text we will introduce many of the vocabulary terms and basic concepts that you will need to know. In Sections III and IV we will use this new vocabulary to discuss the many chemical substances and processes that are essential to life.

section II

a chemical background

chapter 2

Matter

Learning Objectives

By the time you have finished this chapter you should be able to:

1. Define *mass,* and identify which of two given objects has the greater mass.

2. Define the following terms: element, compound, atom, and molecule.

3. Describe the difference between a compound and a mixture.

4. State the difference between homogeneous and heterogeneous mixtures, and give three examples of each.

5. Describe the difference between a chemical change and a physical change.

6. State the units of length, mass, volume, and temperature in the metric and SI systems.

7. Convert measurements between the metric and English systems.

8. Define *density* and *specific gravity* and calculate these quantities when told the mass of a given volume of a substance.

When Judy Bering finally returned to work, her secretary was delighted. Judy had been out for 10 days following an auto accident, and work had certainly piled up while she was gone. Happily, Judy now felt fine and wasted no time getting started on the many jobs waiting for her. Three days later, however, her secretary noticed a sudden change in Judy's work habits—she was away from her desk much more often than usual, and seemed to be worried about something. Judy finally admitted that ever since she had awakened that morning she had needed to go to the bathroom about every half-hour, and had felt terribly thirsty all day long. This intense thirst and frequent urination continued for two more days before Judy finally called her doctor for an appointment.

Upon learning of Judy's symptoms, the doctor requested a laboratory test of the specific gravity of her urine over a period of time. The initial specific gravity of Judy's urine was measured to be 1.003 (the normal range is 1.005 to 1.030), indicating that her urine was "lighter," or more dilute, than it should be. Over the course of the test Judy was not allowed to drink any liquids, and yet the specific gravity of her urine did not increase as it normally should.

This indicated to the doctor that Judy might be suffering from a disease called diabetes insipidus. Although Judy had suffered these symptoms for only a few days, her diabetes insipidus actually resulted from the concussion she had received in the auto accident. That blow to her head had damaged a small organ called the pituitary gland, located just under the brain. This damaged gland had eventually stopped producing a substance called vassopressin (also often called antidiuretic hormone, or ADH), which causes the kidneys to reabsorb water from the urine. Without this substance in her blood, Judy's kidneys were not functioning properly, and were producing 15 to 20 liters of urine each day (compared with a normal level of 1 to 2 liters per day).

A person suffering from diabetes insipidus who does not constantly replace the water lost through urination will undergo dehydration. Fortunately, this condition is treatable, and Judy was given a nasal spray that replaces the vassopressin not supplied by the posterior pituitary gland.

Laboratory tests such as those given to Judy require many different types of measurements to accurately analyze the various chemicals and fluids in the body. This chapter begins to define the vocabulary that is needed to carefully describe and to measure the substances around us.

Matter

2.1 What Is Matter?

The physical world in which we live is made up of matter. **Matter** is defined as anything that has mass and occupies space. Of course that definition doesn't do you much good unless you know what mass is. You probably have a general feeling about the concept of mass, and would certainly be able to tell which has greater mass—a brick or a feather. The **mass** of an object is a measure of how hard it is to start the object moving, or how hard it is to change its speed or direction once it is moving. For example, a bowling ball is harder to push than a balloon because the bowling ball has a greater mass. The mass of an object is constant no matter where in the universe it is found (Figure 2.1). Imagine a bowling ball and a balloon both floating around weightlessly in the cabin of an orbiting spacecraft. The balloon would bounce harmlessly off the instrument panel, but the bowling ball could badly damage the delicate equipment.

2.2 Composition of Matter

Matter is composed of extremely small particles called **atoms.** The diameter of an atom is about eight-billionths of an inch (0.00000002 cm, or 2×10^{-8} cm—see Appendix 1 if you are confused by this notation). It is very difficult to imagine anything so small. To give you an example, a single page of this textbook is about 500,000 atoms thick.

The last 100 years have seen a tremendous body of knowledge develop about the atom, its structure, and the principles governing its behavior, even though no one has actually seen a single atom. Scientists, however, have been able to build elaborate models of atomic structure that fit experimental data very closely. Recently, very fuzzy pictures of heavy atoms have been made with the use of a specially designed electron microscope (Figure 2.2).

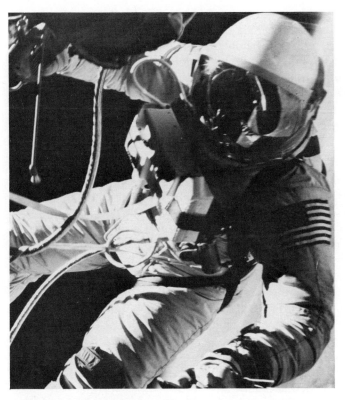

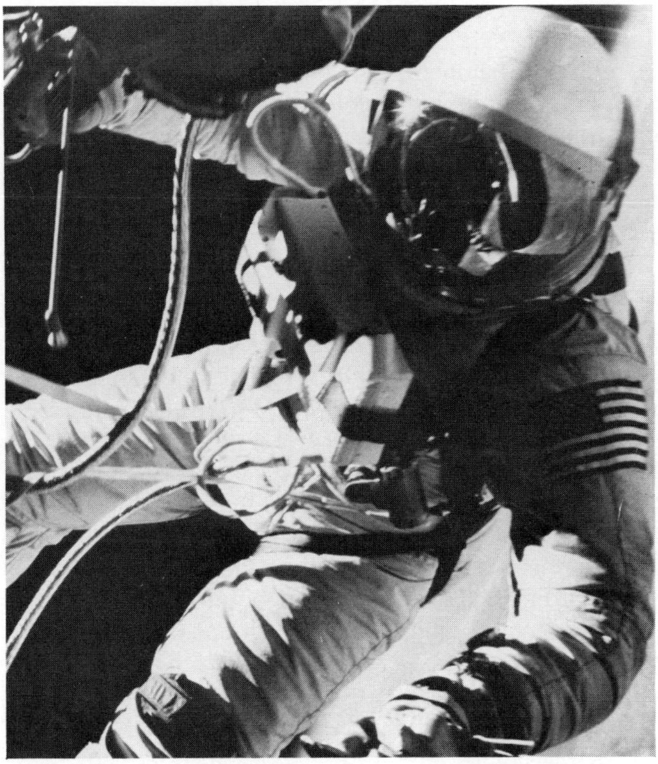

Figure 2.1 Even though this astronaut is weightless in outer space, he has the same mass as on earth. (Courtesy NASA)

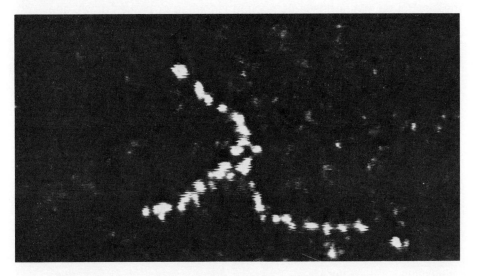

Figure 2.2 Chains of thorium atoms separated by organic molecules were formed on a thin carbon film and photographed with a high resolution scanning electron microscope. The strings of white spots are the chains of thorium atoms. (Courtesy A. V. Crewe. From A. V. Crewe, R. B. Park, J. Biggins, *Science 168,* Figure 4 © 1970 by the American Association for the Advancement of Science)

The Greeks pictured the atom as being indivisible, but the work of many scientists over the last 80 years has shown that the atom is made up of smaller particles. Dozens of subatomic particles have now been identified, but only three are important for our discussions: the **proton,** the **neutron,** and the **electron.** It is the number of these particles and the way in which they are arranged that give each atom its particular chemical properties. We will discuss these important subatomic particles in greater detail in Chapters 5 and 7.

2.3 Classes of Matter: Elements, Compounds, and Mixtures

Any living or nonliving matter can be classified as either an element, a compound, or a mixture (Figure 2.3). An **element** is a pure substance that cannot be broken down into simpler substances by ordinary chemical processes. There are 106 known elements, the heaviest of which are synthetically produced.

An **atom** is the smallest unit of an element having the properties of that element. When two or more atoms are joined together, a **molecule** is formed. Atoms of more than one element can combine to form a **compound,** which has properties that are totally different from those of the elements from which it was formed (Figure 2.4). For instance, two atoms of the element hydrogen, which is a gas that can burn, will react with one atom of the element oxygen, which is a gas we breathe, to form the compound called water. But water, as you know, won't burn, and no living organism can stay alive by breathing water. (You might think that fish breathe water, but they only filter water through their gills to remove the oxygen that is dissolved in the water.)

Compounds can be broken down to simpler substances by chemical means. But no matter how a compound is formed or broken down, one fact will always be true: **A compound is composed of specific elements in a definite proportion by weight.** This is what makes a compound different from a mixture, and is known as the Law of Definite Proportions.

A **mixture** can be made up of two or more substances (elements or

Figure 2.3 Matter can be classified as either a pure substance or a mixture.

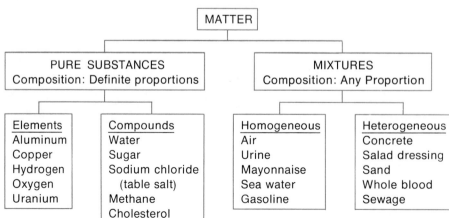

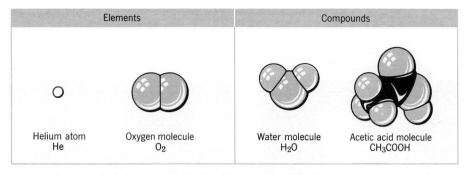

Elements		Compounds	
Helium atom He	Oxygen molecule O_2	Water molecule H_2O	Acetic acid molecule CH_3COOH

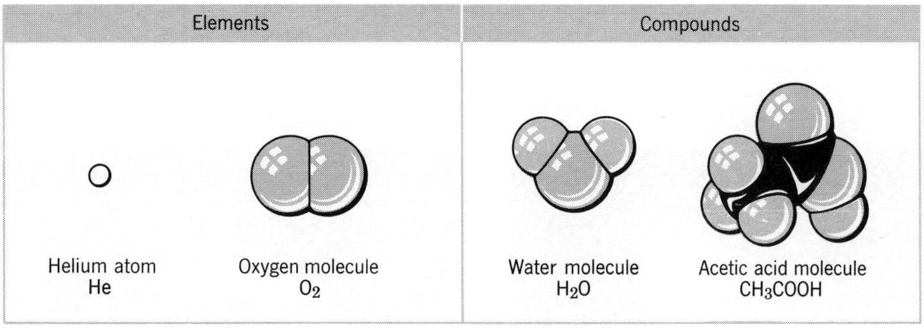

Figure 2.4 Elements such as helium exist as single atoms, but other elements such as oxygen exist in molecular form. Compounds, such as water and acetic acid, exist as molecules containing atoms of more than one element.

compounds) mixed together in any proportion. When two substances combine to form a compound, they form a new substance having different properties. But in a mixture all the substances keep their own individual properties.

Sugared coffee, for example, is a mixture of sugar and coffee; the proportion of sugar to coffee in your cup depends upon how sweet you like your coffee. The air that we breathe, the water that we drink, the ground on which we walk, and the gasoline that we put into our cars, are each mixtures of various substances. Mixtures may be **homogeneous,** meaning they are so uniform that you can't tell one part from another, or **heterogeneous,** meaning that one part may be different from another. For example, a well-mixed cup of coffee with sugar in it is a homogeneous mixture. It will have the sugar molecules evenly distributed throughout the coffee, and each sip will taste the same. On the other hand, coffee with sugar in it that has not been totally dissolved and stirred is an example of a heterogeneous mixture. You would certainly be able to tell the difference between the first few sips of coffee and the last.

2.4 Names and Symbols for the Elements

The table on the inside back cover of this book lists the known elements together with their symbols. The symbol of an element is a shorthand way to indicate one atom of that element. The symbol given to an element is the first letter in the Latin or Latinized name of that element; for example, H for hydrogen, C for carbon, and O for oxygen. If more than one element has a name starting with the same letter, the first two letters of the Latin name can be used for the symbol, with the first letter capitalized and the second in lower case. For example, Co for cobalt and Ca for calcium. Note that because chlorine and chromium share the same first two letters, these two elements have the symbols Cl and Cr, respectively. In a few cases, the Latin name of an element is different from the common English name. For example, the symbol Fe for iron comes from the Latin name ferrum, the symbol Ag for silver comes from the Latin name argentum, and the symbol Pb for lead comes from the Latin name plumbum.

Figure 2.5 Physical changes (a and b) involve a change in form, whereas chemical changes (c and d) involve a change in the basic nature of the substance (a, b, c, Mimi Forsyth/Monkmeyer; d, Hugh Rogers/Monkmeyer).

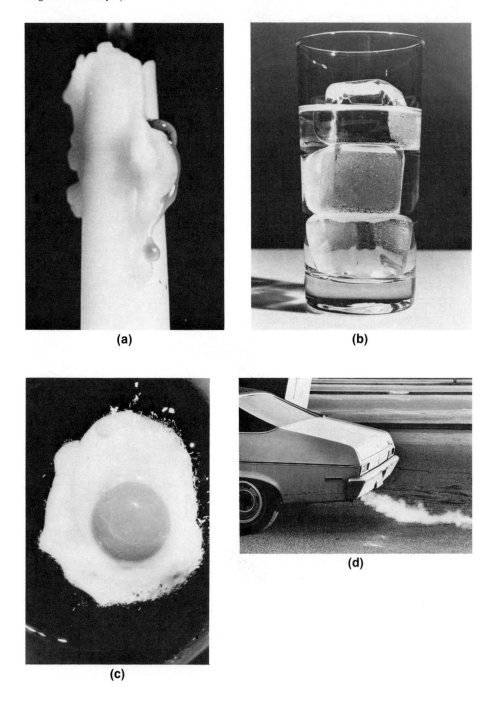

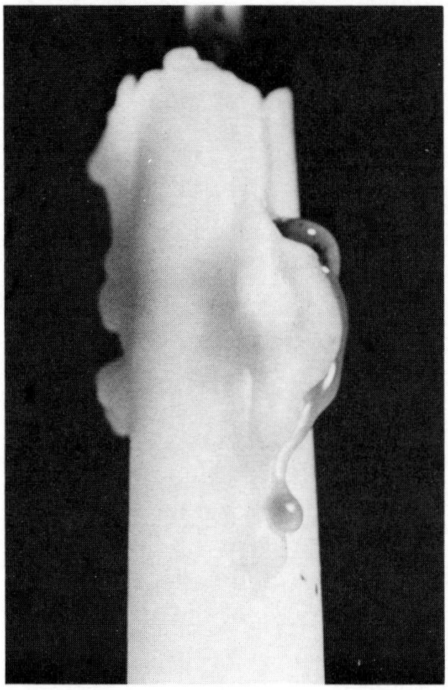

(a)

(b)

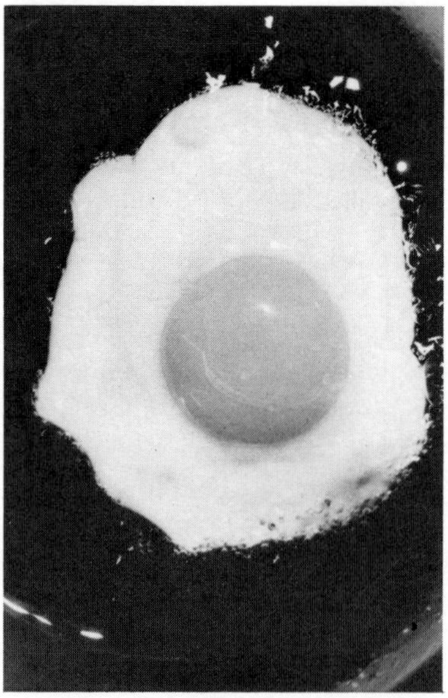

(c)

(d)

2.5 Physical and Chemical Changes

In this book we will be studying two types of changes that matter can undergo: physical change and chemical change. Common examples of substances undergoing physical change include ice melting in a glass of water, wax melting on a candle, sugar dissolving in a cup of coffee, hair drying under a hair dryer, and fingernails being trimmed with a nail clipper. A **physical change** is one in which a substance changes form, but keeps its chemical identity. For example, water still keeps all of its chemical properties whether it is in the form of ice, water, or steam. When ice melts to form water, the water has simply undergone a physical change and can be returned to its previous state by placing it in a freezer. Similarly, wax dripping down a candle undergoes a physical change, turning from a liquid to a solid as it cools. However, the wax keeps the same chemical identity during this physical change (Figure 2.5).

Chemical changes are occurring when you fry an egg, allow your bicycle to rust in the rain, accelerate your car from a stoplight, digest your dinner, or run a 100-meter dash. In a **chemical change,** the starting materials (reactants) are "used up," and different substances (products) are formed in their place. Obviously a fried egg has a different appearance and taste from a raw one, and when it cools on your plate it does not return to the raw state. Similarly, if you were to collect the products that come out of the exhaust pipe of a car, they would have none of the properties of the gasoline that went into the engine. The basic difference to remember, therefore, is that **a physical change involves only a change of form, whereas a chemical change involves a basic change in the chemical composition of the substances involved.**

Measuring Matter

2.6 The Metric System

Scientists throughout the world have long used the metric system for the measurement of matter. In 1960 a revised set of units based on the metric system was proposed. This set of units is known as the International System of Units (or SI units), and their use is slowly being adopted around the world. Table 2.1 lists the basic units of the SI system. Some SI units will be used in this book, but we will also use the more common metric units in many of our discussions.

Using the familiar English system of measurements, we all have learned that 4 quarts make a gallon, 12 inches make a foot, and 16 ounces make a pound. This is a complicated system of measurements to use, because there is no consistent relationship between the number of smaller units needed to make up a larger unit of measure. By contrast, the great advantage of the metric system is that all units of measure are related to their subunits by multiples of 10, and standard prefixes are used to indicate the number of multiplications or divisions by 10 that are required. The idea of such a numerical prefix is not new to you; we all know that the prefix tri-,

Table 2.1 Base Units in the SI System

Quantity	Name	Symbol
Length	meter	m
Mass	kilogram	kg
Time	second	s
Temperature	kelvin	K
Amount of substance	mole	mol
Electric current	ampere	A
Luminous intensity	candela	cd

such as found in the words tricycle, tripod, or trio, indicates that the object being named has three of something. In the same way, the prefix kilo- used in the metric system means 1000. One kilometer, then, is 1000 meters, and one kilogram is 1000 grams. Table 2.2 shows some of the prefixes used in the metric system. Table 2.3 shows the relationship between the English system of measurements and the metric system. The back overleaf of the text contains a list of other common conversion factors.

The use of decimals in the metric system makes calculations much easier to perform than in the English system. In sections 2.7 to 2.10, examples will be given to show conversions between units in the metric system and the English system. If you have trouble following these calculations, turn to Appendix 2 for a more detailed explanation.

2.7 Length

The unit measure of length in the SI system is the **meter.** (*Metre* has been chosen as the preferred international spelling for this unit, but *meter* is the spelling that has been used in English-speaking countries and is the spelling we will use in this book.) This unit was defined in 1790 as one

Table 2.2 Some Prefixes Used in the Metric System

Prefix	Symbol	Multiple	Example
kilo-	k	$1000 = 10^3$	kilometer, km = 1000 meters
hecto-	h	$100 = 10^2$	hectometer, hm = 100 meters
deka-	da	$10 = 10^1$	dekameter, dam = 10 meters
			meter, m (basic unit)
deci-	d	$0.1 = 10^{-1}$	decimeter, dm = 0.1 meter
centi-	c	$0.01 = 10^{-2}$	centimeter, cm = 0.01 meter
milli-	m	$0.001 = 10^{-3}$	millimeter, mm = 0.001 meter
micro-	μ	$0.000001 = 10^{-6}$	micrometer, μm = 0.000001 meter

Table 2.3 Some Relationships between Units in the Metric System and the English System of Measure

	Metric Units	Conversion Factor	English Units
Length	Meter	1 m = 3.3 ft	Feet
Mass	Kilogram	1 kg = 2.2 lb	Pound
Volume	Liter	1 *l* = 1.06 qt	Quart

ten-millionth of the distance from the north pole to the equator. Although this standard has been redefined twice since that time, its length has remained pretty much unchanged. One meter is equal to 39.37 inches and is, therefore, just slightly longer than a yard (Figure 2.6).

A kilometer is the unit that is used to measure large distances, such as the distance between cities (1 kilometer = 0.621 mile). There are 1000 meters in a kilometer. Units commonly used that are smaller than a meter are the millimeter and the centimeter.

$$\begin{aligned} 1 \text{ centimeter (cm)} &= 10 \text{ millimeters (mm)} \\ 1 \text{ meter (m)} &= 100 \text{ cm} \\ 1 \text{ m} &= 1000 \text{ mm} \end{aligned}$$

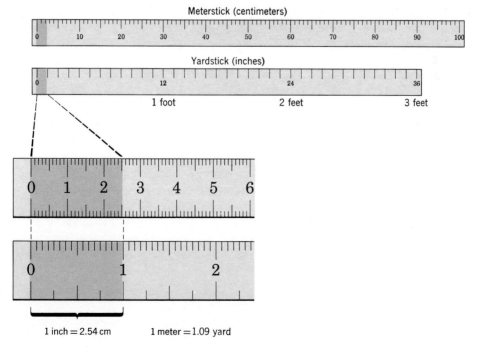

Figure 2.6 A meter is just slightly longer than a yard, and an inch is about 2.5 centimeters.

Example 2-1

1. How many centimeters are there in 2.5 meters?

$$2.5 \text{ meters} = (?) \text{ centimeters}$$

The relationship between meters and centimeters is 1 m = 100 cm. Using this conversion factor,

$$2.5 \cancel{\text{m}} \times \frac{100 \text{ cm}}{1 \cancel{\text{m}}} = 2.5 \times 100 \text{ cm} = 250 \text{ cm}$$

2. A newborn baby measures 18.2 inches in length, but this length must be written in centimeters on its hospital chart. How many centimeters long is the baby?

$$18.2 \text{ inches} = (?) \text{ centimeters}$$

From Figure 2.6, the relationship between inches and centimeters is 1 in = 2.54 cm. Using this conversion factor,

$$18.2 \cancel{\text{in}} \times \frac{2.54 \text{ cm}}{1 \cancel{\text{in}}} = 18.2 \times 2.54 \text{ cm} = 46.2 \text{ cm}$$

2.8 Mass

The unit measure of mass in the metric system is the **gram.** The SI standard that defines this mass is a block of platinum metal weighing exactly one kilogram (1000 grams) that is kept in a vault in France by the International Bureau of Weights and Measures. Units of mass commonly used are the kilogram, gram, and milligram (Figure 2.7*a*).

$$1 \text{ kg} = 1000 \text{ g}$$
$$1 \text{ g} = 1000 \text{ mg}$$
$$1 \text{ kg} = 2.2 \text{ lb}$$

The term **weight** is probably more familiar to you than mass, but there is a difference between these terms. Weight is a measure of the force or attraction of gravity on an object, but the mass of an object does not depend on gravitational attraction. For example, because the moon's gravitational pull is roughly one-sixth that of the earth, an astronaut who weighed 180 pounds on the earth would weigh only 30 pounds on the moon. But the astronaut would have the same mass in either location. It is common and accepted practice, however, to use the terms *mass* and *weight* interchangeably and we will do so in this book.

Example 2-2

1. A doctor prescribes a dose of 0.1 g of medication. How many 25 mg tablets are required to fill the prescription?

This problem must be solved in two steps:

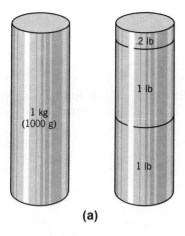

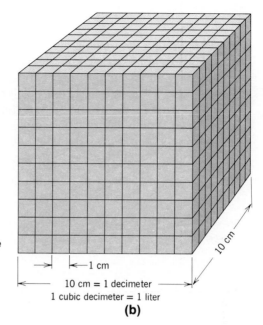

(a)

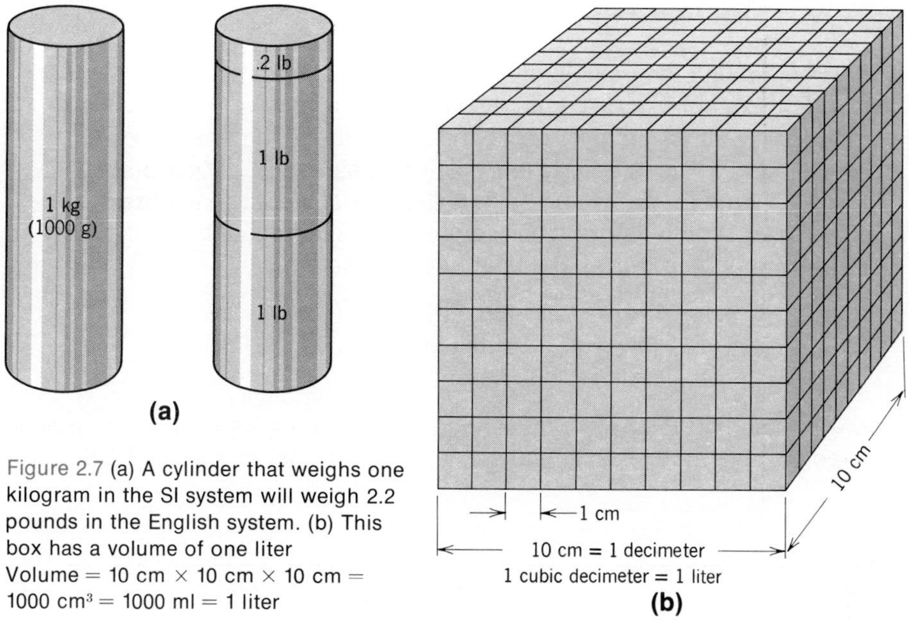

Figure 2.7 (a) A cylinder that weighs one kilogram in the SI system will weigh 2.2 pounds in the English system. (b) This box has a volume of one liter
Volume = 10 cm × 10 cm × 10 cm = 1000 cm³ = 1000 ml = 1 liter

→ ←1 cm
← 10 cm = 1 decimeter →
1 cubic decimeter = 1 liter
(b)

(a) First, 0.1 g = (?) mg

The relationship between grams and milligrams is

$$1 \text{ g} = 1000 \text{ mg}$$

Using this conversion factor,

$$0.1 \text{ g} \times \frac{1000 \text{ mg}}{1 \text{ g}} = 0.1 \times 1000 \text{ mg} = 100 \text{ mg}$$

(b) Then, 100 mg = (?) tablets

We are told that 1 tablet = 25 mg, so

$$100 \text{ mg} \times \frac{1 \text{ tablet}}{25 \text{ mg}} = \frac{100}{25} \text{ tablets} = 4 \text{ tablets}$$

2. Labels on drugs given to infants often list the dosage per kilogram of body weight. What is the weight in kilograms of a 16.5 lb baby?

$$16.5 \text{ lb} = (?) \text{ kg}$$

From Table 2.3, the relationship between pounds and kilograms is

$$1 \text{ kg} = 2.2 \text{ lb}$$

Using this conversion factor,

$$16.5 \text{ lb} \times \frac{1 \text{ kg}}{2.2 \text{ lb}} = \frac{16.5}{2.2} \text{ kg} = 7.5 \text{ kg}$$

2.9 Volume

You will often find the volume of an object stated in terms of some unit of length. For example, to calculate the volume of a box we might multiply the length × width × height. The SI unit of volume is the cubic meter, abbreviated m^3. This is a fairly large unit of measure. For most practical purposes, therefore, we will use the metric unit of volume—the **liter**—which is only slightly larger than a quart. (The international spelling is *litre,* but in this book we will use the English spelling).

$$1\ m^3 = 1000\ \text{liters}\ (l)$$
$$1\ \text{liter} = 1000\ \text{milliliters (ml)}$$
$$1\ \text{ml} = 1\ \text{cubic centimeter (cm}^3\ \text{or cc)}$$

You will often find laboratory measuring devices such as syringes or pipets labeled in either cubic centimeters or milliliters (Figures 2.7*b* and 2.8).

Example 2-3

1. A bottle of antibiotic contains 0.075 liters. How many 5 ml doses does it contain?

   This problem must be solved in two steps:

(a) (b) (c) (d) (e)

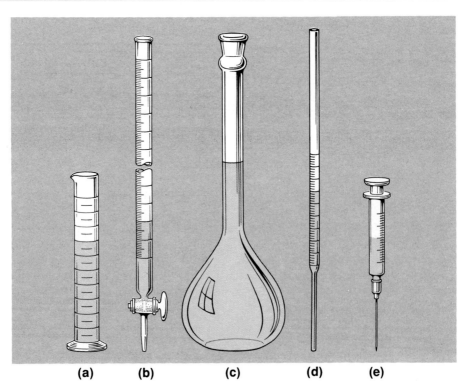

Figure 2.8 This laboratory equipment is commonly used to measure volume: (a) graduated cylinder, (b) buret, (c) volumetric flask, (d) pipet, and (e) syringe.

(a) First, 0.075 liters = (?) milliliters
The relationship between liters and milliliters is

$$1\ l = 1000\ \text{ml}$$

Using this conversion factor,

$$0.075\ \cancel{l} \times \frac{1000\ \text{ml}}{1\ \cancel{l}} = 0.075 \times 1000\ \text{ml} = 75\ \text{ml}$$

(b) Then, 75 ml = (?) doses
We are told that 1 dose = 5 ml, so

$$75\ \cancel{\text{ml}} \times \frac{1\ \text{dose}}{5\ \cancel{\text{ml}}} = \frac{75}{5}\ \text{doses} = 15\ \text{doses}$$

2. A patient excretes 2.65 quarts of urine. How many liters of urine is this?

$$2.65\ \text{qt} = (?)\ \text{liters}$$

The relationship between quarts and liters is

$$1.06\ \text{qt} = 1\ \text{liter}$$

Using this conversion factor,

$$2.65\ \cancel{\text{qt}} \times \frac{1\ l}{1.06\ \cancel{\text{qt}}} = \frac{2.65\ l}{1.06} = 2.50\ l$$

2.10 Temperature

Many instruments have been developed to measure the temperature of a substance. The instrument you are probably most familiar with is the mercury thermometer, a thin graduated glass tube with a mercury containing bulb on the end. As this bulb is heated, the mercury expands and rises in the tube. The particular units marked on the glass tube depend upon the temperature scale used. We are used to seeing temperatures expressed in the **Fahrenheit (°F)** scale, which is part of the English system of measurement. On this scale, the freezing point of water is 32°F and the boiling point of water is 212°F (a difference of 180 degrees). In the metric system, the **Celsius (°C)** temperature scale is used. On this scale the freezing point of water is 0°C and the boiling point is 100°C (a difference of 100 degrees) (Figure 2.9 and Table 2.4). A third temperature scale is often used by scientists, and is the one belonging to the SI system; this is the **Kelvin (K)** temperature scale. (Note that Kelvin temperatures are written as K, not °K.) The freezing point of water on the Kelvin scale is 273.15 K and the boiling point is 373.15 K. As with the Celsius scale, this is a difference of 100 degrees. You may wonder how a number such as 273.15 was chosen for the freezing point of water. This number was picked so that zero degrees on the Kelvin scale indicates the lowest temperature it is theoretically possible to reach. Zero degrees Kelvin is known as **absolute zero.** The following examples show how to make conversions between these temperature scales.

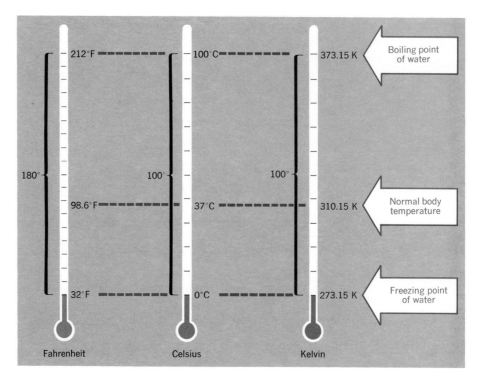

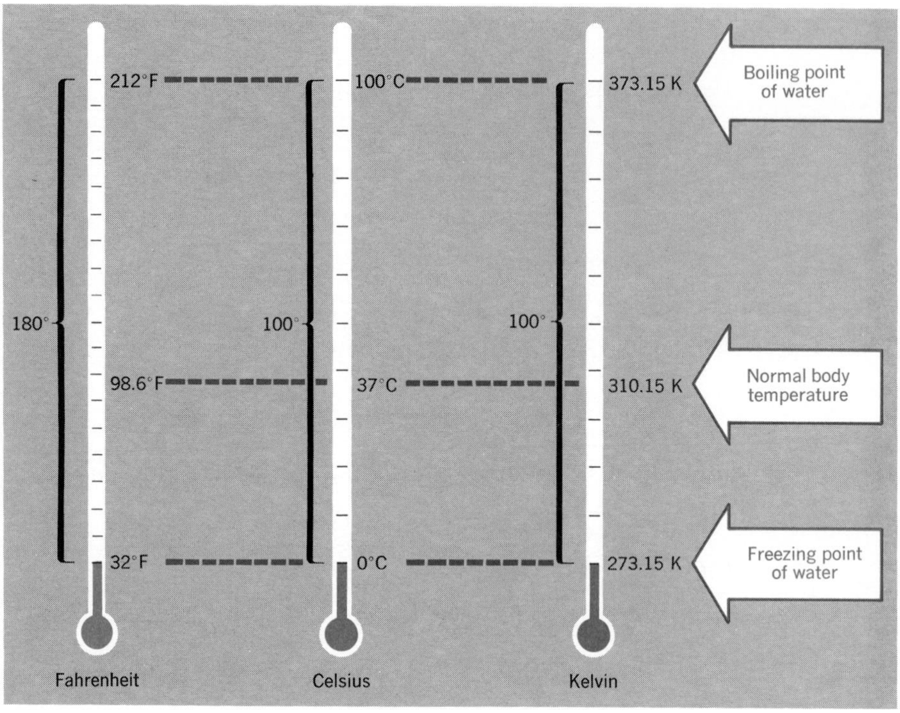

Figure 2.9 Relationships between the Fahrenheit, Celsius, and Kelvin temperature scales.

Table 2.4 A Comparison of Temperatures on the Fahrenheit and Celsius Scales

	Fahrenheit	Celsius
A cold winter day	−5°F	−21°C
Freezing point of water	32°F	0°C
Room temperature	77°F	25°C
Normal body temperature	98.6°F	37°C
A hot summer day	100°F	38°C
Boiling point of water	212°F	100°C

Example 2-4

1. To convert temperatures between the Celsius and Fahrenheit temperature scales, use either of the two following relationships:

$$°C = \frac{5}{9}(°F - 32°) \qquad \text{or} \qquad 1.8(°C) = (°F) - 32°$$

(a) 68°F = (?) °C

Substituting, for example, in the first relationship shown above, we get

$$°C = \frac{5}{9}(68° - 32°)$$

$$°C = \frac{5}{9}(36°)$$

$$°C = 20°$$

(b) 37.0°C = (?)°F

Substituting, for example, in the second relationship shown above, this time we get

$$1.8\,(37.0°) = °F - 32°$$

$$66.6° = °F - 32°$$

$$66.6° + 32° = °F$$

$$98.6° = °F$$

2. To convert temperatures between the Celsius and Kelvin temperature scales, use the following relationship (we will round off our conversion to the nearest whole number):

$$K = (°C) + 273$$

(a) 20°C = (?) K
Using the relationship for this conversion, we get

$$K = 20 + 273$$

$$K = 293$$

(b) 345 K = (?)°C
Again using the above relationship,

$$(°C) + 273 = 345$$

$$°C = 345 - 273$$

$$°C = 72$$

2.11 Density and Specific Gravity

We have seen that the measurement of specific gravity can be a useful clinical tool, such as in the diagnosis of the disease diabetes insipidus. However, before we can perform measurements of specific gravity, we must understand the concept of density.

The **density** of a substance is defined as the mass of that substance per unit of volume (most often expressed as g/cc).

$$\text{Density} = \frac{\text{mass (weight)}}{\text{volume}}$$

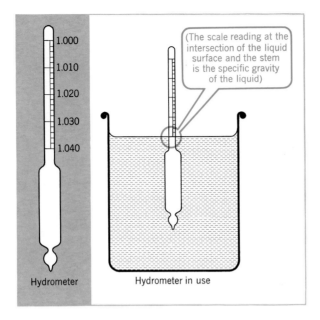

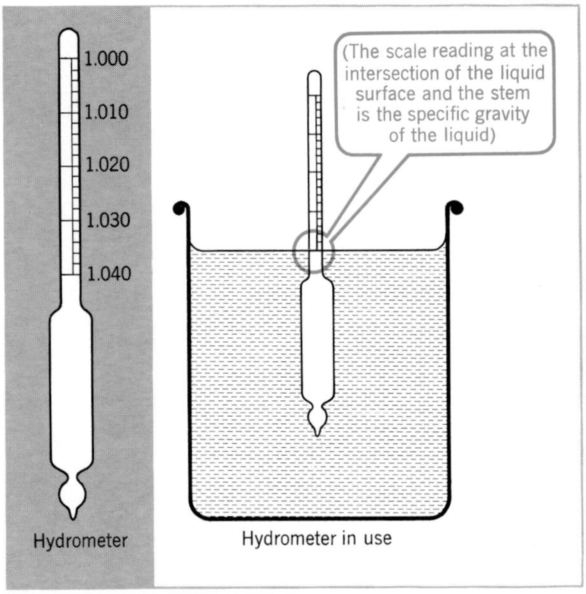

Figure 2.10 This hydrometer is calibrated to measure the specific gravity of liquids that are more dense than water.

The density of lead is 11.3 g/cc, of iron 7.9 g/cc, calcium 1.5 g/cc, water 1.0 g/cc, and alcohol 0.79 g/cc.

Example 2-5

1. Calculate the density of nickel if a 5.0 cc sample weighs 44.5 g.

$$\text{Density of nickel} = \frac{44.5\ \text{g}}{5.0\ \text{cc}} = 8.9\ \text{g/cc}$$

2. What volume would a 42.5 g sample of a liquid occupy if its density is 1.7 g/cc?

$$42.5\ \text{g} = (?)\ \text{cc}$$

The value for the density can be used as a unit factor to solve this problem.

$$42.5\ \cancel{\text{g}} \times \frac{1\ \text{cc}}{1.7\ \cancel{\text{g}}} = 25\ \text{cc, or 25 ml of liquid}$$

Specific gravity is the ratio of the mass of a substance to the mass of an equal volume of water (measured at the same temperature). More simply, specific gravity compares the density of a substance with the density of water at that temperature, and is most commonly used to compare the density of other liquids to that of water.

$$\text{Specific gravity} = \frac{\text{density of sample}}{\text{density of water (1.000 g/cc)}}$$

For example, carbon tetrachloride has a specific gravity of 1.59 and, therefore, is more dense than water. Alcohol, with a specific gravity of 0.79, is less dense than water.

The specific gravity of a liquid can be measured with a **hydrometer,** a weighted bulb-shaped instrument which has a scale calibrated in specific gravities at a given temperature (Figure 2.10). The specific gravity of urine is measured with a **urinometer,** which is a hydrometer calibrated for the specific gravities of urine. The specific gravity of urine increases with the amount of solid waste present in the urine, and normal readings range from 1.005 to 1.030. Low urine specific gravity may indicate kidney damage or diabetes insipidus, whereas high values may be caused by diabetes mellitus or dehydration.

Example 2-6

A 50.0 ml sample of urine weighs 50.5 grams. (a) What is the density of the urine? (b) What is its specific gravity? (c) Is the urine more or less dense than water?

(a) $\text{Density} = \frac{\text{mass}}{\text{volume}}$

$= \frac{50.5 \text{ g}}{50.0 \text{ cc}}$ (1 ml = 1 cc)

$= 1.01$ g/cc

(b) $\text{Specific gravity} = \frac{\text{Density of sample}}{\text{Density of water}}$

$= \frac{1.01 \text{ g/cc}}{1.00 \text{ g/cc}}$

$= 1.01$

(c) The urine is slightly more dense than water.

Chapter Summary

Matter is anything that has mass and occupies space. The mass of an object is a measure of how hard it is to change its speed or direction of movement. Matter can be divided into two groups: pure substances and mixtures. Pure substances are either elements, which can't be broken into simpler substances by ordinary chemical means, or compounds, which are composed of two or more elements in definite proportions by weight. An atom is the smallest unit of an element having the properties of that element. A molecule contains two or more atoms joined together. Mixtures are made up of two or more substances combined in any proportion. Mixtures can be homogeneous, meaning uniform throughout, or heterogeneous, meaning nonuniform. Matter can undergo physical changes, which are changes in form, or chemical changes, which result in changes in the chemical composition of the substances involved.

Scientists measure matter using units in the SI system, a modified metric system. All units in the SI system are related to their subunits by multiples of 10, and prefixes are used to indicate the power of 10 required. The unit of length is the meter and the unit of mass is the kilogram. The unit of volume is the cubic meter in the SI system, and the liter in the metric system. Temperature is measured on the Kelvin temperature scale (SI) or the Celsius temperature scale (metric). The density of a substance is the mass of that substance per unit of volume. The specific gravity of a substance compares the density of the substance with the density of water.

Exercises and Problems

1. Is air matter? Why or why not?
2. Give two examples of (a) an element, (b) a compound.
3. What is the difference between an atom and a molecule?
4. Give two examples of common foods in your kitchen that are

   (a) Homogeneous
   (b) Heterogeneous

5. Label each of the following as either a physical or chemical change.

   (a) A log burning
   (b) Milk souring
   (c) Salt dissolving in water
   (d) Ski goggles fogging
   (e) Snow melting
   (f) A piece of chalk breaking

6. Make the following conversions (review Appendix 2).

   (a) 400 g to kg
   (b) 1500 m to km
   (c) 10 cm to mm
   (d) 1.5 kg to g
   (e) 15 ml to *l*
   (f) 100 mm to m
   (g) 0.02 g to mg
   (h) 2.4 *l* to ml
   (i) 0.3 m to cm
   (j) 67 cm to m
   (k) 125 mg to g
   (l) 12 m to cm
   (m) 5.4 m to ft
   (n) 1050 yd to m
   (o) 76.2 cm to m
   (p) 85 km to mi
   (q) 5.0 pints to *l*
   (r) 15.6 *l* to qt
   (s) 2.0 gal to *l*
   (t) 6.0 oz to g
   (u) 105 g to lb

7. For each of the following, choose the answer that most closely applies.

   (a) a pencil weighs 5 mg or 5 g or 5 kg?
   (b) a man weighs 90 mg or 90 g or 90 kg?
   (c) an aspirin tablet weighs 400 mg or 400 g or 4 kg?
   (d) a coffee cup contains 25 ml or 250 ml or 2.5 *l*?
   (e) a teaspoon contains 5 ml or 50 ml or 0.5 *l*?
   (f) a quart of milk contains 10 ml or 0.1 *l* or 1 *l*?
   (g) the length of a pencil is 15 mm or 15 cm or 15 m?
   (h) the length of a football field is 92 cm or 92 m or 92 km?

8. Make the following conversions:

(a) 71°F to °C
(b) 90.0°C to °F
(c) 151 K to °C
(d) −11°C to °F
(e) −49°F to °C
(f) 302°F to K

9. For each of the following three liquids calculate (1) the density, (2) the specific gravity, (3) the weight of 250 ml of the liquid, (4) the volume in ml of 1.0 g of the liquid.

(a) liquid A:	100 ml weighs 79 g
(b) liquid B:	1 liter weighs 1.1 kg
(c) liquid C:	500 ml weighs 490 g

10. A prescribed injection of a drug is 1.5 ml. If the syringe is graduated in cubic centimeters, to what mark on the syringe would you draw the drug?

11. Wine in the United States is sold in bottles marked 4/5 of a quart. How many liters of wine are there in 4/5 of a quart? How many ml?

12. A sign on the freeway outside San Francisco reads "Los Angeles, 412 miles." How many kilometers would that be?

13. The daily dosage for ampicillin to treat an ear infection is 100 mg per kilogram of body weight. What is the daily dose for a 22 lb baby with an ear infection?

14. One of the most deadly of the poisonous mushrooms is the amanita, or death angel.

(a) If 50 g of amanita mushrooms are enough to kill a 180 lb man, what is the lethal dose of these mushrooms calculated in milligrams per kilogram of body weight?
(b) Using the lethal dose you have just calculated, how many grams of amanita mushrooms would be sufficient to kill a 120 lb woman? a 90 lb teenager?

15. To treat acne, dermatologists wipe the patient's face with a sponge cooled to 77 K by liquid nitrogen. What is the temperature of the sponge (a) in °C (b) in °F?

16. A weather forecaster on the television news states that it is 22°C in Paris and 32°C in Rome. What are the temperatures in Paris and Rome in °F?

17. (a) What is the density of mercury if 15.0 ml weighs 204 g?
(b) What is the weight in kilograms of 250 ml of mercury?

18. Would you expect gasoline to have a specific gravity greater than or less than water? Explain your answer.

19. A patient suffering from symptoms resembling those of diabetes mellitus excretes urine with a specific gravity of 1.002. (a) Does this fact support the diagnosis? Explain why. (b) What would be the weight of a 100.0 ml sample of this urine?

chapter 3

Energy

Learning Objectives

By the time you have finished this chapter, you should be able to:

1. Define *energy*.

2. Describe the difference between kinetic energy and potential energy, and give several examples of each.

3. State which has greater kinetic energy: water or steam.

4. Define a *calorie* and use experimental data to calculate the number of calories in a food sample.

5. Describe five types of electromagnetic energy, and list them according to increasing wavelength.

Jerry and Sue invited their neighbors John and Rosemary to sail with them the morning of July 2 for a cruise to one of the outer islands. The sea was calm, the air was a warm 75°F, and the weather was clear and sunny. They were having a wonderful time until, about 15 miles from their destination, the wind died. Because they were to meet friends for dinner that evening on the island, Jerry decided to start the auxiliary motor. Suddenly an explosion ripped through the boat, and the engine burst into flames. Jerry quickly ordered everyone to don life jackets, and tried unsuccessfully to put out the fire. As the flames spread, Jerry and John tore the seat planking from the boat and threw it overboard for flotation, and everyone jumped into the water. The boat's radio had been destroyed in the explosion, so the four could only quietly cling to the planking and hope that their friends would soon start a search effort.

Their friends did indeed worry when the boat became overdue, and alerted the Coast Guard. A rescue boat was sent out, and the four neighbors were discovered floating near the debris of the burnt boat. Although it had been only four hours since the accident occurred, Rosemary was dead and Jerry died before he reached the hospital. Sue and John were in serious condition, but both survived thanks to the immediate treatment they received.

All four had suffered from hypothermia, a name that describes the lowering of the body's inner temperature. Hypothermia can be fatal when body temperature drops as few as 6 degrees below the normal level of 98.6°F (This is equivalent to a drop of 3 degrees Celsius from the normal body temperature of 37°C). Surprisingly, a person can survive a temperature drop of 40 to 50°F in the hands and feet, but only a small drop in the temperature of the body core can cause death. Because heat flows from a warm region to a colder region, hypothermia can occur under ordinary conditions even when the air temperature is relatively mild. Being wet increases the flow of heat from the body, because water will conduct heat about 240 times better than still air.

Hypothermia begins as soon as the body starts to lose energy faster than it can be produced. The hands and feet are affected first. A drop of only 3°F in body temperature will reduce manual dexterity to the point that one is unable to perform the basic tasks necessary for survival. In addition, as the temperature of the body is lowered, the brain becomes numb. Individuals become confused and irrational. Such an effect was apparent when Rosemary, after two hours in the water, decided that she was going to swim the 15 miles to shore. Jerry had to swim after her and physically drag her back to the rest of the group. This vigorous activity exhausted both of them, critically using up their decreasing supplies of energy.

The speed with which hypothermia develops varies considerably with the energy reserves of the individual, and the nature of the survival situation. However, anything that one can do to prevent heat loss is extremely important. The proper choices of clothing, insulation, or shelter, and the minimization of muscular activity will all help prevent the onset of hypothermia.

The need for an individual to maintain a stable body temperature in the face of possible dangerous environments is only one example of the complex energy interactions between living systems and their environments. An understanding of these interactions requires an examination of the different types of energy that are involved, and the energy changes that can occur.

Energy

3.1 What Is Energy?

Energy has become a widely discussed topic, and the "energy crisis" has affected all of us. A great deal has been said and written about apparent shortages in the world's supply of petroleum, and great controversies have arisen over alternative sources of energy such as coal, solar, and nuclear power. Even though you have felt the effects of our nation's energy supply problems, you may have only a vague notion of precisely what energy is. Some days you may wake up feeling energetic, ready to get things accomplished—and the tasks you have completed by the end of the day will have resulted from your expenditure of energy. It is precisely the ability to accomplish something—the capacity to do work—that defines **energy.** A rushing stream, a rock poised at the top of a hill, the gasoline in a car, the muscles in an arm, each has the capacity to cause change, to do work on an object. Therefore, each is said to possess energy.

Although scientists define energy as the capacity to do work, work can mean different things to different people. Painting a fence, shoveling snow, or chopping logs may be work to some people, but fun and relaxation to others. For this reason, scientists have given the term *work* a mathematical meaning that does not depend on an individual's opinion. **Mechanical work** is defined to be the application of a force through a distance. Force is a push or pull on an object that causes the object to start moving, or to change speed or direction once it is moving. If you push a car and set it into motion, you have done work on the car. However, if you push a car but are unable to get it to move, you have not done any mechanical work.

From this definition, we see that we can't measure the energy contained in an object until it has been transferred from that object to another one. Using the above example, the energy expended by your muscles can be measured only when this energy is transferred from your muscles to the car that you are pushing.

3.2 Kinetic Energy

Energy can take many forms, but it is often convenient to label energy as either energy of motion or energy of position. The name applied to energy of motion is **kinetic energy.** A car traveling at 20 mph, a rock hurtling through the air, or a pot of boiling water all have energy by virtue of motion. That is, each possesses kinetic energy. The moving car could do a great deal of damage if it hit a parked car, the rock could break a window, and the boiling water could cook vegetables. Although it is obvious that both the car and the rock have motion and, therefore, possess kinetic energy, you may wonder about the boiling water. The rock and the car are large objects whose motion we can see, but if we had a "super" microscope that could let us see at the level of atoms and molecules, we would realize that all matter is in constant motion. The molecules of water boiling in the pot are moving extremely rapidly and, for this reason, possess a great deal of kinetic energy (Figure 3.1).

Kinetic energy can be measured, and its value will depend upon the mass of the particle and the particle's speed or velocity. Algebraically,

$$\text{Kinetic energy} = \frac{1}{2} \times \text{mass} \times (\text{velocity})^2$$

$$\text{KE} = \frac{1}{2} \times \text{m} \times \text{v}^2, \quad \text{or } \frac{1}{2}\text{mv}^2$$

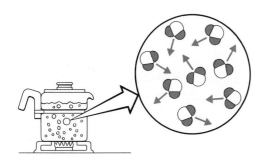

Figure 3.1 We can use the kinetic energy of the rapidly moving molecules in boiling water to do work—such as cooking vegetables.

Figure 3.2 Which ball would you rather have hit you? (Right, M.E. Newman/Woodfin Camp).

Although you may not have thought of it in this way before, you already have a good general feeling for the kinetic energy possessed by an object. For example, you would probably rather have a five-year-old throw a baseball at your stomach than have a major league pitcher try it. In each case the baseball would have the same mass, but would be traveling at a different speed and, therefore, would have different kinetic energy. The ball thrown by the child would be traveling at a much slower speed; it would have less kinetic energy and would do less damage to your tissues (Figure 3.2). As a second example, would you rather have your parked car hit by a bicycle going 5 mph or a dump truck going 5 mph? In this example, the speed is the same, but the much larger mass of the dump truck gives it a much larger kinetic energy and a much greater capacity to do work—that is, to crumple your fender.

3.3 Temperature

A baseball, a car, and a dump truck are all objects that we can weigh, whose speed we can determine, and whose kinetic energy we can then calculate. But how can we determine the kinetic energy of a particle that we can't see, such as an atom or a molecule? The temperature of a substance allows us to measure the kinetic energy of its particles. Scientifically, **temperature** is a measure of average kinetic energy. The term *average kinetic energy* is used because all molecules, like all people, are not alike (Figure 3.3). At any temperature, some molecules will be moving very rapidly, some very slowly, and the majority somewhere in between. But just as we can talk about the average behavior or performance of a group of students, so can we refer to the average behavior of a group of molecules or atoms. As the average kinetic energy of a group of molecules increases, so does the temperature of that substance. From experience, you know that the hotter a pan of water, the more severely it can burn your hand, or the faster it can cook vegetables. You now understand that the hotter water has a higher average kinetic energy and, therefore, has a greater ability to do work in the form of damaging tissues or cooking vegetables.

3.4 Potential Energy

A rock hurtling through the air has kinetic energy; when it strikes a window, it can break it. But a rock perched on the top of a cliff also possesses energy; it has the capacity to do work. If it fell off the cliff onto a passing car, it could do severe damage. The rock on the top of the cliff is said to possess **potential energy,** or energy of position. Such energy is not in use, but it is stored and has the capacity to do work when it is converted to other forms of energy. For example, the water stored behind a dam possesses potential energy. When released in a controlled fashion, it can drive turbines to produce electrical energy or, if released in an uncontrolled fashion by collapse of the dam, the water could display its energy by destroying everything in its path (Figure 3.4).

Energy Sources

It is convenient to classify the energy around us according to the source of that energy. Such categories include electrical energy, chemical energy, nuclear energy, gravitational energy, heat energy, mechanical energy, and electromagnetic energy.

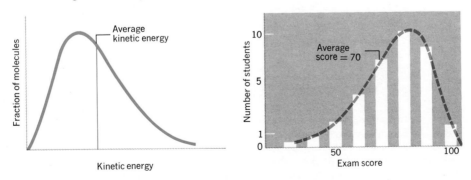

Figure 3.3 At any one temperature each gas molecule will have a specific kinetic energy just as each student will have a specific score on an exam. However, just as there is an average score for the exam, there is an average kinetic energy for the gas molecules, and temperature is a measure of this average kinetic energy.

Figure 3.4 The water stored behind a dam possesses potential energy that, when released in a controlled fashion, can drive turbines to produce electrical energy (top) or, when released in an uncontrolled fashion, can cause death and destruction. Such a disaster occurred when a dam collapsed on Buffalo Creek in West Virginia in 1972 (bottom). (Top, Courtesy Bureau of Reclamation, Pacific Northwest Region; bottom, Wide World Photos)

3.5 Heat Energy

Heat is a form of energy with which you are quite familiar. You may never have thought of heat as doing work, but hot gases produced from burning gasoline push the pistons that power your car, and hot steam turns the turbines in a generating plant to power your electric appliances.

Heat is energy that is transferred from one place to another because of a difference in temperature. We measure heat in units called calories. A **calorie (cal)** is the amount of energy that must be transferred to one gram of water in order to raise its temperature exactly one degree Celsius. For example, raising the temperature of a cup of water (250 grams) from room temperature (25°C) to boiling (100°C) to make a cup of coffee requires 18,750 calories of energy. To make such a calculation we need to use the following equation:

$$\begin{aligned}\text{calories} &= \text{grams of water} \times \text{temperature change (denoted } \Delta t\text{)}\\ &= \text{g} \times \Delta t\end{aligned}$$

So in this case,

$$\begin{aligned}\text{cal} &= 250\ \text{g} \times (100^\circ - 25^\circ)\\ &= 250\ \text{g} \times 75^\circ\\ &= 18{,}750\ \text{cal}\end{aligned}$$

A calorie is actually a very small unit of energy, so we often find it more convenient to talk in terms of 1000 calories, called a kilocalorie. A **kilocalorie (kcal)** is the amount of heat energy required to raise the temperature of 1000 g (about 1 quart) of water one degree Celsius. The food Calorie (note the capital C) that you have heard so much about, is actually a kilocalorie. That 1 ounce bag of potato chips that you had for a snack contains 160 food Calories, or 160 kilocalories. To avoid confusion, in this book we will always refer to food energy content in terms of kilocalories.

The instrument that is used to determine the calorie content of potato chips or other substances is called a **calorimeter,** and is shown in Figure 3.5. The sample to be tested is placed in the inner chamber, and is then burned completely. The energy released from this burning warms the water in the surrounding container, and the number of calories transferred can be calculated from the rise in water temperature.

Example 3-1

What is the average calorie content of a peanut, if the temperature of 1000 grams of water in a calorimeter increases by 50°C when 10 peanuts are burned?

$$\begin{aligned}\text{calories} &= \text{grams} \times \Delta t\\ &= 1000\ \text{g} \times 50^\circ\\ &= 50{,}000\ \text{cal produced by 10 peanuts}\end{aligned}$$

$$\frac{50{,}000\ \text{cal}}{10\ \text{peanuts}} = \frac{5000\ \text{cal}}{1\ \text{peanut}} = \frac{5\ \text{kcal}}{1\ \text{peanut}}$$

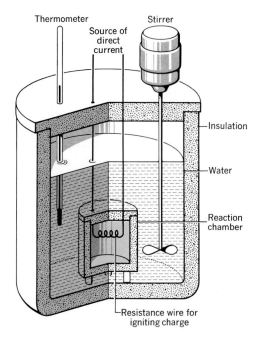

Figure 3.5 A calorimeter consists of a reaction chamber, where the sample is burned, surrounded by water. The water, in turn, is contained in an insulated unit constructed so that the water can be stirred and the temperature change of the water can be measured.

The minimum amount of energy required daily to maintain the basic continuous processes of living in the body at rest is called the **basal metabolism rate (BMR),** and varies for each individual. A woman weighing 121 lb (55 kg) will require about 1400 kcal, and a man weighing 143 lb (65 kg) about 1600 kcal. Any calories that are consumed in excess of that amount either must be used to supply energy for the work an individual does, or will be deposited as fat. About 3,500 excess kilocalories will produce one pound of body fat. Table 3.1 gives examples of the energy used in various activities.

3.6 Electromagnetic Energy

Light represents another form of energy that always surrounds us. The light that we see is only a tiny part of an entire range of electromagnetic energy, from low-energy radio waves to very high-energy X rays and

Table 3.1 Energy Expenditures for Everyday Activities

Activity	Kcal per Hour per Pound of Body Weight	Activity	Kcal per Hour per Pound of Body Weight
Watching TV	0.6	Dishwashing	1.0
Eating a meal	0.7	Walking (3 mph)	1.5
Standing	0.8	Playing ping-pong	2.7
Driving a car	1.0	Running	4.0
Typing rapidly	1.0	Swimming (2 mph)	4.5

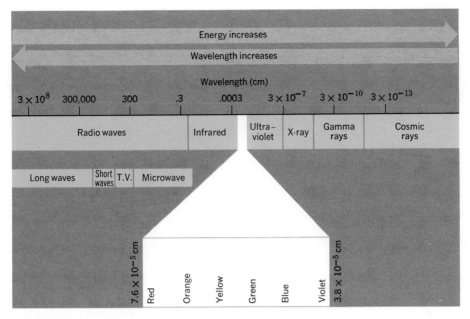

Figure 3.6 The electromagnetic spectrum. Visible light makes up a very small part of this spectrum.

gamma rays. What we see as white light is actually made up of the whole spectrum of colors that appears in a rainbow. The location of this visible spectrum in the **electromagnetic spectrum** is shown in Figure 3.6.

Light waves may be compared with ocean waves, with their crests and troughs. The **wavelength** of light (represented by the Greek letter lambda, λ) is the distance from one crest to the next crest or from trough to trough, or from middle to middle for that matter (Figure 3.7). Each wavelength of light corresponds to a specific level of energy. The longer the wavelength, the lower the energy of the light. This relationship can be expressed mathematically by the equation

$$E = \frac{k}{\lambda}$$

where E is the energy of the electromagnetic radiation, and k is a constant related to the speed of light. Look at the visible spectrum shown in Figure 3.6. Visible light with the highest energy and the shortest wavelength is found in the purple region, whereas red light has the lowest energy and the longest wavelength.

Microwave Radiation

You're certainly familiar with the use of long-wavelength, low-energy radiation to send radio and television signals. Slightly shorter wavelength, or **microwave,** radiation is becoming increasingly popular as a means of rapid cooking. Microwaves interact with food by setting the molecules within the food into motion, causing the food to heat uniformly. This is quite different from conventional cooking, in which the molecules on the

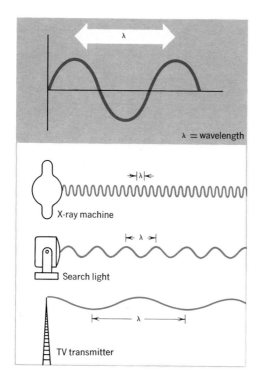

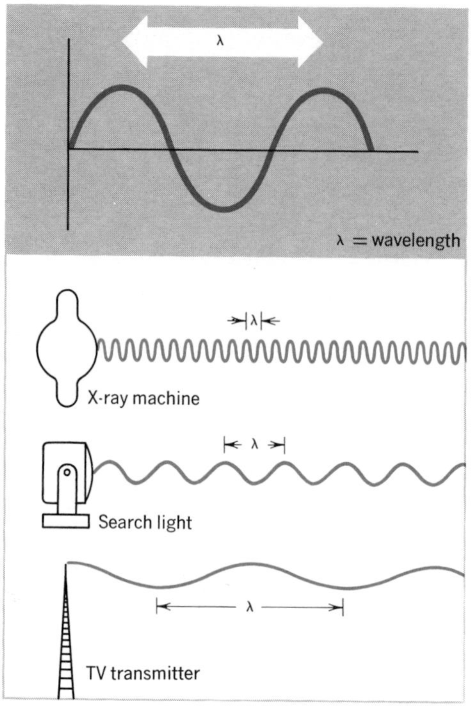

Figure 3.7 Light travels in waves. The wavelength of light can be defined as the distance from crest to crest. Each wavelength is associated with a particular energy: the longer the wavelength, the lower the energy.

outside of the food are heated first, and the heat is then transferred gradually to the molecules in the center of the food. Because microwave ovens uniformly heat all of the food at the same time, cooking time is greatly reduced. You will often find warning signs posted in areas where microwave ovens are being used, because these cooking devices can interfere with the operation of electronic cardiac pacemakers.

Infrared Radiation

Radiation in the **infrared (IR)** range cannot be seen, but can be felt as heat and can be detected by a thermometer. Infrared radiation from the sun warms us, and light fixtures that give off this radiation are used to keep food warm in restaurants. Although infrared radiation doesn't affect ordinary photographic film, special film can detect the infrared radiation given off by warm objects. Such film has been used to locate thermal pollution from power plants and to map the density of vegetation in certain regions (Figure 3.8). Two species of snakes have infrared sensors that can detect low-density infrared radiation given off by their prey. Using these sensors, the snakes can pinpoint and capture their prey in the dark, and can locate hiding places having comfortable living temperatures.

Ultraviolet Radiation

Ultraviolet (UV) light has shorter wavelengths and, therefore, higher energy than visible light. It is penetrating radiation, and can harm living tissue by causing changes in certain biological systems or by causing irreparable damage to cells. Fortunately, most of the ultraviolet radiation coming to the earth from the sun is absorbed by the ozone layer in the upper atmosphere, and never reaches us. Because some wavelengths of ultraviolet light are very effective in killing bacteria, many sterilizing units use this type of radiation. It is perhaps surprising, then, that UV light is also quite important to our bodies. It causes the production within the skin of vitamin D, a compound necessary for the prevention of rickets.

Unfortunately, too much ultraviolet exposure can produce the familiar burn we call sunburn. Prolonged exposure to the sun over the years can cause wrinkling of the skin, thick warty growths on the skin (a condition called keratosis) and, in some cases, skin cancer. The cells in the skin do have some protection against damage by ultraviolet light. First, the dead horny layer of cells on the surface of the skin absorbs some of the UV light. Second, exposure to the sun triggers cells in the skin to produce more of the skin pigment called melanin, which acts to further block the ultraviolet light. This increase in protective skin pigment, of course, is what we call a suntan. However, an effective tan takes three to five days to develop and, even then, reduces the penetration of UV light by only one-half. To gain additional protection, you need to apply sun screens containing chemicals that prevent the most damaging shorter-wavelength UV light from penetrating the skin.

X Rays and Gamma Rays

X rays and **gamma rays** have shorter and shorter wavelengths, higher energies, and greater penetrating power. X rays originate from specially constructed X-ray tubes, whereas gamma rays come from natural sources. The shortest X rays can pass through steel walls, and gamma rays can pass through a 10-inch thick lead plate. Because X rays will not pass as easily through bones and teeth as they will through tissue, they are a very useful diagnostic tool for the medical and dental professions. However, penetration of cells by high-energy radiation such as X rays or gamma rays can disrupt normal chemical processes, causing cells to grow abnormally and to die. Because effects of such radiation build up over time, dosages and repeated exposure to X rays must be carefully controlled. Cancer cells are more sensitive to radiation than normal cells, so X rays and gamma rays can be used to treat certain forms of cancer.

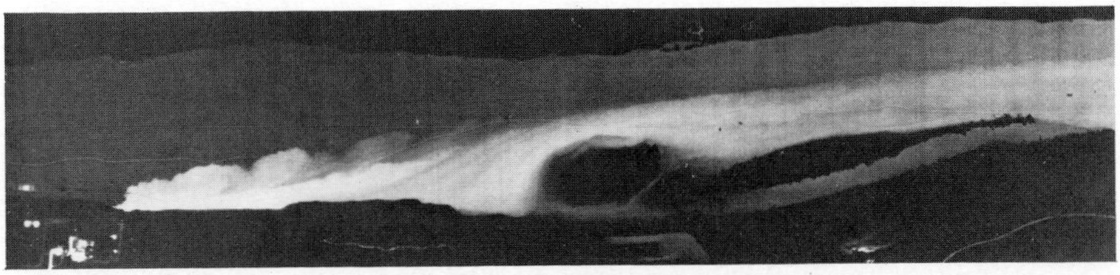

Figure 3.8 Infrared-sensitive film was used to record the heat discharge from a power plant on the Connecticut River. The power plant and an oil tanker with its hot engine room is at the lower left, and the glowing cloud running down the bank is the warm water being discharged from the cooling system of the power plant. (Courtesy Environmental Analysis Department-HRB-Singer, Inc.)

Medical radiation equipment is designed to control the exact degree of penetration of the radiation and to concentrate the radiation on the cancerous area to minimize the damage to normal tissue.

Conservation of Energy*

3.7 First Law of Thermodynamics

Energy can be converted from one form to another. For example, electrical energy is converted into heat energy in our furnaces, stoves, and toasters, and into light energy and heat energy in our lamps. Chemical energy is converted into heat energy when gas is burned in stoves or furnaces, or into mechanical energy and heat energy when gasoline is burned in car engines. Although energy can be converted from one form to another, it must all be accounted for. This is a statement of the **First Law of Thermodynamics,** which says that energy can neither be created nor destroyed, but only changed in form. In other words, the total amount of energy at the end of a reaction or process must equal the total amount of energy at the beginning. You know that gasoline is burned to provide energy for powering automobiles. However, you might be surprised to learn that only about 20% of this energy is used in actually moving the car. Where does the rest of the energy go? (The First Law of Thermodynamics says that it cannot just disappear.) In this particular case, the rest is lost as heat energy—*lost* in the sense that this energy does not do any useful work.

In the same way that gasoline provides energy for a car, the food that we eat provides energy for our cells. But in this case too, the conversion of food into energy that our cells can use is not 100% efficient. For example, about 44% of the energy that can be transferred from a molecule of sugar is converted into energy that our cells can use to do work. The rest ends up as heat energy that helps to maintain body temperature. (Actually, a transformation of energy in any type of system will result in the production of some waste heat.) Much to many people's unhappiness, our bodies must obey the First Law of Thermodynamics. If we take in more food energy than our bodies need, this excess energy doesn't just disappear, but is stored within our tissues, mainly in the form of unwanted fat. To get rid of this excess stored energy, we must take in less food energy than our bodies require so that our cells will start using the fatty tissue as a source of energy to maintain body activities.

3.8 Entropy

You might be wondering why we need a constant supply of new energy from the sun to maintain life on earth if energy is conserved in all natural processes. Although the First Law of Thermodynamics tells us that the total quantity of energy remains unchanged, it tells us nothing about the

---

* *This section is optional and may be skipped without loss of continuity.*

quality of that energy. Only concentrated forms of energy can be used to do work, and work is essential to the maintenance of living organisms. The energy from the sun reaches us in the very concentrated form of light energy, but this energy is quickly converted to heat, a less concentrated form of energy.

If you think about it for a moment, you will realize that all naturally occurring (or spontaneous) processes tend to go in one direction. Left alone, water always runs downhill, heat flows from a hot to a cold object, gases flow from regions of high pressure to regions of low pressure, and people grow old. We would be quite shaken if we were to see water flowing uphill on its own, but we still might ask what it is that prevents naturally occurring processes from reversing direction. The answer to this question is fairly obvious when the final state of the system is at a lower energy level than the initial state. Water flows downhill from higher to lower potential energy, and heat flows from a state of higher to lower kinetic energy. It is quite reasonable that there should be a general tendency for all substances to reach a state of lower energy.

But there are other spontaneous processes with which you are familiar that don't seem to depend upon the energy of the system. For example, if you were to put a drop of ink into a glass of water, the ink would immediately begin to spread throughout the liquid, eventually giving the water a uniform tint (Figure 3.9). This process certainly occurs in a single direction. You would be astonished if you ever saw a glass of tinted liquid suddenly change so that all of the color moved through the liquid to form a single concentrated spot of dye. Yet, it is impossible to point out any change of energy taking place in this process.

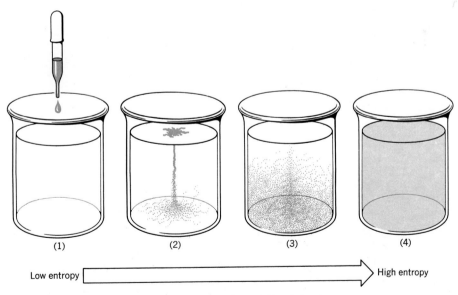

Figure 3.9 Ink dropped into a glass of water will quickly change from a state of low entropy (the droplet) to a state of high entropy (spread throughout the water).

This example leads us to conclude that there is another factor besides energy levels that determines the direction of spontaneous processes. That factor is the amount of randomness or disorder of the system. The term used to describe the disorder of a system is **entropy.** The more random and disordered a situation is, the greater its entropy. All spontaneous reactions go toward a condition of greater randomness, greater disorder, and greater entropy. We can now see that the drop of ink, which was originally in a small compact drop, will spontaneously spread throughout the liquid into a random and disordered arrangement of ink molecules, which increases the entropy of the system. The driving force behind the spreading of the drop of ink is the tendency for the entropy of the system to increase.

The **Second Law of Thermodynamics** states that the entropy of the universe is increasing. This law is interwoven through all of science and into each of our lives. Even the writers of nursery rhymes had an intuitive feeling for the Second Law when they wrote, "All the king's horses and all the king's men couldn't put Humpty together again." You now see that the breaking of an egg is an irreversible process that leads to increased disorder of the egg parts and, therefore, greater entropy (Figure 3.10).

By now it may have occurred to you that there are some naturally occurring processes that can be reversed. Water can be pumped uphill to a storage reservoir, heat can be pumped from cold to hot areas in a refrigerator, and air can be pumped from low pressure to high pressure in a bicycle tire. The key word in each of these examples is *pump*. Each requires that energy be added; that is, work must be done. This means that

Figure 3.10 The writer of this nursery rhyme had an intuitive understanding of the concept of entropy.

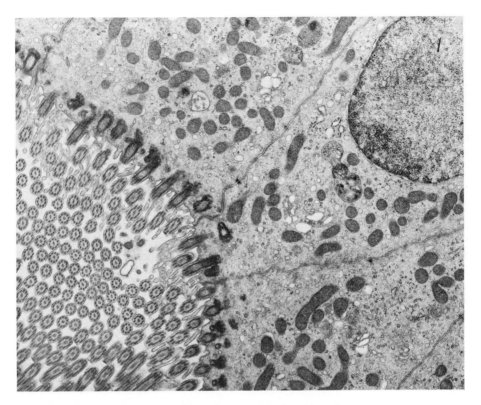

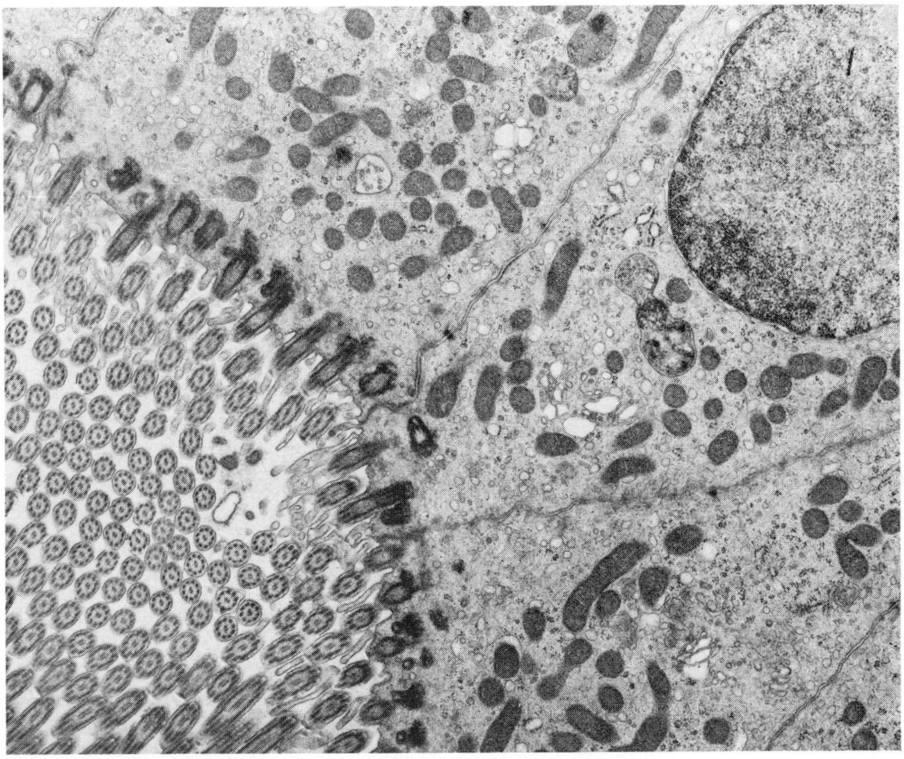

Figure 3.11 This electron micrograph of a cell from the retina of a rabbit illustrates the highly organized, complex structures found in living organisms. To function normally, the cell must constantly expend energy to counteract the natural tendency toward increased entropy and the breakdown of these complex structures. (Courtesy Perkin-Elmer Corporation)

each of these seemingly "reversed" processes must be coupled with, or related to, another reaction in which energy is produced. Electrical energy is needed to power the water pump or to run the refrigerator, and chemical energy is needed to give our muscles the strength to push a bicycle tire pump. But in the reactions that are required to produce this energy, entropy is increased. So, if we now look at both of the reactions required to pump the water uphill (that is, the process of actually pumping the water uphill and the process of generating the energy necessary to do this pumping), we would find that the total entropy of the combined processes has increased.

An awareness of entropy is very important in understanding the functioning of living organisms. Such organisms are composed of cells, which are highly complex structures. Therefore, there is a natural tendency for these cells to break down and increase the entropy of the system (Figure 3.11). In order to maintain the structure and functioning of the living system, energy must be added to counteract the natural drive toward increased entropy. We will see that this energy, which comes from the food that organisms eat, actually originates on the sun. If we want to consider the coupled reactions of this system, we see that the life processes that build up and maintain the complicated structures in living organisms depend upon energy-producing reactions on the sun. It is rather amazing to realize that each of our bodies maintains its highly complex structures and functions only at the expense of huge entropy increases on the far-off sun.

Chapter Summary

Energy is the capacity to cause a change or to do work on an object. Kinetic energy is energy of motion, and temperature is a measure of the average kinetic energy of the particles of a substance. Potential energy is stored energy or energy of position. There are many types of energy. Heat is energy that is transferred from one place to another because of a difference in temperature, and can be measured in calories. Electromagnetic radiation is energy that travels in waves. The shorter the wavelength, the higher the energy of the radiation. Radiation of all wavelengths has many uses, and can have both beneficial and harmful effects on living organisms.

Exercises and Problems

1. Which one in each of the following pairs has the greater kinetic energy?
   (a) 100 g of water or 100 g of steam
   (b) A car moving at 25 mph or at 50 mph
   (c) A football player or a sprinter, both running 100 yards in 12 seconds.

2. Which one in each of the following pairs has the greater potential energy?
   (a) Snow in the mountains or ocean water
   (b) A pendulum at the top or the bottom of its swing
   (c) A resting bow and arrow or a drawn bow and arrow

3. How many calories are there in:
   (a) 3.5 Kilocalories
   (b) 125 Food Calories
   (c) 0.01 Kcal
   (d) 0.25 Kilocalories

4. Calculate the number of calories transferred in each of the following sets of conditions.

	Grams of water	*Initial temperature*	*Final temperature*
(a)	25 g	20°C	27°C
(b)	50 g	24°C	32°C
(c)	1000 g	25°C	76°C
(d)	500 g	45°C	23°C

5. Which light contains more energy: orange light or yellow light?

6. List the following in order of (a) increasing energy and (b) increasing wavelength:
   X rays, short waves, ultraviolet light, yellow light, microwaves

7. Discuss the potential and kinetic energy of a skier as she:
   (a) stands in the lift line
   (b) travels to the top of the lift
   (c) stands at the top of the mountain
   (d) skis down the hill to the bottom

8. When four cashews are burned in a calorimeter, the temperature of the 1000 g of water changes from 25° to 69°C. How many kilocalories of energy are transferred from one cashew? How many food Calories?

9. How many kilocalories of energy are transferred from one gram of butter if burning 10 grams of butter raises the temperature of 1000 grams of water in a calorimeter from 20° to 90°C?

10. In which place would you probably get a more severe sunburn on a clear day: skiing in the Rocky Mountains or swimming at a Delaware beach? Explain your answer.

11. Ultraviolet light can be classed into two groups: UV-A with longer wavelengths and UV-B with shorter wavelengths. Which type of ultraviolet light do you think is more likely to cause skin cancer? Give a reason for your answer.

*12. In a nuclear power plant, only 40% of the energy released in the nuclear reaction is converted to electrical energy. Why doesn't this violate the First Law of Thermodynamics?

*13. Describe the changes in entropy of the molecules of a sugar cube as it dissolves in water.

---

* *Problems marked with an asterisk are from optional sections in the chapter.*

chapter 4

The Three States of Matter

Learning Objectives

By the time you have finished this chapter, you should be able to:

1. State the difference between the two principal types of solids.

2. Define the *melting point* and *boiling point* of a substance.

3. Describe the changes that occur at the molecular level as an ice cube is warmed from −5°C to 110°C.

4. Describe the five properties of ideal gases stated by the kinetic molecular theory.

5. State, in your own words, five laws of gas behavior, and give everyday examples of each.

6. Perform calculations using Boyle's law, Charles' law, and the general gas law.

Cliff Brown was nearing retirement at the age of 65, but was starting to wonder if he'd ever be able to enjoy the fishing and hunting he'd been looking forward to. Over the past five years he had been increasingly bothered by a shortness of breath and continual coughing. He knew that his heavy cigarette smoking was to blame for many of these symptoms, but had never been able to cut down to less than two packs a day. But now his condition seemed to get worse—he was trying to recover from a bad cold, but felt extremely tired all the time and was finding it more and more difficult to breathe.

After seeing his doctor, Cliff was sent to the hospital to have a series of tests done on his pulmonary system (the parts of his body involved in lung function). Chest X rays showed that his lungs were enlarged and his diaphragm flattened. Other tests revealed that his ability to exhale air was well below normal, and that the movement of oxygen (O_2) into his blood and carbon dioxide (CO_2) out of the blood was not occurring at the proper rate. Blood tests verified this imbalance of oxygen and carbon dioxide in his blood. Oxygen in his blood was measured at an abnormally low tension, or pressure, of 40 mm Hg (normal is 80 to 100 mm Hg), and carbon dioxide was measured at a very high tension of 70 mm Hg (normal is 40 mm Hg). The diagnosis was clear: Cliff was suffering from emphysema, one of the chronic obstructive pulmonary diseases (COPD).

Cliff was admitted to the hospital for treatment that would help ease the imbalance of these two gases in his blood and would reduce the amount of work required for him to breathe. This therapy provided Cliff with air to breathe that had a concentration of oxygen slightly higher than that found in normal room air. This extra oxygen was administered to Cliff at a very slow rate of 2 liters per minute because a much higher rate could actually have been quite dangerous for him. To understand this, you have to know that the brain keeps track of the concentration of carbon dioxide in the blood. High blood levels of carbon dioxide cause the brain to increase the rate of breathing. However, in an emphysema patient who has chronically high blood levels of carbon dioxide, the brain undergoes a change to become sensitive to blood O_2 levels instead of blood CO_2 levels. Low concentrations of oxygen in the blood of these persons will trigger more rapid respiration. Therefore, if oxygen is administered to an emphysema patient at a high rate, the blood oxygen level will rapidly increase and the brain will not receive a signal for the need to breathe. This means that the patient may actually stop breathing, and could easily die.

Cliff remained in the hospital for a week. His upper respiratory infection was cleared up with antibiotics, and the oxygen tension in his blood was increased to an adequate level. At the end of the week Cliff was taken off the oxygen treatment and his blood gases were monitored for 48 hours. This check revealed that Cliff's disease had progressed so far that he was unable to maintain an adequate oxygen tension in his blood by his own breathing efforts. To allow Cliff to return to his home, he was supplied with a home oxygen unit containing liquid oxygen. This machine had long tubes that enabled Cliff to breathe oxygen as he moved around the house; he also had a portable unit that he could fill with oxygen and carry over his shoulder when he wanted to leave the house for several hours.

You might wonder how the oxygen that Cliff needed to breathe could be stored as a liquid for his use. Liquid and gas are two of the three possible states, or forms, in which matter can exist. The other state, of course, is solid. In this chapter we will closely examine these different states of matter to see how they differ and how they can be converted from one to another. We will then be able to discuss some of the physical laws that describe the behavior of matter when it is in the gaseous state.

States of Matter

4.1 Solids

As we have just stated, matter may be found in the form of a gas, a liquid, or a solid (Figure 4.1). However, there are really two types of solids, differing in the arrangement of the particles that make them up. An

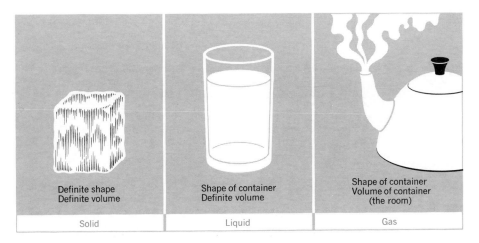

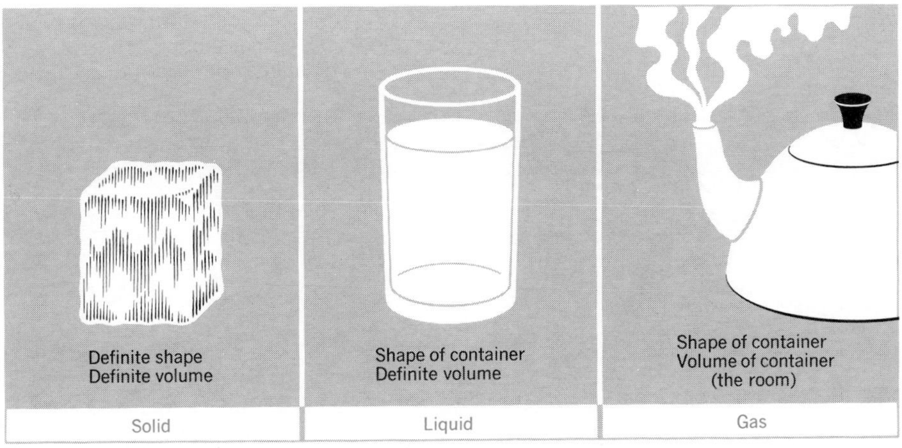

Figure 4.1 The three states of matter: water can exist as a solid, liquid, or gas.

amorphous solid, such as window glass or clay, is one in which the particles are arranged in a noncrystalline, or disordered, fashion. A **crystalline solid,** such as quartz or table salt, is one in which the particles are arranged in a highly ordered fashion (Figure 4.2). There are many different solids that are found as crystals, some of which you can easily identify—such as crystals of quartz or water crystals in a snowflake. Particles in a crystalline solid have specific positions and fixed neighboring particles, giving the solid a rigid structure. It is important to realize,

Figure 4.2 Clay (left) is an example of an amorphous solid, whose particles are arranged in a disordered fashion. A quartz crystal (right) is an example of a crystalline solid, whose particles are arranged in a specific pattern. (Left, Kathy Bendo; Right, Ward's Natural Science Establishment.)

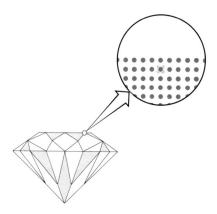

Figure 4.3 The particles in a solid are very closely packed together, but they are still able to vibrate around a fixed point.

however, that these particles are not totally motionless. They move back and forth and up and down, or vibrate around a fixed point (Figure 4.3).

When the temperature is low, these vibrations are not enough to destroy the attraction between the particles of the solid. But if energy is added in the form of heat, the kinetic energy of the particles increases. Their vibrations become more and more violent until the solid finally begins to break apart, or melt. The specific temperature at which this occurs is called the **melting point** of the solid. The greater the attraction between the particles of a solid, the higher the melting point. When a solid has reached its melting point, additional heat does not go into increasing the average kinetic energy of the particles—that is, into increasing the temperature—but goes entirely into melting the solid. The amount of energy that must be added to change a solid to a liquid at the melting point is called the **heat of fusion** (expressed in cal/g). The heat of fusion of water is 80 cal/g. (This same amount of energy will be given off when the liquid again becomes a solid.) When the solid is completely melted, the temperature of the substance will again begin to rise as heat is added (Figure 4.4).

4.2 Liquids

The particles in a liquid are not held together as tightly or as rigidly as they are in a solid. Although the particles are fairly close together, they can move from place to place by slipping past one another (Figure 4.5). This allows a liquid to take on the shape of its container, even though the volume of the liquid does not change.

Although a liquid will flow, its particles are still strongly attracted to one another. The strength of this attraction determines two properties of the liquid: its viscosity and its surface tension. **Viscosity** is a measure of how easily the liquid flows. The higher the attraction between the particles of the liquid, the greater the viscosity or syrupy nature of the liquid. Molasses is more viscous than water, which in turn is more viscous than gasoline. **Surface tension** is the resistance of particles on the surface of the liquid to expansion of the liquid. Water has a high surface tension, whereas gasoline has a low surface tension. Both the viscosity and the surface

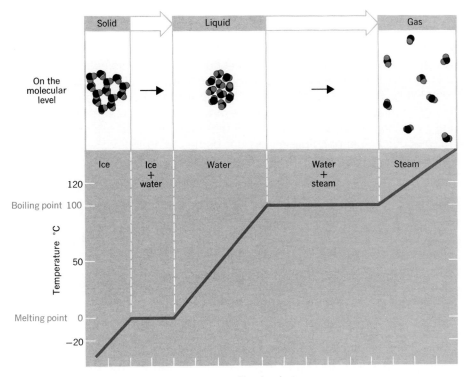

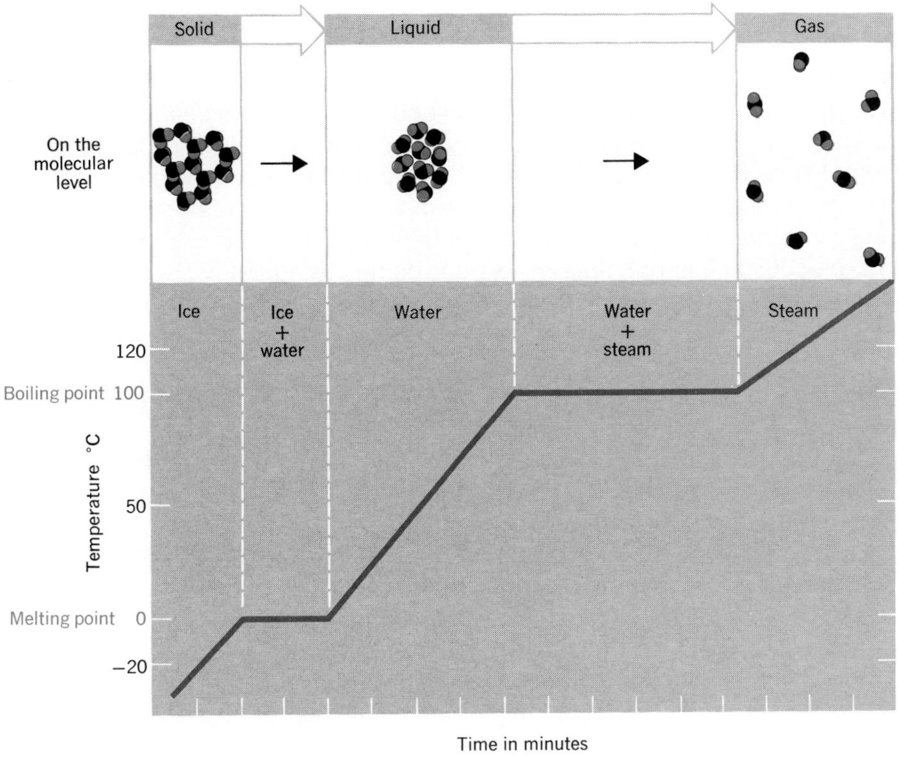

Figure 4.4 This heating curve illustrates the phase changes of water from solid to liquid to gas as heat is added at a constant rate. The temperature at which ice changes to water is the melting point of water, and the temperature at which water changes to steam is the boiling point.

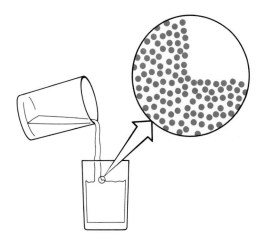

Figure 4.5 A liquid consists of particles that are fairly close together and can move from place to place by slipping past one another.

tension of a liquid will decrease with an increase in temperature (that is, as the kinetic energy of the particles increases).

Very energetic particles near the surface of a liquid can break away from the surface. If these particles don't collide with air molecules and return to the surface of the liquid, they will escape from the liquid completely. This is the process called **evaporation.** Because the very energetic molecules are lost through evaporation, the average kinetic energy of the particles left behind will be lower. This means that the liquid will be cooler. This principle allows our bodies to rid themselves of heat through the evaporation of sweat. Similarly, when alcohol is applied to the skin, the cool feeling is caused by the rapidly evaporating alcohol molecules, which leave behind a cooler liquid on the skin. Note that as more heat is added to a liquid the kinetic energy of the particles will increase, which will then increase the rate of evaporation.

Adding enough heat to a liquid will increase the motion of the particles so much that violent collisions will occur within the liquid, and bubbles of vapor (gas) will be formed. This is the process called boiling. The specific temperature at which boiling occurs under conditions of normal atmospheric pressure is the **normal boiling point** of a substance. At this temperature, additional energy added to the substance goes into pulling the particles apart—that is, changing the liquid to a gas (see again Figure 4.4). The amount of energy required to change a substance from a liquid to a gas at its boiling point is called the **heat of vaporization** (expressed in cal/g). The heat of vaporization of water is 539 cal/g. When all of the substance has become a gas, the temperature will again increase as heat is added.

4.3 Gases and the Kinetic-Molecular Theory

Particles in the gaseous state are moving very rapidly (almost 1000 miles per hour at room temperature) in a completely random fashion (Figure 4.6). The physical properties of gases can be explained by the **kinetic-molecular theory of gases.** First, this theory states that gases are made up of very small particles with a great deal of space between them. This fact explains why it is easier to run through air than through water. Second, the particles of gas travel at high speeds in straight lines until they collide with one another or with the sides of the container. This explains why gas molecules will move throughout the container, taking on the shape of the container. Third, the collisions that occur are completely elastic; that is, a gas particle

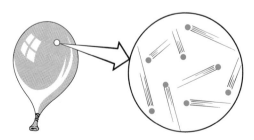

Figure 4.6 A gas consists of widely separated particles which are moving very rapidly in a random, chaotic fashion.

doesn't lose any energy in a collision, and bounces off at its original speed. Fourth, there is no attraction or repulsion between the particles of a gas. And fifth, the kinetic energy of the gas particles changes with temperature. When heated, the particles move faster and faster. When cooled, the particles move slower and slower until, theoretically, a temperature is reached when all motion stops. This temperature, called absolute zero, is equal to −273.15°C or 0 K.

The kinetic-molecular theory describes the behavior of an *ideal* gas. Actually, real gases may not always behave as described, especially under conditions of very high pressure or very low temperature. But under less extreme conditions, real gases seem to behave like ideal gases.

Laws of Gas Behavior

A great deal of study has led scientists to formulate laws describing the behavior of gases. Although you may not know the precise mathematical statement of these laws, you certainly are familiar with their general form from everyday experience. For example, an aerosol can will explode when heated, and a tire may blow out when driven at high speeds on a very hot day (because increasing the temperature of a gas increases its pressure). When you push down on a bicycle pump, the plunger goes down in the air cylinder (because increasing the pressure on a gas decreases its volume). When you accidentally burn the cookies you were baking, the whole house soon smells like burned cookies (because a gas will diffuse to all regions of a container). When you open a bottle of carbonated beverage, you hear a hissing sound (because the solubility of a gas in a liquid depends upon the pressure).

4.4 Units of Pressure

Before we can discuss the exact mathematical forms of the laws mentioned above, we need to take a look at how pressure is measured. The **pressure** exerted by a gas is nothing more than the collisions of the billions of gas particles against the sides of the container. Pressure is defined as a force that is exerted per unit of area, and may be measured in any of the following three units:

Atmosphere (atm)

One atmosphere of pressure is equivalent to 14.7 pounds per square inch, which is the average pressure exerted by the earth's atmosphere at sea level. One atmosphere of pressure will support a column of mercury 760 mm high (Figure 4.7). **Standard atmospheric pressure** is defined to be 1 atm.

Millimeters of Mercury (mm Hg)

This has long been the most common unit used to express pressure. 1 mm Hg = $\frac{1}{760}$ atm; this is a very small amount of pressure.

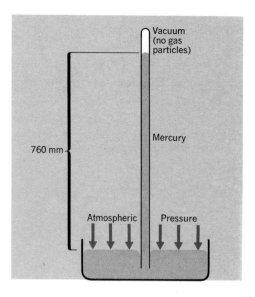

Figure 4.7 The mercury barometer is an instrument used to measure atmospheric pressure. The height of the column will vary from place to place and from time to time as weather conditions affect the atmospheric pressure. At sea level the atmosphere will support a column of mercury about 760 mm high, but at an altitude of three and one-half miles the column would be only about 380 mm high.

Torr (torr)

1 mm Hg is also called a torr, which honors Evangelista Torricelli, the inventor of the mercury barometer. Therefore, 1 mm Hg = 1 torr, and 1 atm = 760 torr.

4.5 Boyle's Law (the Relationship between Pressure and Volume)

In the seventeenth century, the British chemist Robert Boyle discovered that the volume of a gas varies inversely with its pressure if the temperature is kept constant (Figure 4.8). In other words, if you increase the pressure, the volume decreases. Decrease the pressure and the volume increases. We can use the kinetic-molecular theory to explain Boyle's law. When the volume of a gas is decreased, there is less space for the particles to

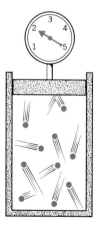

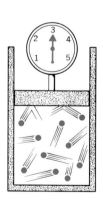

Figure 4.8 Boyle's law states that when the volume of a gas decreases, the pressure will increase (when the temperature of the gas remains constant).

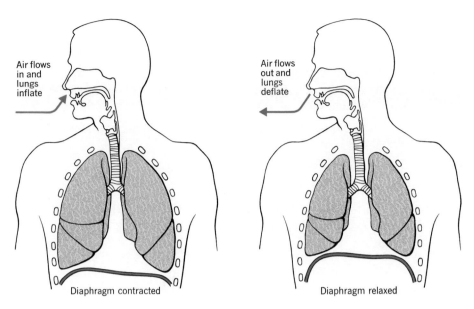

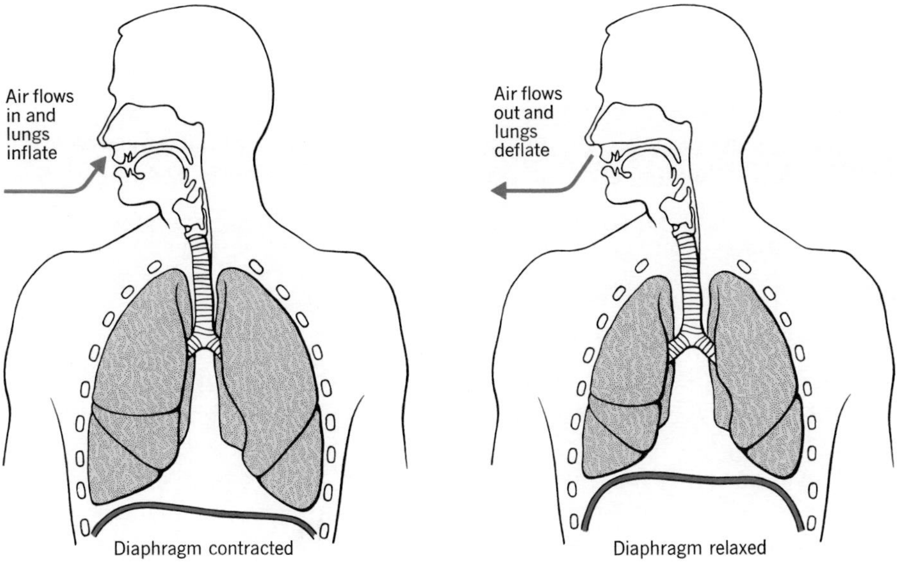

Figure 4.9 Breathing is an example of Boyle's law in action.

move around in, so they will collide with the sides of the container much more often. This will appear as an increase in pressure.

An illustration of Boyle's law at work in our bodies is provided by examining how we breathe (Figure 4.9). Your lungs are located in the thoracic cavity, surrounded by the ribs and a muscular membrane called the diaphragm. To inhale, the diaphragm contracts and flattens out, increasing the volume of the thoracic cavity. This lowers the air pressure in the cavity below that of the atmosphere, causing air to flow into the lungs. To exhale, the diaphragm relaxes and pushes up into the thoracic cavity, increasing the pressure above that of the atmosphere and causing air to flow out of the lungs.

The breathing of patients with polio, paralysis, or other respiratory difficulties is often aided with a respirator. When the pressure inside the respirator is decreased, air enters the lungs and forces the diaphragm to flatten. When the pressure in the respirator is increased, the volume of the lungs is decreased and air is pushed out, allowing the diaphragm to move up into the thoracic cavity.

We have just described Boyle's law in words, but this inverse relationship can also be stated mathematically as follows:

$$PV = k \qquad (k \text{ is a constant})$$

And, when given two conditions of temperature and pressure,

$$P_1V_1 = P_2V_2$$

where P_1 and V_1 are the starting conditions of pressure and volume, and P_2 and V_2 are the final conditions. The following examples show ways to think through various types of problems, as well as how to use the equation for Boyle's law:

Example 4-1

1. Imagine that a gas is inside a cylinder with a movable piston. If the volume of the gas is 3 liters and the pressure is 760 mm Hg, what would be the volume if the pressure were increased to 1140 mm Hg?

   (a) The first step is to identify the quantities that are stated in the problem.

   $P_1 = 760$ mm Hg $\quad V_1 = 3$ liters
   $P_2 = 1140$ mm Hg $\quad V_2 = ?$

   (b) Now let's think about the problem. Will the starting volume of 3 liters increase or decrease if the pressure is increased from 760 mm Hg to 1140 mm Hg? Boyle's law states that the volume of a gas will decrease if the pressure is increased. So we must multiply the starting volume by a fraction or ratio of the pressures that will decrease the volume.

   $$3\ l \times \frac{760\ \cancel{\text{mm Hg}}}{1140\ \cancel{\text{mm Hg}}} = 2\ l$$

   (c) Now let's check our answer using the equation for Boyle's law:

   $$P_1V_1 = P_2V_2$$

   $$760 \text{ mm Hg} \times 3\ l = 1140 \text{ mm Hg} \times V_2$$

   $$\frac{760\ \cancel{\text{mm Hg}} \times 3\ l}{1140\ \cancel{\text{mm Hg}}} = V_2$$

   $$2\ l = V_2$$

2. A 2.5 liter tank contains oxygen at a pressure of 44 atm. What pressure would this amount of oxygen exert in a 55 liter tank?

   (a) Identify the given quantities:

   $V_1 = 2.5$ liters $\quad P_1 = 44$ atm
   $V_2 = 55$ liters $\quad P_2 = ?$

   (b) Reason the problem through. Will the starting pressure increase or decrease if the volume is increased from 2.5 to 55 liters? We know from Boyle's law that if the volume is increased, the pressure will decrease. So we must multiply the starting pressure by a ratio of volumes that will decrease the pressure.

   $$44 \text{ atm} \times \frac{2.5\ \cancel{l}}{55\ \cancel{l}} = 2.0 \text{ atm}$$

(c) Using the Boyle's law equation, we have

$$44 \text{ atm} \times 2.5\ l = P_2 \times 55\ l$$

$$\frac{44 \text{ atm} \times 2.5\ \not{l}}{55\ \not{l}} = P_2$$

$$2.0 \text{ atm} = P_2$$

4.6 Charles' Law (the Relationship between Volume and Temperature)

In the early nineteenth century, the French physicist Jacques Charles discovered that the volume of a gas varies directly with Kelvin temperature when the pressure is constant. That is, when the temperature of a gas increases the volume increases (Figure 4.10). On the molecular level, increasing the temperature means increasing the kinetic energy of the gas particles. Because these particles will now be moving faster, they will collide with the sides of the container much more often. The only way that the pressure can remain constant under these circumstances is to increase the volume of the container. Similarly, when gas particles are cooled, they slow down. In order for the pressure to remain constant, the volume must decrease so that the gas particles can hit the sides of the container just as often as before.

Charles' law can be expressed mathematically by the following direct relationship:

$$V = kT \qquad (k \text{ is a constant})$$

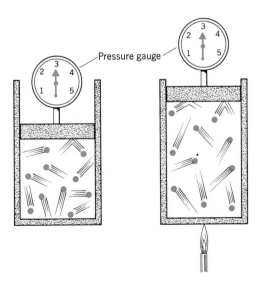

Figure 4.10 Charles' law states that when the temperature of a gas is increased, the volume of the gas will increase (when the pressure remains constant).

And, when given two conditions of volume and temperature,

$$\frac{V_1}{T_1} = \frac{V_2}{T_2}$$

where V_1 and T_1 are the starting conditions of volume and temperature (in degrees Kelvin), and V_2 and T_2 are the final conditions.

Example 4-2

1. Imagine that a gas occupies a volume of 2.0 liters at a temperature of 27°C. To what temperature in degrees Celsius must the gas be cooled to reduce its volume to 1.5 liters?

   (a) Identify the known quantities:

   $V_1 = 2.0$ liters $\quad T_1 = 300$ K (remember that the temperature *must* be in degrees Kelvin, $K = °C + 273$)

   $V_2 = 1.5$ liters $\quad T_2 = ?$

   (b) Reason the problem through. From Charles' law we know that to decrease the volume of a gas the temperature must also decrease. So we must multiply the starting temperature by a ratio of the volumes that will decrease the temperature.

   $$300 \text{ K} \times \frac{1.5 \not{l}}{2.0 \not{l}} = 225 \text{ K}$$

   The answer in degrees Celsius would be

   $$225 = °C + 273$$
   $$°C = 225 - 273 = -48°C$$

   (c) Or use the Charles' law equation

   $$\frac{2.0\ l}{300 \text{ K}} = \frac{1.5\ l}{T_2}$$

   $$2.0\ l \times T_2 = 1.5\ l \times 300 \text{ K}$$

   $$T_2 = \frac{1.5\ l \times 300 \text{ K}}{2.0 \not{l}} = 225 \text{ K} = -48°C$$

2. Imagine that you are skiing on a beautiful, clear day with a temperature outside of −3°C. The cold air that you breathe is warmed to a body temperature of 37°C as it travels to your lungs. If you inhaled 400 ml of air at −3°C, what volume would it occupy in your lungs? (Assume that the pressure is constant.)

   (a) Identify the known quantities:

   $T_1 = -3 + 273 = 270$ K $\quad V_1 = 400$ ml

   $T_2 = 37 + 273 = 310$ K $\quad V_2 = ?$

(b) Reason the problem through. The temperature of the gas is increasing as it enters your lungs, so from Charles' law we know that the volume must also increase. Therefore, we must multiply the starting volume by a ratio of temperatures that will increase the volume.

$$400 \text{ ml} \times \frac{310 \cancel{\text{K}}}{270 \cancel{\text{K}}} = 459 \text{ ml}$$

(c) Or, using the Charles' law equation:

$$\frac{400 \text{ ml}}{270 \text{ K}} = \frac{V_2}{310 \text{ K}}$$

$$\frac{400 \text{ ml} \times 310 \cancel{\text{K}}}{270 \cancel{\text{K}}} = V_2$$

$$459 \text{ ml} = V_2$$

4.7 General Gas Law

We can combine the laws stated by Charles and Boyle to write a general mathematical equation that will allow us to predict gas behavior under various different conditions of pressure, temperature, and volume:

$$\frac{P_1V_1}{T_1} = \frac{P_2V_2}{T_2}$$

Example 4-3

A cylinder contains 250 ml of carbon dioxide at 27°C and 810 torr. What volume would the carbon dioxide occupy at standard temperature and pressure (1 atm, or 760 torr, and 0°C)?

(a) Identify all the known quantities:

$P_1 = 810$ torr $T_1 = 27 + 273 = 300$ K $V_1 = 250$ ml
$P_2 = 760$ torr $T_2 = 0 + 273 = 273$ K $V_2 = ?$

(b) We can solve this problem by reasoning in two steps. First, notice that the pressure is decreasing. From Boyle's law we know that this should make the volume increase. So we multiply the volume by a ratio of pressures that would increase the volume. Second, notice that the temperature is decreasing. From Charles' law we know that a decrease in temperature causes the volume to decrease. So the ratio of temperatures to use is the one that would decrease the volume.

$$250 \text{ ml} \times \frac{810 \text{ torr}}{760 \text{ torr}} \times \frac{273 \text{ K}}{300 \text{ K}} = 242 \text{ ml}$$

(c) Or, using the combined gas law equation, we have

$$\frac{810 \text{ torr} \times 250 \text{ ml}}{300 \text{ K}} = \frac{760 \text{ torr} \times V_2}{273 \text{ K}}$$

$$\frac{810 \text{ torr} \times 250 \text{ ml} \times 273 \text{ K}}{300 \text{ K} \times 760 \text{ torr}} = V_2$$

$$242 \text{ ml} = V_2$$

Another way to combine the various relationships we've been discussing is given by the equation of state of an ideal gas, also known as the ideal gas law:

$$PV = nRT$$

where R is a number called the universal gas constant, and n is the number of gas particles expressed in a unit called the mole (to be discussed in Chapter 9).

4.8 Graham's Law of Diffusion

In the nineteenth century the Scottish chemist Thomas Graham noticed that different gases diffuse (mix with other particles) at different rates. In particular, he observed that lighter (less dense) gases diffuse more rapidly than heavier gases at the same temperature. Mathematically, Graham's law states that a gas diffuses at a rate inversely proportional to the square root of the mass of its particles:

$$\text{Diffusion rate} = \frac{k}{\sqrt{\text{mass}}}$$

This means, for example, that you would detect the smell of the lighter gases produced by a burned pot roast before the heavier gases. And, if one gas were four times heavier than another gas, you would detect the lighter gas in half the time that it would take to detect the heavier gas ($\sqrt{4} = 2$).

4.9 Henry's Law (Solubility of Gases)

When you open a bottle of carbonated beverage and listen to the hiss of the escaping carbon dioxide, you are observing Henry's law in action. At the beginning of the nineteenth century the English chemist William Henry discovered that the solubility of a gas in a liquid at a given temperature is directly proportional to the pressure of that gas on the liquid. That is, the higher the pressure, the more gas that will dissolve in a liquid when the temperature does not change. Carbonated beverages are bottled under

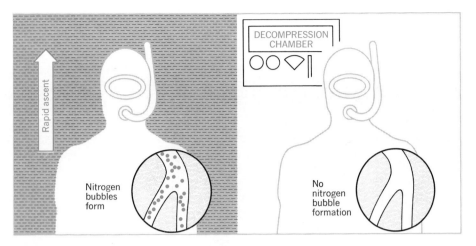

Figure 4.11 The bends is a disorder resulting from a change in the solubility of nitrogen caused by a rapid change in pressure.

high pressure, and when you open the cap, you reduce the pressure in the bottle. This, then, lowers the solubility of the carbon dioxide and permits the gas to escape from the beverage.

Another example of a change in solubility which is caused by a change in pressure is the bends, a disorder that occurs when deep-sea divers are brought to the surface too quickly (Figure 4.11). Deep-sea divers breathe air (mainly nitrogen and oxygen) under high pressure, which increases the solubility of the nitrogen in the blood. If the diver is brought to the surface too quickly, the sudden drop in pressure causes the dissolved nitrogen to leave the blood and form tiny bubbles (air emboli), which cause severe pain in the arms, legs, and joints. The only cure for the bends is slow decompression, which allows the dissolved nitrogen to escape slowly without forming bubbles. Divers who must work for long periods at great pressure breathe a mixture of helium and oxygen instead of air. Helium, like nitrogen, has no dangerous effects on the body, and has the advantage of being only about 40% as soluble in the blood as nitrogen. Also, because helium is a much lighter gas than nitrogen, Graham's law tells us that helium will diffuse from the bloodstream into the lungs at a higher rate than will nitrogen. As a result, the use of helium rather than nitrogen greatly reduces the risk of the bends.

4.10 Dalton's Law of Partial Pressures

We have seen that the pressure of a gas comes from the collisions of the gas particles against the walls of the container. There are two ways we can increase the frequency of collisions and, therefore, increase the pressure. One way is to increase the temperature of the gas. This will increase the kinetic energy of the gas particles and the number of collisions that occur. The second way is to increase the number of particles of gas in the container.

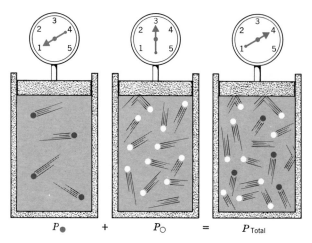

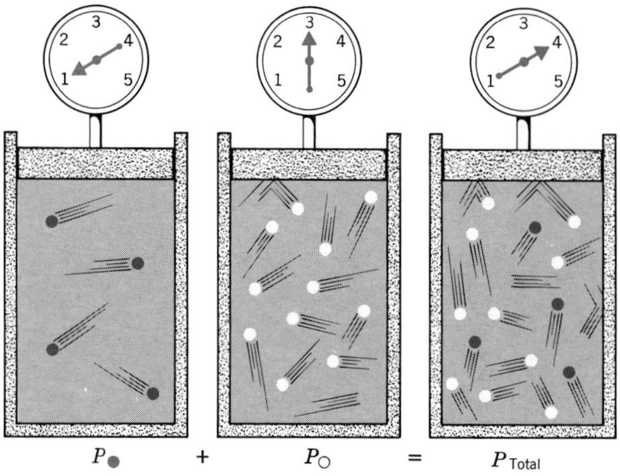

Figure 4.12 Dalton's law states that the total pressure in a container will be equal to the sum of the partial pressures of the gases in that container.

The pressure exerted by a gas at constant temperature does not depend on the type of gas particles, but instead depends directly on the number of particles of gas that are present. For example, we can double the pressure in a container by adding an equal number of particles of the same gas or of a different gas (Figure 4.12). In a mixture of gases, the pressure exerted by each gas is called the **partial pressure, P,** of that gas, and depends only upon the number of particles of that gas present. Dalton's law states that the total pressure of a mixture of gases is equal to the sum of the partial pressures of each of the gases in the mixture. For a mixture of four gases, *A, B, C,* and *D,*

$$P_{\text{total}} = P_A + P_B + P_C + P_D \tag{1}$$

The earth's atmosphere is a mixture of nitrogen, oxygen, argon, and other gases found in small amounts. By Dalton's law then,

$$P_{\text{atmosphere}} = P_{N_2} + P_{O_2} + P_{Ar} + P_{\text{other}} \tag{2}$$

Example 4-4

1. What is the partial pressure of oxygen in the air if $P_{N_2} = 593.0$ mm Hg, $P_{Ar} = 7.0$ mm Hg, and $P_{\text{other}} = 0.2$ mm Hg, when the atmospheric pressure is measured at 760.0 mm Hg?

   Substituting into equation (2), we get

$$760.0 \text{ mm Hg} = 593.0 \text{ mm Hg} + P_{O_2} + 7.0 \text{ mm Hg} + 0.2 \text{ mm Hg}$$
$$P_{O_2} = 760.0 \text{ mm Hg} - 600.2 \text{ mm Hg}$$
$$P_{O_2} = 159.8 \text{ mm Hg}$$

2. A cylinder contains a mixture of oxygen and nitrous oxide (N_2O) which is used as an anesthetic. The pressure gauge on the tank reads 1.20 atm. If the partial pressure of the oxygen is 137 torr, what is the partial pressure of the nitrous oxide?

The total pressure is 1.20 atm = 1.20 × 760 torr
= 912 torr

$$P_{N_2O} + P_{O_2} = P_{total}$$
$$P_{N_2O} + 137 \text{ torr} = 912 \text{ torr}$$
$$P_{N_2O} = 912 \text{ torr} - 137 \text{ torr}$$
$$= 775 \text{ torr}$$

4.11 The Diffusion of Respiratory Gases

The diffusion of respiratory gases (oxygen and carbon dioxide) within our bodies is directly related to the partial pressures of these gases (Figure 4.13). These gases will diffuse from a region of higher partial pressure to a region of lower pressure. For example, venous blood entering the lungs from the tissues has been depleted of its oxygen supply and is carrying the waste product carbon dioxide from the cells. Oxygen will diffuse from the lungs (P_{O_2} = 104 mm Hg) to the blood (P_{O_2} = 40 mm Hg), and carbon

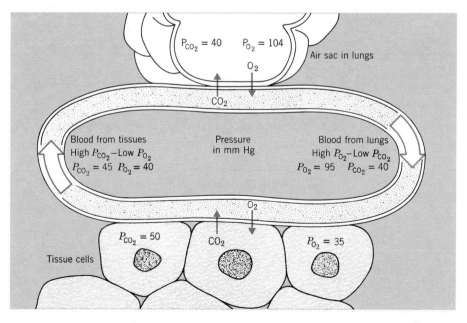

Figure 4.13 The movement of oxygen and carbon dioxide within the blood and tissues depends on the partial pressure of each gas.

dioxide will diffuse from the blood (P_{CO_2} = 45 mm Hg) to the lungs (P_{CO_2} = 40 mm Hg) to be exhaled. The oxygen, most of which is held by the carrier molecule hemoglobin, is then transported by the arterial blood to the tissues. Tissue cells are constantly using oxygen, so the partial pressure of oxygen (or the oxygen tension) in the cells is low. Because the arterial blood has a higher oxygen tension, the oxygen diffuses from the blood (P_{O_2} = 95 mm Hg) to the tissues (P_{O_2} = 35 mm Hg). Carbon dioxide, which is produced in the tissues by the cells, will diffuse from the cells (P_{CO_2} = 50 mm Hg) to the bloodstream (P_{CO_2} = 45 mm Hg) to be carried back to the lungs.

Chapter Summary

Matter can exist in three states: solid, liquid, and gas. Solids have a rigid structure, with definite shape and volume. Liquids have a fixed volume, but take on the shape of the container. Gases have no fixed volume or shape. Adding energy to a solid increases the kinetic energy of the particles until the solid structure breaks down, or melts. The temperature at which the solid melts to form a liquid is called the melting point, and the amount of energy that must be added to convert one gram of the substance from solid to liquid is called the heat of fusion. Adding energy to a liquid will increase the kinetic energy of the liquid particles until the liquid boils. The temperature at which this occurs under conditions of normal atmospheric pressure is called the boiling point. The amount of energy needed to convert a gram of liquid to gas is called the heat of vaporization.

The kinetic-molecular theory of gases describes the physical properties of an ideal gas as follows:

1. Gases are made up of very small particles, with a great deal of space between them.
2. The particles travel at high speeds in straight lines.
3. The collisions of the particles with each other and with the container are completely elastic.
4. There is no attraction or repulsion between the particles.
5. The kinetic energy of the particles increases with the temperature.

Boyle's law states that if the pressure of a gas is increased, the volume will decrease when the temperature is constant. Charles' law states that the volume will increase when the temperature increases, if the pressure remains constant. Graham's law states that the lighter the gas, the faster the rate of diffusion. Henry's law says that the higher the pressure of a gas over a liquid, the greater the solubility of the gas in the liquid. The pressure of a gas depends upon the number of particles of gas in the container, not the nature of the particles. Dalton's law states that the total pressure of the gases in a container is equal to the sum of the partial pressures of each gas. The partial pressure of a gas is defined as the pressure that the gas would exert if it were the only gas in the container.

Exercises and Problems

1. Complete the following conversions:
   (a) 5.0 atm = ________ mm Hg (c) 70 torr = ________ mm Hg
   (b) 190 mm Hg = ________ atm (d) 2.7 atm = ________ torr
2. The element bromine has a melting point of −7°C and a boiling point of 59°C. Draw a heating curve for bromine (similar to Figure 4.4), and label the following:
   (a) Melting point
   (b) Boiling point
   (c) Region A—bromine is a gas
   (d) Region B—bromine is a liquid
   (e) Region C—bromine is a solid
   (f) Region D—bromine exists as solid and liquid
   (g) Region E—bromine exists as liquid and gas
3. Describe in your own words what happens on the molecular level when water is cooled from 110°C to −5°C.
4. How much energy will be given off when 5 grams of water are frozen at 0°C?
5. At room temperature ammonia is a gas, water is a liquid, and sugar is a solid. Which substance has the strongest attraction between its particles? Which has the weakest?
6. Which substance in each of the following pairs is more viscous?
   (a) Motor oil or gasoline
   (b) Water or honey
   (c) Water at 10°C or water at 70°C
7. State in your own words the kinetic-molecular theory of gas behavior.
8. Propose a reason why real gases do not obey the kinetic molecular theory under conditions of very low temperature.
9. Everyone has observed that if the sun comes out immediately after a rain shower, the puddles dry up much faster than if it remains cloudy. Explain, on the molecular level, why this occurs.

10. Why does splashing your face with water cool you on a hot day?

11. Which would cause a more severe burn: water at 100°C or steam at 100°C? Explain your answer.

12. If there is a danger of a frost in late spring, grape growers will sprinkle their vineyards with water to protect their grapes from freezing. Suggest a reason why this is effective.

13. Rooms that must be kept sterile are often kept at a pressure slightly higher than atmospheric pressure. Explain how this procedure helps to keep out dust and microorganisms.

14. When the barometer reads 720 mm Hg, a sample of oxygen occupies a volume of 250 ml. What volume will the sample occupy when the barometer reads 750 mm Hg?

15. A 1.0 liter sample of gas is collected under a pressure of 912 mm Hg. What will the volume of the gas be at standard atmospheric pressure?

16. A diver collecting samples at a depth of 100 meters exhales a bubble having a volume of 100 ml. The pressure at this depth is 11 atm. What will the volume of the bubble be when it reaches the surface of the ocean (assuming the water temperature is constant)?

17. An aerosol hair spray is sold in a 500 ml can that is pressurized with propellant to 4.8 atm at 20°C. The can has been constructed to withstand an internal pressure of 8.0 atm. What is the maximum Fahrenheit temperature that the can will withstand before it explodes?

18. If a sample of nitrogen occupies 505 ml at −23°C, what volume will it occupy at 23°C?

19. If a sample of ammonia occupies a volume of 1.2 liters at 45°C, to what temperature in °C must the gas be lowered to reduce its volume to 1.0 liter?

20. What volume will a gas occupy at 27°C and 760 mm Hg, if it occupies 2.8 liters at 30°C and 909 mm Hg?

21. When collected under the conditions of 27°C and 800 mm Hg, a gas occupies 400 ml. To what temperature must the gas be cooled to reduce its volume to 320 ml when the pressure falls to 720 mm Hg?

22. Chlorine gas molecules are about four times heavier than ammonia molecules, and have a distinctly different smell. If equal quantities of these gases are released at the same time on the far side of the laboratory, which odor would you detect first?

23. Assume that you have two flasks containing equal amounts of water into which equal amounts of carbon dioxide have been introduced. The pressure in one container is 1 atm and in the other container it is 2.5 atm. Which container would have more carbon dioxide dissolved in the water? Give the reason for your answer.

24. The pressure in a bottle containing a mixture of oxygen and nitrogen is 1.00 atm. The partial pressure of the oxygen is 76 mm Hg. What is the partial pressure of the nitrogen?

25. A cyclopropane-oxygen mixture can be used as an anesthetic. If the partial pressure of cyclopropane is 255 torr and the partial pressure of oxygen is 855 torr, (a) What is the total pressure in the tank? (b) What is the total pressure expressed in atmospheres? (c) Which gas has a greater number of molecules in the tank?

chapter 5

The Atom and Radioactivity

Learning Objectives

By the time you have finished this chapter, you should be able to:

1. Describe the structure of an atom, listing three subatomic particles, their relative mass, their charge, and their location in the atom.
2. Define *atomic number.*
3. Define *mass number.*
4. Given the atomic number and mass number of any element, indicate the number of protons, neutrons, and electrons in the atom.
5. Describe how isotopes of an element differ.
6. Define *radioactivity.*
7. Describe the three types of radiations given off by radioactive material.
8. Explain *half-life.*
9. Define *nuclear transmutation.*
10. Define *atomic fusion* and *atomic fission.*
11. Explain the difference between a breeder reactor and a burner reactor.

"Dawn, July 16, 1945, 5:29:35 Mountain War Time. The countdown had reached Zero minus ten seconds.

"In those milliseconds before the most awesome weapon devised by man created its first terrifying sunrise, the only sound on the desert wastes of southern New Mexico was the mating buzz of a colony of spadefoot toads. And, if one strained for it, the distant drone of a B-29 bomber . . .

"A pinprick of a brilliant light punctured the darkness, spurted upward in a flaming jet, then spilled into a dazzling cloche of fire that bleached the desert to a ghastly white. It was precisely 5:29:45 A.M. . . .

"For a fraction of a second the light in that bell-shaped fire mass was greater than any ever produced before on earth. Its intensity was such that it could have been seen from another planet. The temperature at its center was four times that at the center of the sun and more than 10,000 times that at the sun's surface. The pressure, caving in the ground beneath, was over 100 billion atmospheres, the most ever to occur at the earth's surface. The radioactivity emitted was equal to one million times that of the world's total radium supply.

"No living thing touched by that raging furnace survived. Within a millisecond the fireball had struck the ground, flattening out at its base and acquiring a skirt of molten black dust that boiled and billowed in all directions. Within twenty-five milliseconds the fireball had expanded to a point where the Washington Monument would have been enveloped. At eight tenths of a second the ball's white-hot dome had topped the Empire State Building. The shock wave caromed across the roiling desert . . .

"For a split second after the moment of detonation the fireball, looking like a monstrous convoluting brain, bristled with spikes where the shot tower and balloon cables had been vaporized. Then the dust skirt whipped up by the explosion mantled it in a motley brown. Thousands of tons of boiling sand and dirt swept into its maw only to be regurgitated seconds later in a swirling geyser of debris as the fireball detached itself from the ground and shot upward. As it lifted from the desert, the sphere darkened in places, then opened as fresh bursts of luminous gases broke through its surface.

"At 2,000 feet, still hurtling through the atmosphere, the seething ball turned reddish yellow, then a dull blood-red. It churned and belched forth smoking flame in an elemental fury. Below, the countryside was bathed in golden and lavender hues that lit every mountain peak and crevasse, every arroyo and bush with a clarity no artist could capture. At 15,000 feet the fireball cleaved the overcast in a bubble of orange that shifted to a darkening pink. Now, with its flattened top, it resembled a giant mushroom trailed by a stalk of radioactive dust. Within another few seconds the fireball had reached 40,000 feet and pancaked out in a mile-wide ring of greying ash. The air had ionized around it and crowned it with a lustrous purple halo. As the cloud finally settled, its chimney-shaped column of dust drifted northward and a violet afterglow tinged the heavens above Trinity."*

Our knowledge of nuclear energy has rapidly expanded since that awesome morning in July, 1945, and the technology that has been developed for the use of radioactive materials is becoming more and more important in improving standards of human life—from medical diagnosis and treatment, to prevention of food spoilage, to the production of useful energy for our society.

What makes a substance radioactive? How is the tremendous energy released in a nuclear reaction produced? Why is it that radioactivity can both cause and cure cancer? Are there important dangers from increased use of radioactive materials in power generation, industrial processes, and medical technology? It is important that we each know the answers to these questions, for our local and national governments continually make policy decisions concerning nuclear energy that will affect us and many future generations.

** Excerpted from* Day of Trinity *by Lansing Lamont. Copyright © 1965 by Lansing Lamont. Reprinted by permission of Atheneum Publishers.*

Atomic Structure

5.1 The Parts of the Atom

To answer these questions we need to take a more detailed look at the structure of the atom. In Chapter 2 we stated that atoms are composed of many types of particles, but that we would be focusing our attention on three such particles: protons, neutrons, and electrons. **Protons** and **electrons** are electrically charged particles, whereas the **neutron** is neutral (that is, it has no charge). A proton is assigned the smallest unit of positive charge (+1) that will just cancel the negative charge on an electron (−1). Interestingly, the terms *positive* and *negative,* which are used to describe the opposing effects of electrical charges, were first used by Ben Franklin more than 50 years before the discovery of the electron and proton. A proton will repel other protons (charges that are alike repel one another), and will attract electrons (unlike charges attract one another). See Table 5.1.

Table 5.1 Subatomic Particles

Name of Particle	Location in the Atom	Charge	Symbol	Relative Mass (amu)
Proton	Nucleus	+1	p, ${}^{1}_{1}H$	1
Electron	Around the nucleus	−1	e, e^{-1}, ${}^{0}_{-1}e$	$\frac{1}{1837}$
Neutron	Nucleus	0	n, ${}^{1}_{0}n$	1

An atom consists of a small dense **nucleus** that contains protons and neutrons. Because we have just said that protons repel one another, you might wonder what holds the nucleus together. Although nobody completely understands nuclear forces, the neutrons seem to play an important role in binding the positive protons together. The electrons are found in the region surrounding the nucleus, but for the most part an atom is just empty space. To give you an idea of the relative positions of the subatomic particles in an atom, suppose you were in a large baseball stadium. If we were to let a flea on second base represent the nucleus of the atom, the nearest electron would be found somewhere in the top deck of the stands (Figure 5.1).

5.2 Atomic Number

The special characteristic that determines which element an atom represents is the number of protons in that atom. The **atomic number** of an element is the term used to describe the number of protons in the nucleus of any atom of that element. (The atomic numbers of the elements are listed on the inside back cover of the book.) Because the electrical charge on a proton just cancels the electrical charge on an electron, we can see

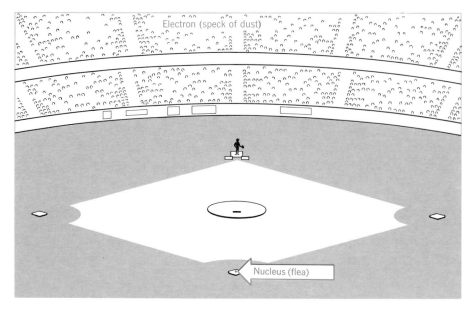

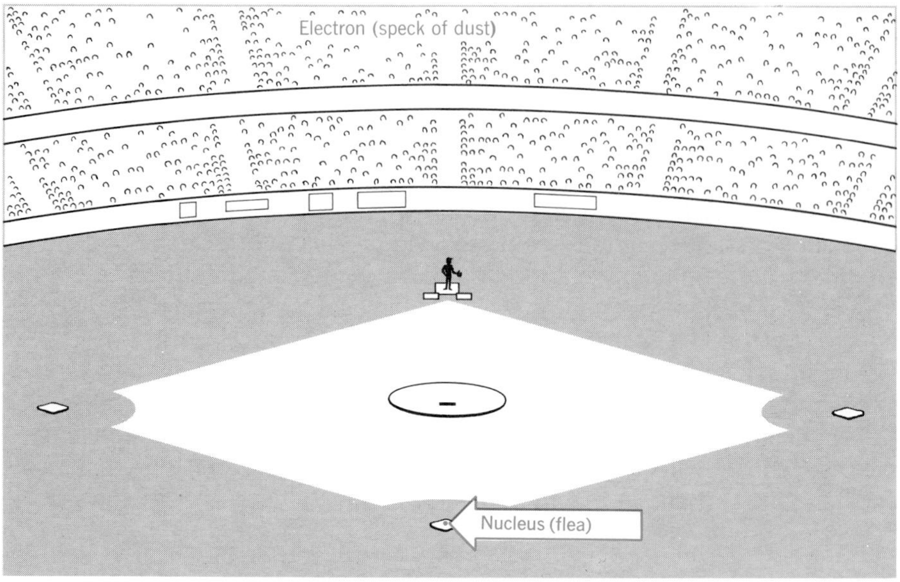

Figure 5.1 An atom is mostly empty space. If we imagine a flea on second base to be the nucleus of an atom, the nearest electron would be a speck of dust somewhere in the top deck of the stands. (From *Chemistry, An Experimental Science*. Used with the permission of CHEM Study.)

that the atomic number of an element also tells us how many electrons there are in a neutral atom of that element. For example, the element sodium has an atomic number of 11. Therefore, each neutral atom of sodium will have 11 protons in the nucleus and 11 electrons surrounding the nucleus. But, no matter how many electrons (or neutrons) there may be, the identity of an element is always determined by the number of protons in its nucleus—that is, by its atomic number.

5.3 Mass Number

Protons, electrons, and neutrons are extremely small particles. Protons and neutrons have a mass of 1.7×10^{-24} grams; electrons are even lighter, having a mass only $\frac{1}{1837}$ that of a proton. In fact, the mass of the electrons in an atom is so small in comparison with the mass of the protons and neutrons that it is ignored when calculating the mass of an atom. We define the **mass number** of an atom to equal the number of protons plus the number of neutrons in the nucleus of the atom.

$$\text{Mass number} = \text{Protons} + \text{Neutrons}$$

or

$$M = p + n$$

Chemists often use shorthand methods to express the mass number and the atomic number of an atom. For example, an atom of carbon (C) has six protons and six neutrons. The atomic number (Z) is 6, and the mass number (M) is 12. Chemists may refer to atoms of carbon having the mass number of 12 as carbon-12, ^{12}C or $^{12}_{6}$C, where the upper number is the mass number and the lower number is the atomic number.

Example 5-1 _____

1. State the number of protons, neutrons, and electrons in a neutral atom of each of the following.

 (a) $^{14}_{6}$C
 (b) Cobalt-60
 (c) ^{235}U

 (a) Using the table on the inside back cover of the book, we can identify this element as carbon. The symbol $^{14}_{6}$C tells us that the mass number is 14 and the atomic number is 6. The atomic number gives both the number of protons and the number of electrons in a neutral atom.

 $$p = 6 \quad \text{and } e^- = 6$$

 We know that $\qquad M = p + n$
 therefore $\qquad 14 = 6 + n$
 $\qquad\qquad 8 = n$

 (b) Looking up cobalt, we find that the atomic number is 27. The mass number is 60. From the atomic number we then have that

 $$p = 27 \quad \text{and } e^- = 27$$
 $$60 = 27 + n$$
 $$33 = n$$

 (c) U is the symbol for the element uranium, which has an atomic number of 92 and a mass number of 235.
 From the atomic number,

 $$p = 92 \quad \text{and } e^- = 92$$

 From the mass number,

 $$235 = 92 + n$$
 $$143 = n$$

2. State three ways of writing the symbol for an atom of iodine with 78 neutrons.

 From the inside back cover we learn that the symbol for iodine is I and its atomic number is 53. Therefore, the mass number for such an atom is

 $$M = 78 + 53 = 131$$

 The three different notations would be

 Iodine-131, ^{131}I, and $^{131}_{53}$I

5.4 Isotopes

In the early nineteenth century, John Dalton developed an atomic theory based on the idea that each atom of a given element is exactly alike. It was not until 100 years later that Fredrick Soddy proved this theory was wrong by showing that the element neon consisted of not one, but two types of atoms. Some of these neon atoms had a mass number of 20, and some had a mass number of 22. (A third type of neon atom with the mass number of 21 also exists.) Neon has an atomic number of 10, so each atom of neon must have 10 protons in the nucleus. Therefore, these two different types of neon atoms must have different numbers of neutrons in their nuclei. Soddy invented the term **isotopes** to describe atoms of an element containing different numbers of neutrons in the nucleus (Figure 5.2). For neon, then, the isotopes are:

	Atomic Number	Mass Number	Number of *p*	Number of *n*
Neon-20	10	20	10	10
Neon-21	10	21	10	11
Neon-22	10	22	10	12

Most elements have two or more naturally occurring isotopes; in fact, tin (Sn) has 10! Only 22 elements have but one type of atom and, therefore, have no naturally occurring isotopes.

5.5 Atomic Weight

Because atoms are so small and light, it is impossible to measure the mass of a small number of atoms. However, it is possible to measure the relative masses of atoms of different elements. A scale based on these relative measurements has been developed, with the carbon-12 isotope being arbitrarily assigned a value of exactly 12 **atomic mass units (amu).** If an

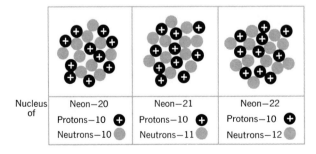

Figure 5.2 The element neon has three types of atoms that differ in the number of neutrons in the nucleus. These atoms are isotopes of neon.

Table 5.2 The Relative Abundance of the Isotopes of Several Elements

Isotope	Percent Natural Abundance	Isotope	Percent Natural Abundance
Hydrogen-1	99.99%	Silicon-28	92.21%
Hydrogen-2	0.01%	Silicon-29	4.70%
		Silicon-30	3.09%
Carbon-12	98.89%		
Carbon-13	1.11%	Chlorine-35	75.53%
		Chlorine-37	24.47%
Nitrogen-14	99.63%		
Nitrogen-15	0.37%	Zinc-64	48.89%
		Zinc-66	27.81%
Oxygen-16	99.76%	Zinc-67	4.11%
Oxygen-17	0.04%	Zinc-68	18.57%
Oxygen-18	0.20%	Zinc-70	0.62%
Fluorine-19	100.00%		
		Bromine-79	50.54%
		Bromine-81	49.46%

isotope of another element were $\frac{1}{2}$ as heavy as the carbon-12 isotope, then its mass in atomic mass units would be $\frac{1}{2} \times 12$, or 6; if an isotope were 2.84 times as heavy as the carbon-12 isotope, its mass would be $2.84 \times 12 = 34.08$ amu.

Because most elements have at least two naturally occurring isotopes, a sample of any of these elements will contain a mixture of the different isotopes. Therefore, the exact mass of any sample will depend upon the share, or percent abundance, of each isotope in that particular sample. Table 5.2 lists the percent abundance of the naturally occurring isotopes of several elements. We would expect the mass of a random sample of any element to be an average of the masses of its isotopes, weighted by the relative abundance of those isotopes. The name given to such a weighted average is **atomic weight.** That is, the atomic weight of an element is the weighted average of the masses of the naturally occurring isotopes of the element, expressed in atomic mass units. The atomic weight of each of the elements is listed on the inside back cover of this book.

Example 5-2

The calculation of a weighted average is not a very difficult procedure. In fact, teachers often use such a procedure in assigning final grades in a course. For example, suppose a teacher were to tell you that each of two mid-terms would determine 25% of your final grade, and the final exam would be worth 50% of the final grade. If you received 76 on your first mid-term, 64 on your second mid-term, and 90 on your final exam, your grade for the course would be 80. This final score is calculated by multiplying each grade by its percent weight in the final grade, and then adding up these products.

$$76 \times 25\% = 76 \times 0.25 = 19$$
$$64 \times 25\% = 64 \times 0.25 = 16$$
$$90 \times 50\% = 90 \times 0.50 = \underline{45}$$
$$80$$

A similar procedure is followed in calculating the atomic weight of an element. The mass of each isotope is multiplied by its percent abundance, and the products are then added up to give the atomic weight. For example, boron has two isotopes, boron-10 and boron-11, whose percentage abundances are 19.6% and 80.4%, respectively. The atomic weight of boron is, then,

$$10.0 \text{ amu} \times 19.6\% = 10.0 \times 0.196 = 1.96$$
$$11.0 \text{ amu} \times 80.4\% = 11.0 \times 0.804 = \underline{8.844}$$
$$10.804 \text{ or } 10.8 \text{ amu}$$

Radioactivity

5.6 What Is Radioactivity?

Some atoms have nuclei that are unstable. This is especially true of elements that have high atomic numbers and, therefore, a high number of positive protons in the nucleus (remember, charges that are alike repel one another). Such an unstable nucleus will decay; that is, it will "spit out" particles, leaving behind a new nucleus called a daughter nucleus. A daughter nucleus may or may not be stable. An unstable daughter nucleus will decay again, and this process will continue until a stable daughter nucleus is formed. Such a series of decays is called a **decay series** or **disintegration series** (Figure 5.3). The decay of a nucleus can give rise to several different forms of radiation (Table 5.3). **Radioactivity** is the term used to describe the giving off, or emission, of such radiation from samples of certain elements or their compounds.

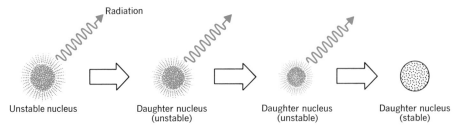

Figure 5.3 Radioactivity is the term used to describe the emission of radiation from an unstable nucleus.

Table 5.3 Natural Radioactive Radiation

Radiation	Composition	Charge	Symbol	Penetration (Rays Stopped by)
Alpha	Helium nucleus	+2	α, ${}^{4}_{2}\text{He}$	Piece of paper
Beta	Electrons	−1	β, ${}^{0}_{-1}\text{e}$	Piece of wood
Gamma	High-energy rays similar to X rays	0	γ	High-density concrete

5.7 Alpha Radiation (α Particles)

One way in which a nucleus can become more stable is by giving off alpha radiation. **Alpha radiation** consists of streams of alpha particles, each of which is made up of two protons and two neutrons (the nucleus of a helium atom, ${}^{4}_{2}\text{He}$). By giving off an alpha particle, the atomic number of the nucleus is reduced by two, and the mass number is reduced by four. A well-known source of alpha radiation is the most abundant isotope of uranium, uranium-238, which decays by giving off an alpha particle to form an atom of thorium-234. A shorthand way of representing this decay is as follows:

$$ {}^{238}_{92}\text{U} \longrightarrow {}^{234}_{90}\text{Th} + {}^{4}_{2}\text{He} $$

The starting unstable nucleus is shown on the left-hand side of the arrow, and the products resulting from the radioactive decay of this nucleus are shown on the right-hand side. To be sure that this shorthand statement is correctly written, we must check that the number of protons and neutrons on one side of the arrow is equal to the number of protons and neutrons on the other side. In other words, the sum of the mass numbers on each side of the arrow must be equal, and the sum of the atomic numbers on each side of the arrow must also be equal. For the radioactive decay above,

$$ \begin{aligned} \text{Mass number:} \quad & 238 = 234 + 4 \\ \text{Atomic number:} \quad & 92 = 90 + 2 \end{aligned} $$

By the way, the thorium atom produced by the decay of uranium-238 is itself unstable, and will decay to form a new nucleus.

Alpha particles are the largest particles emitted by radioactive substances, and have very little penetrating power. Even when traveling through air they lose energy very quickly through collisions with air molecules, and will stop within a few inches. They can be stopped by a piece of paper, and cannot penetrate even the dead layer of cells on the surface of your skin. An intense external dose of alpha radiation, however, would produce a burn on the skin. And alpha particles can do a great deal of damage if they are emitted inside the body, which might result from inhaling or swallowing an alpha emitter.

5.8 Beta Radiation (β Particles)

Beta radiation, like alpha radiation, consists of streams of particles. In this case, the particle is an electron ($_{-1}^{0}e$) that is produced inside the nucleus. During beta decay, a neutron in the nucleus is changed into a proton and an electron. This electron (or β particle) is ejected from the nucleus, leaving behind a daughter nucleus having the same mass number as the original atom, but a different atomic number. For example, the thorium atom that is produced by the alpha decay of uranium-238 is a beta emitter.

$$^{234}_{90}\text{Th} \longrightarrow {}^{234}_{91}\text{Pa} + {}_{-1}^{0}e$$

Beta particles are 7000 times smaller than alpha particles and, therefore, have much more penetrating power. (Consider the ability of a very fine needle to penetrate your skin as compared with a basketball.) Beta particles can pass through a piece of paper, but will be stopped by a piece of wood. Beta radiation can penetrate the dead outer layer of your skin; it will be stopped within the skin layer, causing damage to the skin tissue and making it appear burned (Figure 5.4). As is the case with alpha particles, beta particles hitting the skin from the outside cannot penetrate to internal organs, but the effect on internal organs can be severe if a beta emitter is taken internally.

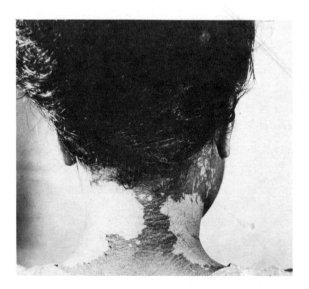

Figure 5.4 The burned area on the neck of this Rongelap native was caused by accidental exposure to over 2000 rads of beta radiation from radioactive fallout. (Courtesy Brookhaven National Laboratory)

5.9 Gamma Radiation (γ Rays)

Gamma rays are not particles, but are high-energy radiation similar to X rays (you might look again at the energy spectrum shown in Figure 3.6). Quite often the daughter nucleus produced by an alpha or beta emitter will be in a high-energy, or excited, state. It can release this energy in the form of gamma radiation to become more stable. The release of gamma rays, therefore, often occurs together with alpha or beta radiation. For example, radium-226 has a radioactive nucleus that releases alpha and gamma radiation when it decays.

$$^{226}_{88}\text{Ra} \longrightarrow {}^{222}_{86}\text{Rn} + {}^{4}_{2}\text{He} + \gamma$$

Because they have such high energy, gamma rays easily pass through paper and wood, but can be stopped by lead blocks or thick concrete walls. Gamma rays will completely penetrate the human body, causing cellular damage as they pass through (Figure 5.5).

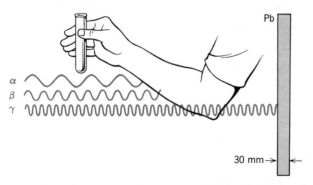

Figure 5.5 Each type of radiation has a different penetrating power. Alpha particles can penetrate about 0.05 mm into the outer layer of skin. Beta particles can penetrate between 0.06 and 4 mm into the skin, depending upon their energy. Gamma radiation, however, will pass right through the arm and will be stopped only by a block of lead 30 mm thick.

Example 5-3

1. Determine the symbol of the element that belongs in the following decay.

$$^{3}_{1}\text{H} \longrightarrow \quad (?) \quad + {}^{0}_{-1}\text{e}$$

The sum of the mass numbers on both sides of the arrow must be equal, so the mass number of the unknown element is

$$3 = M + 0$$
$$M = 3$$

The sum of the atomic numbers on both sides of the arrows must be equal, so the atomic number of the unknown element is

$$1 = Z + (-1)$$
$$Z = 1 + 1 = 2$$

From the back inside cover of the book, we find that the element having an atomic number of 2 is helium, He.
The complete decay, therefore, is

$$^{3}_{1}\text{H} \longrightarrow {}^{3}_{2}\text{He} + {}^{0}_{-1}e$$

2. Polonium-214 is an alpha and gamma emitter that decays to form an isotope of lead. Write the shorthand representation of this decay.

(a) Use the inside back cover and Table 5.3 to help write the necessary symbols

Polonium-214 $= {}^{214}_{84}\text{Po}$
Alpha particle $= {}^{4}_{2}\text{He}$
Gamma ray $= \gamma$
Lead $= {}^{M}_{82}\text{Pb}$ (we have to determine the mass number of the lead)

(b) We can now write a representation of this decay, and use it to determine the mass number of the lead.

$$^{214}_{84}\text{Po} \longrightarrow {}^{M}_{82}\text{Pb} + {}^{4}_{2}\text{He} + \gamma$$

The sum of the mass numbers on each side of the arrow must be equal, therefore,

$$214 = M + 4$$
$$M = 210$$

(c) The correct representation is

$$^{214}_{84}\text{Po} \longrightarrow {}^{210}_{82}\text{Pb} + {}^{4}_{2}\text{He} + \gamma$$

5.10 Half-Life

Each radioactive substance has its own special rate of decay—that is, the number of emissions, or disintegrations, that occur each minute. This rate of decay is expressed by a number called the half-life of the isotope. The **half-life ($t_{1/2}$)** of an isotope is the length of time it takes for one-half of the atoms in a given sample to undergo radioactive decay. This means that after one half-life, one-half of a sample of the radioactive isotope will have decayed to form a new substance, and one-half will remain. After two half-lives, one-half of one-half (or one-fourth) will remain. After three half-lives, one-half of one-fourth (or one-eighth) will remain (Figure 5.6). For example,

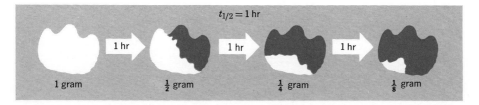

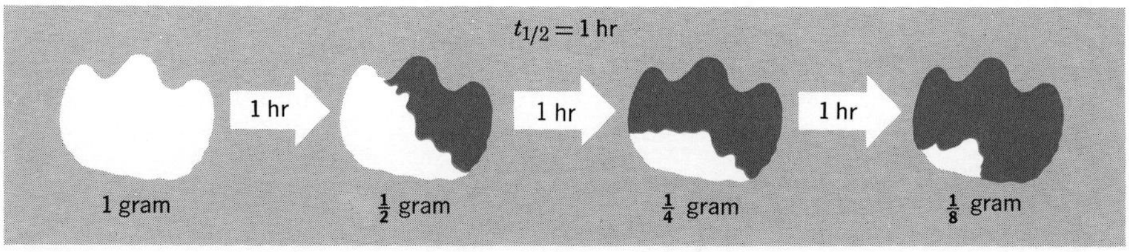

Figure 5.6 After one half-life, one-half of a sample of the radioactive element will have decayed to form a new element and one-half will remain. After two half-lives, one-fourth of the original sample will remain among the decay products.

nitrogen-13 has a half-life of 10 minutes. If you have 1 gram of nitrogen-13, 10 minutes later 0.5 grams of nitrogen-13 will have decayed to form carbon-13, and 0.5 grams of nitrogen-13 will remain.

As another example, an isotope of the element technetium is widely used in medical diagnosis. This isotope has a half-life of 6 hours, which is quite favorable for purposes of diagnosis but which requires the laboratory to constantly renew its supply. If there were 1 gram of this isotope in the laboratory at 6:00 P.M. on Friday, only 0.25 grams would remain for use on Saturday morning (two half-lives later). Continuing in this way, only 1 milligram would still be found among the decay products when the lab opened on Monday morning (Figure 5.7).

Half-lives of the different isotopes can be as short as fractions of a second, or as long as billions of years. The half-life of an isotope is an indication of how stable the isotope is. Most artificially produced radioisotopes are highly unstable and have very short half-lives. Table 5.4 gives some examples of isotopes and their half-lives.

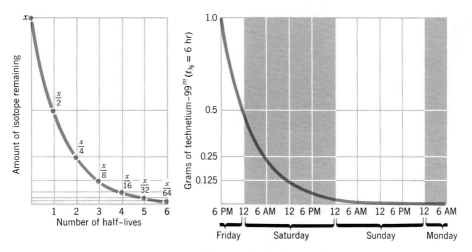

Figure 5.7 The half-life of an element is the amount of time it takes for one-half of the atoms in a given sample to undergo radioactive decay. The half-life of technetium-99^m is 6 hours. If you had a one gram sample of technetium-99^m at 6 PM, Friday, you would have only one-half of a gram of technetium-99^m at 12 midnight. By 6.00 AM Monday morning, only one milligram (0.001 g) would be left.

Table 5.4 Some Radioactive Isotopes and Their Half-Lives

Element	Isotope	Half-Life	Radiations Given Off
Hydrogen	$^{3}_{1}H$	12 years	Beta
Carbon	$^{14}_{6}C$	5730 years	Beta
Phosphorus	$^{32}_{15}P$	14 days	Beta
Potassium	$^{40}_{19}K$	1.28×10^{9} years	Beta and gamma
Cobalt	$^{60}_{27}Co$	5 years	Beta and gamma
Strontium	$^{90}_{38}Sr$	28 years	Beta
Technetium	$^{99m}_{43}Tc$	6 hours	Gamma
Iodine	$^{131}_{53}I$	8 days	Beta and gamma
Cesium	$^{137}_{55}Cs$	30 years	Beta
Polonium	$^{214}_{84}Po$	1.6×10^{-4} seconds	Alpha and gamma
Radium	$^{226}_{88}Ra$	1600 years	Alpha and gamma
Uranium	$^{235}_{92}U$	7.1×10^{8} years	Alpha and gamma
	$^{238}_{92}U$	4.5×10^{9} years	Alpha
Plutonium	$^{239}_{94}Pu$	24,400 years	Alpha and gamma

Example 5-4

Phosphorus-32 is often used to study chemical reactions in biological research. The half-life of ^{32}P is 14 days. If a research laboratory received 500 mg of phosphorus-32, how many milligrams of ^{32}P would remain for use after 70 days?

(a) If 1 half-life equals 14 days, then 70 days would be

$$70 \cancel{\text{days}} \times \frac{1 \text{ half-life}}{14 \cancel{\text{days}}} = 5 \text{ half-lives}$$

(b) Using the graph in Figure 5.7, we see that after 5 half-lives the amount of ^{32}P remaining would be $\frac{X}{32}$, where X is the original amount of the isotope. In this case, $X = 500$ mg.

$$^{32}P \text{ remaining} = \frac{500 \text{ mg}}{32} = 15.6 \text{ mg}$$

The half-lives of radioactive elements can be very useful tools for discovering the age of archaeological objects. Carbon-14, which has a half-life of 5730 years, is used to date once-living material. This carbon isotope is created in the earth's upper atmosphere when nitrogen atoms are bombarded by cosmic rays (which are streams of particles pouring into the atmosphere from the sun and outer space). The procedure of carbon-14 dating assumes that the ratio of carbon-14 to the stable carbon-12 isotope remains constant in living organisms. When such an organism dies, the total amount of carbon-12 it has accumulated during its life becomes fixed and will not change. However, half of the carbon-14 it has accumulated will be gone in 5730 years. Therefore, measuring the ratio of carbon-14 to carbon-12 in material that was once alive is a fairly accurate way to date objects that have died within the last 40,000 years.

Although carbon-14 is very useful for dating once-living objects, isotopes with much longer half-lives must be used to date geologic periods in the earth's history. Uranium-238 decays through a decay series shown in Figure 5.8 to produce the stable lead-206 isotope. The half-life of uranium-238 is 4.5 billion years, so that if an original sample of rock

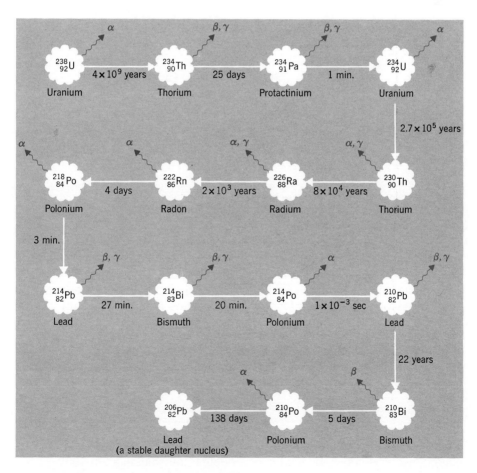

Figure 5.8 The decay series of uranium-238.

contained 1 gram of uranium-238, after 4.5 billion years only 0.50 grams of ^{238}U will remain, the rest having decayed to form ^{206}Pb. Using uranium/lead radioactive dating of rocks, scientists have estimated the age of the planets in the solar system to be approximately 4.5 billion years.

As another example of very long-term dating, potassium-40 decays to form either the calcium-40 isotope or the argon-40 isotope. A measurement of the ratio of potassium-40 to argon-40 showed the jawbone of a human-like creature found in Tanzania to be 5.5 million years old. Until this dating, the earliest human was believed to exist only 2.2 million years ago.

5.11 Nuclear Transmutation

A **nuclear transmutation** is a reaction in which a high-speed particle collides with a nucleus to produce a different nucleus. Some transmutations occur in nature, and many are produced in laboratories. For example, carbon-14 is produced in the upper atmosphere, and can be produced in laboratories, by bombarding nitrogen-14 with neutrons.

$$\underset{\text{Projectile}}{^{1}_{0}n} \quad + \quad \underset{\text{Target}}{^{14}_{7}N} \quad \longrightarrow \quad ^{14}_{6}C \quad + \quad \underset{\text{Proton}}{^{1}_{1}H}$$

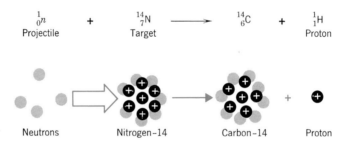

| Neutrons | Nitrogen–14 | Carbon–14 | Proton |

Carbon-14 is a beta emitter that can be substituted into biological compounds and traced as it travels through a living system.

$$^{14}_{6}C \longrightarrow ^{14}_{7}N + ^{0}_{-1}e$$

By using carbon dioxide "labeled" with carbon-14, Melvin Calvin, in the 1940s, was able to obtain a detailed picture of the chemical pathways of photosynthesis (the process that plants use to make sugar molecules from carbon dioxide and water).

A nuclear transmutation was produced in the laboratory for the first time in 1919 by Ernest Rutherford, who bombarded nitrogen gas with alpha particles.

$$^{14}_{7}N \quad + \quad ^{4}_{2}He \quad \longrightarrow \quad ^{18}_{9}F$$

The fluorine nucleus produced in this transmutation is very unstable and rapidly decays to form oxygen-17 and a proton.

$$^{18}_{9}F \quad \longrightarrow \quad ^{17}_{8}O \quad + \quad \underset{\text{Proton}}{^{1}_{1}H}$$

This experiment led to the discovery of the proton. The existence of the neutron also was shown by means of a nuclear transmutation. In 1932, James Chadwick bombarded beryllium-9 with alpha particles, causing the following transmutation:

$$^{9}_{4}\text{Be} + ^{4}_{2}\text{He} \longrightarrow ^{12}_{6}\text{C} + ^{1}_{0}n$$
$$\text{Neutron}$$

The neutrons that were produced in this transmutation had enough energy to cause additional nuclear reactions in nuclei with which they collided.

Before 1940, the uranium atom was the heaviest atom known. However, the invention of the cyclotron and other nuclear accelerators allowed scientists to create very high-energy projectiles, and in 1940, E. M. McMillan and P. H. Abelson produced neptunium (^{93}Np) by bombarding uranium with a stream of high-energy deuterons (the nucleus of hydrogen-2, $^{2}_{1}\text{H}$).

$$^{238}_{92}\text{U} + ^{2}_{1}\text{H} \longrightarrow ^{239}_{92}\text{U} + ^{1}_{1}\text{H}$$

$$^{239}_{92}\text{U} \xrightarrow[t_{1/2} = 23.5 \text{ min}]{} ^{239}_{93}\text{Np} + ^{0}_{-1}e$$

The neptunium produced in this way is a radioactive beta emitter, and decays to form plutonium-239. Plutonium is an alpha emitter with a half-life of 24,400 years, and is an important fuel used in atomic reactors and bombs.

$$^{239}_{93}\text{Np} \xrightarrow[t_{1/2} = 2.33 \text{ days}]{} ^{239}_{94}\text{Pu} + ^{0}_{-1}e$$

Example 5-5 _____

Insert the bombarding particle for the following nuclear transmutation.

$$^{238}_{92}\text{U} + (?) \longrightarrow ^{246}_{98}\text{Cf} + 4\,^{1}_{0}n$$

The sum of the mass numbers on both sides of the arrow must be equal, so the mass number of the bombarding particle is

$$238 + M = 246 + 4\,(1) \quad \text{[because there are 4 neutrons produced]}$$
$$M = 250 - 238$$
$$= 12$$

The sum of the atomic numbers on both sides of the arrow must be equal, so the atomic number of the bombarding particle is

$$92 + Z = 98 + 4\,(0)$$
$$Z = 98 - 92 = 6$$

The element with an atomic number of 6 is carbon. Therefore, the bombarding particle is $^{12}_{6}\text{C}$.

The Nucleus and Energy

Nuclear power has been acclaimed by many scientists as a solution to the world's energy crisis. There is no doubt that as we use up our store of fossil fuels, something must take its place to supply the energy needs

of our society. Although nuclear fuel may be an answer, great care and control must be exercised in expanding the use of nuclear power to minimize its impact on the environment and on future generations. In order to participate in decisions regarding the use of this nuclear technology, it is important that you understand how such a vast amount of energy can be produced from the nuclei of atoms, and what hazards may be involved.

5.12 Nuclear Fission

When bombarded with neutrons, the nuclei of several isotopes (^{235}U, ^{233}U, and ^{239}Pu) are capable of breaking apart or undergoing **fission** to form smaller, more stable nuclei. For example,

$$^{1}_{0}n + {}^{235}_{92}U \longrightarrow {}^{236}_{92}U \longrightarrow {}^{139}_{56}Ba + {}^{94}_{36}Kr + 3\,{}^{1}_{0}n + \text{Energy}$$

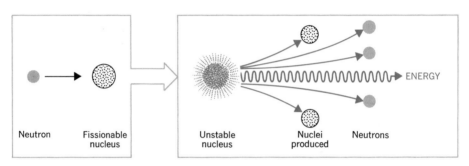

| Neutron | Fissionable nucleus | Unstable nucleus | Nuclei produced | Neutrons |

This fission process yields great amounts of energy. When 1 kilogram of ^{235}U or ^{239}Pu is fissioned, the amount of energy released is equivalent to 20,000 tons of TNT. Atomic bombs dropped on Japan at the end of World War II contained about 1 kilogram of fissionable material.

In nuclear fission, an atom of fissionable fuel (such as uranium-235) is struck by a neutron and breaks into two small fragments that fly apart at high speeds. The kinetic energy of these fast-moving particles is converted into heat as they collide with surrounding molecules. In addition to these two fragments, two or three neutrons are released, and these neutrons can react with other ^{235}U nuclei. If this process continues, a **chain reaction** occurs which, if uncontrolled, could result in an explosion (Figure 5.9). However, this chain reaction cannot occur unless there are at least a minimum number of fissionable nuclei (called the **critical mass**) present in one place.

5.13 Nuclear Reactors

Burner Reactors
In 1942, Enrico Fermi used a converted squash court at the University of Chicago as a place to build the first atomic reactor. Today, several hundred reactors are being built in various sizes and designs; however, the basic components of all nuclear reactors are the same. These components include:

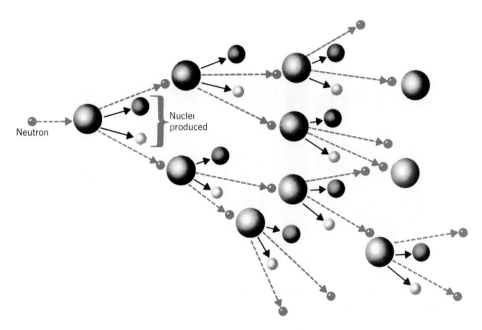

Figure 5.9 A nuclear chain reaction. When a nucleus undergoes fission, two or three neutrons are produced, which are capable of reacting with other nuclei. If a critical mass of fissionable nuclei is present, a chain reaction will occur.

1. The fuel, which contains significant amounts of one of the fissionable isotopes (^{235}U, ^{233}U, or ^{239}Pu), of which ^{235}U is the only naturally occurring isotope.
2. A moderator, such as graphite or water, which slows down the neutrons produced in the fission process.
3. Control rods made of cadmium or boron steel, which will absorb neutrons and which can be moved in and out of the reactor to control the rate of the reactions.
4. A heat transfer fluid such as water, liquid sodium, or pressurized gas, which removes the heat from the reactor core and transfers it to a steam generating system.
5. Shielding, which consists of an internal thermal shield to protect the walls of the reactor from radiation damage, and an external biological shield of high-density concrete to protect the workers from radiation.

The exact reactor design will depend upon its use, which could be electric power generation, the production of plutonium-239, the propulsion of ships

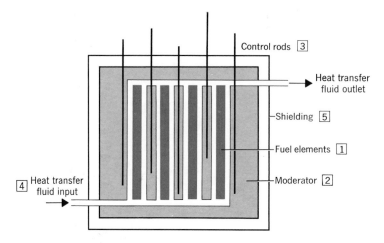

Figure 5.10 The core of a nuclear reactor.

or rockets, or the production of heat for desalinization, drying, evaporation, or other industrial uses (Figures 5.10 and 5.11).

Breeder Reactors

In burner reactors, the ^{235}U fuel is used up as energy is produced. This uranium isotope is the only naturally occurring fissionable fuel, and is found only in extremely small amounts in uranium ore. Because the world's supply of this fuel may be relatively quickly exhausted, the true promise of nuclear power currently hinges on the development of **breeder reactors.** A breeder reactor is one in which more fissionable fuel is produced than is used up in the heat generation process.

A breeder reactor uses uranium-238 and thorium-232, both of which are relatively abundant isotopes. By positioning them so that they are bombarded by some of the neutrons produced in the fission of uranium-235, new fissionable isotopes are formed.

$$\underset{\text{Fertile isotope}}{{}^{238}_{92}\text{U}} + {}^{1}_{0}\text{n} \longrightarrow {}^{239}_{92}\text{U} \xrightarrow[t_{1/2} = 24 \text{ min.}]{\nearrow {}^{0}_{-1}\text{e}} {}^{239}_{93}\text{Np} \xrightarrow[t_{1/2} = 2.3 \text{ days}]{\nearrow {}^{0}_{-1}\text{e}} \underset{\text{Fissionable isotope}}{{}^{239}_{94}\text{Pu}}$$

$$\underset{\text{Fertile isotope}}{{}^{232}_{90}\text{Th}} + {}^{1}_{0}\text{n} \longrightarrow {}^{233}_{90}\text{Th} \xrightarrow[t_{1/2} = 22 \text{ min.}]{\nearrow {}^{0}_{-1}\text{e}} {}^{233}_{91}\text{Pa} \xrightarrow[t_{1/2} = 27 \text{ days}]{\nearrow {}^{0}_{-1}\text{e}} \underset{\text{Fissionable isotope}}{{}^{233}_{92}\text{U}}$$

If the technology of breeder reactors can be sufficiently developed, the supplies of fertile uranium-238 and thorium-232 will assure a production of energy equal to hundreds of times the world's fossil fuel reserves.

Nuclear Wastes

One of the critical problems in the use of reactors for energy production is the question of safety. As we will see in the next chapter, exposure to

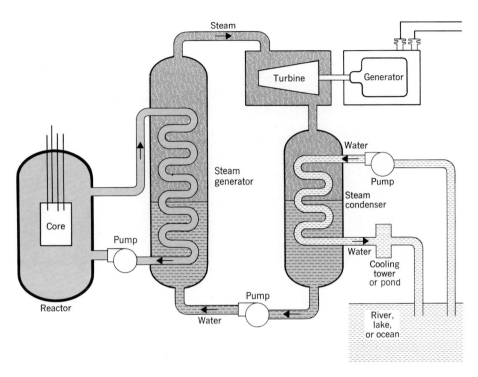

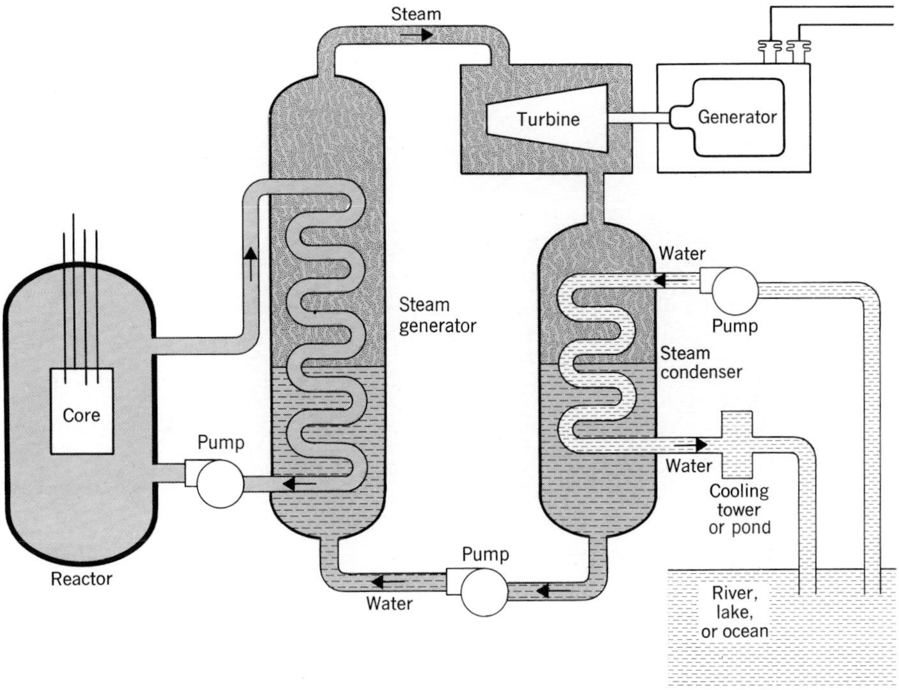

Figure 5.11 A schematic drawing of a nuclear power plant.

large amounts of radioactive material (or even small amounts over long periods of time) can have damaging effects on living organisms. Reactor safety systems must be able to protect against malfunctions that could result in accidents ranging from a small leak of radioactive gas, to a major release of radioactive materials that would contaminate an entire surrounding area. Long-lived radioactive wastes are produced in the fission process, and must be safely stored for long periods of time—at least 20 half-lives. For example, strontium-90, a common and biologically dangerous waste (with a relatively short half-life of 28.1 years), should be kept in safe storage for about 600 years. Various alternatives for waste disposal continue to be carefully studied. Technology is being developed to trap radioactive gases and to store them until they have decayed, to concentrate liquid wastes, and to dispose of highly radioactive wastes in solid form (Figure 5.12). But there is still a great deal of controversy over how much risk we are all being exposed to from nuclear reactors and from the wastes they produce.

5.14 Nuclear Fusion—A Captured Sun

Enormous amounts of energy are given off when small nuclei (such as those of hydrogen, helium, or lithium) combine to form heavier, more stable nuclei. This process is called **nuclear fusion.** Nuclear fusion reactions are the source of the energy released by the sun and by the

Figure 5.12 These huge million-gallon tanks shown under construction will store radioactive waste materials. Before storage, the highly radioactive liquid wastes are converted to solid salt cakes to reduce the volume. The double-walled carbon steel tanks pictured here will be encased with 3 feet of concrete, and then will be completely surrounded with up to 40 feet of dirt. (Courtesy Energy Research and Development Administration)

hydrogen bomb. The fusion reaction occurring on the sun involves the combination of four hydrogen atoms, and releases four times as much energy per gram of fuel as does a fission reaction. The following fusion reaction is the one that is most likely to be used here on earth because it occurs at the lowest temperature and because deuterium can be obtained from the sea at a very low cost.

$$\underset{\text{Deuterium}}{{}^{2}_{1}\text{H}} + \underset{\text{Tritium}}{{}^{3}_{1}\text{H}} \longrightarrow {}^{4}_{2}\text{He} + {}^{1}_{0}n + \text{Energy}$$

| Deuterium | Tritium | Helium | Neutron | Energy |

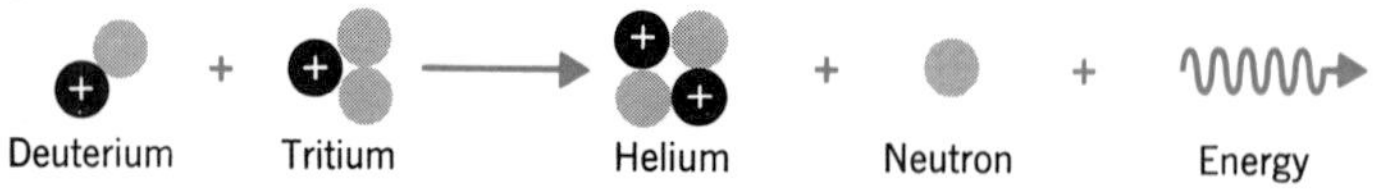

The use of nuclear fusion is proving to be much more difficult than the development of nuclear fission because temperatures of 100 million degrees Celsius or more are required for the fusion reaction to occur. If the technology can be developed to control a fusion reaction, however, this process has great advantages over fission as a source of the world's energy supply. It would be more efficient, the fuel would be cheap and almost inexhaustible, there is no possibility of runaway accidents from a melting of the reactor core (as could occur in fission reactors), and there is very little radioactive waste. Tritium, which is one of the least toxic radioactive substances, may be produced, but it can be returned to the system as a fuel.

Chapter Summary

The atom consists of a small, dense nucleus containing positive protons and neutral neutrons, and a region surrounding the nucleus containing negative electrons. The atomic number of an element indicates the number of protons in the nucleus of each atom of the element. The mass number is the sum of the number of protons and neutrons in the nucleus of an atom. Atoms of the same element that differ in the number of neutrons in their nuclei are called isotopes. The atomic weight of an element is a weighted average of the masses of the naturally occurring isotopes of that element, expressed in atomic mass units.

The nuclei of atoms of elements with high atomic numbers are often unstable, and will emit particles to form a new daughter nucleus. This process is called radioactive decay. There are several types of natural radioactive radiation, but the three most common are alpha, beta, and gamma radiation. Each radioactive substance has its own characteristic rate of decay, called the half-life of the isotope. Nuclear transmutation, the collision of a high-speed particle with a nucleus to produce a new nucleus, occurs in nature and has been used in the laboratory to produce elements with atomic numbers greater than 92.

When bombarded with neutrons, the nuclei of several elements are capable of undergoing fission to produce smaller, more stable nuclei and large amounts of energy. The energy generated in a nuclear fission reaction can be produced in a controlled fashion in a nuclear reactor. The fusion of several small nuclei into a heavier, more stable nucleus also generates large amounts of energy, but the technology has not yet been developed to release this energy in a controlled fashion.

Exercises and Problems

1. Complete the following chart:

	Subatomic Particle	*Location*	*Charge*	*Relative Mass*
1.	________	______	+1	______
2.	________	______	______	______
3.	________	______	______	$\frac{1}{1837}$

2. Complete the following chart (assume the atom is neutral)

	Atomic Number	Symbol	Mass Number	Number of p	e^-	n
(a)	3		7			
(b)		S				16
(c)	20					20
(d)	35		80			
(e)		Sn				66
(f)		Hg	200			

3. What is the atomic number, mass number, and symbol for the following atoms?
   (a) An atom with 8 protons and 8 neutrons
   (b) An atom with 90 protons and 142 neutrons
   (c) An atom with 47 protons and 60 neutrons

4. Calculate the atomic number and the mass number, and use two different shorthand methods to denote an atom of each of the following elements.

Name	*No. of p*	*No. of n*
a. chlorine (Cl)	17	18
b. cobalt (Co)	27	33
c. hydrogen (H)	1	2

5. State the number of protons, neutrons, and electrons in a neutral atom of each of the five isotopes of zinc (Table 5.2).

6. Using the information with which you answered question 5, describe how isotopes differ.

7. What causes a substance to be radioactive?

8. What is the difference between the three most common radiations given off by radioactive material?

9. What is the meaning of the term *nuclear transmutation*? Give an example.

10. What is the difference between atomic fission and atomic fusion? Which process is presently being used in power plants?

11. Why is the breeder reactor preferable to the burner reactor for meeting the world's long-range energy requirements?

12. Fill in the correct symbol, including the mass number and atomic number.
    (a) $^{144}_{60}Nd \longrightarrow {}^{140}_{58}Ce + (?)$

    (b) $(?) \longrightarrow {}^{90}_{38}Sr + {}^{0}_{-1}e + \gamma$

    (c) $(?) \longrightarrow {}^{206}_{82}Pb + {}^{4}_{2}He$

    (d) $^{214}_{82}Pb \longrightarrow {}^{214}_{83}Bi + (?) + \gamma$

(e) $^{150}_{64}Gd \longrightarrow (?) + ^{4}_{2}He$

(f) $^{234}_{91}Pa \longrightarrow (?) + ^{0}_{-1}e + \gamma$

13. Insert the correct bombarding particle for the following nuclear transmutations.

(a) $^{27}_{13}Al + (?) \longrightarrow ^{30}_{15}P + ^{1}_{0}n$

(b) $^{252}_{98}Cf + (?) \longrightarrow ^{257}_{103}Lw + 5\ ^{1}_{0}n$

(c) $^{46}_{20}Ca + (?) \longrightarrow ^{47}_{20}Ca$

14. Strontium-90 is a beta emitter with a half-life of 28 years.
    (a) Write the shorthand representation for this decay.
    (b) Draw a half-life graph for the decay of 100 grams of strontium-90.
    Use your graph to answer the following:
    (c) How long would it take for $\frac{3}{4}$ of the sample to decay?
    (d) How many grams would be left after 4 half-lives?
    (e) How many grams would remain after 252 years?
    (f) How many half-lives would it take to have less than 1 gram remaining?

15. It has been suggested that radioactive wastes be stored for at least 20 half-lives. How long would that be for cesium-137, a radioactive waste produced in nuclear power plants?

16. If you had $10 to gamble in a slot machine in Las Vegas, but could bet only half of it the first hour, and half of the remainder in each succeeding hour, at the end of what hour would you have $1.25 left? (Neglect any earnings.) How is this problem similar to the half-life of a radioactive substance?

17. Radon-222 is an alpha particle emitter which decays to polonium-218. Write the shorthand representation of this radioactive decay.

18. Bismuth-210 is a beta emitter which decays to form an isotope of polonium. Write the shorthand representation of this radioactive decay.

19. When the thorium-230 isotope decays, it gives off an alpha particle and gamma radiation. Write the shorthand representation of this radioactive decay.

20. Calculate the atomic weight of chlorine to three digits (see Table 5.2).

chapter 6

Radioactivity and the Living Organism

Learning Objectives

By the time you have finished this chapter you should be able to:

1. Describe two ways in which ionizing radiation can be harmful to living tissue.
2. Define the following terms: curie, rad, rem, LET.
3. Describe two ways of detecting radioactivity.
4. List three sources of background radiation.
5. Describe two ways radioactive substances can help in medical diagnosis.
6. Describe three different techniques for using radioactive materials in medical therapy.

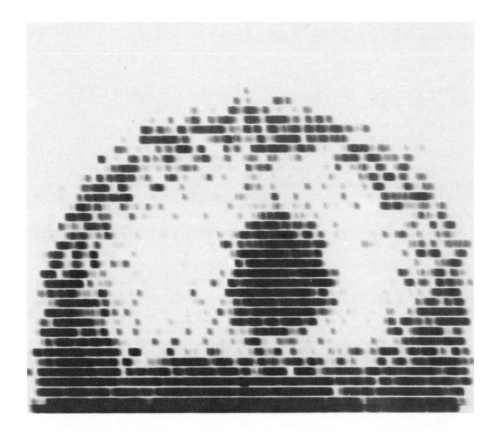

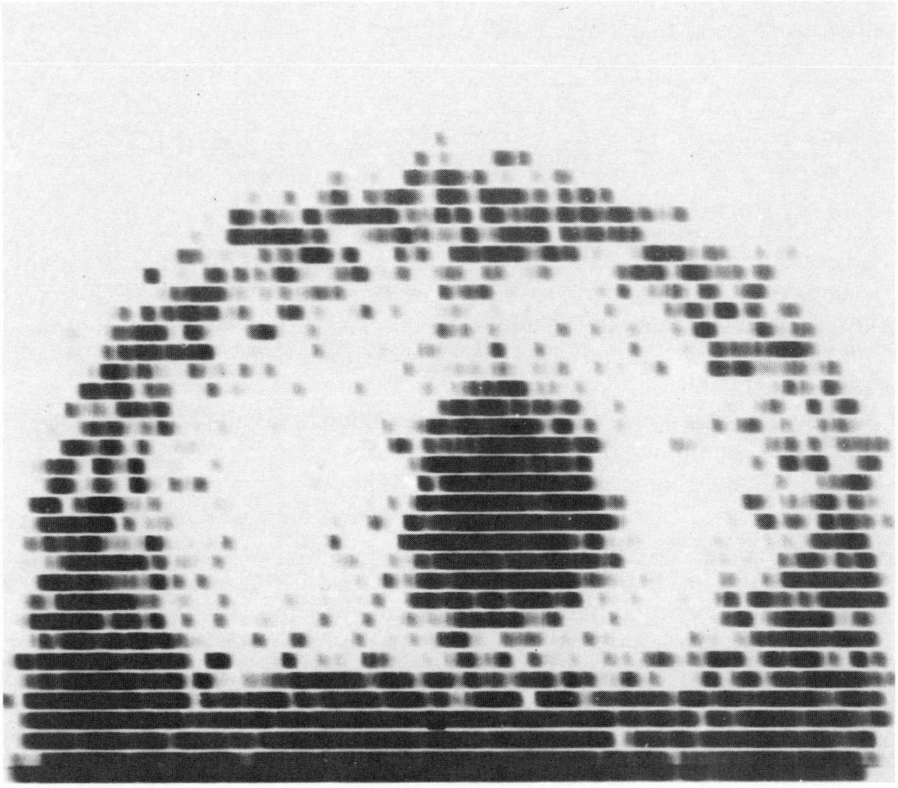

George Downing, a 34-year-old advertising executive, tried to ignore the familiar throbbing in his head. For weeks he had been suffering constant headaches that made his work impossible and his life miserable. His doctor had performed a long series of routine examinations and tests, but nothing unusual had been found. At his doctor's suggestion, George was now on his way to a New York hospital to have a brain image done by the Department of Radiology.

Thirty minutes before the procedure was to begin, George was given an intravenous injection of a small amount of radioactive material. He was positioned under an instrument called a gamma camera, which detects the high-energy rays emitted by the radioactive element. The camera recorded these rays on a sensitized photographic film. In less than an hour the doctor was able to take three images, one from the front of George's head and one from each side. After it was developed, the photographic film showed that the radioactive material had concentrated in one spot in George's brain. This indicated to the doctor that a tumor was present. Such a diagnosis is based on the fact that a brain tumor will damage the protective blood-brain barrier, allowing the radioactive material to leak into the brain in the region of the tumor. When this occurs, the brain image will show a concentrated accumulation of the radioactive substance in the region of the tumor. Several days later George underwent surgery to remove the tumor. Using all three images, the surgeon was able to pinpoint the tumor exactly and to remove it successfully.

The Effects of Radiation on Living Cells

6.1 Ionizing Radiation

The use of radioactive materials is becoming more and more of a standard procedure in many areas of medical diagnosis. Such materials are also playing important roles in the treatment of disease. It may seem just a bit contradictory, then, to realize that these same materials can also critically damage and kill living organisms.

Why are radioactive substances potentially so harmful? It is because alpha, beta, gamma, and cosmic radiation, when they hit living tissue, are each capable of producing unstable and highly reactive charged particles called ion pairs. For this reason, these radiations are called **ionizing radiation.** In addition, such radiation can transfer so much energy to the molecules in living tissue that these molecules will vibrate apart, forming high-energy, uncharged fragments called free radicals. Free radicals are even more reactive than ions; they are capable of pulling other molecules apart, and can completely disrupt living cells. Ions and free radicals may recombine with one another (resulting in little damage), or may combine with other molecules to form new substances foreign to the cell and, therefore, potentially dangerous (Figure 6.1). The new substances produced in this way are often highly reactive chemically, but are not themselves radioactive.

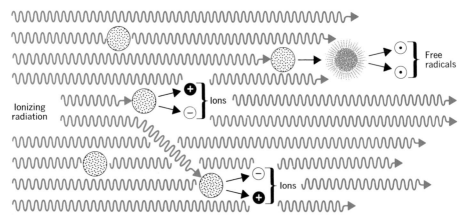

Figure 6.1 When ionizing radiation hits living tissue, it is capable of producing unstable and highly reactive charged particles called ions, and even more reactive uncharged particles called free radicals.

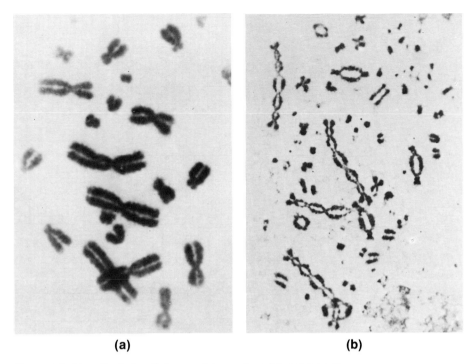

Figure 6.2 The effect of radiation on living cells. (a) A normal cell nucleus containing 23 pairs of chromosomes, composed of DNA molecules and protein. (b) Ionizing radiation has caused the chromosomes in this cell nucleus to double and triple abnormally. (Courtesy Argonne National Laboratory)

Focusing, for a moment, on an individual cell of the body, there are two ways in which ionizing radiation can cause great damage. First, the damage may be caused by direct action—a direct hit on a biologically important molecule that can cause the molecule to split into biologically useless fragments. The most vital molecule in a living cell is the DNA molecule, which carries all the "blueprints" necessary for the cell to divide and reproduce. If the DNA is destroyed, the cell cannot divide, and will die. When such cells die without replacement, the entire irradiated tissue will eventually die. Furthermore, if this tissue is essential to the organism, the entire organism may die prematurely.

Even if the DNA molecule is only damaged, and not destroyed completely, it may cause the cell to divide abnormally into new cells with altered DNA. Such cells are known as **mutant** cells. A mutant cell may have its DNA so altered that it is no longer under the body's control. The cell may then begin to grow and divide in an uncontrolled fashion, destroying the normal cells around it. Cells that behave in this manner are called **cancerous** or **malignant** (Figure 6.2).

Studies on survivors of the atomic bombs dropped on Hiroshima and Nagasaki (as well as other cases) have shown that exposure to high doses of ionizing radiation can cause cancer in humans. The possibility of developing cancer from lower dosages is still under study. Because the effects of radiation exposures build up over time, however, it seems most sensible to avoid exposure as much as possible.

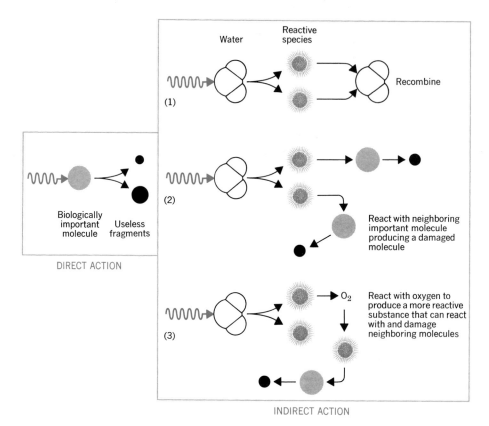

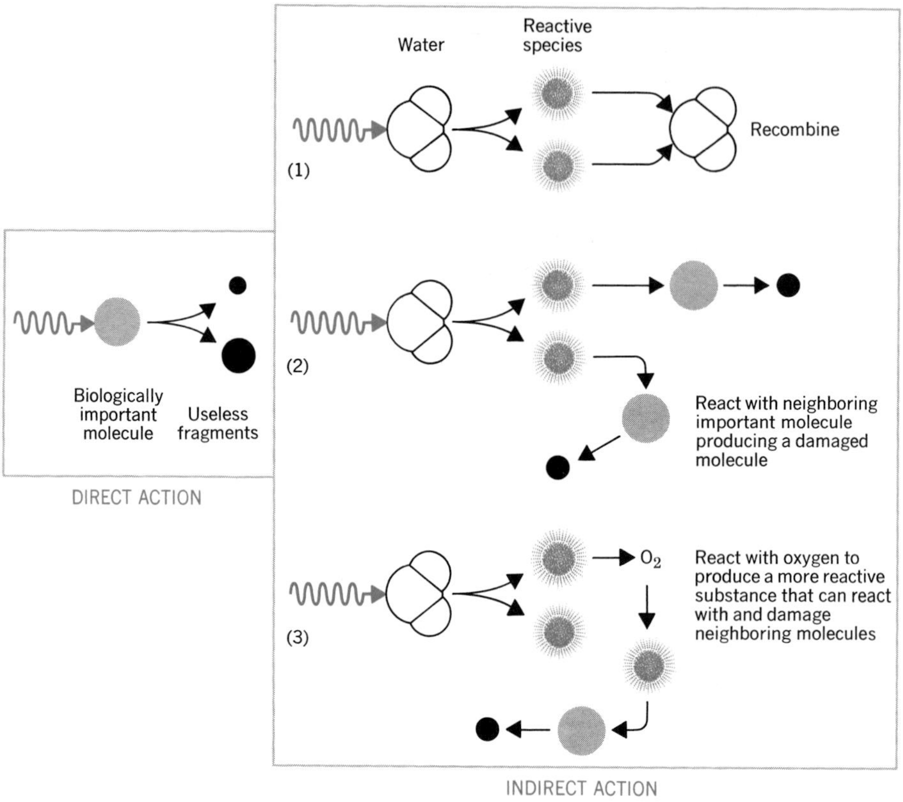

Figure 6.3 Ionizing radiation can damage living tissue by direct action, such as a direct hit on a biologically important molecule, or by indirect action, such as reacting with water to produce highly reactive substances which can then react with biologically important molecules.

Direct hits of ionizing radiation on biologically important molecules are not the only way that radiation can damage cells. Ionizing radiation can also cause damage through indirect action. Animal cells are about 80% water, and ionizing radiation can cause water molecules in living tissue to form positive and negative ions, or to form highly reactive free radicals. Such free radicals can recombine to form water (H_2O, which would be harmless), can combine to form hydrogen (H_2, which can be tolerated in small amounts by living cells), or can combine to form hydrogen peroxide (H_2O_2, which is a highly toxic substance). This may be the reason that radiation sickness resembles hydrogen peroxide poisoning in many respects. Free radical fragments can also react with oxygen in the cell to produce a free radical that is even more undesirable than hydrogen peroxide (Figure 6.3).

6.2 Radiation Dosage

There are several scientific units used to measure various aspects of ionizing radiation. These different units help answer specific questions

Table 6.1 Units Used to Measure Ionizing Radiation

Unit	Measures	
curie (Ci)	Rate of disintegration	1 Ci = 3.7×10^{10} disintegrations per second
rad (D)	Absorbed dose of radiation	1 rad = absorbed radiation that liberates 100 ergs of energy/gram of tissue
rem	Absorbed dose of radiation	1 rem = absorbed dose that will produce same biological effect as 1 rad of therapeutic X ray
LET	Energy transferred to tissue by absorbed dose	LET = amount of energy transferred to tissue per unit of path length

about the nature of the radioactive source and its effects on living tissue. The following pages will describe a few of these terms: the curie, the rad, the rem, and the LET (Table 6.1).

The first aspect of ionizing radiation which one might wish to describe is the activity level of the source of that radiation. That is, how often do disintegrations occur per unit of time? The unit used to describe the activity of a radioactive source is the **curie (Ci).** One curie is equal to 3.7×10^{10} disintegrations per second, which is the rate of disintegration of one gram of radium-226. It is important to realize that this unit doesn't tell us anything about the type of radiation produced, or the effect that this radiation will have on tissue or other matter. It describes only the rate at which radiation is produced. In radiation therapy, hospitals often use a sample of cobalt-60 as their source of gamma radiation. If a source is rated at 1.4 curies, it will deliver $1.4 \times 3.7 \times 10^{10}$, or 5.18×10^{10} disintegrations per second. A 2 Ci cobalt-60 source will deliver to a patient twice as many gamma rays per unit of time as a 1 Ci cobalt-60 source.

To study the effect of ionizing radiation on living tissue, we must try to measure the amount of energy that has been absorbed by the tissue. The dose actually absorbed by a tissue will be influenced by many factors, including the nature of the radioactive source, the energy of the radiation, the distance from the source to the tissue, the nature of the tissue itself, and the duration of exposure. The **rad** (**r**adiation **a**bsorbed **d**ose) is the unit used to measure the amount of energy that is absorbed by irradiated tissue. Regardless of the type of radiation, one rad is an absorbed dose of radiation that results in the transfer of 100 ergs of energy to a gram of the irradiated tissue. Most mammals will not survive a dose of 1000 rads of radiation to the entire body (this means the delivery of 1000×100 ergs, or 100,000 ergs of energy per gram of the mammal's tissue). Now, the erg is a very small unit of energy—100,000 ergs equal 0.0024 calorie. Why is such a small amount of energy all that is needed for a killing dose of radiation? As we discussed earlier, it is the ability of ionizing radiation to change important biological molecules through direct or indirect action which results in the death of cells.

Scientists have discovered that the absorption of one rad of different types of ionizing radiation will produce different effects on identical biological systems. For example, it takes twice the dose of cobalt-60 gamma radiation as it does a specific level of X rays to produce the same disruption in a certain type of pollen grain. Therefore, in order to study the effects of different types of radiation on humans, a unit called the **rem** (**r**oentgen **e**quivalent **m**an) was devised. For each type of radiation, the rem is defined as the absorbed dose of radiation which will produce the same biological effect as one rad of therapy X rays. When a dose is stated in rems, there is no need to specify the type of radiation, because the biological effects will be the same for each type. For example, 700 rems of any type of radiation is a dose sufficient to kill 50% of a population of rats within 30 days. (The lethal dose for 50% of the exposed individuals within 30 days is often abbreviated LD_{50}/30 days. See Table 6.2.) One rad of beta radiation has a rem of 1, but one rad of alpha radiation has a rem of 10 to 20.

Table 6.2 LD_{50}/30-Day Dose Values in Rems for Four Mammals

Animal	Rem
Dog	300
Man	500 (estimated)
Mouse	600
Rat	700

Why should one rad of different types of radiation have such different effects on living tissue? One way in which these radiations are different is the manner in which the energy of the radiation is deposited in the tissues. For example, alpha particles and neutrons (because they are relatively heavy particles) "crash" through tissues, producing dense clusters of ions along fairly short paths. X rays and gamma rays, on the other hand, "zip" through tissues, leaving isolated ion pairs at scattered points along a much longer path. This difference is described by the **linear energy transfer (LET),** or the average energy released per unit of path length, of a radiation. High LET radiation (neutrons and α particles) produces closely packed groups of ion pairs, and does most of its damage by direct action. When a biologically important molecule is hit by high LET radiation, the damage is usually so great that it can't be repaired at all. By contrast, low LET radiation (β particles, X rays, and γ rays) leaves behind sets of ion pairs well spread out through the tissue, and does most of its damage through indirect action. As a result, damage to biologically important molecules caused by sublethal doses of low LET radiation is much more likely to be repaired by the cell than is damage by high LET radiation.

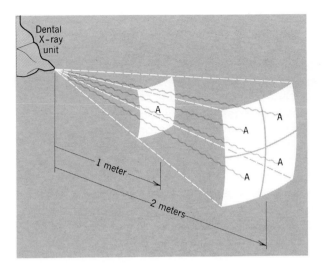

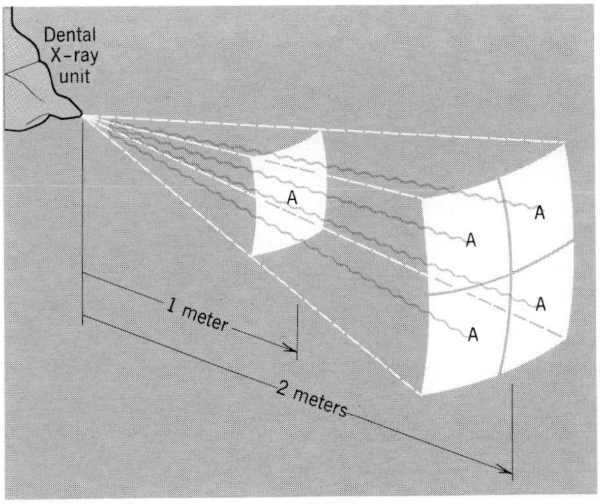

Figure 6.4 An illustration of the inverse square law. The number of radiations hitting area A at one meter from the radioactive source will hit an area four times larger at two meters from the source. Therefore, the intensity of the radiation at two meters is one-fourth its intensity at one meter.

6.3 Protection Against Radiation

In Chapter 5, we saw that various kinds of radiation differ in their penetrating power. Alpha and beta radiation have low penetrating power and are easily stopped; X rays and gamma rays have high penetrating power, and require lead or thick concrete to be stopped. Because lead is such an effective shield, doctors and technicians often use lead aprons to protect parts of the patient's body from radiation during medical or dental X rays.

In addition to the use of shielding, another important safety factor is the distance from the source of the radiation. The farther you are from the source, the lower the intensity of the exposure. Because the rays spread out in a cone shape as they move away from the source, the number of radiations striking any object will decrease as the object moves away (Figure 6.4). In more mathematical language, the intensity of radiation on a given surface area decreases with the square of the distance from the source. Scientists call such a relationship an **inverse square law.**

Even with good shielding, the most sensible safety precaution is to minimize your exposure to ionizing radiation. The living cell has an amazing ability to repair much of the damage from radiation exposure, but the harmful effects of radiation keep building up over time.

6.4 Detecting Radiation: Radiation Dosimetry

You cannot hear, feel, taste, or smell low levels of radiation. (Do you feel anything when you have a dental X ray?) Because such radiation is potentially dangerous, various instruments and detection systems have been designed to measure radiation dose and exposure. Depending on the particular instrument, radiation dosage may be measured by recording the amount of ionization in a gas, the amount of chemical reaction produced, the heat produced when the radiation gives off energy, or the light produced in a luminescent material.

A **Geiger-Müller** counter is a widely used radiation detecting instrument. The detector consists of a gas-filled ionization chamber in the shape of a tube. Radiation that enters the tube will produce ions in the gas, allowing an electrical current to flow. This current will either generate a click in a speaker or will turn a numerical counter to indicate the number of radiations that have entered the tube (Figure 6.5). The Geiger-Müller counter is sensitive to beta radiation. Most gamma rays will pass through the gas without affecting it, and all but the most energetic alpha particles will be stopped before even entering the tube. Although the counter will indicate the number of particles entering the tube, it tells us nothing about the energy of those particles.

The **scintillation counter** is the most widely used instrument for determining both the number of radiations and the dose rate at the same time. Incoming radiations strike a surface that has been coated with special chemicals, producing tiny flashes of light. The number of these flashes and

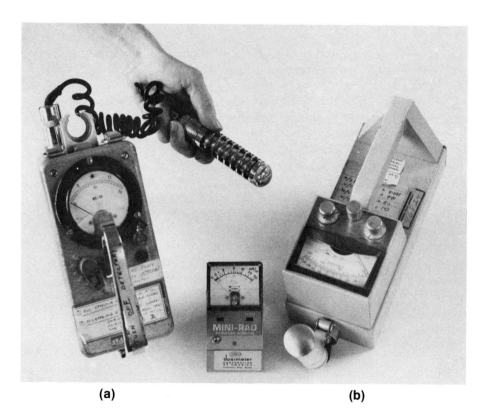

(a) (b)

Figure 6.5 (a) and (b): Two portable Geiger-Müller counters frequently used in laboratories. The small dosimeter in the middle can be worn on a person's belt.

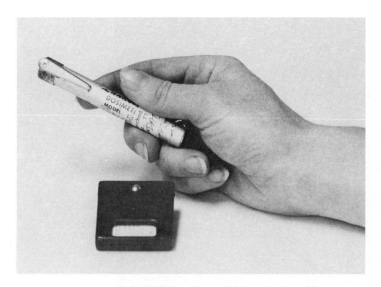

Figure 6.6 A pocket dosimeter and a film badge, both of which can be worn on a pocket and will detect any significant radiation exposure.

their brightness (which depends on the energy of the radiation) is electronically detected and converted to electric pulses that can be counted, measured, and recorded.

We have seen that it is important to carefully monitor the exposure of researchers, X-ray technicians, nurses, and others who work with ionizing radiation. To make this as easy as possible, small portable devices have been designed to accurately measure radiation dose. Two very commonly used devices are **pocket dosimeters** and **film badges** worn by the individual. The dosimeter has a built-in scale indicating the radiation exposure of the person wearing it. The film in the badge must be developed to obtain a reading. The degree of darkening in the resulting negative indicates the total amount of radiation to which the wearer has been exposed (Figure 6.6).

Photographic film is also used to pinpoint the position of radioactive material in living tissue. This technique, known as **autoradiography,** involves introducing radioactive elements or compounds into material which is then exposed to a photographic emulsion. The emulsion is developed after an exposure time that varies with the different materials being tested. This technique has been used for such purposes as determining the location of calcium-45 in a leaf, or of strontium-90 in the body of a mouse (Figure 6.7). On a microscopic level, this technique can help pinpoint the specific location of a radioactive chemical in a living cell.

6.5 Background Radiation

There is no way that we can totally escape exposure to ionizing radiation. In fact, the average human body undergoes several hundred thousand radioactive disintegrations per minute as a result of the natural

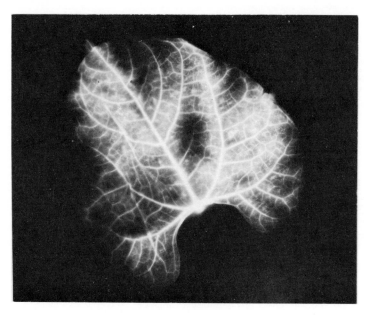

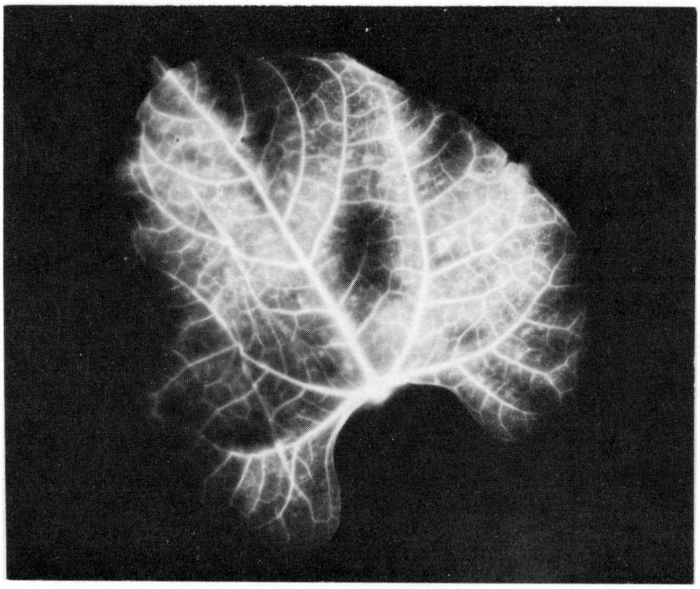

Figure 6.7 An autoradiogram of the leaf of a bean plant that had absorbed calcium-45 through its root system. (Courtesy Victor Arena. From V. Arena, *Ionizing Radiation and Life,* The C. V. Mosby Co., St. Louis, 1971)

radioisotopes (mostly potassium-40) found in the body. Seventy of the 350 naturally occurring isotopes are radioactive, and small amounts of radioactive materials are found in the soil we walk on, the food we eat, the water we drink, and the air we breathe.

On the average, we each receive a very small dose (0.211 rems, or about 200 millirems) of background radiation per year. Sixty-three percent of this radiation is from outer space and natural materials in the soil, water, and air; 35% comes from medical radiation such as medical and dental X rays; and 2% is from radioactive fallout and pollution from nuclear power plants (Table 6.3). We cannot control the natural background radiation, but we can control radiation exposure from other sources. Very few people would argue that we should stop all uses of radioactive materials, for these materials are playing a large role in improving living conditions around the world. But when considering decisions that might increase the level of background radiation, it becomes important to weigh the benefits to humanity against the risk. Such decisions are especially difficult because so little is known about the long-term effects on health of very low levels of radiation.

Radioisotopes and Medicine

One area in which the benefits of ionizing radiation appear to far outweigh the risk is the use of radioisotopes in medical diagnosis, treatment, and research.

Table 6.3 Summary of the Annual Per Capita Radiation Dose in the United States

Radiation Source	Dose (Millirem/Year, 1970)	
Environmental		
Natural		130
Cosmic rays	(45)	
Terrestrial radiation		
External (Soil, potassium-40, decay products of uranium and thorium)	(60)	
Internal (mainly potassium-40)	(25)	
Global fallout		4
All other (Nuclear reactors, fuel processing, nuclear tests, etc. By the year 2000 it will equal about 0.5 mrem/year)		0.06
Medical		
Diagnostic		72
Radiography	(58)	
Fluoroscopy	(14)	
Radiopharmaceutical (By the year 2000 it will equal about 16 mrem/year)		2
Occupational		0.8
Miscellaneous (TV, consumer products, air transport)		2.7
	Total	211

Data from "Estimates of Ionizing Radiation Doses in the United States, 1960-2000," Environmental Protection Agency, Rockville, Maryland, 1972.

6.6 Medical Diagnosis

Major advances in medical diagnosis have been made possible by being able to trace the paths taken by specific chemical compounds in living systems. **Radioactive tracers** are chemicals that contain radioactive atoms and that have the same chemical nature and behavior as the compounds they are meant to trace. A living system treats the tracer just as it would the normal compound, but the radioactive atoms make it possible to follow the compound's path through the system.

In 1923 Georg von Hevesy was the first to use a radioactive tracer in his study of the uptake of lead-212 in plants. Artificially produced

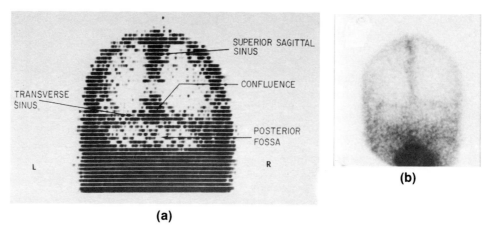

Figure 6.8 (a) A normal posterior brain scan taken with a rectilinear scanner. (b) A normal posterior brain image taken with a gamma camera. (Courtesy Sheldon Baum, M.D., from *Basic Nuclear Medicine,* S. Baum and R. Bramlet, Appleton-Century-Crofts, New York, 1975)

radioisotopes were first used in 1936, when Joseph G. Hamilton and Robert S. Stone of the University of California at Berkeley used radioactive sodium produced in a cyclotron to study the uptake and excretion of sodium. Since then, the use of natural and artificially produced radioisotopes has greatly increased. In 1970 about 8 million "atomic cocktails" of some 30 different radioisotopes were administered to patients in diagnostic and therapeutic treatments.

In early tracer diagnostic procedures, the movement of the radioisotope was followed with a geiger counter. The 1950s saw the development of instruments called scanners. These devices have a sensitive measuring head that moves slowly back and forth across the patient's body to detect the radioactivity. After several minutes to one-half hour, a picture can be put together from the various passes of the scanner (Figure 6.8a). In recent years computer programs have been developed to analyze the scanning data in order to produce much clearer and sharper pictures. The gamma camera, which does not have to move across the patient, is a new instrument now being widely used. This camera will record all the radiation coming from a certain region of the patient, and will produce an image in several seconds to a few minutes (Figure 6.8b and 6.9).

6.7 Radioisotopes Used in Diagnosis

Some radioisotopes and their medical uses are listed in Table 6.4, but a few of these isotopes are worth discussing in greater detail.

Technetium-99^m

Over the past 20 years many different radioisotopes have been used to produce images of various body organs. However, technetium-99^m, because

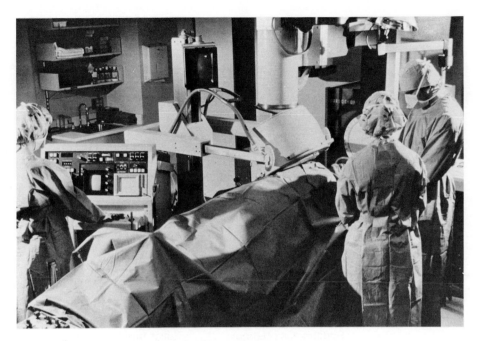

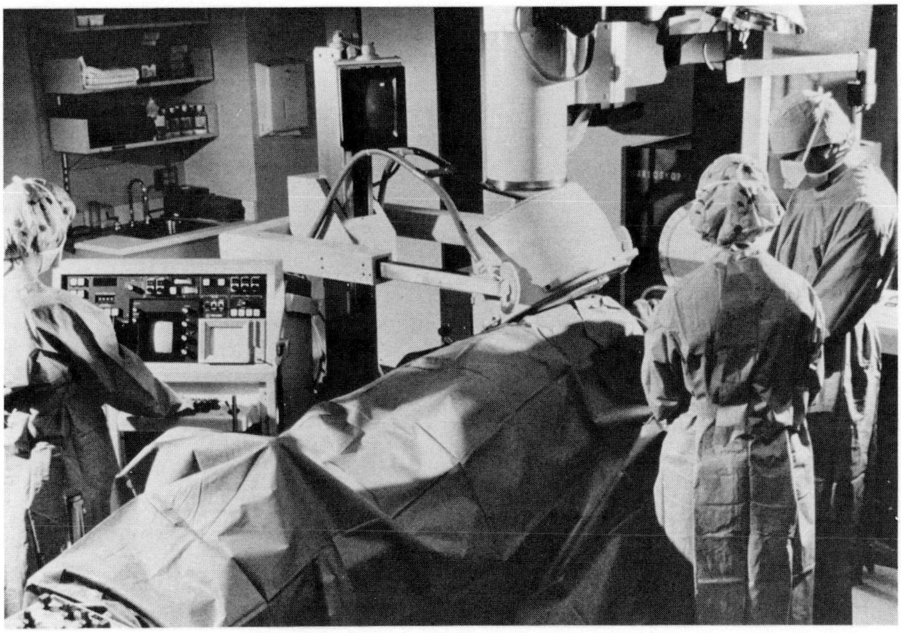

Figure 6.9 A gamma camera in use. (Spectrum One: Nuclear Cardiology System, courtesy of Ohio-Nuclear, Inc.)

of its more suitable chemical and physical properties, is now replacing many of these other radioisotopes in common diagnostic procedures. Technetium is one of four elements lighter than uranium which do not occur naturally and, therefore, must be artificially produced. It is obtained from the decay of molybdenum-99.

$$^{99}_{42}\text{Mo} \xrightarrow[t_{1/2} = 67\text{ hr}]{} {}^{99m}_{43}\text{Tc} + {}^{0}_{-1}\text{e} + \gamma$$

The *m* after the mass number means that the isotope is **metastable,** or in an energy state higher than normal. When it decays, technetium-99^m releases this excess energy as gamma rays and forms the lower-energy isotope technetium-99.

$$^{99m}_{43}\text{Tc} \xrightarrow[t_{1/2} = 6.02\text{ hr}]{} {}^{99}_{43}\text{Tc} + \gamma$$

The advantages of technetium-99^m over other isotopes are (1) the energy of the gamma rays are best for detection by the cameras now in use, (2) no alpha or beta radiations are given off (which would increase the absorbed dose to the patient), and (3) the 6-hour half-life allows enough time for the isotope to localize in the body after injection, and yet allows the isotope to be administered in amounts that limit the radiation exposure to that of a comparable X-ray procedure.

Technetium 99^m was the radioactive isotope used to detect the brain tumor described at the beginning of this chapter. Various technetium

Table 6.4 Some Radioisotopes and their Uses in Medical Diagnosis

Radioisotope	Compound Given	Diagnostic Use
Technetium-99^m	Sodium pertechnetate	Brain and thyroid scanning
	Technetium sulfate colloid	Liver, spleen, and bone marrow scanning
	Labeled serum albumen	Cardiac blood pool, placenta, and lung scanning
	Technetium diphosphonate	Bone scanning
	Technetium gluconate	Kidney function and scanning
Iodine-131	Labeled serum albumen	Blood volume determination, cardiac output and blood pool scanning, lung scanning, placenta localization
	Labeled fats	Fat absorption
	Sodium iodohippurate	Kidney function and scanning
	Sodium rose bengal	Liver function and scanning
Gallium-67	Gallium citrate	Tumor scanning
Thallium-201	Thallous chloride	Cardiac scanning
Xenon-133	Xenon gas	Lung function and cerebral blood circulation
Chromium-51	Sodium chromate	Red blood cell survival, blood volume, and spleen scanning
Cobalt-57	Labeled vitamin B_{12}	Vitamin B_{12} absorption
Iron-59	Iron citrate	Iron metabolism
Mercury-203	Chlormerodrin	Kidney and brain scanning
Selenium-75	Selenomethionine	Pancreas scanning

compounds are used in other diagnostic procedures. For example, a compound of technetium and sulfur is taken up by the normal cells of the liver, spleen, and bone marrow. However, areas in which cancerous cells are growing will not take up this compound, and will appear as areas of decreased activity (Figure 6.10). In a similar manner, dead or abnormal tissue in the kidneys will not absorb and concentrate technetium gluconate, so diseased areas of the kidney can be located on the camera image.

Technetium diphosphonate is used to detect whether breast cancer cells have spread, or **metastasized,** to the bones. Bone cells will absorb the diphosphonate and incorporate it into the structure of the bone in those places where bone tissue is being destroyed by the cancer cells. In this case the highest radioactivity will be found in areas of the metastasized cells (Figure 6.11).

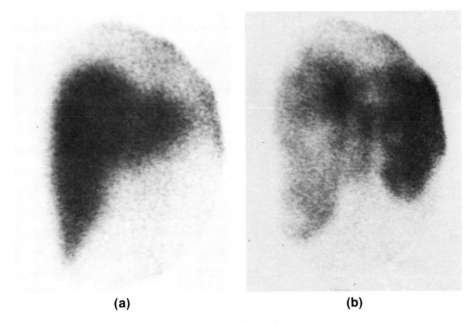

(a) (b)

Figure 6.10 Liver images taken 10 minutes after an intravenous injection of 8 millicuries of technetium-99^m sulfur colloid. (a) A normal image showing the liver on the left and the spleen on the right. (b) An abnormal image showing white areas in the liver, which are metastasized cells from colon cancer. These abnormal cells don't pick up the colloid. The spleen is greatly enlarged, which indicates that it is trying to take over the functions of the damaged liver. (Courtesy Good Samaritan Hospital, Corvallis, Oregon)

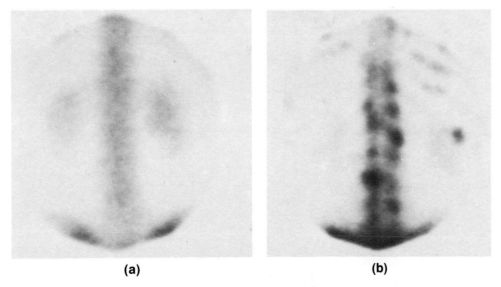

(a) (b)

Figure 6.11 Bone images taken three hours after an intravenous injection of 20 millicuries of technetium-99^m diphosphonate: (a) normal, (b) showing the metastasis of breast cancer cells. (Courtesy Good Samaritan Hospital, Corvallis, Oregon)

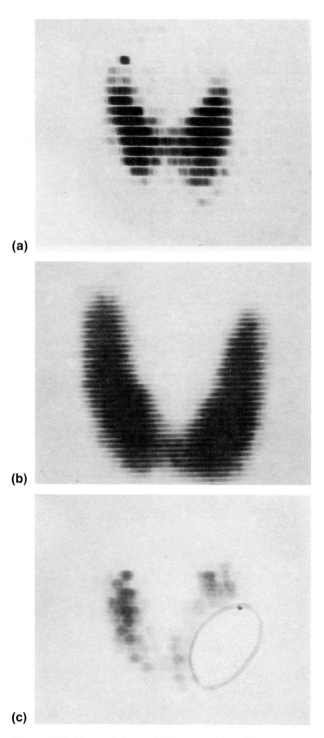

(a) (b) (c)

Figure 6.12 These pictures of (a) a normal thyroid, (b) an enlarged thyroid, and (c) a cancerous thyroid were taken by a linear photo scanner using iodine-131. (Courtesy U.S. Energy Research and Development Administration, Technical Information Center, Oak Ridge, Tennessee)

Technetium-99^m, in several forms, is also allowing doctors to determine the location and extent of both old and new damage to the heart. Such procedures can help the doctor to decide what therapy the patient should or should not receive. Radioactive technetium (as with calcium) deposits in the mitochondria of heart cells that are irreversibly damaged. Thallium-201, which is produced in a cyclotron, is being used together with technetium to diagnose heart damage. Thallium, unlike technetium, acts like potassium and concentrates in normal heart muscle. It can be used, for example, to determine whether a patient who enters the hospital with chest pains, but with few other symptoms of a heart attack, has a deficiency in the blood supply to the heart that might result in a future heart attack. Both of these radioisotopes can be administered intravenously, and their use is safer, more comfortable, and less expensive than the previous technique of inserting a catheter through a vein into the heart.

Iodine-131

Early studies using iodine as a tracer made use of iodine-128, which has a half-life of 25 minutes. In 1938, searching for a tracer with a longer half-life, Hamilton was able to produce iodine-131 with the use of a cyclotron. This isotope is a beta and gamma emitter with a half-life of eight days, and is a very versatile tracer.

The greatest value of radioiodine lies in the measurement of thyroid function. Thyroxin, an iodine compound manufactured in the thyroid, is released in the blood to help control the use of nutrients by the body. To test the functioning of the thyroid gland, a patient can be given a small amount of iodine-131 in the form of the iodide ion. Because any iodine in the body will be concentrated in the thyroid gland, the doctor can monitor the amount of radioiodine in the patient's blood to determine the rate at which the thyroid takes up the iodine. The uptake of radioiodine is monitored over a 24-hour period, giving the doctor an indication of how well the thyroid gland is functioning. Too rapid an uptake means the patient is suffering from an overactive or hyperthyroid; too slow an uptake means the patient could be suffering from an underactive or hypothyroid. Scans taken of the thyroid will give pictures such as those in Figure 6.12.

Liver function can be measured using a chemical dye called Rose Bengal which has been tagged with iodine-131 and injected into a vein. The liver normally removes the dye from the bloodstream and transfers it via the bile duct to the intestines for excretion. The speed with which the dye is removed can be followed by detectors monitoring the liver, small intestines, and bloodstream. This test, which can be quickly performed, could help save the life of an auto accident victim brought into the emergency room with suspected liver damage. Or this test could help a physician determine without additional surgery whether, for example, the artificial bile duct connecting the liver to the intestines of a three-year-old is still functioning properly (Figure 6.13).

Chromium-51

Chromium, in the form of sodium chromate, attaches to red blood cells. This property makes chromium useful in several types of tests. Red blood cells tagged with chromium-51 can be used in measuring blood volume to

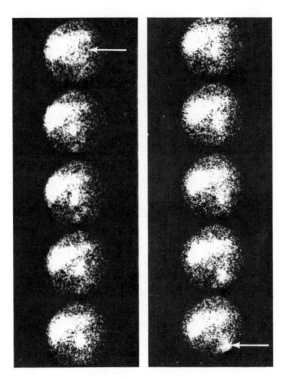

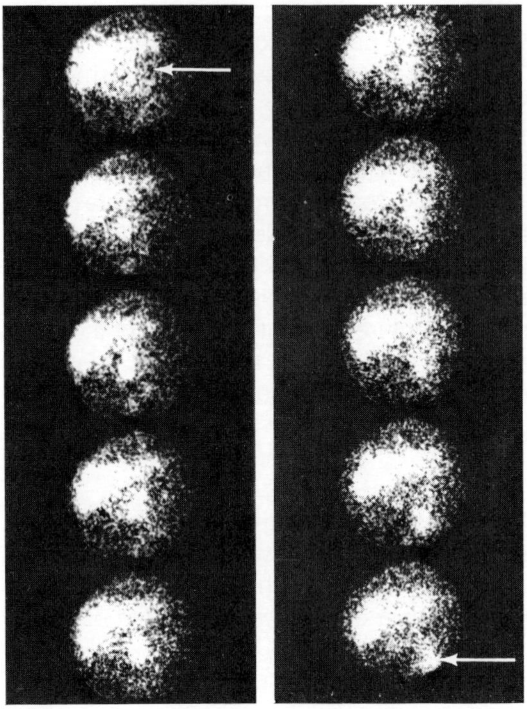

Figure 6.13 A three-year-old child who had been born without a bile duct, and who had had an artificial duct surgically created, was displaying symptoms indicating that this duct may have become blocked. These pictures are time-lapse photographs made with a scintillation camera one hour after the child was injected with Rose Bengal dye tagged with iodine-131. The small light area, seen moving downward and to the right in these pictures, showed that the artificial duct was still open. (Courtesy U.S. Energy Research and Development Administration)

determine whether transfusions are needed in cases involving bleeding, burns, or surgical shock. These cells can be used to measure the lifetimes of red blood cells, aiding in the detection of certain types of anemia. A monitor placed over the heart can detect radioactive red blood cells and, therefore, can determine the blood flow rate through the heart. Obstetricians use red blood cells tagged with chromium-51 to find the exact location of the placenta in a pregnant woman; in a condition known as placenta previa, the placenta may be in such a position that fatal bleeding could occur. (Radioiodine was originally used for this procedure, but red blood cells tagged with chromium-51 have the advantage of not being transferred to the fetus.)

Phosphorus-32

In many types of tumors, phosphorus is found in abnormally high concentrations. This fact often allows doctors to tell the difference between normal and cancerous cells. Radioactive phosphorus can be used to help locate such tumors, but because phosphorus-32 gives off only beta radiation, a counter must be very close to the tissue in which the phosphorus is present. Phosphorus-32 finds its greatest use in the detection of skin cancers and in brain surgery when the cancerous tissue is very hard to distinguish from normal tissue. If the patient is given ^{32}P, the surgeon can measure the radioactivity of the brain cells during surgery to determine which tissues to remove.

6.8 Mammography

Mammography is a technique used for the early detection of breast cancer. In this technique, regular X-ray equipment is used to make an exposure. However, instead of producing the image on film by chemical processing, the picture appears on paper through a photoelectric process similar to a Xerox copy (Figure 6.14). This process produces a much clearer image of soft tissues than does X-ray techniques, and allows for the detection of breast cancers at a highly curable stage, before they could be detected by any other means.

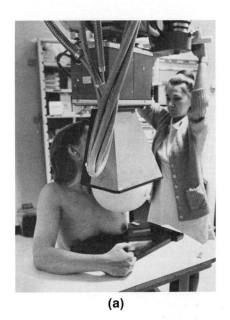

(a) (b)

Figure 6.14 (a) A mammogram being taken. (b) A mammogram showing a breast tumor. (a, Josephus Daniels/Photo Researchers; b, courtesy Dr. W. Lloyd, Good Samaritan Hospital, Corvallis, Oregon)

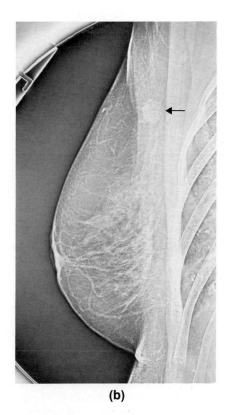

6.9 Radiation Therapy

Cancer cells, by their very nature, divide much more rapidly than do normal cells and are, therefore, more sensitive to radiation. This is the key fact that allows doctors to use radiation in killing cancerous tissue while leaving normal tissue relatively unharmed.

Teletherapy is the term used to denote the use of high-intensity radiation to destroy cancerous tissue that can't be removed by surgery. In such therapy, radiation from an X-ray machine, a cobalt-60 source, or a particle accelerator is focused into a thin beam and directed at the cancerous tissue. In the treatment of internal cancers, the patient is carefully positioned so that the tumor is at the center of a large circular framework. The source of radiation can be rotated about the patient so that the radiation will produce minimal skin tissue damage, but yet will be constantly focused on the tumor cells (Figure 6.15).

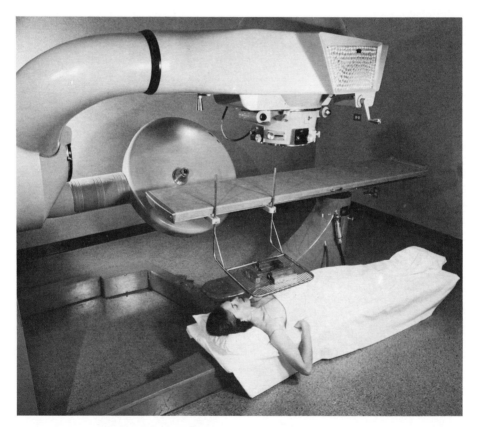

Figure 6.15 This cobalt-60 unit containing a 3000-curie source can be rotated around the stationary patient, allowing for many possible treatment positions. This patient is undergoing irradiation of the portions of the neck and chest outlined by the lines painted on her. The tray above the patient contains blocks of lead that absorb gamma rays and control the field of treatment. (Courtesy the University of Texas at Houston, M.D. Anderson Hospital & Tumor Institute)

Brachytherapy is the medical procedure of inserting a radioisotope by means of a needle, or in the form of a seed, into the area to be treated. Radium-226 and strontium-90 are two isotopes often used in this way for the treatment of skin and eye cancers. Seeds containing radon-222, gold-198, or iridium-192 are sometimes implanted in the tumors of patients with advanced inoperable cancer in order to slow the tumor growth.

Radiopharmaceutical therapy involves administering radioisotopes in chemical forms designed so that they will be concentrated in specific areas of the body. For example, because radioiodine is concentrated in the thyroid, it can be used to destroy thyroid cells in the treatment of an overactive or hyperthyroid. Phosphorus-32 is used to treat the disease polycythemia vera, which causes an abnormal increase in the number of red blood cells. There is no known cure for this disease, but the radiophosphorus destroys the cells that produce red blood cells. This will slow down the formation of red blood cells and will temporarily alleviate the symptoms of the disease.

Certain types of cancerous tumors cause malignant effusions, which are large accumulations of excess fluids in the chest and abdominal cavities. To treat this condition, gold-198 in a suspension of very finely divided particles is introduced into the abdominal cavity. Gold-198 is chemically inert and nontoxic, and cannot pass through the abdominal wall. It has a half-life of 2.7 days, and emits beta and gamma radiation which is absorbed by tissues in the immediate neighborhood. This treatment destroys the cancerous cells and stops the oozing of fluid.

Chapter Summary

Radiations given off by radioactive materials are called ionizing radiation because they are capable of producing ions in material through which they pass. Ionizing radiation can cause damage to living tissue in two ways: through direct damage to biologically important molecules, or by reacting with water to form chemically reactive particles that can disrupt cellular functions. Radiation is measured in several different units such as the curie, the rad, the rem, and the LET. These various units of measure give specific information about the nature of the radioactive source and its effect on living organisms. Various instruments such as the Geiger-Müller counter, scintillation counter, dosimeter, and film badge, can be used to detect the presence of, or the amount of exposure to, ionizing radiation. To protect yourself from the dangerous effects of ionizing radiation, you should try to minimize your exposure to such radiation, stay as far as possible from the source, and use lead or other shielding when working with ionizing radiation. However, we cannot entirely eliminate our exposure to radiation. The amount of radiation occurring naturally in the environment is called background radiation.

Although ionizing radiation in large doses can be harmful to living tissue, it has proven extremely useful in the diagnosis and treatment of disease. Radioisotopes can be used to determine blood volume and cardiac output, to analyze liver, kidney, spleen, and thyroid function, and to locate tumors or cancerous lesions in the brain, bones, and many other tissues. The ionizing radiation given off by radioactive elements can then be used to destroy cancerous cells and tumors.

Exercises and Problems

1. Explain how ionizing radiation can be both the cause of cancer and the means of stopping or slowing its growth.
2. Beta radiation is low LET radiation. Explain how beta radiation can damage living cells. What are the chances of a cell recovering from a dose that is sublethal to the organism?
3. Alpha radiation is high LET radiation. Explain how its effects on living cells are different from beta radiation. Is a cell more likely to recover from sublethal doses of alpha or beta radiation?
4. You are working in a laboratory which you fear has been contaminated with a radioactive beta emitter. How might you try to detect this contamination?
5. A radiation therapy cobalt-60 source is rated at 1.7 Ci. How many disintegrations will it deliver in 15 seconds?
6. Both the rad and the rem are units used to describe the absorbed dose of radiation. What is the difference between these two units?
7. While running an experiment, you expose a group of 200 mice to 600 rem of radiation. How many mice would you expect to die within 30 days?
8. You are setting up a dental office in a building that is being remodeled. The contractor asks you for your preferences regarding the placement of the operating controls for the X-ray machine. As the dentist, what would be your order of preference? Give the reasons for your order.
   First possibility: On a wall 5 feet from the examining chair, with convenient access to the patient.
   Second: On a wall 5 feet from the patient but separated from the patient by a concrete dividing wall.
   Third: On a wall 11 feet from the patient.

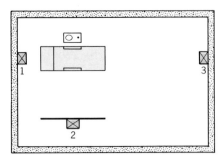

9. Would you be exposed to more background radiation living in Denver or San Francisco? Give reasons to support your answer.

10. If a woman's mother has had breast cancer, the woman has a two to three times greater chance of developing breast cancer than a woman whose mother has not had breast cancer. Why is it important for such a woman to have a mammogram after the age of 35?

11. After giving her patient a complete examination, a doctor feels that the patient may be suffering from an abnormal thyroid. How might radioisotopes be used to confirm the doctor's diagnosis?

12. Strontium-90 decays as follows:

$$^{90}_{38}\mathrm{Sr} \longrightarrow {}^{90}_{39}\mathrm{Y} + {}^{0}_{-1}e$$

Iridium-192 decays as follows:

$$^{192}_{77}\mathrm{Ir} \longrightarrow {}^{192}_{78}\mathrm{Pt} + {}^{0}_{-1}e + \gamma$$

Assume that a patient is suffering from an inoperable cancer in his nose. Which of the two above radioisotopes is better for treating this tumor? Give a reason for your choice.

chapter 7

Electron Configuration and the Periodic Table

Learning Objectives

By the time you have finished this chapter, you should be able to:

1. Describe Bohr's model of the atom.

2. Indicate the maximum number of electrons possible in energy levels 1 through 7.

3. Explain the difference between an atom in the ground state and an excited atom.

4. Explain, on an atomic level, what is happening in a luminous material.

5. Indicate the electron configuration by energy level for each of the elements with atomic number less than 18.

6. Define *periodicity.*

7. Use the periodic table to state the symbol, atomic number, atomic weight, and electron configuration of any element.

8. Given one element in a period or chemical family, pick out another member of that same period or family.

9. State the number of valence electrons in an atom of any representative element.

10. Identify on a periodic table the elements that are metals, nonmetals, representative elements, transition metals, and metalloids.

11. Draw the electron dot diagram for any representative element.

12. Predict how atomic size changes across a period and down a group.

13. Define *ionization energy,* and predict how it will change across a period or down a group.

14. Define *electron affinity,* and predict how it will change across a period or down a group.

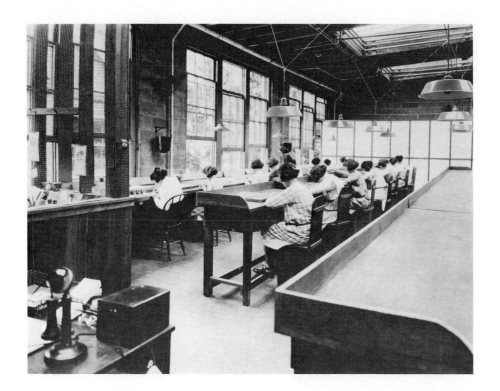

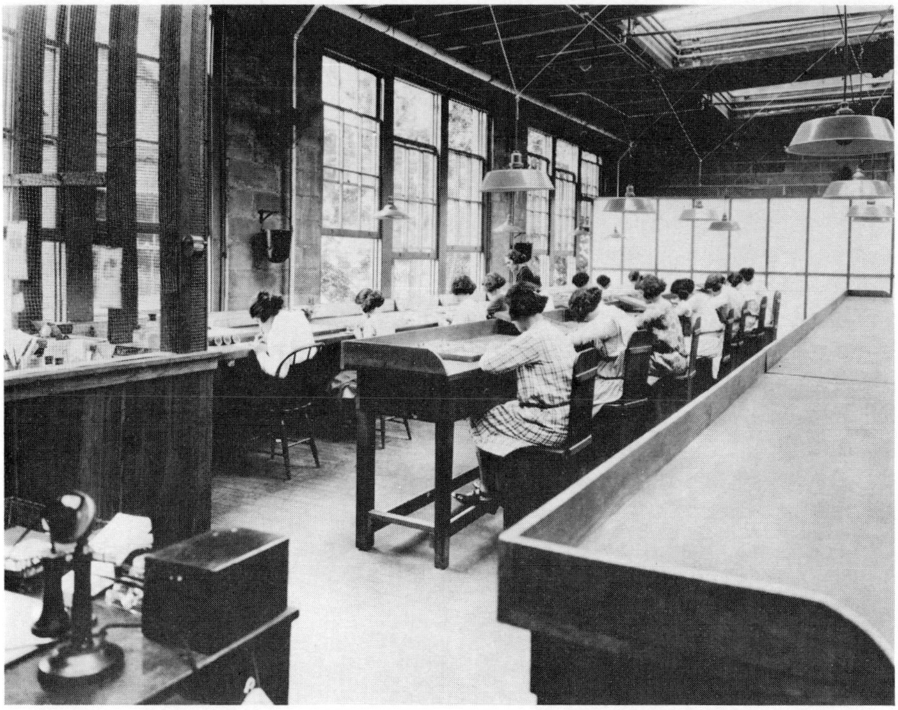

In 1918 Ruth Adams, then 16 years old, was excited to have finally found a job. She was hired by the Radium Luminous Materials Corporation of Orange, New Jersey, to apply luminous paint to watch dials and instrument dials. As you can imagine, this was painstaking work that required great precision. To help her get started, the other women showed Ruth how to keep a fine point on her brush by turning the bristles on her tongue and lips. The paint she used was invented by the president of the company, and contained phosphorescent zinc sulfide with a small amount of radium and adhesive added.

At that time, neither Ruth nor anyone else realized the dangers that radioactive material posed to human tissues. However, it wasn't long until the possible harm of internal exposure to radium was very clearly brought to the public's attention. In 1925 the *New York Times* reported that five watch dial painters had died, and ten others had been stricken with "radium necrosis," a general breakdown of the bone tissues. But still, so little was known about the nature of radium and its effects on human tissues that the company, when brought to trial, defended itself with the incredible claim that small quantities of radium were actually beneficial to health!

The watch dial painters who died in the 1920s suffered symptoms that included severe anemia, tumors of the mouth and nose, and inflammation of the bone marrow in the jaw or other bone structures. Although many of the women who worked in this industry during that time are still living and in good health, many others died after developing bone cancer 20 to 30 years later. Ruth Adams worked in the watch dial factory for six years before leaving to get married. She showed no ill effects until some 25 years later, when she died of bone cancer which had spread throughout her body.

The disabilities or deaths of the women who were exposed to radium resulted from destruction of their bone tissue, especially the tissue of the bone marrow, which produces the blood cells for the body. You might wonder what special property of radium makes it so damaging to the bones of the body, as opposed to other tissues. We will see that radium (Ra) has chemical properties which are very similar to those of calcium (Ca), which is the main component of bones and teeth. Unfortunately, the body is unable to tell the difference between the toxic radium and the essential calcium. Therefore, both elements are deposited in the bones, where the ionizing radiation produced by the radium does its damage.

We have stated that radium and calcium have similar chemical properties. These properties are determined for each element by the arrangement of electrons surrounding the nucleus of the atom. Therefore, a good way to start this chapter is to take a closer look at the arrangement of electrons in the various elements.

Electron Configuration

In an earlier chapter we described an atom of any element as containing a small, dense, positive nucleus surrounded by negative electrons. You will remember that a neutral atom contains the same number of electrons as there are protons in the nucleus. Therefore, the number of electrons in a neutral atom of an element will be equal to the atomic number of that element. Because protons and electrons are too small to be seen with even the most powerful microscopes, we must depend on various theoretical models to describe the arrangement of the electrons around the nucleus.

7.1 The Bohr Model

One of the first models of the atom, called the planetary model, was developed in the early 1900s by Niels Bohr, who based his theory on the experimental data of Ernest Rutherford. Over the past 75 years this theory has been modified quite a bit, but it is helpful for us to understand Bohr's simple model. He pictured the atom as consisting of a small, dense nucleus surrounded by electrons traveling in orbits, similar to planets traveling around the sun (Figure 7.1). In his model the electrons could occupy only certain positions around the nucleus. Each of these positions corresponded to a certain energy value, so these energy positions were called **energy levels** or **energy shells.** Electrons closest to the nucleus

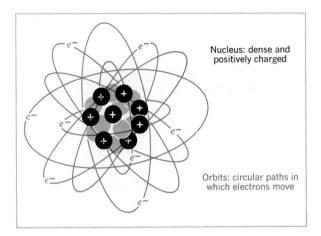

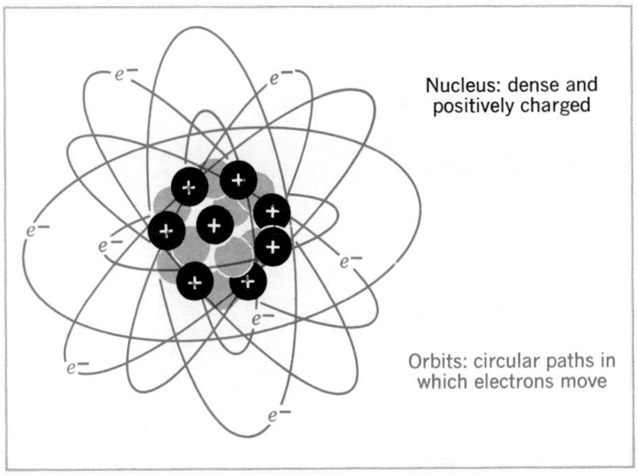

Figure 7.1 Bohr's planetary model of the atom has a dense, positively charged nucleus, with electrons moving in circular paths, called orbits, around the nucleus.

are in the lowest possible energy states, and electrons have higher and higher energy values as they are found farther from the nucleus.

We might compare this idea of electron energy levels with the potential energy you possess as you climb the stairs to the balcony of a theater. At the bottom of the stairs you are at the lowest possible energy position. You must use a specific amount of energy to climb up one stair, and once there you then possess greater potential energy relative to the ground floor. Also, there is no way that you can stand between two steps—you must use a specific amount of energy to get from one step to another step, and you are not able to stop part way in between (Figure 7.2). Similarly, an electron is not permitted to exist in between any of the specific energy levels of the atom. To move from one energy level to another an electron must gain (or lose, depending upon whether it is moving away from or closer to the nucleus) an amount of energy exactly equal to the energy difference between the two levels.

The energy levels within an atom are labeled with numbers, with the first energy level being the one closest to the nucleus. As the number of the energy level increases, it is found farther from the nucleus and contains electrons having greater and greater energy. Each energy level can contain only a certain maximum number of electrons, but in nature this maximum number is, in most cases, never realized (Table 7.1).

Table 7.1 Energy Levels in the Atom

Energy level number (n)	1	2	3	4	5	6	7
Maximum number of electrons allowed in theory ($2n^2$)	2	8	18	32	50	72	98
Maximum number of electrons actually found in nature	2	8	18	32	32	18	8

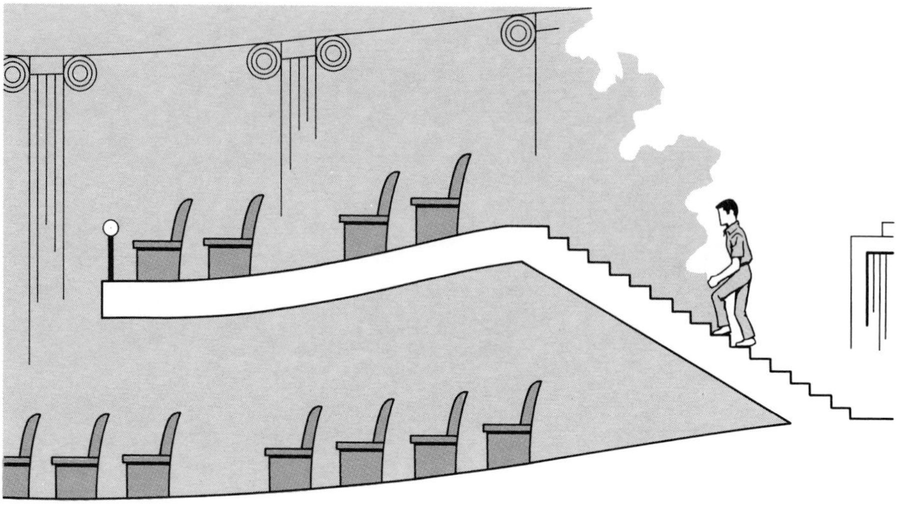

Figure 7.2 We can compare the energy possessed by electrons in an atom to the potential energy that is gained as you climb the stairs to a theater balcony. The higher you go, the more potential energy you possess. Also, you must exert a specific amount of energy to climb each stair, and there is no way to stand between steps.

7.2 Electron Configuration of the First Twenty Elements

We have learned that each neutral atom of a given element has a specific number of electrons equal to the atomic number of that element. For the first twenty elements, these electrons will occupy positions in energy levels according to the following rules.

1. The number of electrons in a neutral atom of an element is equal to the atomic number.
2. Beginning with energy level 1, the electrons will fill in the lowest possible energy positions that are available. When any level has been filled, electrons go on to fill the next level. For example, sodium has atomic number 11 and, therefore, has 11 electrons. Two of these electrons will be in the first energy level, which fills the first level. The second level will be filled with the next 8 electrons. This leaves one remaining electron, which will be found in the third energy level.
3. For the first 18 elements, this pattern is closely followed. (See Table 7.2.) But for elements 19 and 20, the nineteenth and twentieth electrons do not go in the third level, even though this level has room for more than 8 electrons. Instead, these electrons are found in the fourth energy level.

Table 7.2 Electron Configuration of Elements with Atomic Numbers under 21

Element	Atomic Number	Energy Level Electron Configuration 1	2	3	4	5	6	7
Hydrogen, H	1	1						
Helium, He	2	2						
Lithium, Li	3	2	1					
Beryllium, Be	4	2	2					
Boron, B	5	2	3					
Carbon, C	6	2	4					
Nitrogen, N	7	2	5					
Oxygen, O	8	2	6					
Fluorine, F	9	2	7					
Neon, Ne	10	2	8					
Sodium, Na	11	2	8	1				
Magnesium, Mg	12	2	8	2				
Aluminum, Al	13	2	8	3				
Silicon, Si	14	2	8	4				
Phosphorus, P	15	2	8	5				
Sulfur, S	16	2	8	6				
Chlorine, Cl	17	2	8	7				
Argon, Ar	18	2	8	8				
Potassium, K	19	2	8	8	1			
Calcium, Ca	20	2	8	8	2			

In order to know how to assign electrons for elements having atomic numbers greater than 20, additional rules are required. Fortunately, for our study of living organisms it is not too important that we understand these rules, because the elements that make up the majority of living matter have atomic numbers less than 21.

7.3 Formation of Ions

In a normal atom, all of the electrons will be found in the lowest possible energy levels. Such an atom is said to be in the **ground state.** If the atom absorbs energy (such as heat or light) from an external source, some electrons might jump to a higher energy level. When this occurs, each

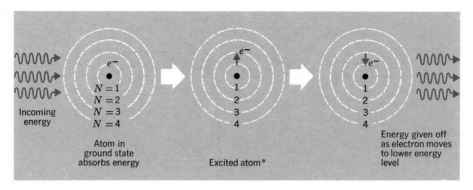

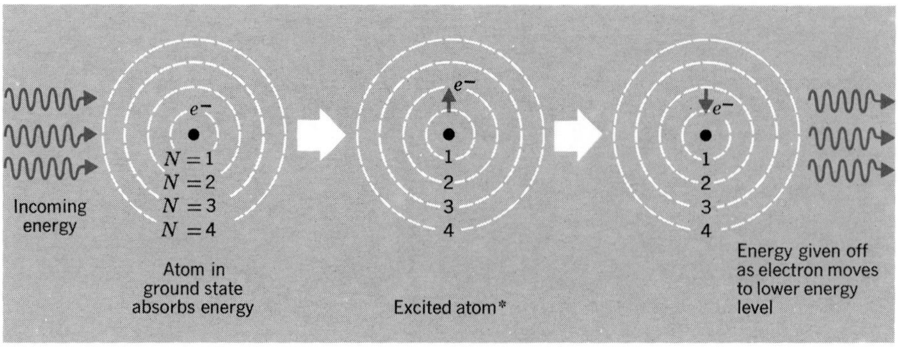

Figure 7.3 Electrons can move to energy levels farther from the nucleus by absorbing energy, thus forming an excited atom. These electrons may then return to lower energy levels by giving off energy, usually in the form of light.

such electron absorbs an amount of energy exactly equal to the energy difference between the two levels. An atom having one or more electrons in a higher than normal energy state is called an **excited atom.** Excited atoms are unstable, and in such atoms electrons will drop back to lower energy levels by giving off energy, usually in the form of radiant energy (UV, IR, or light waves). The light given off by an atom as an electron drops from one energy level to another has a specific energy or wavelength corresponding to the energy difference between the two levels (Figure 7.3). If enough energy is supplied to an atom, one of its electrons may jump completely away from the atom, leaving the atom with one less electron than it has protons. This new particle is then no longer a neutral atom, but is rather a positively charged **ion.** Ionizing radiation, such as gamma radiation, can give an atom enough energy that one or more of its electrons break completely away. This, then, explains why one of the effects of ionizing radiation on living cells is to produce positive ions.

7.4 Luminescence

When an element or compound becomes excited by absorbing energy from an external source, and then gives off that energy as visible light, it is said to be **luminescent.** There are two kinds of luminescence: fluorescence, in which the element or compound stops giving off light as soon as the energy being supplied to the compound is stopped; and phosphorescence, in which the element or compound continues to give off light for a short time after the incoming energy has stopped. There are many examples of luminescence with which you are familiar. Laundry whiteners and brighteners make use of fluorescent dyes which become attached to clothing in the laundering process. These compounds absorb energy from the ultraviolet light in sunlight, and then give off light in the visible region. Through both reflection and fluorescence, then, the clothing gives off more visible light than falls on it—this makes the clothing appear "brighter". Red reflective tape, which is often attached to bicycles and cars to increase their visibility, contains compounds that absorb nonred light and give off the red-orange light we see (Figure 7.4). The luminous dial you may have

Figure 7.4 The reflective trim on these bicycle tires contains a fluorescent material, greatly increasing the visibility of these bicycles at night. (Courtesy 3M Company)

on your watch contains tiny amounts of radioactive material mixed with a powdered phosphor. The radioactive material releases a small but steady amount of energy, which is absorbed by the atoms in the phosphor. When the electrons in these atoms return to the ground state, they release the visible light which you see as a faint glow on the watch dial.

Living organisms make use of luminescence for sexual attraction, protection, and hunting. The firefly uses bioluminescence to attract a mate. This insect uses chemical energy to excite electrons in a special compound that produces the firefly's blink as the electrons return to the ground state. Green plants use the reverse of this process in the production of food: energy from sunlight excites electrons in molecules of the plant's green pigment, called chlorophyll. As these electrons return to the ground state the energy is not released in the form of light, but instead is finally trapped as chemical energy in a molecule of sugar. This process of using the sun's energy to produce a sugar molecule from carbon dioxide and water is called photosynthesis, and will be studied in detail in Chapter 17.

* 7.5 The Quantum Mechanical Model of the Atom

The simple planetary model of the atom developed by Bohr did not explain some of the experimental data that was later collected, and in the late 1920s Erwin Schrödinger, P. A. M. Dirac, Werner Heisenberg, and others developed a new model of the atom called the quantum mechanical model. The model is based on fairly complicated mathematical concepts, and we will look at only some of the ways this theory changed the idea of the atom developed by Bohr.

* This section is optional and can be skipped without loss of continuity.

Although the Bohr atom described electrons orbiting around the nucleus, scientists found it impossible to accurately pinpoint the position and speed of an electron. Therefore, the electron's position could not be described as following a definite path. Instead, scientists could only specify regions in which there is a large probability of finding one or two electrons. These probability regions are known as atomic **orbitals.** Each energy level was found to have energy sublevels consisting of one or more atomic orbitals.

Atomic Orbitals

Each orbital is a particular region around the nucleus of the atom in which there is a high probability of finding an electron with a specific energy. An orbital can contain zero, one, or two electrons.

The first energy level contains one atomic orbital, labeled the 1*s* orbital. The probability region described by this orbital is a sphere with its center at the nucleus (Figure 7.5). The second energy level contains four atomic orbitals. One of the orbitals is spherical in shape, and is called the 2*s* orbital. The probability regions of the other three orbitals are shaped like dumbbells that cross each other at right angles (Figure 7.5). These orbitals, called the 2*p* orbitals, have slightly more energy than the 2*s* orbital.

Each higher energy level will contain one *s* and three *p* orbitals. In addition to these, there are two other types of orbitals: the *d* orbitals, which appear in groups of five starting from energy level 3, and the *f* orbitals, which appear in groups of seven starting from energy level 4. We will not discuss the shapes of the *d* and *f* orbitals in this textbook.

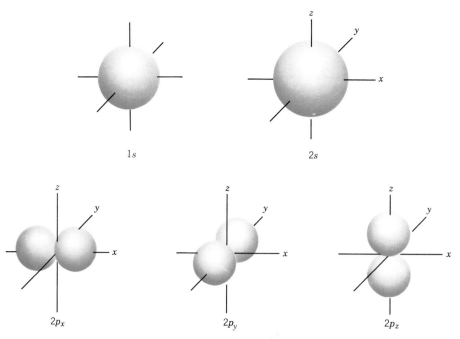

Figure 7.5 Two-dimensional drawings representing the probability regions (areas in which it is most likely to find an electron) for the 1*s*, 2*s*, and 2*p* orbitals.

Table 7.3 The Electron Orbital Configurations for the Elements With Atomic Numbers from 11 to 18

Element	Atomic number	Electron orbital configuration 1s	2s	$2p_x$	$2p_y$	$2p_z$	3s	$3p_x$	$3p_y$	$3p_z$
Na	11	↑↓	↑↓	↑↓	↑↓	↑↓	↑	—	—	—
Mg	12	↑↓	↑↓	↑↓	↑↓	↑↓	↑↓	—	—	—
Al	13	↑↓	↑↓	↑↓	↑↓	↑↓	↑↓	↑	—	—
Si	14	↑↓	↑↓	↑↓	↑↓	↑↓	↑↓	↑	↑	—
P	15	↑↓	↑↓	↑↓	↑↓	↑↓	↑↓	↑	↑	↑
S	16	↑↓	↑↓	↑↓	↑↓	↑↓	↑↓	↑↓	↑	↑
Cl	17	↑↓	↑↓	↑↓	↑↓	↑↓	↑↓	↑↓	↑↓	↑
Ar	18	↑↓	↑↓	↑↓	↑↓	↑↓	↑↓	↑↓	↑↓	↑↓

The order in which electrons fill these orbitals is shown in Figure 7.6. In addition to filling the lowest possible energy level, there are several other rules that describe how electrons are found in orbitals. First, an

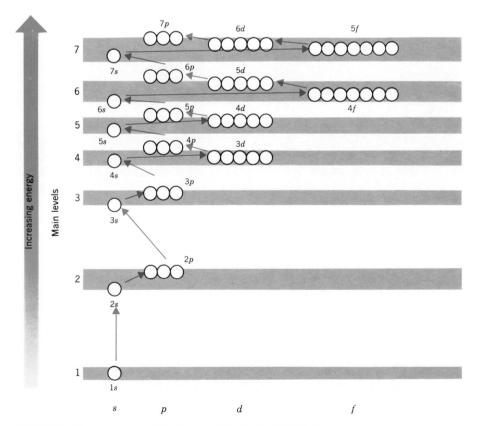

Figure 7.6 The order in which electrons fill atomic orbitals is indicated by following the arrows, beginning with the 1s orbital.

orbital will hold no more than two electrons, and can hold two electrons only if they are spinning in opposite directions. Electrons behave as though they are spinning around an axis, and scientists use arrows — ↑ for clockwise and ↓ for counterclockwise — to indicate the direction of this spin. Second, electrons will not pair up in an orbital if there is another orbital of the same energy available. They pair up only when all orbitals of the same energy have at least one electron in them. Table 7.3 shows the electron orbital configuration of elements with atomic numbers 11 to 18.

Example 7-1 _____

1. What is the electron orbital configuration of carbon, C?

 Carbon's atomic number is 6, so a neutral atom of carbon contains 6 electrons. Using Figure 7.6 as a guide, we can fill in the orbitals. The 1s orbital is filled with 2 electrons, leaving 4 electrons still to be assigned. Next, the 2s orbital is filled with 2 electrons. These electrons will pair up in the 2s orbital before filling the 2p orbitals because the 2p orbitals have slightly more energy than the 2s orbital. The next electron is found in the $2p_x$ orbital. The last electron will not pair up in the $2p_x$, but will instead be found in a different 2p orbital having the same energy.

 $$1s \quad 2s \quad 2p_x \quad 2p_y \quad 2p_z$$

 Carbon (↑↓) (↑↓) (↑) (↑) (○)

2. What is the electron orbital configuration of calcium, Ca?

 Calcium has an atomic number of 20 and, therefore, a neutral atom of calcium will have 20 electrons. Using Figure 7.6 as a guide, we can again fill in the orbitals. The 1s is filled with 2 electrons. The 2s is filled with 2 electrons and the three 2p orbitals are filled with 6 electrons, making a total of 10 electrons that we so far have placed. Ten electrons remain. The 3s orbital is filled with 2 electrons, and the three 3p orbitals with 6 electrons. We now have only 2 electrons remaining. You will notice from Figure 7.6 that the 4s orbital has less energy than the 3d orbital. Therefore, the 4s will be filled first with two electrons, completing the electron orbital configuration for calcium.

 $$1s \quad 2s \quad 2p_x \quad 2p_y \quad 2p_z \quad 3s \quad 3p_x \quad 3p_y \quad 3p_z \quad 4s$$

 Calcium (↑↓) (↑↓) (↑↓) (↑↓) (↑↓) (↑↓) (↑↓) (↑↓) (↑↓) (↑↓)

 Note that the 3d orbitals will be filled in elements having atomic numbers 21 through 30, and the 4p orbitals will then be filled in the elements having atomic numbers 31 through 36.

The Periodic Table of Elements

7.6 Mendeleev's Periodic Table

When discussing radium we mentioned that the chemical properties of each element depend upon that element's electron configuration. It would certainly be quite a chore to memorize the electron configuration of all 106 chemical elements. But take heart; there is a way of writing down the chemical elements so that their similarities are easily seen. This method makes use of the **periodicity,** or repeating nature, of the chemical properties of the elements, which was first noticed by the Russian scientist Dmitri Mendeleev in 1869. He had devised a table showing each of the then-known elements arranged according to their atomic weight, and noticed that the chemical properties of the elements seemed to show periodic relationships. That is, various similarities in chemical properties repeated themselves among the elements as he read down the table. However, Mendeleev also realized that there seemed to be gaps in this table which would have to be filled in if the periodic relationships were to hold exactly. He wisely left these spaces blank, suspecting that these gaps represented elements that had not yet been discovered. About 45 years later it was found that if Mendeleev's table were slightly rearranged so that the elements were listed by atomic number rather than by atomic weight, this would eliminate some of the irregularities left in the table. The resulting arrangement of elements is known as the **periodic table of the elements,** and modern-day periodic tables can be drawn so as to give a large amount of information about each element. A periodic table is shown on the inside front cover of this book.

Let's examine the periodic table closely. You will see that each element appears in a box containing specific information about that element. In this case the box lists the symbol of the element, the atomic number, the atomic weight, and the electron configuration for the main energy levels.

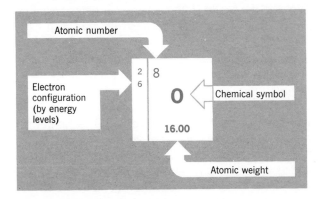

7.7 Periods and Families

Each horizontal row of the periodic table is called a **period.** The periods are numbered from one to seven corresponding to the seven energy levels

Table 7.4 Electron Configurations of Three Chemical Families

Group Number	Chemical Family	Element	Atomic Number	Electron Configuration 1	2	3	4	5	6	7
I	Alkali metals	Lithium	3	2	1					
		Sodium	11	2	8	1				
		Potassium	19	2	8	8	1			
		Rubidium	37	2	8	18	8	1		
		Cesium	55	2	8	18	18	8	1	
		Francium	87	2	8	18	32	18	8	1
II	Alkaline earth metals	Beryllium	4	2	2					
		Magnesium	12	2	8	2				
		Calcium	20	2	8	8	2			
		Strontium	38	2	8	18	8	2		
		Barium	56	2	8	18	18	8	2	
		Radium	88	2	8	18	32	18	8	2
VII	Halogens	Fluorine	9	2	7					
		Chlorine	17	2	8	7				
		Bromine	35	2	8	18	7			
		Iodine	53	2	8	18	18	7		
		Astatine	85	2	8	18	32	18	7	

of an atom which can contain electrons. This means that any element in, say, period four will have its outermost electrons located in the fourth energy level of the electron shell. For example, find sodium (symbol Na, atomic number 11) on the table. Sodium has only one electron in its outermost energy level and, because sodium appears on period three of the periodic table, this indicates that sodium contains that one electron in the third energy level.

Each column of the periodic table is called a **group** or **chemical family.** Mendeleev noticed that the various members of a family show similar chemical behavior. We will see that it is the number of electrons in the outermost energy level of an atom which determines its chemical behavior. These outermost energy level electrons are called **valence electrons;** each member of a chemical family has the same number of valence electrons (Table 7.4). Notice the columns on the periodic table labeled with the Roman numerals I through VII (on some periodic tables these are identified as the A-group families). These seven families and group 0 are called the **representative elements** (Figure 7.7). The Roman numeral above these columns indicates the number of valence electrons for each member of the family. For example, the members of group II are beryllium (Be), magnesium (Mg), calcium (Ca), strontium (Sr), barium (Ba), and radium (Ra). Each neutral atom of these elements will contain two valence electrons. These two electrons will be located on the fourth energy level in an atom of calcium, and on the seventh energy level in an atom of radium.

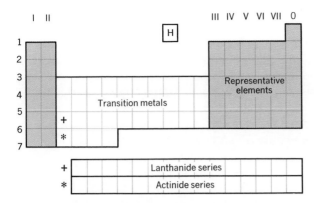

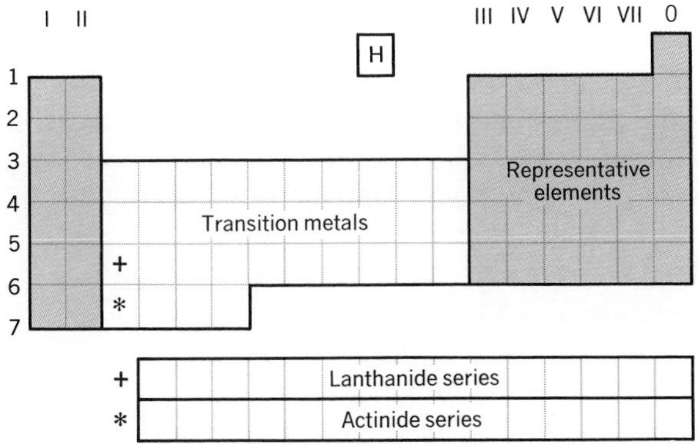

Figure 7.7 The location of the representative elements, the transition metals, and the metals of the lanthanide and actinide series on the periodic table.

The elements of group 0 are called the **noble gases.** Some of these gases can be forced to form compounds with oxygen or with members of the halogen family (the elements in group VII), but otherwise are extremely unreactive. Each noble gas, with the exception of helium (He), has an electron configuration of eight electrons in the outermost energy level, and the chemical stability of these elements comes from this stable arrangement of electrons. Although helium contains only two electrons, these valence electrons completely fill the first energy level of this atom, giving helium its stability. At the present time, chemists still cannot completely explain the great stability given to the noble gases by their particular arrangement of valence electrons.

We have said that members of the same chemical family will show similar chemical behavior. For example, chlorine (Cl_2) and iodine (I_2), both members of the halogen family, are highly effective in killing bacteria, and are used as disinfectants and antiseptics, respectively. Germanium (Ge) and silicon (Si), members of group IV, are both widely used as semiconductors in transistors. Hard water problems can be caused by either calcium (Ca^{2+}) or magnesium (Mg^{2+}) ions because both of these group II elements undergo similar chemical reactions with soap. Two radioactive wastes of special concern are strontium-90 and cesium-137. Strontium, like radium, is in group II. Therefore, it will have chemical properties similar to calcium. Strontium-90 in the atmosphere falls onto the grass, is eaten by cows, and is consumed by humans when they drink milk. As with radium, the human body will mistake strontium-90 for calcium, and will deposit it in bones and teeth. Cesium is in group I, and will behave in the body in much the same way as sodium, an element that is essential for many body functions.

You have probably noticed that hydrogen is separated slightly from the other elements on the periodic table, indicating that hydrogen actually belongs in a chemical family of its own. It has chemical properties similar to elements of group I and to elements of group VII.

7.8 Metals and Nonmetals

Various properties of the elements allow them to be classified as **metals** or **nonmetals,** and the periodic table can be divided into two large sections based on this classification (Figure 7.8). Metals are generally shiny, dense, and easily worked. They have high melting points, and are good conductors of electricity. Nonmetals, on the other hand, tend to be of low density and brittle when solid. Most nonmetals have low melting points (that is, are liquids or gases at room temperature), and are poor conductors of electricity.

The heavy, jagged line on the periodic table divides those elements that are metals from the nonmetals, with the metals lying on the left of the line and clearly making up the majority of elements. However, there is no sharp distinction between the elements having metallic properties and those having nonmetallic properties; the elements next to the jagged line may exhibit properties of both metals and nonmetals. Such elements are known as **metalloids** or **semimetals.** For example, arsenic (symbol As, atomic number 33) is a brittle gray nonmetal. When freshly cut, however, it has a bright metallic shine. Arsenic acts as a metal in forming compounds with oxygen and chlorine, but acts as a nonmetal in other chemical reactions. Aluminum (symbol Al, atomic number 13) is a silvery metal, but it also has some nonmetallic properties.

The metals of the periodic table can be further classified. The elements of group I are soft metals which can be cut with a knife. They react very rapidly with water, and are called **alkali metals.** Because of their high reactivity, they are found only in a combined form in nature. The elements of group II are harder metals which react less rapidly with water, and are called the **alkaline earths.** The three rows of ten elements in the middle of the table are called **transition metals** (or the B-group elements), and they often show similar chemical properties not only within groups, but also along periods. This similarity in chemical behavior results from the fact that as we move along a period across the transition metals,

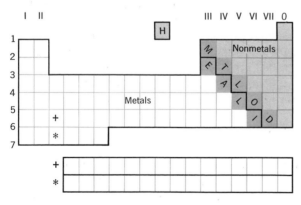

Figure 7.8 The location of the metals, nonmetals, and metalloids on the periodic table.

electrons are not being added to the outermost energy level of the atoms, but rather are being added to the next-to-the-outermost level (Figure 7.7). (The 3*d* orbitals are being filled in the transition metals of period 4, and the 4*d* orbitals in the transition metals of period 5, etc.) The outermost energy level of each of the transition metals contains only one or two electrons. For example, iron (Fe), cobalt (Co), and nickel (Ni), located on the fourth row of the periodic table, all have two electrons on the same outermost energy level. Although they contain different total numbers of electrons, they are very similar in chemical behavior, and are often grouped together for this reason.

We have seen that the same number of electrons in the outermost energy level of the transition metals masks the difference in the number of electrons in the next-to-outermost level. This is doubly true for the **lanthanide** and **actinide** series, the two rows of 14 elements at the bottom of the table. The differences in the electron arrangements of these elements are hidden beneath two identical outer shell electron configurations (Figure 7.7). (Electrons in these elements are filling the *f* orbitals—the 4*f* in the lanthanide metals and the 5*f* in the actinide metals.) The naturally occurring elements in these two rows were once known as the rare earths because they appeared to exist only in very small quantities. These very similar elements were finally separated and identified using a technique called chromatography. It is now known that many of these elements are, in fact, more abundant than either gold or silver.

7.9 Electron Dot Diagrams

We have stated that the chemical behavior of an atom is related to its electron configuration, especially to the arrangement of its valence electrons. There are several ways to represent the electron configuration of an element. Tables 7.2 and 7.3 show two ways to write out electron configurations, and Figure 7.9 illustrates two types of diagrams often used. In the rest of this book we will be focusing our attention on the valence electrons of elements, and it will be convenient for us to use electron dot diagrams to represent an element's configuration of valence electrons. As you may guess from the name, in these diagrams a dot is used to represent each valence electron, and these dots are placed around the symbol of the element. On each side of the element's symbol there is room for two dots.

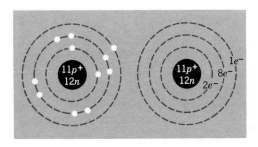

Figure 7.9 Schematic drawings of the sodium atom.

Beginning on any side, the dots are first placed singly, then paired for elements having more than four valence electrons.

Let's look at the electron dot diagrams for the elements in period 2. Remember that the group number of a representative element tells you the number of valence electrons for each member of that group or family.

Element	Symbol	Atomic Number	Group	Valence *e*	Electron Dot Diagram
Lithium	Li	3	I	1	Li·
Beryllium	Be	4	II	2	Be:
Boron	B	5	III	3	·B:
Carbon	C	6	IV	4	:C:
Nitrogen	N	7	V	5	·N: (with dots above and below)
Oxygen	O	8	VI	6	·O: (with dots above and below)
Fluorine	F	9	VII	7	:F: (with dots above and below)
Neon	Ne	10	0	8	:Ne: (with dots above and below)

7.10 The Periodic Law

Various physical and chemical properties of atoms vary periodically with their atomic number. That is, a particular pattern of properties will be repeated as the atomic number increases. This fact is often referred to as the **periodic law.** An examination of three of these periodic properties will help us to better understand the chemical behavior of the elements, especially the different chemical behavior of metals and nonmetals.

Atomic Size

One property of the elements which shows a periodic relationship is the size of their atoms. From Figure 7.10 you can see that, with only a few exceptions, the atomic radius decreases as you move across a period of the periodic table (for example, from atomic numbers 3 to 9.) The group I and group VII elements are labeled on the graph, and this will help you see that as you move down a group or chemical family the atomic radius increases as the atomic number increases. One result of this pattern is that the metallic elements on the left-hand side of the table will

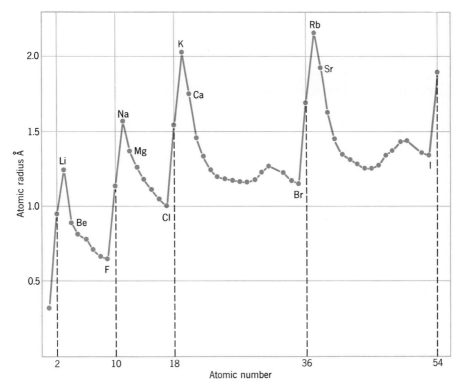

Figure 7.10 Graph of the periodic trends of atomic radii. With a few irregularities, atomic radii decrease across a row and increase as the atomic numbers increase down a chemical family.

have large atomic radii, and the nonmetals on the upper right-hand part of the table will have small radii.

Ionization Energy

A second important property of an element is its ionization energy. The **ionization energy** of an element is the amount of energy that must be added to an atom to remove one electron from its outermost energy level. Put another way, the ionization energy indicates how strongly the nucleus attracts the valence electrons. The stronger the attraction for these valence electrons, the greater is the ionization energy. It is easy to see that the size of an atom will affect how strongly the nucleus attracts the valence electrons. In small atoms the nucleus and the valence electrons are close together and, therefore, the electrical attraction is strong. In larger atoms the outermost energy level is farther from the nucleus and, therefore, the attraction for the valence electrons is weaker. This means that the ionization energy will increase across a period as the radius decreases. On the other hand, as we move down any family the atomic radii are increasing. Therefore, the attraction of the nucleus will become weaker and the ionization energy will decrease (Figure 7.11).

Notice that the metals generally have lower ionization energies than

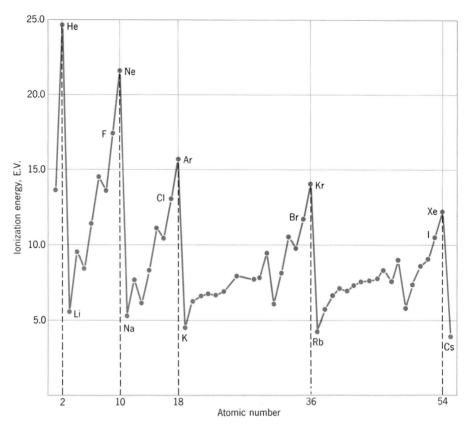

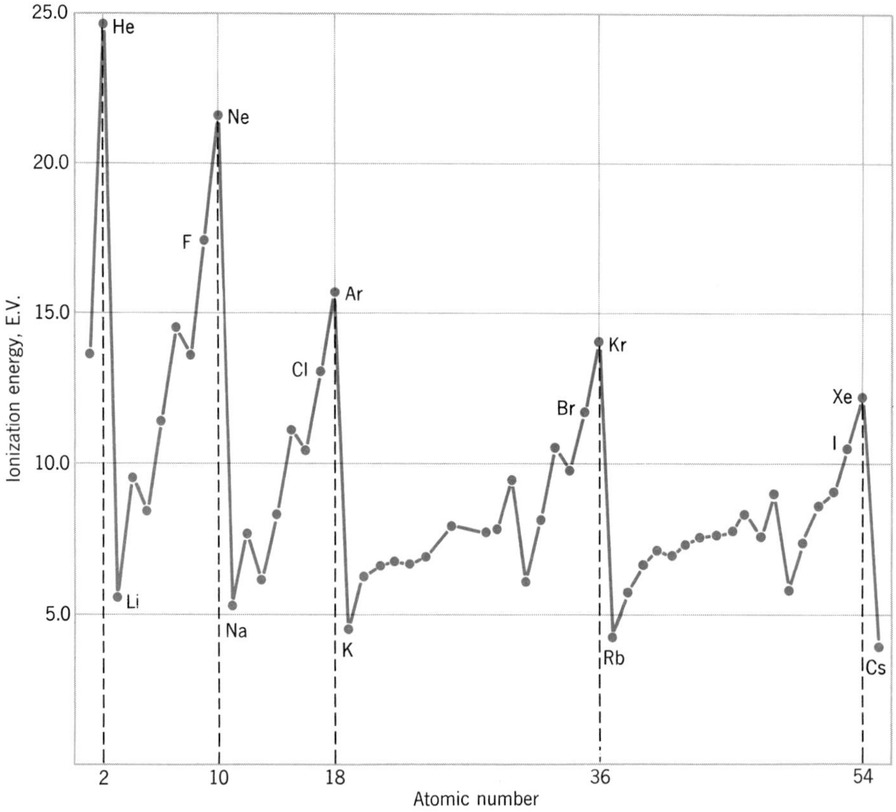

Figure 7.11 Graph of the periodic trends of ionization energies. Ionization energies tend to increase across a row and to decrease down a chemical family.

do the nonmetals, so that it is easier to remove a valence electron from a metal than from a nonmetal. Also, Figure 7.11 shows that the noble gases, with their very stable configuration of electrons, all have very high ionization energies. This particular electron configuration somehow makes it extremely difficult to remove an electron from the outermost energy level of the atom.

Electron Affinity

A third property showing a periodic relationship among the elements is electron affinity. The **electron affinity** of an element is the amount of energy which is released when an electron is added to a neutral atom of that element. This property indicates the ability of an atom to attract additional electrons. The larger the electron affinity, the larger the attraction of the atom for additional electrons. Determination of electron affinities is a complex process, and values are not yet available for all the elements. However, enough values have been determined to show trends in electron affinity (Figure 7.12). Metals have low electron affinities and nonmetals have high electron affinities. As we might expect from their stable electron configurations, the noble gases have low electron affinities.

I	II	III	IV	V	VI	VII	0
H 0.75							He (-0.22)
Li 0.62	Be (-2.5)	B 0.86	C 1.27	N 0.0	O 1.46	F 3.34	Ne (-0.30)
Na 0.55	Mg (-2.4)	Al (0.52)	Si 1.24	P 0.77	S 2.08	Cl 3.61	Ar (-0.36)
K 0.50	Ca (-1.62)	Ga (0.37)	Ge 1.20	As 0.80	Se 2.02	Br 3.36	Kr (-0.40)
Rb 0.49	Sr (-1.74)	In 0.35	Sn 1.25	Sb 1.05	Te 1.90	I 3.06	Xe (-0.42)
Cs 0.47	Ba (-0.54)	Tl 0.5	Pb 1.05	Bi 1.05	Po (1.8)	At (2.8)	Rn (-0.42)
Fr (0.46)	Ra						

Figure 7.12 Periodic table of electron affinities (in electron volts; values in parentheses are not determined experimentally). [From "Experimental Values of Atomic Electron Affinities," by E. C. M. Chen and W. E. Wentworth, *Journal of Chemical Education,* 52, 488 (1975).]

Chapter Summary

The Bohr model pictures the atom as consisting of a small, dense, positive nucleus surrounded by negative electrons traveling in orbits around the nucleus. The electrons cannot occupy just any position around the nucleus, but can be found only in certain allowed energy regions called energy levels. The energy level closest to the nucleus contains electrons with the least amount of energy. To move from one energy level to another an electron must absorb or give off energy exactly equal to the energy difference between the two energy levels. Each energy level can hold no more than a certain number of electrons, and specific rules are followed in determining the electron configuration of an atom. If an atom absorbs energy from an outside source, an electron in that atom can move from a lower energy level to a higher energy level. Such an atom is said to be excited. If the atom then gives off energy in the form of visible light, it is said to be luminescent. If enough energy is absorbed by an atom, one of its electrons may jump completely away from the atom, forming a positively charged ion.

The elements show periodicity, or a repeating nature, in their chemical and physical properties. Three such properties are atomic size, ionization energy, and electron affinity. If the elements are arranged according to their atomic numbers on a periodic table, members of each vertical column, called a group or family, will display similar properties. Each member of a family will have the same number of valence, or outer energy level, electrons. Elements on the periodic table can be divided into metals and nonmetals and can be further classified as representative elements, transition metals, lanthanide metals, and actinide metals.

Exercises and Problems

1. Describe the Bohr model of the atom.
2. What is the maximum number of electrons found on the fourth energy level? On the sixth?
3. Explain, on the atomic level, why a watch dial glows in the dark.
4. Fill in the following table.

	Symbol	Atomic Number	Atomic Weight	Electron Configuration
(a) Lithium				
(b) Nitrogen				
(c) Neon				
(d) Magnesium				
(e) Aluminum				
(f) Chlorine				

5. What is the electron configuration of a ground state atom containing 16 electrons?
6. State the number of valence electrons in a neutral atom of each of the following elements.

   a. Rubidium
   b. Indium
   c. Phosphorus
   d. Krypton
   e. Beryllium
   f. Silicon
   g. Bromine
   h. Selenium

7. Identify the following as (1) either a metal, metalloid, or nonmetal and (2) either a representative element or transition metal.

   a. Strontium
   b. Selenium
   c. Iron
   d. Germanium
   e. Copper
   f. Fluorine
   g. Silicon
   h. Potassium

8. Draw the electron dot diagrams for each of the following atoms.

   a. Cesium
   b. Germanium
   c. Calcium
   d. Neon
   e. Arsenic
   f. Aluminum
   g. Sulfur
   h. Iodine

9. For each of the following pairs predict which element has (1) the larger radius, and (2) the larger ionization energy.

a. Na and P
b. C and O
c. Li and Rb
d. As and F
e. Ne and Xe
f. N and Sb
g. Sr and Si
h. Fe and Br

10. For each of the following pairs, predict which element has the higher electron affinity.

    (a) Na and Cl (b) Cl and I (c) Cl and Ar

11. Which has a higher ionization energy?

    a. A sodium atom, or a sodium ion with a charge of +1?
    b. A magnesium ion with a charge of +1, or a magnesium ion with a charge of +2?

    Give a reason for your answers.

12. Which has the higher electron affinity, a chlorine atom or a chloride ion with a charge of −1?
    Give a reason for your choice.

13. Arrange the following elements in order from the most metallic to the least metallic: Sulfur, Chlorine, Silicon, Phosphorus.

14. Look at the following electron configuration for a magnesium atom, atomic number 12.

energy level, *n*	1	2	3	4	5
electron configuration	2	8	1	1	

    Is this atom in the ground state? Why or why not?

15. Data for the nuclei of three atoms are shown below. Which two atoms would have similar chemical properties?

    Atom A contains 4 protons and 5 neutrons
    Atom B contains 8 protons and 8 neutrons
    Atom C contains 12 protons and 12 neutrons

    Give a reason for your choice.

16. In the early 1940s, zinc beryllium silicate was used in the manufacture of fluorescent lights. The use of this substance has been banned because workers who came in contact with it developed a disease called "berylliosis". This disease resulted from the disruption of the normal chemical pathways of a specific element in the body. From your knowledge of the periodic table, state which element you think this was.

17. You are running a carefully controlled experiment that keeps track of the use of sodium ions by a specific one-celled organism. You discover that your culture water has been contaminated by trace amounts of the following ions: potassium ion, barium ion, iodide ion. Do you think that any of these ions would interfere with your data? Why or why not? Design an experiment that would support your conclusion.

chapter 8

Combinations of Atoms

Learning Objectives

By the time you have finished this chapter, you should be able to:

1. State the octet rule.

2. Describe an ionic bond and an ionic compound.

3. Write the formula for a binary ionic compound when given its name, and write the name of the compound when given its formula.

4. Describe a covalent bond and a covalent compound.

5. Describe a single, double, and triple covalent bond.

6. Draw the electron dot diagram for a covalent compound formed from the representative elements.

7. Write the formula for a binary covalent compound when given its name, and write the name of the compound when given its formula.

8. Write the formula for a compound containing a polyatomic ion when given its name, and write the name of the compound when given its formula.

9. Define *electronegativity* and describe how the electronegativity of the elements changes on the periodic table.

10. Describe the difference between a polar and a nonpolar covalent bond.

11. Predict whether a simple molecule will be polar or nonpolar.

12. Describe the hydrogen bond.

A light wind was blowing as the afternoon freight train rumbled through a small town in southeastern Louisiana. The train was just starting to gather speed about two miles outside of town when a freight car suddenly jumped the tracks, pulling 18 cars with it. One of those 18, a tank car containing 30 tons of liquid chlorine (Cl_2), lay on its side with a huge gash ripped open. The chlorine, which had immediately vaporized to a greenish-yellow gas, now poured out of the tank car and was carried by the breeze back toward the town.

The Harrison family lived in a nearby farmhouse, unaware of the approaching danger. Within minutes of the train accident the irritating odor of chlorine began to fill the house. Mrs. Harrison suddenly found she was having difficulty breathing, and her two small children began retching and vomiting. Her husband came running from the barn with his eyes streaming and quickly loaded his family into their car. As they rushed off, Mrs. Harrison noticed that her 11-month-old son Randy was having a great deal of trouble breathing. Their desperate drive to the hospital was accompanied by the sound of the volunteer fire department's siren screaming out the emergency signal, and the sight of the sheriff's car helping to evacuate people from nearby farms.

The deadly cloud of chlorine gas forced nearly 1,000 people to flee from their homes, offices, and schools. Hundreds of farm animals died from exposure to the gas, and no one could live in several square miles of the countryside for several days. Fifty people were treated at the hospital for severe irritation caused by the chlorine, and ten, including the Harrison family, were hospitalized with critical poisoning. Unfortunately, Randy died from the effects of the chlorine gas before his family could even reach the hospital.

Chlorine, at room temperature, is a greenish-yellow gas that has a characteristic irritating and suffocating odor. In low concentrations it will irritate mucous membranes and the respiratory system, and in high concentration will cause difficulty breathing—leading in extreme cases to death from suffocation.

Another dangerous substance is the element sodium (Na). This element is an alkali metal which is so highly reactive that it is never found in a pure state in nature. (When isolated in the pure form, however, it is a soft, silvery metal that can easily be cut with a knife.) Great care must be taken in handling sodium to be sure it is kept away from water. When it does contact water, sodium reacts extremely vigorously, releasing hydrogen gas (H_2) that can be ignited by the heat from this reaction.

Chlorine and sodium, then, are both highly reactive elements that are potentially dangerous to living tissue. However, suppose you drop a piece of sodium into a container of chlorine gas, and warm the container. You will soon see a white powder start to form—a chemical reaction is taking place. The product of this reaction is sodium chloride (NaCl), the substance more commonly known as table salt. But table salt has none of the properties of the reactants, sodium and chlorine. In fact, sodium chloride is an essential part of our diets. It plays an important role in maintaining the proper amount of water in our cells and tissues, and is needed for the contraction of muscles and the transport of nerve impulses.

Sodium chloride is a chemical compound, a homogeneous substance produced by the reaction of different elements. This chemical reaction, like many other spontaneous processes in nature, results in the formation of a more stable substance. In this case, the elements sodium and chlorine are both relatively unstable, and will react spontaneously (the addition of heat will only speed up the reaction) to form sodium chloride, which is very stable.

Ionic Bonds and Ionic Compounds

8.1 Transfer of Electrons

About 1920, W. Kossel and G. N. Lewis noticed that the elements in groups I through VII enter into chemical combinations that involve the loss, gain, or sharing of electrons in such a way as to end up with a total of eight valence, or outer energy level, electrons. (You might remember that this is the electron configuration of the noble gases, which are extremely stable.) The tendency of these elements to attain an outer octet, or eight valence electrons, is called the **octet rule.** Although there are many exceptions to this rule among the heavier elements, we will find it quite useful in predicting the composition of chemical compounds

$Na \longrightarrow Na^{+} + e^{-}$

$e^{-} + Cl \longrightarrow Cl^{-}$

Figure 8.1 In the formation of sodium chloride, a sodium atom will lose one electron to a chlorine atom, producing the positive sodium ion Na^{+} and the negative chloride ion Cl^{-}.

produced from the reactions between lighter elements (atomic numbers 1 to 22).

There are two ways in which chemical elements can attain a stable octet of electrons: by gaining or losing electrons, or by sharing electrons. The first of these ways, the transfer of electrons from one atom to another, causes electrically neutral atoms to become ions—and the force of attraction between these ions is called an **ionic bond.** Such a transfer occurs when an element such as chlorine, which has a very strong attraction for additional electrons (that is, high electron affinity), reacts with an element such as sodium, which has a weak attraction for its valence electron (that is, low ionization energy). If an atom of sodium loses its single valence electron, it will form a positive sodium ion, Na^{+}. Notice that a sodium ion has 10 electrons—the same number as an atom of neon (Ne), the noble gas closest to sodium on the periodic table. Similarly, if an atom of chlorine gains an electron (such as might be lost from a sodium atom), it will form a negative chloride ion, Cl^{-}. Such a chloride ion will have 18 electrons, the same as the closest noble gas, argon (Ar) (Table 8.1 and Figure 8.1).

In general, we can expect that elements with few valence electrons (the metals in groups I, II, III) will lose electrons when they react with elements that have almost eight valence electrons (the nonmetals in groups VI and VII). The ions formed by such a transfer are attracted to each other (remember that unlike charged particles attract), and it is this attraction between ions that forms the ionic bond. When, for example, we talked about dropping a chunk of sodium into a container of chlorine gas, many billions of atoms were involved in this reaction (that is, in the transfer of electrons). The attraction formed between the positive and negative ions causes these ions to be grouped into an orderly three-dimensional pattern

Table 8.1 Electron Configurations of Atoms and Ions

	Atomic Number	Electron Configuration 1	2	3		Atomic Number	Electron Configuration 1	2	3
Sodium, Na	11	2	8	1	Chlorine, Cl	17	2	8	7
Sodium ion, Na^{+}	11	2	8		Chloride ion, Cl^{-}	17	2	8	8
Neon, Ne	10	2	8		Argon, Ar	18	2	8	8

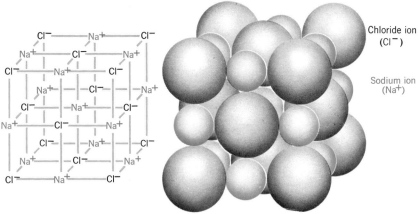

Figure 8.2 A sodium chloride crystal consists of sodium and chloride ions in a closely packed arrangement. (Photo courtesy American Museum of Natural History)

called a **crystal lattice.** This entire large grouping of ions is then called an **ionic compound** (Figure 8.2). In a crystal of sodium chloride each sodium ion is surrounded by six chloride ions, and each chloride ion is surrounded by six sodium ions. Because there is one sodium ion, Na^+, for every chloride ion, Cl^-, a crystal of sodium chloride is electrically neutral. And we will find this always to be true—the ratio of ions in the crystal lattice of an ionic compound will always result in an electrically neutral compound.

An ionic compound contains no specific molecules that we can point out. No ion is attracted exclusively to another ion, but rather each ion is attracted to all the ions surrounding it. You can see that the ionic bond is not a thing or a substance, but is simply the force of attraction between oppositely charged ions.

8.2 Electrovalence

The electrical charge carried by an ion is a measure of the ion's valence, or combining capacity. When this valence results from a transfer of electrons, it is called **electrovalence** or **ionic valence.** Elements that form ionic compounds have been assigned electrovalence numbers that indicate the kind of charge (positive or negative), and the amount of charge (the number of electrons gained or lost), on their ions.

To obtain a stable octet of electrons, elements in group I of the periodic table will lose one electron, forming ions with a charge of +1. Therefore, the electrovalence of the elements in group I is +1. Elements in group II and group III have electrovalences of +2 and +3, respectively. Group VI elements have an electrovalence of −2, and group VII elements have −1. Elements in groups IV and V rarely form ions, so we will not concern ourselves with their electrovalence numbers. Some elements can have more than one electrovalence number; that is, they may appear in different compounds as ions having different charges. The electrovalence numbers of some common ions are listed in Table 8.2. In our study of living organisms we will be discussing compounds containing these elements, so it is important that you become familiar with their electrovalence numbers.

Table 8.2 Electrovalence Numbers of Some Common Ions

Name of Ion	Electrovalence Number of Its Common Ion	Symbol of Ion	Name of Ion	Electrovalence Number of Its Common Ion	Symbol of Ion
Lithium ion	+1	Li^+	Bromide ion	−1	Br^-
Potassium ion	+1	K^+	Chloride ion	−1	Cl^-
Silver ion	+1	Ag^+	Fluoride ion	−1	F^-
Sodium ion	+1	Na^+	Iodide ion	−1	I^-
Barium ion	+2	Ba^{2+}	Oxide ion	−2	O^{2-}
Calcium ion	+2	Ca^{2+}	Sulfide ion	−2	S^{2-}
Magnesium ion	+2	Mg^{2+}			
Zinc ion	+2	Zn^{2+}			
Aluminum ion	+3	Al^{3+}			
Copper(I) ion (Cuprous)	+1	Cu^+	Iron(II) ion (Ferrous)	+2	Fe^{2+}
Copper(II) ion (Cupric)	+2	Cu^{2+}	Iron(III) ion (Ferric)	+3	Fe^{3+}

8.3 Naming Binary Ionic Compounds

Positive ions carry the same name as their parent element. For example, Na^+ is the sodium ion and Ca^{2+} is the calcium ion. If an element can form more than one type of positive ion, the different electrovalence numbers are indicated by Roman numerals after the name of the element. For example, Fe^{3+} is the iron(III) ion and Cu^+ is the copper(I) ion. As shown in Table 8.2, an older practice of naming ions calls these two ions the ferric ion and the cuprous ion, respectively. You will find that laboratory instruction manuals and labels on bottles of chemicals often make use of this older practice of naming ions and compounds. The negative ion is named by using a prefix taken from the name of the parent element, and then adding the suffix -ide. For example, Br^- is the bromide ion and O^{2-} is the oxide ion.

Ionic compounds that contain only two different elements are called binary ionic compounds, and are named for the ions from which they are formed. The name of the positive ion always appears first. For example, NaBr is sodium bromide and CuO is copper(II) oxide. Binary ionic compounds may contain more than one atom of an element, but are still named the same way. For example, $CaCl_2$ is calcium chloride and Na_2O is sodium oxide. But what do we mean when we write a formula such as $CaCl_2$?

8.4 Writing Chemical Formulas

Chemical formulas are a shorthand way of representing chemical compounds. The formula contains two pieces of information: it tells us the types of atoms or ions, and the ratio of these atoms or ions in the compound. The type of atom is indicated by the symbol of the element, and the ratio of atoms is shown by subscripts following the symbol.

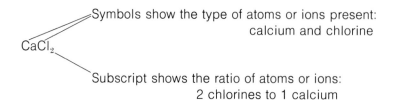

Symbols show the type of atoms or ions present:
calcium and chlorine

$CaCl_2$

Subscript shows the ratio of atoms or ions:
2 chlorines to 1 calcium

As we mentioned before, all ionic compounds are electrically neutral. The ratio of ions used in the chemical formula is the lowest set of whole numbers that will give us such electrical neutrality. This means that the formula for calcium chloride would not be expressed as, say, $Ca_{1/2}Cl$ or Ca_6Cl_{12}. Notice that the charges on the ions are not written in the chemical formula.

Example 8-1 _____

1. Name the following ionic compounds:

 (a) KBr Using Table 8.2, we see that K represents the potassium ion and Br represents the bromide ion. The name is potassium bromide.

 (b) MgI_2 Mg represents the magnesium ion and I represents the iodide ion. The name is magnesium iodide.

 (c) Cu_2O We see from Table 8.2 that there are two electrovalences for copper: $+1$ and $+2$. Because all ionic compounds are electrically neutral, the total charge of the two copper ions must exactly balance the charge of the oxide ion. The electrovalence of the oxide ion is -2 and, therefore, the electrovalence of each copper ion must be $+1$. The name of this compound is copper(I) oxide.

2. Write the chemical formula for the following compounds:

 (a) *Sodium fluoride.* From Table 8.2 we see that the sodium ion is Na^+ and the fluoride ion is F^-. Therefore, to maintain electrical neutrality we need one sodium ion for every fluoride ion; the ratio will be 1:1. The correct formula is NaF (notice that the subscript "1" is not written, and the charges on the ions are not included).

 (b) *Magnesium chloride.* From Table 8.2 we see that the magnesium ion is Mg^{2+} and the chloride ion is Cl^-. Therefore, to maintain electrical neutrality we need two chloride ions for each magnesium ion; the ratio is 1:2. The correct formula is $MgCl_2$ (the subscript "2" refers only to the chloride ion).

 (c) *Iron(II) oxide.* From Table 8.2 we see that the iron(II) ion is Fe^{2+} and the oxide ion is O^{2-}. Therefore, to maintain electrical neutrality we need one iron(II) ion for every oxide ion. The ratio will be 1:1. The correct formula is FeO.

 (d) *Iron(III) oxide.* From Table 8.2 we see that the iron(III) ion is Fe^{3+} and the oxide ion is O^{2-}. Therefore, to maintain electrical neutrality we need two iron(III) ions (a total charge of $+6$) for every three oxide ions (a total charge of -6); the ratio is 2:3. The correct formula is Fe_2O_3 (*Hint:* An easy rule to help you write binary formulas is to "crisscross" valence numbers.)

 $$Fe^{\underset{\longrightarrow}{③}+} O^{\underset{\longrightarrow}{②}-} \longrightarrow Fe_2O_3$$

 (e) *Magnesium oxide.* From Table 8.2 we see that the magnesium ion is Mg^{2+} and the oxide ion is O^{2-}. Using the crisscross method we have

 $$Mg^{\underset{\longrightarrow}{②}+} O^{\underset{\longrightarrow}{②}-} \longrightarrow Mg_2O_2$$

 But this formula is not the lowest whole number ratio of atoms, so the correct answer is MgO.

Covalent Bonds and Covalent Compounds

8.5 Sharing Electrons

Different elements that have about the same attracting power for their electrons cannot reach stability by pulling electrons away from one another to form an ionic bond. Rather, these elements must reach stability by sharing electrons, and forming a **covalent bond.** A covalent bond results when two positive nuclei attract the same electrons, thus holding the two nuclei close together. When two or more atoms share electrons through covalent bonds, a single (electrically neutral) unit called a **molecule** is formed. Covalent compounds are composed of molecules, which in turn are composed of atoms held together by covalent bonds. We will see that covalent bonds are normally formed between nonmetallic elements (remember that nonmetals are elements having high ionization energies and, therefore, strong attraction for their valence electrons).

Some nonmetallic elements exist in nature not as atoms, but as **diatomic molecules**—two atoms of the element covalently bonded together (Table 8.3). Chlorine, for example, exists as the diatomic molecule, Cl_2. Each chlorine atom has seven valence electrons, and needs one more electron to reach the stable electron configuration of argon. Two chlorine atoms can share a pair of electrons (thus forming a covalent bond) to reach this stable octet. A shorthand way of representing a covalent bond is to draw a dash between the chemical symbols of the elements involved (Figure 8.3).

Table 8.3 Elements That Are Found as Diatomic Molecules

Hydrogen	H_2	Fluorine	F_2
Nitrogen	N_2	Chlorine	Cl_2
Oxygen	O_2	Bromine	Br_2
		Iodine	I_2

Hydrogen is another element that is found as a diatomic molecule, H_2. Each hydrogen atom has one valence electron, and needs one more electron to reach the stable electron configuration of the closest noble gas, helium. Perhaps you can now picture how two hydrogen atoms are able to share a pair of electrons to become stable (Figure 8.4).

8.6 Multiple Bonds

Often a stable octet of electrons can be attained only if more than one pair of electrons is shared between two nuclei. A single shared pair of electrons results in a **single covalent bond,** which we have represented by a dash between the symbols of the elements. A **double bond** is formed when two

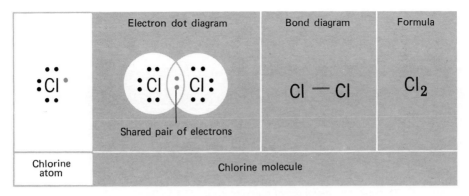

Figure 8.3 The diatomic molecule of chlorine contains a single covalent bond.

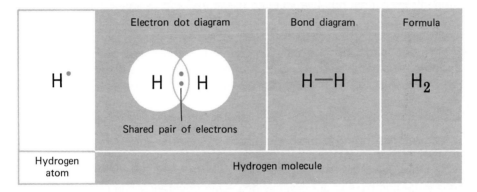

Figure 8.4 The diatomic molecule of hydrogen contains a single covalent bond.

pairs of electrons are shared between two nuclei, and is represented by two dashes, ═. Carbon dioxide (CO_2) is an example of a compound containing double covalent bonds. The carbon atom must share two pairs of electrons with each oxygen atom to reach a stable octet of electrons (Figure 8.5).

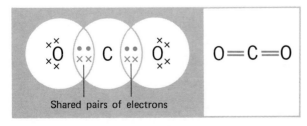

Figure 8.5 The carbon dioxide molecule contains two double bonds, resulting in an octet of electrons around each atom.

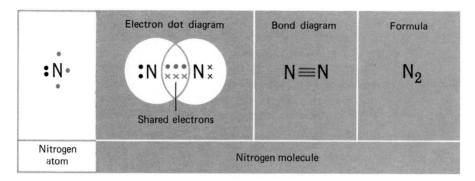

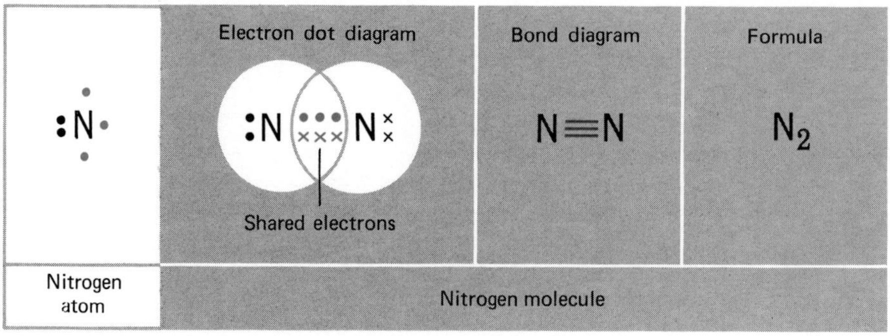

Figure 8.6 The diatomic molecule of nitrogen contains a triple covalent bond.

The element nitrogen is found in nature as a diatomic molecule, N_2. Each nitrogen atom has five valence electrons, and needs to share three electrons to reach a stable octet. This happens when each of two nitrogen atoms shares three electrons with the other atom. These three shared pairs of electrons form a **triple bond,** represented by three dashes (Figure 8.6). Quadruple bonds, which would be four pairs of electrons shared between two atoms, are structurally impossible and, therefore, are never found.

8.7 Covalence Numbers

The **covalence number** of an element indicates the number of electrons an atom of that element will share in a covalent compound. Because ions are not formed in covalent compounds, we will not specify signs for covalence numbers in our discussion. Some covalence numbers are shown in Table 8.4. For example, carbon's covalence number is 4; in each

Table 8.4 Covalence Numbers of Some Common Elements

Group	Element	Symbol	Electron Dot Diagram	Common Covalence* Number
	Hydrogen	H	H·	1
IV	Carbon	C	·Ċ· (4 single dots)	4
V	Nitrogen	N	·N̈· (5 dots)	3
	Phosphorus	P	·P̈· (5 dots)	3
VI	Oxygen	O	·Ö: (6 dots)	2
	Sulfur	S	·S̈: (6 dots)	2
VII	Chlorine	Cl	·C̈l: (7 dots)	1
	Bromine	Br	·B̈r: (7 dots)	1
	Iodine	I	·Ï: (7 dots)	1

* Some of these elements have more than one covalence number.

$$O\!=\!\overset{3}{\underset{2}{C}}\!=\!O \qquad H\overset{4}{\underset{H}{\underset{3}{C}}}\overset{2}{\cdots}H \qquad H\overset{H}{\underset{H}{\underset{4}{C}}}\overset{1}{-}\overset{H}{\underset{H}{\underset{4}{C}}}_{3} \qquad H\overset{4}{\underset{3}{C}}\overset{1}{=}\overset{4}{\underset{3}{C}}\overset{H}{\underset{H}{H}} \qquad H\overset{4}{-}C\overset{}{\underset{1,2,3,}{\equiv}}C\overset{4}{-}H$$

Figure 8.7 In each of the carbon compounds shown, the carbon atom has four pairs of electrons associated with it.

compound it forms, carbon must share four pairs of electrons. Several carbon compounds are shown in Figure 8.7, and you will notice that there are four covalent bond lines attached to each carbon atom.

Example 8-2 _____

1. Write the electron dot diagram for the following compounds:

 (a) *Water, H_2O.* Oxygen has a covalence number of 2, which means it needs to share two electrons to become stable. Hydrogen has a covalence number of 1, and needs to share one electron. Therefore, the oxygen atom will share one electron with each hydrogen. Using electron dot diagrams to illustrate this, we have

$$\begin{array}{c} H \\ H\!:\!\ddot{O}\!: \end{array}$$

 (b) *The compound that is formed between carbon and bromine.* Carbon has a covalence number of 4, and must share four electrons to become stable. Bromine has a covalence number of 1, so carbon must share an electron with four atoms of bromine. Using electron dot diagrams, we have

$$\begin{array}{c} :\ddot{Br}: \\ :\ddot{Br}\!:\!\ddot{C}\!:\!\ddot{Br}: \\ :\ddot{Br}: \end{array}$$

2. Draw the bond diagram for the compound formed between carbon and sulfur.

 Carbon has a covalence number of 4 and sulfur has a covalence number of 2. Therefore, a carbon atom will share two electrons with one sulfur atom and two electrons with another sulfur atom to become stable. The bond diagram is $S\!=\!C\!=\!S$.

8.8 Polar Covalent Bonds

Imagine two identical atoms sharing a pair of electrons in a covalent bond. We would expect the electrons to be shared absolutely equally between the two identical nuclei, resulting in the center of positive charge and the

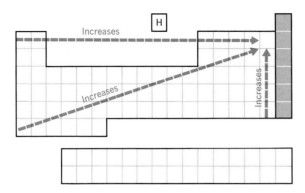

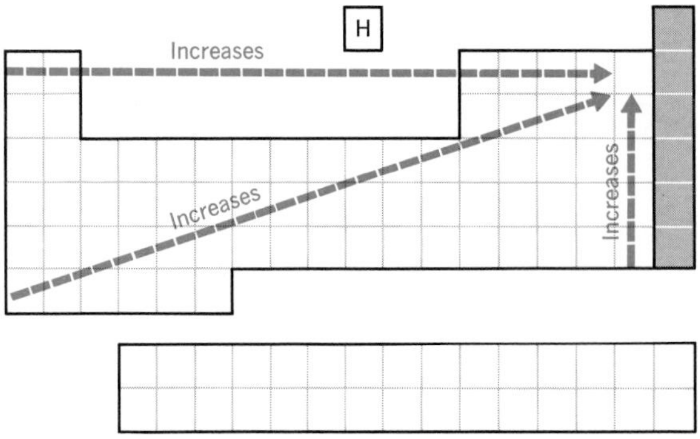

Figure 8.8 A periodic table showing the trends in electronegativity among the elements.

center of negative charge occurring at the same place along the bond; this is called a **nonpolar covalent bond** (see Figure 8.9a).

An atom of each element has a specific tendency to attract a shared pair of electrons, and the strength of this attraction is known as **electronegativity.** In each row of the periodic table the element with the smallest radius (and needing the fewest number of electrons to become stable) is the most electronegative; among all the elements, fluorine is the most electronegative. The electronegativity of the elements increases to the right along a row, and decreases from top to bottom down each group, as shown in Figure 8.8.

Now imagine that two different atoms are sharing electrons. We would expect that one atom might attract the electrons more strongly than the other, so the electrons will be found closer to the more electronegative (or electron-attracting) atom. Therefore, in such a bond the center of positive charge is not found in the same place as the center of negative charge, resulting in an unequal distribution of charge (see Figure 8.9b). This unequal distribution of charge results in the formation of an electric dipole.

Type of bond	Nonpolar covalent	Polar covalent	Partial covalent Partial ionic	Ionic
Center of positive and negative charge	+ −	+ −	+ −	+ −
Electro–negativity difference	0	Increases		

Figure 8.9 The bonding continuum. There is no absolute break between ionic and covalent bonding, but rather a continuous range from nonpolar covalent bonding to ionic bonding.

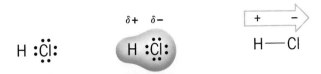

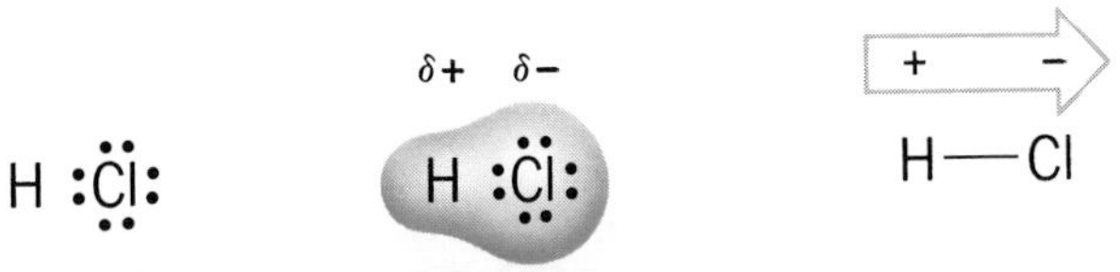

Figure 8.10 Schematic diagrams of the polar covalent bond in a hydrogen chloride molecule. The Greek letter delta (δ) indicates a partial charge.

The molecule has two electric poles, just as a magnet has two magnetic poles. This kind of bond is called a **polar covalent bond.**

For example, consider hydrogen chloride, HCl. Chlorine has a greater electron affinity than hydrogen and is the more electronegative element, so the shared electrons will be pulled closer to the chlorine nucleus in the hydrogen-chlorine bond. This will give the chlorine end of the bond a partial negative charge, and the hydrogen end of the bond a partial positive charge (Figure 8.10). In general then, we would expect the negative part of a polar covalent bond to be in the region of the more electronegative atom, and the positive part to be in the region of the less electronegative atom.

Atoms of the same element will form nonpolar covalent bonds, with centers of positive and negative charges coinciding. Atoms of unlike elements which have differences in electronegativity will result in the centers of electrical charges moving apart. When the difference in electronegativity becomes very large, the electrical charges will be completely separated to the two opposite ends of the bond—but this is exactly the description of the ionic bond that we discussed earlier. That is, we can imagine a continuous range from nonpolar covalent bonding to complete ionic bonding, with different compounds appearing at specific places all along this continuum (Figure 8.9). Even though there is no clear break between ionic bonding and polar covalent bonding, it is still convenient to try to classify compounds as either ionic or covalent. In general, bonds formed between metals and nonmetals will be ionic, and bonds formed between nonmetals will be covalent (they will be nonpolar covalent when the electronegativity of the atoms is the same, and polar covalent when one atom has greater electronegativity than the other).

8.9 Naming Binary Covalent Compounds

Covalent compounds formed from two nonmetallic elements are called binary covalent compounds, and are named according to the following rules.

1. In a compound formed from two nonmetals, the element that occurs earlier in the following list is written and named first in the formula (generally this is the element with the lower electronegativity).

   B, Si, C, P, N, H, S, I, Br, Cl, O, F

2. The second element is named by using a prefix taken from the parent element, and adding the suffix -ide.

3. Prefixes are added to the name of each element to indicate the number of atoms of each element in the molecule.

Mono–	1	Tetra–	4
Di–	2	Penta–	5
Tri–	3	Hexa–	6

The prefix *mono* is usually left out, except when it helps us distinguish between two different compounds. For example, CO is carbon monoxide and CO_2 is carbon dioxide.

Example 8-3

Name the following compounds:

(a) HCl According to the list in rule (1), the hydrogen should be named first. There is just one atom of each element in the formula, so the name is hydrogen chloride.

(b) CCl_4 From rule (1), carbon should be named first. There are four chlorine atoms, so the prefix *tetra* must be added to chloride. The name is carbon tetrachloride.

(c) S_2Cl_2 Sulfur is the element that should be named first. There are two atoms of each element, so the prefix *di* must be added to the name of each. The name is disulfur dichloride.

8.10 Writing Formulas for Covalent Compounds

To write the formula for a binary covalent compound, follow rule (1) to decide which element is written first. The subscripts for each element are indicated by the prefixes given in the name. Remember that the prefix *mono* is generally not stated.

Example 8-4

Write the formulas for the following compounds:

(a) *Hydrogen iodide.* The name indicates that there is one atom of each element in the formula, HI.

(b) *Sulfur trioxide.* The name indicates that there is one sulfur atom and three oxygen atoms in the molecule. The formula is SO_3.

(c) *Diphosphorus pentoxide.* The name indicates that there are two atoms of phosphorus and five atoms of oxygen in the molecule. The formula is P_2O_5.

8.11 Polyatomic Ions

A **polyatomic ion** is a group of covalently bonded atoms that, as a group, carries an electrical charge, but which is so stable that it will go through most chemical reactions as a unit and won't come apart. Some polyatomic ions and their electrovalence numbers are listed in Table 8.5.

Table 8.5 Some Common Polyatomic Ions

Name of the Ion (Common Name)	Formula	Electrovalence
Ammonium ion	NH_4^+	+1
Acetate ion	$C_2H_3O_2^-$	−1
Carbonate ion	CO_3^{2-}	−2
Hydrogen carbonate ion (Bicarbonate)	HCO_3^-	−1
Chlorate ion	ClO_3^-	−1
Chromate ion	CrO_4^{2-}	−2
Dichromate ion	$Cr_2O_7^{2-}$	−2
Cyanide ion	CN^-	−1
Hydroxide ion	OH^-	−1
Nitrate ion	NO_3^-	−1
Nitrite ion	NO_2^-	−1
Permanganate ion	MnO_4^-	−1
Peroxide ion	O_2^{2-}	−2
Phosphate ion	PO_4^{3-}	−3
Monohydrogen phosphate ion	HPO_4^{2-}	−2
Dihydrogen phosphate ion	$H_2PO_4^-$	−1
Sulfate ion	SO_4^{2-}	−2
Hydrogen sulfate ion (Bisulfate)	HSO_4^-	−1
Sulfite ion	SO_3^{2-}	−2
Hydrogen sulfite ion (Bisulfite)	HSO_3^-	−1

Ionic compounds containing polyatomic ions are so common and play such a central role in our daily lives that it is important for you to become familiar with the names and formulas of the polyatomic ions shown in Table 8.5. Examples of such familiar ionic compounds are sodium hydrogen carbonate (sodium bicarbonate) $NaHCO_3$, which is baking soda, and magnesium hydroxide $Mg(OH)_2$, which is a common antacid.

Many ionic compounds consist of a metal ion and a polyatomic ion. These compounds are named by placing the name of the metal first and the name of the polyatomic ion second. For example, Na_2CO_3 is sodium carbonate and $KHSO_4$ is potassium hydrogen sulfate or potassium bisulfate. Whenever you see a name of a compound that doesn't end in *-ide,* this tells you that a polyatomic ion is involved. (Note, however, that the reverse is not true—compounds whose names end in *ide* might contain polyatomic ions. The names of compounds containing OH^-, CN^-, and O_2^{2-} all end in *-ide*).

The formulas for compounds containing polyatomic ions follow the rules given for ionic compounds, except that when more than one polyatomic ion appears in the formula, the symbol of the polyatomic ion is put in parentheses, with the subscript following the parentheses.

Example 8-5

Write the formulas for the following compounds:

(a) *Magnesium hydroxide.* The electrovalence of the magnesium ion is +2 and the electrovalence of the hydroxide ion is −1. Therefore, to maintain neutrality we need one Mg^{2+} ion and two OH^- ions. The formula is written $Mg(OH)_2$, where the 2 indicates two of the polyatomic ion OH^-.

(b) *Calcium phosphate.* The electrovalence of the calcium ion is +2 and that of the phosphate ion is −3. By the crisscross method, we have

$$Ca^{2+} \quad PO_4^{3-} \longrightarrow Ca_3(PO_4)_2$$

The 4 inside the parentheses refers only to the oxygen, but the 2 that appears outside the parentheses indicates two of the polyatomic ion PO_4^{3-}.

8.12 Shapes of Molecules

The properties and behavior of a molecular compound depend not only on the combination of atoms that make it up, but also on the shape or arrangement of these atoms in the molecule. The properties affected by the shape of the molecule range from the odor of the compound to the role that the molecule plays in regulating the chemical reactions in living organisms. For example, all psychedelic drugs seem to affect one particular site in the brain. One current theory is that this site is sensitive

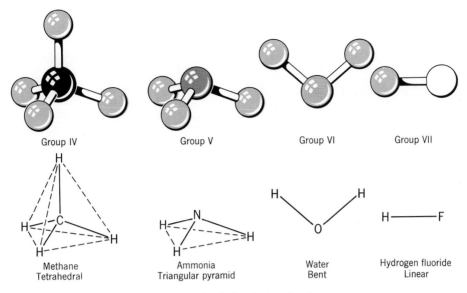

Group IV Group V Group VI Group VII

Methane Ammonia Water Hydrogen fluoride
Tetrahedral Triangular pyramid Bent Linear

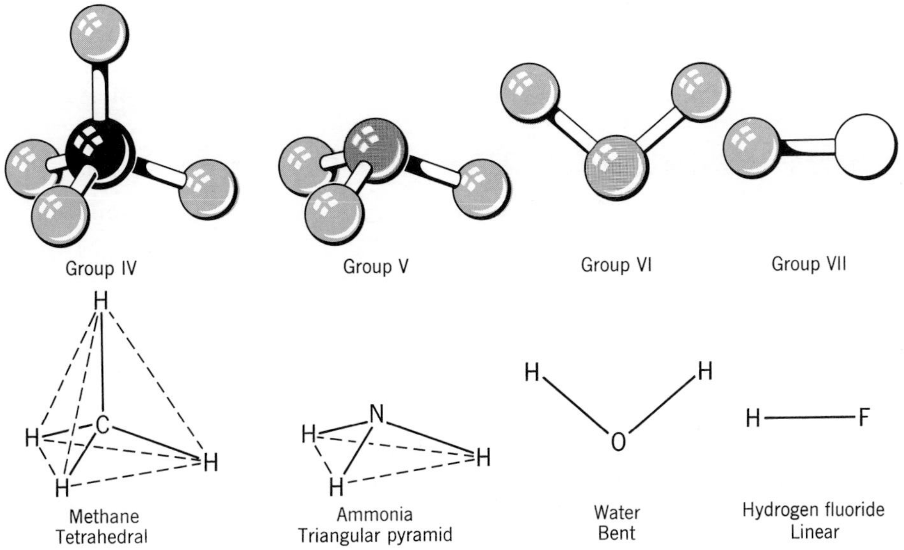

Figure 8.11 The shapes of some covalently bonded molecules.
The dotted line is an outline of the shape formed by each molecule.

to, or will "recognize," a specific three-dimensional shape. According to the theory, the molecules of the psychedelic drugs each have particular regions that closely resemble this three-dimensional shape and, therefore, will each trigger hallucinogenic symptoms.

We have seen before that elements within the same group or family will have similar chemical properties. Such similarities are also true of the arrangement of atoms that are covalently bonded to nonmetals in groups IV through VII. If all bonds to the central atom are single bonds, the arrangement around atoms of elements in group IV will be tetrahedral in shape, group V will be triangular pyramidal, group VI bent, and group VII linear (Figure 8.11).

8.13 Polar and Nonpolar Molecules

A molecule containing many different polar bonds can itself be polar or nonpolar, depending upon the shape of the molecule. As is the case with molecular shape, the polarity or nonpolarity of a molecule also plays a large part in its behavior and role in living organisms. We will see that polar molecules will be found with other polar molecules, and nonpolar molecules will be found with other nonpolar molecules in living systems.

As with covalent bonds, the centers of positive and negative charges will coincide in a nonpolar molecule, and will not coincide in a polar molecule. It is important to realize that a polar molecule is still electrically neutral, even though there will be a separation of charge in the molecule resulting in regions of positive and negative charge.

Several examples will illustrate how the shape of a molecule can determine whether the molecule will be polar or nonpolar. Note that in each

Table 8.6 Molecular Shape and the Polarity of Molecules

Compound	Formula	Shape	Bond Diagram	Center of Charge	Type of Molecule
Hydrogen chloride	HCl	Linear	$\overset{\delta+}{H}-\overset{\delta-}{Cl}$		Polar
Carbon dioxide	CO_2	Linear	$\overset{\delta-}{O}=\overset{\delta+}{C}=\overset{\delta-}{O}$		Nonpolar
Water	H_2O	Bent	$\overset{\delta+}{H}-\overset{\delta-}{O}-\overset{\delta+}{H}$		Polar

of our examples, however, the bonds that are formed are polar (Table 8.6). Hydrogen chloride (HCl), as we have already seen, is a linear molecule containing one polar bond. The molecule is polar, with the hydrogen at the positive end and the chlorine at the negative end. Carbon dioxide (CO_2) is also a linear molecule, in this case containing two polar double bonds. However, the linear arrangement of atoms in the molecule makes the centers of positive and negative charge coincide, so the entire carbon dioxide molecule is nonpolar. A molecule of water contains three atoms and two polar bonds just as we found in carbon dioxide. But the water molecule has a bent (rather than linear) shape, and the centers of positive and negative charge do not coincide. Therefore, a water molecule will be polar (Figure 8.12).

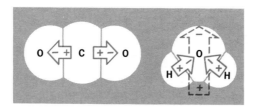

Figure 8.12 Both CO_2 and H_2O contain two polar covalent bonds. But the shape of their molecules causes a molecule of CO_2 to be nonpolar, and a molecule of H_2O to be polar.

8.14 Hydrogen Bonding

Molecules containing hydrogen attached to a highly electronegative atom such as fluorine, oxygen, or nitrogen, will have an intermolecular (between molecules) type of bonding called **hydrogen bonding.** The partially positive hydrogen region on one molecule will be attracted to the partially negative fluorine, oxygen, or nitrogen region on another molecule. Such a hydrogen bond has a strength about one-tenth that of an ordinary covalent bond. In some very large biological molecules, hydrogen bonding may occur between different parts of the same molecule, causing the molecule to bend back on itself (intramolecular hydrogen bonding).

Hydrogen bonding plays an important role in nature. Hydrogen bonds are responsible for the unusual properties of water (such as high melting and boiling points) which make this fluid so important to all living

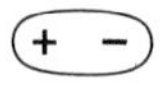

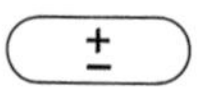

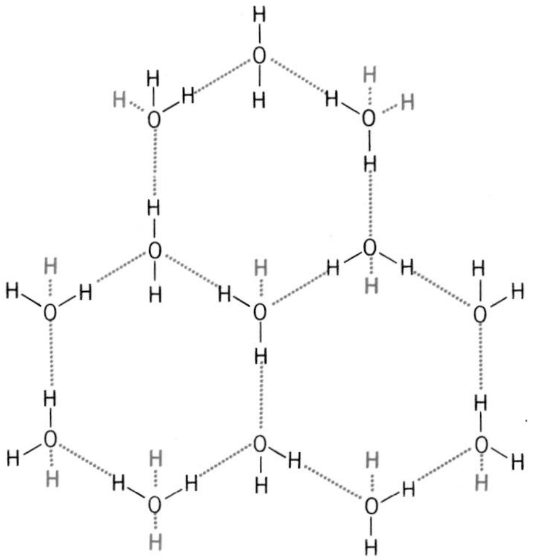

Figure 8.13 The hydrogen bonding in ice. The hydrogen bonds are represented by colored dotted lines.

organisms (Figure 8.13). These properties will be discussed in detail in Chapter 11. Hydrogen bonds also determine the shapes of large biological molecules in living substances and, on a more familiar level, account for the hand-softening property of the lanolin in skin creams, for hard candy getting sticky, and for cotton fabrics taking longer to dry than synthetic fabrics.

Chapter Summary

Elements become more stable by entering into reactions in which they lose, gain, or share electrons so as to attain eight valence electrons. When electrons are transferred from one atom to another, ions are formed. The force of attraction between these ions is called an ionic bond. An ionic compound is a group of ions combined in an orderly fashion called a crystal lattice. In general, ionic bonds form between metals and nonmetals. A covalent bond is formed when electrons are shared between two atoms, creating electrically neutral units called molecules. Two atoms can share two electrons in a single covalent bond, or can share four electrons in a double covalent bond, or six electrons in a triple covalent bond. The tendency for an atom of an element to attract the electrons in a covalent bond is called the electronegativity of that element. When electrons are shared equally between two atoms, a nonpolar bond is formed. If there is a difference in the electronegativity of the two atoms, a polar bond will be formed. Molecules can be polar or nonpolar depending upon the shape of the molecule and the types of bonds in the molecule. Polyatomic ions are groups of covalently bonded atoms which, as a group, carry an electrical charge and stay together as a unit through most chemical reactions. Another type of bonding that is extremely important in living organisms is hydrogen bonding. A hydrogen bond is formed between a hydrogen attached to a highly electronegative atom (such as fluorine, oxygen, or nitrogen) and a fluorine, oxygen, or nitrogen atom on another molecule or on another part of the same molecule.

Exercises and Problems

1. What is the difference between an ionic bond and a covalent bond?
2. What is the difference between the units that make up an ionic compound and the units that make up a covalent compound?
3. State the difference between a single, double, and triple covalent bond. Give examples of each.
4. State the difference between the following:
   (a) an atom and an element
   (b) an atom and an ion
   (c) an atom and a diatomic molecule
   (d) an atom and a polyatomic ion
5. Why is it unlikely that (a) calcium would form an ion with a +1 charge? (b) potassium would form an ion with a +2 charge?
6. Predict whether the bonds formed between the following atoms will be polar or nonpolar. If the bond is polar, indicate which is the more electronegative atom.
   (a) Cl and Cl
   (b) H and Br
   (c) C and N
   (d) S and Cl
   (e) N and O
   (f) P and O
   (g) F and Br
   (h) O and O
7. Predict whether the following molecules will be polar or nonpolar.
   (a) Oxygen difluoride, OF_2
   (b) Fluorine, F_2
   (c) Hydrogen iodide, HI
   (d) Methane, CH_4
   (e) Chloromethane, CH_3Cl
8. Would you expect hydrogen bonds to form between molecules of hydrogen fluoride, HF? Why or why not?
9. Write the correct formulas for the compounds composed of the following pairs of ions.
   (a) K^+, CO_3^{2-}
   (b) Na^+, S^{2-}
   (c) Ca^{2+}, NO_3^-
   (d) Sr^{2+}, S^{2-}
   (e) Cr^{3+}, Cl^-
   (f) Fe^{3+}, HPO_4^{2-}
   (g) Cu^{2+}, $C_2H_3O_2^-$
   (h) Ba^{2+}, SO_4^{2-}
10. Name each of the compounds for which you wrote the formulas in question 9.

11. Write the correct formulas for each of the following sets of compounds.

	Set 1	Set 2	Set 3
(a)	Lithium fluoride	Ammonium carbonate	Iodine pentafluoride
(b)	Potassium sulfide	Calcium bisulfate	Hydrogen selenide
(c)	Magnesium bromide	Magnesium bicarbonate	Dinitrogen tetroxide
(d)	Silver chloride	Lithium phosphate	Hydrogen bromide
(e)	Iron(III) sulfide	Barium nitrate	Boron trichloride
(f)	Barium iodide	Magnesium dihydrogen phosphate	Sulfur dioxide
(g)	Aluminum oxide	Iron(III) sulfate	Carbon disulfide
(h)	Copper(I) oxide	Potassium cyanide	Disulfur dichloride

12. Name the following compounds:
   (a) NH_4I
   (b) PCl_3
   (c) $Ca(OH)_2$
   (d) CBr_4
   (e) $FeCl_2$
   (f) $(NH_4)_2Cr_2O_7$
   (g) O_2F_2
   (h) $Zn(NO_2)_2$
   (i) P_2O_5
   (j) NaH
   (k) HI
   (l) $NaMnO_4$
   (m) Li_2S
   (n) K_2SO_3
   (o) CaC_2
   (p) $Al_2(SO_4)_3$

13. Draw the electron dot diagrams for the following molecules:
   (a) HI
   (b) F_2
   (c) CH_3Cl
   (d) HCN
   (e) OF_2
   (g) PCl_3

14. Draw the bond diagrams for each of the following compounds:
   (a) NH_3
   (b) CCl_4
   (c) HCl
   (d) C_2Br_2
   (e) SF_2
   (f) I_2

15. What is the total number of each type of atom in one unit of each of the following:
   (a) $(NH_4)_3PO_4$
   (b) $Al_2(HPO_4)_3$
   (c) $Ca(C_2H_3O_2)_2$

chapter 9

Chemical Equations and the Mole

Learning Objectives

By the time you have finished this chapter, you should be able to:

1. Write a balanced chemical equation given the reactants and the products.

2. Define a *mole.*

3. Calculate the formula weight for a chemical compound.

4. Calculate the number of moles of a substance given its weight in grams.

5. Calculate the number of grams in a given number of moles of a substance.

6. Use a balanced chemical equation to determine the amount of reactants necessary to produce a given amount of product.

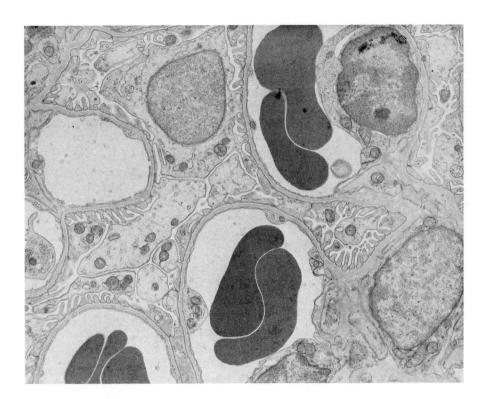

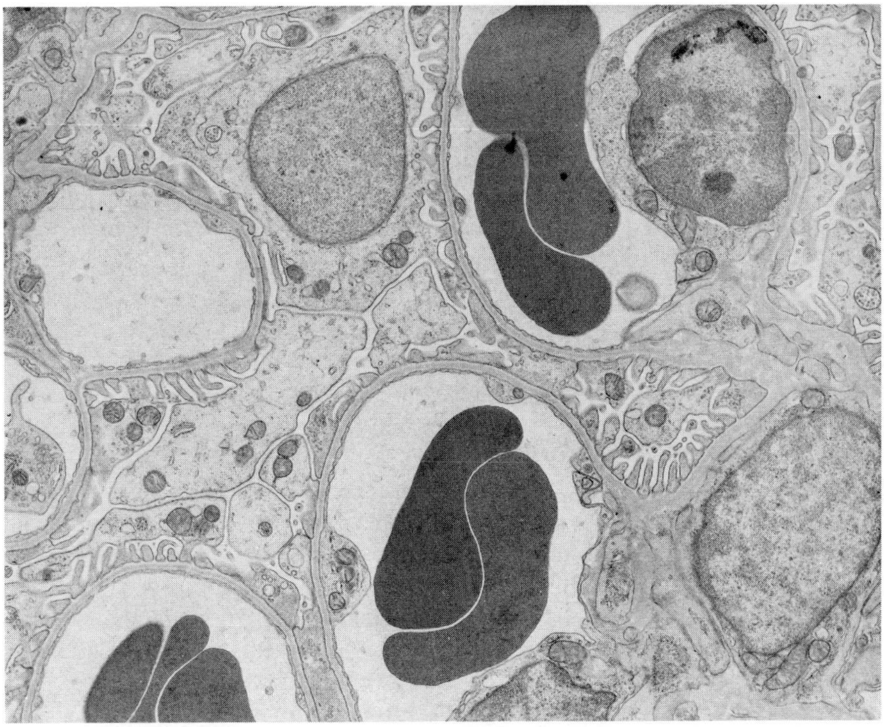

At 45 years of age, Betty Johns felt that she was in perfect health. However, it had been eight years since her last physical examination, and she felt that now was a good time to schedule another checkup. At Betty's exam the doctor measured her blood pressure at the very high level of 180/120. Although she had not suspected it, this meant that Betty was suffering from hypertension, or high blood pressure.

Because initial treatment with a diuretic failed to lower Betty's blood pressure to normal levels, the doctor next prescribed the drug alpha-methyl dopa, which is a muscle relaxant that dilates the small arteries and thereby reduces the blood pressure. However, when Betty returned to the doctor a month later for a follow up exam, she complained of a constant feeling of fatigue and lack of energy. This time the doctor found that Betty's blood pressure was within the normal range at 140/90, but that her hemoglobin level had fallen from a normal level of 13.5 to a very low level of 10. Further tests confirmed the diagnosis that Betty was now suffering from hemolytic anemia, a destruction of red blood cells caused by the drug alpha-methyl dopa. The doctor immediately stopped the use of this drug, and prescribed a different smooth-muscle relaxant for Betty. Within three weeks her hemoglobin level had returned to normal.

It is an unfortunate fact of modern medicine that the treatment given for one disease may very well end up causing another. However, for us the important question at this moment is what exactly caused the destruction of Betty's red blood cells? The alpha-methyl dopa that Betty was taking for her high blood pressure reacts with oxygen in the body to form the compound hydrogen peroxide, H_2O_2. This is usually not a problem because most people have two enzymes—molecules that control the rate of chemical reactions in the body—which protect against the build-up of hydrogen peroxide in the cells by breaking this compound down to form water. Some people, however, are missing one or both of these enzymes, leaving their cells without protection against the harmful effects of hydrogen peroxide.

Betty's red blood cells lacked the selenium-containing enzyme called glutathione peroxidase, which is the main enzyme that destroys hydrogen peroxide in red blood cells. This resulted in a build-up of hydrogen peroxide, which can damage red blood cells in two ways. First, this compound changes the iron found in hemoglobin from Fe^{2+} to Fe^{3+}, which makes the hemoglobin lose its ability to carry oxygen in the blood. Second, the hydrogen peroxide reacts with molecules in the membrane of the red blood cell, resulting in the destruction (hemolysis) of these cells and causing the hemolytic anemia from which Betty suffered.

Each of the processes that we have just described—the breakdown of hydrogen peroxide to water, the change in the iron ion, and the destruction of red blood cells—takes place by means of chemical reactions. But in order for us to study and understand such processes we need to be familiar with the vocabulary and symbols that are used to represent chemical reactions.

Chemical Equations

9.1 Writing Chemical Equations

Chemical equations are a shorthand way of representing what occurs in a chemical reaction. A chemical equation contains the formulas of the starting materials, or **reactants,** separated by an arrow from the resulting materials, or **products.**

$$\underbrace{A + B}_{\text{Reactants}} \longrightarrow \underbrace{C + D}_{\text{Products}}$$

The substances that form the reactants and products may be atoms, ions, molecules, or ion groups. In this chapter we will be concerned mainly with reactions between atoms and molecules. We will study reactions between ions in greater detail in Chapter 11.

Atoms are neither created nor destroyed in a chemical reaction, so the chemical equation representing the reaction must be **balanced.** That is, for each element involved in the reaction the equation must show the same number of atoms on the product side as are found on the reactant side. This means that the total mass of the products will equal the total mass of the reactants, which is known as the Law of Conservation of Mass.

For example, either hydrogen peroxide or water can be formed from hydrogen and oxygen, but, as we have seen, our tissues are greatly affected by which one is formed. Hydrogen peroxide can badly damage cells, whereas water is an essential part of all cells. In writing a chemical equation for the formation of either hydrogen peroxide or water, the reactants or starting materials are oxygen (O_2) and hydrogen (H_2). Although the hydrogen needed for this reaction does not come from the elemental hydrogen molecule H_2, but rather comes from other molecules containing hydrogen, we will just use elemental hydrogen to simplify this introductory discussion:

$$H_2 + O_2 \longrightarrow$$

To complete either equation we must write the correct chemical formulas for the products, water or hydrogen peroxide. We know that water is H_2O. From Table 8.5 we can find that the peroxide ion is O_2^{2-} and, therefore, to maintain electrical neutrality we need two hydrogen ions, H^+. This gives us the correct formula for hydrogen peroxide, H_2O_2. Now we can write the reactants and products of the two equations we have discussed:

(Hydrogen peroxide)	$H_2 + O_2 \longrightarrow H_2O_2$
(Water)	$H_2 + O_2 \longrightarrow H_2O$

Let's check to see if these two equations are balanced. Starting with the equation for the formation of hydrogen peroxide, we must check the number of atoms of each element on both sides of the equation. You can see that there are two hydrogen atoms on the reactant side and two on the product side. There are two oxygen atoms on the reactant side and two on the product side. Therefore, this equation is balanced.

$$H_2 + O_2 \longrightarrow H_2O_2 \qquad (1)$$

Now look at the chemical equation for the formation of water. There are two hydrogen atoms on each side of the arrow, but there are two oxygen atoms on the reactant side and only one on the product side. However, to balance an equation the only numbers that can be changed (after the correct chemical formulas have been written) are the numbers, or coefficients, that appear in front of the formulas of the substances involved. So, to obtain two oxygen atoms on the product side of the equation we must place a coefficient of 2 in front of the formula for water.

$$H_2 + O_2 \longrightarrow 2H_2O$$

Now the oxygen atoms balance, but the hydrogen atoms are unbalanced—there are four hydrogen atoms on the product side and only two on the reactant side. Placing a coefficient of 2 in front of the formula for hydrogen on the reactant side will balance the hydrogens. Now you can check to see that the equation is balanced (Figure 9.1).

$$2H_2 + O_2 \longrightarrow 2H_2O \qquad (2)$$

By comparing these two balanced equations you can now see that the reactions for the formation of water and hydrogen peroxide are actually different: in one reaction 1 molecule of hydrogen combines with 1 molecule of oxygen to form 1 molecule of hydrogen peroxide, and in the other

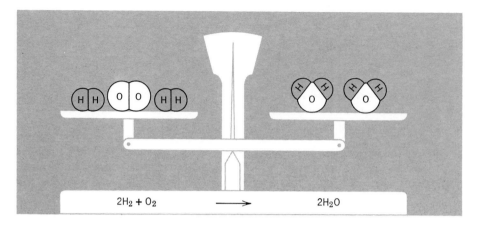

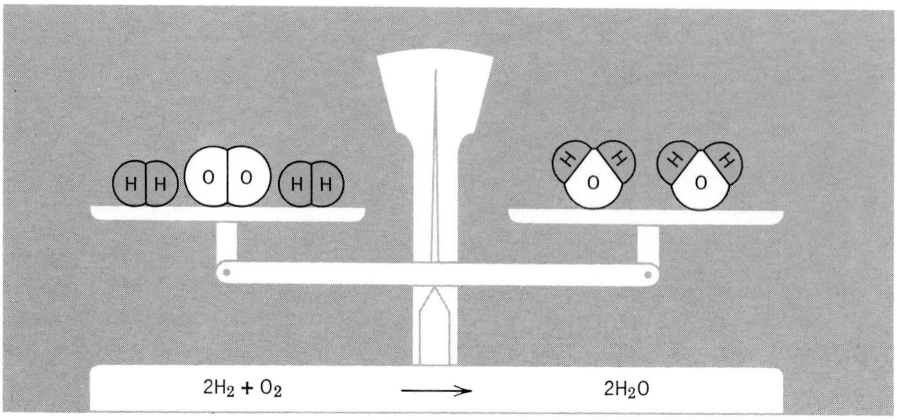

Figure 9.1 When an equation is balanced, the number of atoms of each element on the reactant side will equal the number of atoms of each element on the product side. Therefore, the total mass of the reactants will equal the total mass of the products.

reaction 2 molecules of hydrogen combine with 1 molecule of oxygen to form 2 molecules of water.

9.2 Balancing Chemical Equations

Balancing equations is not really a complicated procedure. The following steps should help you quickly master this skill:

1. First, write the correct chemical formula for each reactant and each product. Once this is done you should never change the formula of a substance; only the coefficients can be changed. Any change of subscript would change the nature of the compound. For example, CO and 2CO represent one and two molecules of the compound carbon monoxide. But CO and CO_2 are entirely different substances, and you wouldn't want to confuse the two. Carbon monoxide is deadly, but carbon dioxide is produced in, and exhaled from, our bodies.

2. Because the balancing of equations involves juggling coefficients, it is often easiest to begin by giving the coefficient 1 to the compound having the most complicated formula.

3. You must then start balancing each of the elements. You will often find it easiest if you leave oxygen until last.

4. Treat polyatomic ions as one unit (that is, just as

though they were a single element) if they remain unchanged in the reaction.

5. When you think the equation is balanced, it is a good idea to check each element again to make sure that it really is balanced.

The following examples illustrate these five steps for balancing equations. After you have read through the examples, try them yourself to see if you can get the same answer. You will find that skill in balancing equations comes with lots of practice.

Example 9-1

1. Balance the following equation:

$$Na + Cl_2 \longrightarrow NaCl$$

Step 1: Already completed.
Steps 2, 3, 4: This is not a balanced equation; there are two chlorine atoms on the reactant side of the equation and only one on the product side. Placing a coefficient of 2 in front of the formula for sodium chloride will balance the chlorine atoms in this equation.

$$Na + Cl_2 \longrightarrow 2NaCl$$

But now there are two sodium atoms on the product side, and only one on the reactant side. Placing a coefficient of 2 in front of the sodium on the reactant side will balance the equation.

$$2Na + Cl_2 \longrightarrow 2NaCl$$

Step 5: We have two atoms of sodium and two atoms of chlorine on each side of the equation.

2. A small amount of the pollutant sulfur trioxide is released into the air when sulfur-containing coal or petroleum is burned. Write the balanced chemical equation for this reaction.

Step 1: When a substance burns, it reacts with oxygen in the air, so the reactants for this chemical reaction are elemental sulfur and oxygen (remember that oxygen exists as a diatomic molecule). The product is sulfur trioxide.

$$S + O_2 \longrightarrow SO_3$$

Steps 2, 3, 4: When the equation is written as above, the sulfurs balance but not the oxygen. In order for the oxygen to balance, we must have the same number of atoms on both sides, and the lowest possible number is 6.

$$S + 3O_2 \longrightarrow 2SO_3$$

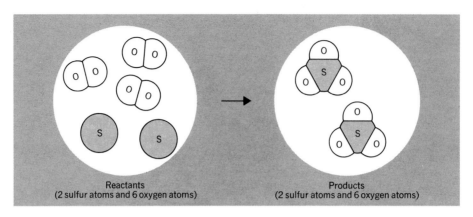

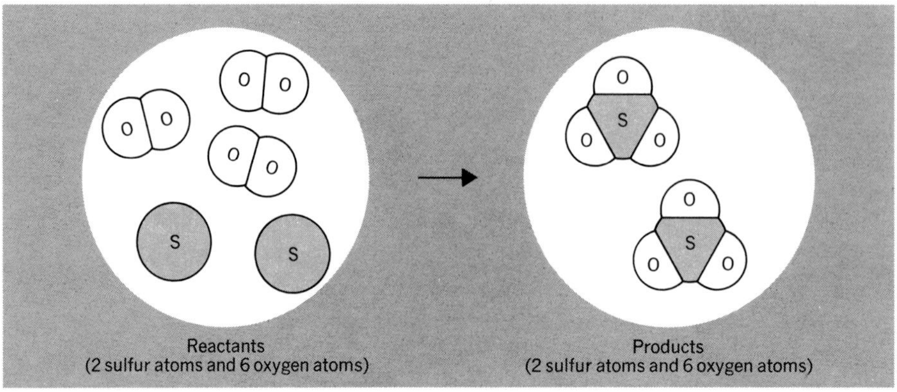

Figure 9.2 The balanced equation for the formation of sulfur trioxide is $2S + 3O_2 \rightarrow 2SO_3$.

Now to balance the sulfur,

$$2S + 3O_2 \longrightarrow 2SO_3$$

Step 5: We have two sulfur atoms on each side and six oxygen atoms on each side (Figure 9.2).

3. When aluminum reacts with sulfuric acid (H_2SO_4), hydrogen gas and aluminum sulfate are produced.

   Step 1: Using Table 8.2 and 8.5, we find that the electrovalence number of aluminum is +3, Al^{3+}, and that of sulfate is −2, SO_4^{2-} The formula for aluminum sulfate, therefore, is $Al_2(SO_4)_3$.

   $$Al + H_2SO_4 \longrightarrow H_{2(g)} + Al_2(SO_4)_3$$

   (g) = gas

   Steps 2, 3, 4: Start with one $Al_2(SO_4)_3$. First balance the aluminum.

   $$2Al + H_2SO_4 \longrightarrow H_{2(g)} + Al_2(SO_4)_3$$

   Next balance the sulfate, SO_4.

   $$2Al + 3H_2SO_4 \longrightarrow H_{2(g)} + Al_2(SO_4)_3$$

   And finally, balance the hydrogens.

   $$2Al + 3H_2SO_4 \longrightarrow 3H_{2(g)} + Al_2(SO_4)_3$$

   Step 5: There are now two atoms of aluminum on each side, six atoms of hydrogen on each side, and three sulfate ions on each side (Table 9.1).

4. One way in which kidney stones develop is from a reaction between calcium ions and phosphate ions in the blood to produce calcium phosphate, the solid that forms the stones (Figure 9.3). Write the equation for this reaction.

Table 9.1 A Tally of Atoms for the Following Balanced Equation
$2Al + 3H_2SO_4 \longrightarrow 3H_2 + Al_2(SO_4)_3$

Symbol of Atom	Number of Atoms on the Reactant Side	Number of Atoms on the Product Side
Al	2	2
H	6	6
S	3	3
O	12	12

Step 1: $Ca^{2+} + PO_4^{3-} \longrightarrow Ca_3(PO_4)_2$

Steps 2, 3, and 4: Start by balancing the calcium ion.

$$3Ca^{2+} + PO_4^{3-} \longrightarrow Ca_3(PO_4)_2$$

Then balance the phosphate ion

$$3Ca^{2+} + 2PO_4^{3-} \longrightarrow Ca_3(PO_4)_2$$

Step 5: There are three calcium ions and two polyatomic phosphate ions on each side of the equation.

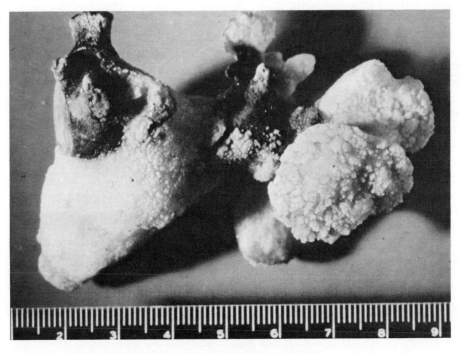

Figure 9.3 These kidney stones, composed mainly of calcium phosphate, were formed in the kidneys of a person with a urinary tract infection. (Courtesy William P. Mulvaney, M.D.)

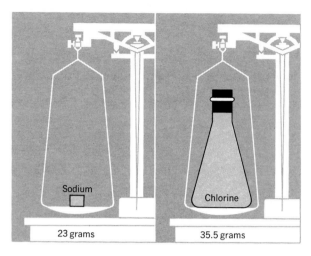

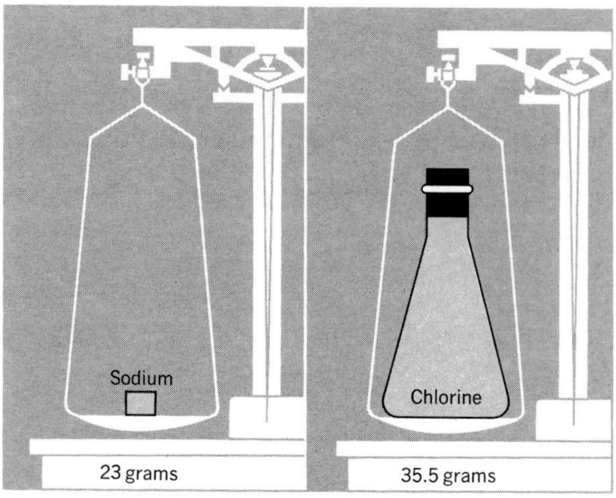

Figure 9.4 Although they do not weigh the same, the number of atoms in this sample of sodium equals the number of atoms in this sample of chlorine.

Chemical Calculations

9.3 The Mole

When sodium (Na) is placed in a container of chlorine (Cl_2), sodium atoms and chlorine atoms will undergo a chemical reaction to produce the compound sodium chloride, NaCl. One sodium atom will react with each chlorine atom—or more precisely, two sodium atoms will react with each chlorine molecule, Cl_2, to produce two sodium chloride ion groups (remember that ionic compounds do not exist as single molecules). If a chemist wanted to produce a certain exact amount of sodium chloride, however, it would be impossible to measure out the sodium and chlorine atoms individually. Atoms are too small to count out one at a time, even millions at a time. Therefore, a unit of measurement had to be invented that would allow us to measure out equal numbers of sodium and chlorine atoms; this unit is the **mole.** The mole is defined as the number of atoms in 12.0000 grams of carbon-12. Because atoms are so small, this number of atoms is very large: 6.02×10^{23}. This number is called **Avogadro's number,** named for the nineteenth-century scientist Amadeo Avogadro who contributed a great deal to our knowledge of atomic weights. A mole, then, contains 6×10^{23} units; these units can be atoms, molecules, ions, electrons, or anything else. A mole of sodium atoms contains 6×10^{23} sodium atoms; a mole of chlorine molecules contains 6×10^{23} chlorine molecules; a mole of garden peas contains 6×10^{23} garden peas (enough to cover our planet and almost ten others like it with a layer of garden peas 4 feet deep*).

** A calculation made by Professors D. H. Andrews and R. J. Kokes of Johns Hopkins University.*

How much does one mole weigh? That depends upon the nature of the particles, just as a dozen lemons, grapefruit, or pumpkins each has a different total weight. **The atomic weight in grams of an element will contain one mole of atoms of that element.** For example, the atomic weight of sodium is 23, so 23 grams of sodium will contain one mole of sodium atoms. Likewise, aluminum has an atomic weight of 27, so 27 grams of aluminum will contain one mole, or 6×10^{23} atoms, of aluminum (Figure 9.4).

It has become common to use the term *mole* to refer to not only Avogadro's number of particles, but also to the number of grams of a substance that will contain 6×10^{23} particles. Therefore, we will refer to 23 grams of sodium as one mole of sodium, and will think of one mole of aluminum as being 27 grams of aluminum.

Example 9-2

Determine the weight of each of the following.

(a) *One mole of magnesium atoms:* From the table on the back inside cover of the book, we find that the atomic weight of magnesium is 24. Therefore, the weight of one mole of magnesium atoms is 24 g.

(b) *3.5 moles of iron:* The atomic weight of iron is 56. Therefore, 3.5 moles will weigh $3.5 \cancel{\text{moles}} \times \frac{56 \text{ g}}{1 \cancel{\text{mole}}} = 196 \text{ g}$.

(c) *0.1 mole of sulfur atoms:* The atomic weight of sulfur is 32. Therefore, 0.1 mole of sulfur atoms will weigh $0.1 \cancel{\text{mole}} \times \frac{32 \text{ g}}{1 \cancel{\text{mole}}} = 3.2 \text{ g}$.

9.4 Formula Weight

When atoms react to form compounds, there is no net gain or loss of weight. The particle that forms, whether it be a molecule or ion group, will have a **formula weight** that is equal to the sum of the atomic weights of all the atoms appearing in its chemical formula.

For example, the formula weight for sodium chloride, NaCl, is 58.5 amu—the atomic weight of one sodium atom (23 amu) plus the atomic weight of one chlorine atom (35.5 amu). The formula weight of carbon tetrachloride, CCl_4, equals the sum of the atomic weight of one carbon atom plus the atomic weight of four chlorine atoms.

$$\begin{aligned} C + (4 \times Cl) &= CCl_4 \\ 12 + (4 \times 35.5) &= 154 \end{aligned}$$

One mole of any substance will have a mass equal to the formula weight of that substance expressed in grams. For example, one mole of chlorine molecules, Cl_2, will weigh 2×35.5 or 71 grams. One mole of the compound sodium chloride, NaCl, will weigh $23 + 35.5$ or 58.5 grams. If you weigh out 58.5 grams of NaCl, which is one mole of table salt, your table salt will contain 6×10^{23} NaCl ion groups.

Example 9-3

1. Calculate the formula weight for each of the following.

   (a) KBr   The atomic weight of potassium is 39 amu, and that of bromine is 80 amu. Therefore, the formula weight of KBr is $39 + 80 = 119$ amu.

   (b) $Ca(OH)_2$ The formula weight will be the sum of the atomic weight of one calcium atom (40 amu) plus two times the weight of the polyatomic ion OH ($16 + 1 = 17$ amu).

$$Ca + (2 \times OH) = Ca(OH)_2$$
$$40 + (2 \times 17) = 74 \text{ amu}$$

   (c) $Mg_3(PO_4)_2$ The formula weight can be calculated as follows:

$$3 \times Mg + (2 \times [P + (4 \times O)])$$
$$3 \times 24 + (2 \times [31 + (4 \times 16)])$$
$$72 + (2 \times 95)$$
$$72 + 190 = 262 \text{ amu}$$

2. What is the weight of one mole of $Ca(OH)_2$?

   One mole of $Ca(OH)_2$ will have a mass equal to the formula weight in grams. Therefore, one mole of $Ca(OH)_2$ will weigh 74 g.

3. How many KBr ion groups are there in 119 g of KCl?

   The weight of one mole of KBr is 119 g. Because one mole contains 6×10^{23} units, 119 g or 1 mole of KBr will contain 6×10^{23} ion groups.

9.5 Solving Problems With the Mole

Problems involving mole calculations can be solved using the same techniques shown in Appendix 2 for problems involving the metric system. Before continuing with this chapter, it might be helpful for you to stop and review Appendix 2.

Now let's do some examples of mole calculations:

Example 9-4

1. A chemist needs 0.45 mole of potassium chlorate, $KClO_3$. How many grams of $KClO_3$ is this?

Step 1: First you must determine the weight of one mole of $KClO_3$. This will give you a relationship between grams and moles that you can use to solve the problem.

$$\text{Formula weight of } KClO_3 = 39.1 + 35.5 + (3 \times 16.0)$$
$$= 39.1 + 35.5 + 48.0 = 122.6$$

Therefore, 1 mole of $KClO_3$ = 122.6 g

Step 2: From the equality in Step 1, we can write two unit factors

$$\frac{1 \text{ mole } KClO_3}{122.6 \text{ g}} \quad \text{or} \quad \frac{122.6 \text{ g}}{1 \text{ mole } KClO_3}$$

Step 3: Our problem asks:

$$0.45 \text{ mole } KClO_3 = (?) \text{ g}$$

Therefore, for our answer to appear in the correct units we must use the second of the two unit factors shown in Step 2.

$$0.45 \cancel{\text{mole}}\ KClO_3 \times \frac{122.6 \text{ g}}{1 \cancel{\text{mole}}\ KClO_3} = 0.45 \times 122.6 \text{ g} = 55 \text{ g}$$

2. If you have a flask containing 9 grams of water, how many moles of water are in the flask?

Step 1: First you will need to find the relationship between grams and moles of water.

Formula weight of $H_2O = (2 \times 1) + 16 = 18$. Therefore, 1 mole of water = 18 grams.

Step 2: From the equality in Step 1, we are able to write two unit factors:

$$\frac{1 \text{ mole of water}}{18 \text{ g}} \quad \text{or} \quad \frac{18 \text{ g}}{1 \text{ mole of water}}$$

Step 3: Our problem asks:

$$9 \text{ grams} = (?) \text{ moles of water}$$

Therefore, for our answer to appear in the correct units we must use the first of the two unit factors shown in Step 2.

$$9 \cancel{\text{g}} \times \frac{1 \text{ mole of water}}{18 \cancel{\text{g}}} = \frac{9}{18} \text{ mole of water}$$
$$= 0.5 \text{ mole of water}$$

3. How many molecules of chlorine gas will there be in a tank containing 7.1 grams of chlorine?

Step 1: To solve this problem we need to establish a relationship between grams of chlorine and molecules of chlorine. We know that

$$1 \text{ mole of } Cl_2 = 2 \times 35.5 \text{ g} = 71 \text{ grams}$$

and

$$1 \text{ mole of } Cl_2 = 6 \times 10^{23} \text{ molecules}$$

Therefore,

$$71 \text{ grams of } Cl_2 = 6 \times 10^{23} \text{ molecules}$$

Step 2: From the equality in Step 1, we can write two unit factors:

$$\frac{71 \text{ grams of } Cl_2}{6 \times 10^{23} \text{ molecules}} \quad \text{or} \quad \frac{6 \times 10^{23} \text{ molecules}}{71 \text{ grams of } Cl_2}$$

Step 3: Our problem asks the following question:

$$7.1 \text{ grams of } Cl_2 = (?) \text{ molecules}$$

In order that our answer appear in the correct units, we use the second of the two unit factors shown in Step 2.

$$7.1 \text{ grams of } Cl_2 \times \frac{6 \times 10^{23} \text{ molecules}}{71 \text{ grams of } Cl_2}$$

$$= \frac{7.1 \times 6 \times 10^{23} \text{ molecules}}{71}$$

$$= 0.1 \times 6 \times 10^{23} \text{ molecules}$$

$$= 6 \times 10^{22} \text{ molecules}$$

9.6 Calculations Using Balanced Equations

A balanced chemical equation contains a great deal of information for the chemist. Not only does it indicate the identity of the reactants and products, but it also tells the relative number of these substances involved in the reaction. Consider the following reaction:

$$2H_2 + O_2 \longrightarrow 2H_2O$$

This equation can be read: "Two hydrogen molecules will react with one oxygen molecule to produce two water molecules," or in quantities with which we can work: "Two moles of hydrogen will react with one mole of oxygen to produce two moles of water."

This, in turn, tells us that 4 grams of hydrogen will react with 32 grams of oxygen to produce 36 grams of water.

Example 9-5

The following equation gives three sets of information. Write out each in sentence form.

$$2Al + 3H_2SO_4 \longrightarrow 3H_2 + Al_2(SO_4)_3$$

(1) Two atoms of aluminum will react with three molecules of hydrogen sulfate (sulfuric acid) to produce three molecules of hydrogen and one aluminum sulfate ion group.

(2) Two moles of aluminum will react with three moles of sulfuric acid to produce two moles of hydrogen and one mole of aluminum sulfate.

(3) 54 grams of aluminum will react with 294 grams of sulfuric acid to produce 6 grams of hydrogen and 342 grams of aluminum sulfate.

Now we can put together all the procedures that we have learned in this chapter—balancing chemical equations and solving problems with the mole—to analyze many different types of problems concerning chemical reactions.

Example 9-6

1. Imagine that you are working in a laboratory and want to obtain 2.4 moles of magnesium chloride, a compound that is produced (along with water) when magnesium oxide is heated with hydrochloric acid, HCl. How many grams of each reactant would you need?

   Step 1: First you need to write a balanced chemical equation for this reaction.

   Hydrochloric acid + Magnesium oxide ⟶ Magnesium chloride + Water

   Unbalanced: $HCl + MgO \longrightarrow MgCl_2 + H_2O$
   Balanced: $2HCl + MgO \longrightarrow MgCl_2 + H_2O$

   We have now written a chemical equation representing the production of one mole of magnesium chloride.

   Step 2: No matter how much magnesium chloride we might want to produce, the important fact to remember is that the proportion of reactants and products involved in the reaction will always remain the same. That is, for example, we will always need two moles of hydrochloric acid for every mole of magnesium chloride produced, and will always obtain one mole of water for every mole of magnesium oxide that reacts. Using the coefficients of the balanced equation, we can write a series of reaction ratios expressing the proportions that will hold between any two of the reactants or products:

   (a) $\frac{\text{2 moles of HCl}}{\text{1 mole of MgO}}$ or $\frac{\text{1 mole of MgO}}{\text{2 moles of HCl}}$

   (b) $\frac{\text{2 moles of HCl}}{\text{1 mole of MgCl}_2}$ or $\frac{\text{1 mole of MgCl}_2}{\text{2 moles of HCl}}$

   (c) $\frac{\text{2 moles of HCl}}{\text{1 mole of H}_2\text{O}}$ or $\frac{\text{1 mole of H}_2\text{O}}{\text{2 moles of HCl}}$

(d) $\frac{1 \text{ mole of MgO}}{1 \text{ mole of } MgCl_2}$ or $\frac{1 \text{ mole of } MgCl_2}{1 \text{ mole of MgO}}$

(e) $\frac{1 \text{ mole of MgO}}{1 \text{ mole of } H_2O}$ or $\frac{1 \text{ mole of } H_2O}{1 \text{ mole of MgO}}$

(f) $\frac{1 \text{ mole of } MgCl_2}{1 \text{ mole of } H_2O}$ or $\frac{1 \text{ mole of } H_2O}{1 \text{ mole of } MgCl_2}$

Step 3: Now we need to go back to the problem to see which of these reaction ratios will help us find the correct answer. The problem asks how many grams of each reactant are needed to produce 2.4 moles of $MgCl_2$. Therefore, we need to choose reaction ratios which will show the relationships between the reactants HCl and MgO, and the product $MgCl_2$. We can then use these ratios to calculate the number of moles of each reactant that will be needed.

$$2.4 \text{ moles of } MgCl_2 \times \frac{2 \text{ moles of HCl}}{1 \text{ mole of } MgCl_2} = 4.8 \text{ moles of HCl}$$

$$2.4 \text{ moles of } MgCl_2 \times \frac{1 \text{ mole of MgO}}{1 \text{ mole of } MgCl_2} = 2.4 \text{ moles of MgO}$$

Step 4: We know how many moles of each reactant are required, but we have been asked to express the answer in grams. Therefore, we must use the appropriate unit factors to convert moles of each reactant into grams.

(a) 4.8 moles of HCl = (?) grams.

$$1 \text{ mole of HCl} = 1.0 + 35.5 = 36.5 \text{ grams}$$

Therefore,

$$4.8 \text{ moles of HCl} \times \frac{36.5 \text{ grams}}{1 \text{ mole of HCl}} = 175 \text{ grams}$$

(b) 2.4 moles of MgO = (?) grams.

$$1 \text{ mole of MgO} = 24 + 16 = 40 \text{ grams}$$

Therefore,

$$2.4 \text{ moles of MgO} \times \frac{40 \text{ grams}}{1 \text{ mole of MgO}} = 96 \text{ grams}$$

So, to produce 2.4 moles of $MgCl_2$ we will need 175 grams of HCl and 96 grams of MgO.

2. Antacid tablets are taken by millions of persons to reduce the discomfort of an upset stomach. The active ingredient in some commercial antacid tablets is magnesium hydroxide, $Mg(OH)_2$, which will react with stomach acid (HCl) to produce magnesium chloride ($MgCl_2$) and water. One popular tablet contains 0.10 grams of $Mg(OH)_2$. How many grams of stomach acid will this tablet neutralize?

Step 1: Write the balanced equation.

Unbalanced: $Mg(OH)_2 + HCl \longrightarrow MgCl_2 + H_2O$
Balanced: $Mg(OH)_2 + 2HCl \longrightarrow MgCl_2 + 2H_2O$

Step 2: Write the reaction ratios that will be needed to solve the problem. In this case we are interested in the relationship between $Mg(OH)_2$ and HCl.

$$\frac{1 \text{ mole of } Mg(OH)_2}{2 \text{ moles of } HCl} \quad \text{or} \quad \frac{2 \text{ moles of } HCl}{1 \text{ mole of } Mg(OH)_2}$$

Step 3: In order to use these reaction ratios, we need to determine how many moles of $Mg(OH)_2$ are found in one tablet.

$$1 \text{ mole of } Mg(OH)_2 = 24 + [2 \times (16 + 1)] = 58 \text{ grams}$$

Therefore,

$$0.10 \cancel{\text{gram}} \times \frac{1 \text{ mole of } Mg(OH)_2}{58 \cancel{\text{grams}}} = \frac{0.10}{58} \text{ mole of } Mg(OH)_2$$

$$= 0.0017 \text{ mole of } Mg(OH)_2$$

Step 4: Using a reaction ratio from Step 2, we can now compute the number of moles of HCl that will be neutralized.

$$0.0017 \cancel{\text{mole of } Mg(OH)_2} \times \frac{2 \text{ moles of } HCl}{1 \cancel{\text{mole of } Mg(OH)_2}} = 0.0034 \text{ mole of } HCl$$

Step 5: To solve the problem, then, all we must do is convert moles of HCl into grams of HCl.

$$1 \text{ mole of } HCl = 1.0 + 35.5 = 36.5 \text{ grams}$$

$$0.0034 \cancel{\text{mole of } HCl} \times \frac{36.5 \text{ grams}}{1 \cancel{\text{mole of } HCl}} = 0.12 \text{ gram}$$

Therefore, 0.10 gram of $Mg(OH)_2$ will neutralize 0.12 gram of HCl.

Chapter Summary

A chemical equation is a shorthand way of indicating what occurs in a chemical reaction. The formulas of the reactants or starting substances are separated from the formulas of the products or resulting substances by an arrow. A balanced chemical equation is one in which the number of atoms of each element on the product side of the equation equals the number of atoms of that element on the reactant side of the equation.

A mole is a unit of measure that allows equal numbers of particles of different substances to be weighed out. A mole contains Avogadro's number, or 6×10^{23}, particles. The atomic weight in grams of an element will contain one mole of atoms of that element. The formula weight of a compound is equal to the sum of the atomic weights of each atom in the formula of that compound. One mole of a compound will have a mass equal to the formula weight of that compound expressed in grams.

Exercises and Problems

1. Calculate the formula weight for each of the following compounds:

Group 1	Group 2	Group 3
(a) Sodium hydroxide	KCl	$Ca(OH)_2$
(b) Calcium chloride	H_2SO_4	$(NH_4)_2SO_4$
(c) Sulfur dioxide	$Mg_3(PO_4)_2$	CH_4
(d) Sodium phosphate	CO_2	HCl
(e) Barium sulfate	$NaHCO_3$	$NaNO_3$
(f) Hydrogen bromide	$KMnO_4$	$AgNO_3$
(g) Boron trifluoride	HNO_3	H_2CO_3
(h) Water	$CuSO_4$	$CaCO_3$

2. What is the weight in grams of each of the following? (Use the formula weights calculated in exercise 1.)

Group 1	Group 2	Group 3
(a) 0.15 mole of sodium hydroxide	4.2 moles of KCl	0.95 mole of $Ca(OH)_2$
(b) 2.5 moles of calcium chloride	1.5 moles of H_2SO_4	4.6 moles of $(NH_4)_2SO_4$
(c) 0.80 mole of sulfur dioxide	0.015 moles of $Mg_3(PO_4)_2$	12.5 moles of CH_4
(d) 0.50 mole of sodium phosphate	5.5 moles of CO_2	0.025 mole of HCl
(e) 3.60 moles of barium sulfate	0.32 mole of $NaHCO_3$	0.62 mole of $NaNO_3$
(f) 0.020 mole of hydrogen bromide	0.075 mole of $KMnO_4$	0.50 mole of $AgNO_3$
(g) 0.125 mole of boron trifluoride	0.10 mole of HNO_3	1.25 moles of H_2CO_3
(h) 0.0010 mole of water	0.060 mole of $CuSO_4$	0.0001 mole of $CaCO_3$

3. Calculate the number of moles in each of the following samples. (Use the formula weights that you calculated in exercise 1.)

Group 1	Group 2	Group 3
(a) 0.012 g of sodium hydroxide	89.4 g of KCl	407 g of $Ca(OH)_2$
(b) 33.3 g of calcium chloride	7.35 g of H_2SO_4	3.3 g of $(NH_4)_2SO_4$
(c) 2.48 g of sulfur dioxide	131 g of $Mg_3(PO_4)_2$	148.8 g of CH_4
(d) 0.82 g of sodium phosphate	198 g of CO_2	2.19 g of HCl
(e) 2.33 g of barium sulfate	67.2 g of $NaHCO_3$	64.6 g of $NaNO_3$
(f) 52.65 g of hydrogen bromide	498.8 g of $KMnO_4$	0.68 g of $AgNO_3$
(g) 122.4 g of boron trifluoride	94.5 g of HNO_3	80.6 g of H_2CO_3
(h) 64.8 g of water	3.19 g of $CuSO_4$	1.25 g of $CaCO_3$

4. Calculate the following:

(a) the number of moles in 3.6×10^{23} molecules of oxygen.
(b) the weight in grams of 1.5×10^{23} molecules of chlorine.

5. Balance the following equations.

(a) $Na + H_2O \longrightarrow NaOH + H_{2(g)}$
(b) $KClO_3 \longrightarrow KCl + O_{2(g)}$
(c) $MnO_2 + HCl \longrightarrow Cl_{2(g)} + MnCl_2 + H_2O$
(d) $C_3H_8 + O_2 \longrightarrow CO_2 + H_2O$
(e) $NH_3 + O_2 \longrightarrow NO + H_2O$
(f) $CO_3^{2-} + H^+ \longrightarrow CO_2 + H_2O$

6. Balance the following equations.

(a) $HCl + Cr \longrightarrow CrCl_3 + H_{2(g)}$
(b) $FeCl_3 + Na_2CO_3 \longrightarrow Fe_2(CO_3)_3 + NaCl$
(c) $PbS + H_2O_2 \longrightarrow PbSO_4 + H_2O$
(d) $C_4H_{10} + O_2 \longrightarrow CO_2 + H_2O$
(e) $CrO_4^{2-} + H^+ \longrightarrow Cr_2O_7^{2-} + H_2O$

7. Write a balanced equation for each of the following reactions:

(a) Hydrogen reacts with bromine to form hydrogen bromide.
(b) Calcium bicarbonate, when heated, breaks apart to form calcium carbonate, water, and carbon dioxide.
(c) Silver nitrate will react with copper to form copper(II) nitrate and silver.

(d) Hydrogen reacts with nitrogen to form ammonia, NH_3.
(e) Methane and chlorine gases will react to form carbon tetrachloride and hydrogen chloride.

8. Use this balanced equation

$$Mg_3N_2 + 6H_2O \longrightarrow 3Mg(OH)_2 + 2NH_3$$

to answer the following questions:

(a) How many moles of $Mg(OH)_2$ would be produced from the reaction of 0.10 mole of Mg_3N_2?
(b) How many moles of $Mg(OH)_2$ and of NH_3 would be produced from the reaction of 500 g of Mg_3N_2?
(c) How many grams of Mg_3N_2 and H_2O must react to produce 0.060 mole of $Mg(OH)_2$?
(d) How many grams of Mg_3N_2 are needed to produce 52.2 g of $Mg(OH)_2$?
(e) What is the maximum number of grams of $Mg(OH)_2$ that can be produced by the reaction of 10.0 g of Mg_3N_2 and 14.4 g of H_2O?

9. Wine is produced by the process of fermentation, in which the sugar in grapes is converted by the action of yeast into ethyl alcohol and carbon dioxide.

$$\underset{\text{Sugar}}{C_6H_{12}O_6} \xrightarrow{\text{yeast}} \underset{\text{Ethyl alcohol}}{2C_2H_5OH} + 2CO_{2(g)}$$

How many kilograms of ethyl alcohol would be produced if all of the 5.00 kilograms of sugar in a batch of grapes were fermented?

10. Aspirin is produced by the reaction of salicylic acid with acetic anhydride.

$$\underset{\text{Salicylic acid}}{C_7H_6O_3} + \underset{\text{Acetic anhydride}}{C_4H_6O_3} \longrightarrow \underset{\text{Aspirin}}{C_9H_8O_4} + C_2H_4O_2$$

How many grams of salicylic acid are required to produce an aspirin tablet that contains 0.33 gram of aspirin?

11. Sulfur trioxide can be produced by the reaction of the air pollutant sulfur dioxide with the oxygen in the air.

(a) Write the balanced equation for this reaction.
(b) How many grams of sulfur dioxide must be released into the air to produce 1.00 kilogram of sulfur trioxide?

12. The pollutant sulfur trioxide is removed from the atmosphere by rain, which reacts with the sulfur trioxide to form sulfuric acid, H_2SO_4. Along major freeways, rain water has been found to be quite acidic and, as a result, is damaging to plant life. For every kilogram of sulfur trioxide that reacts with rain, how many grams of sulfuric acid will be formed?

13. Sodium lauryl sulfate, a detergent, can be prepared by the following reactions:

1. $C_{12}H_{25}OH + H_2SO_4 \longrightarrow C_{12}H_{25}OSO_3H + H_2O$
   (Lauryl alcohol) (Lauryl sulfonic acid)

2. $C_{12}H_{25}OSO_3H + NaOH \longrightarrow C_{12}H_{25}OSO_3Na + H_2O$
   (Lauryl sulfonic acid) (Detergent)

If a day's production of detergent is 11 tons, how many tons of lauryl alcohol will be required?

chapter 10

Reaction Rates and Chemical Equilibrium

Learning Objectives

By the time you have finished this chapter, you should be able to:

1. Define *activation energy.*

2. Draw a potential energy diagram for an endothermic and exothermic reaction.

3. List four factors that will affect the rate of a chemical reaction.

4. Define *catalyst.*

5. Define *chemical equilibrium,* and give two examples.

6. State two ways in which a chemical equilibrium can be disrupted.

7. State Le Chatelier's Principle, and predict the changes that will occur in an equilibrium when a stress is applied.

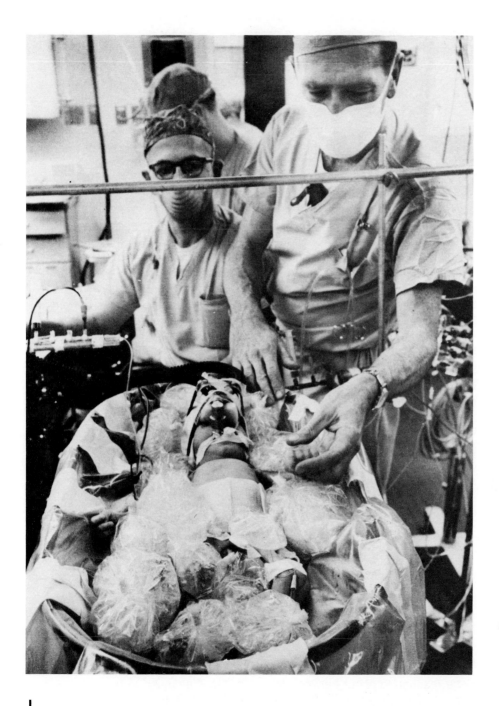

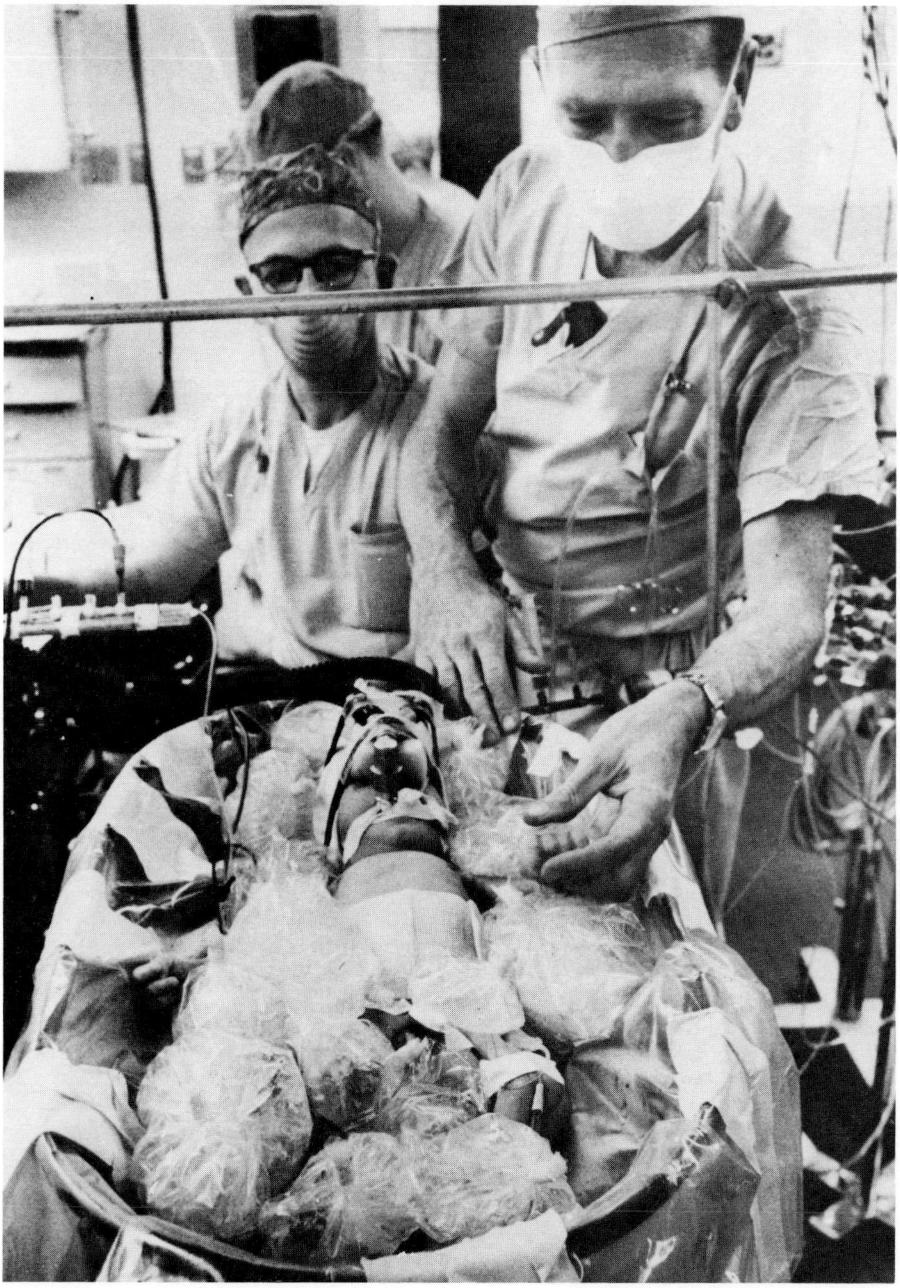

In the delivery room of a large Boston hospital, a baby was born to Jan and Tom Harper. The baby, a boy named Tim, was found to be suffering from a fatal heart defect. In order to survive, Tim had to have surgery quickly.

His chances for survival under normal open-heart surgery techniques were small. The heart-lung machine that is ordinarily used to maintain blood circulation during such an operation is risky for a child so young. The machine damages red blood cells, and the tubes and other equipment that would have to fit into such a small chest cavity would make it difficult for the surgeon to see. But Tim's chances for life were still good, thanks to a new surgical technique called deep hypothermia.

In Chapter 3 we saw that lowering the body temperature could result in death. But doctors now use hypothermia under controlled conditions to save human lives. Just 36 hours old, Tim was taken to surgery and was anesthetized. His body was covered with a plastic blanket which was then covered with cracked ice. When his body temperature had dropped to 77°F (25°C), the surgeon opened Tim's chest and hooked him up to a heart-lung machine that pumped cooled blood through Tim's body until his temperature was lowered to 68°F (20°C). At this point the pump was turned off. Tim's heart was motionless, and his blood circulation had stopped. The surgeons then began the delicate process of rearranging the blood vessels between the heart and the lungs. Less than an hour later, the task was completed. The pump was turned on, and now pumped warmed blood through Tim's body. As his temperature increased, his heart again began to pump blood through the repaired vessels. Tim recovered quickly from the surgery, and has since grown to be a very healthy and active toddler.

But how can the body, especially the brain, survive without any blood circulating for an hour? Under normal conditions the brain undergoes irreparable damage if it is cut off from its oxygen supply for as little as three minutes. However, by lowering the body temperature, doctors are able to slow down the chemical reactions occurring within the brain cells. Under controlled conditions, these chemical processes can be slowed to a point where blood circulation can be stopped for up to an hour.

Controlling the rate of chemical reactions can, therefore, mean the difference between life and death to a young child. It can also mean the difference between a nuclear disaster and the peaceful use of nuclear power to generate electricity. In order to understand how scientists control chemical reactions for our benefit, it is important to study the various factors that influence the rate at which chemical reactions occur.

Rates of Chemical Reactions

10.1 Activation Energy

For a reaction to occur between two particles, they must be brought close enough together for their outer-shell electrons to interact. In fact, they must collide. Not only must they collide, however, but they must do so with enough energy to overcome the repelling forces between the electrons surrounding the two nuclei. The amount of energy necessary for a successful collision is called the **activation energy, E_{act}.** In order for a reaction to occur, molecules must collide with energy at least equal to the activation energy. When such a collision occurs, the molecules form a reactive group called the **activated complex.** This complex of atoms is neither the reactants nor the products. Rather, it is a highly unstable combination of atoms which represents a state of change between the reactants and the products.

It may help to imagine the activation energy as a hill the particles must get over to complete the reaction. Each reaction is associated with a hill of a different height, or different activation energy. Imagine a person trying to roll a bowling ball up such a hill; in most cases the bowling ball will roll only part way up the hill and then roll down again. This is true of a reaction—most collisions between molecules do not occur with enough energy to overcome the activation barrier and to form the activated complex; the molecules simply bounce off each other unreacted. Just as the bowler can only occasionally give the ball enough energy to get up over the hill, so it is only on occasion that the particles in a reaction collide with sufficient energy to overcome the activation energy barrier and eventually form products (Figure 10.1).

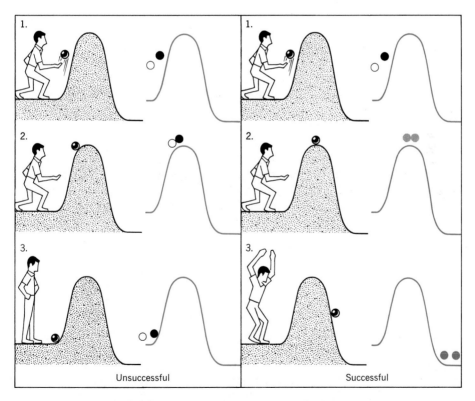

Figure 10.1 Just as a bowling ball will roll over the hill only when it has been given enough energy, so a reaction will occur between two molecules only when they collide with sufficient energy to get over the activation energy barrier.

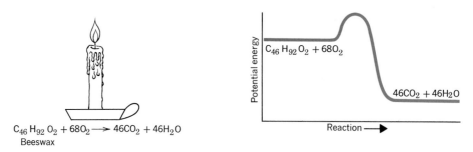

$$C_{46}H_{92}O_2 + 68O_2 \longrightarrow 46CO_2 + 46H_2O$$
Beeswax

Figure 10.2 Burning candle wax is an exothermic reaction in which the products have less potential energy than the reactants.

10.2 Exothermic and Endothermic Reactions

You know that a candle will not spontaneously burst into flame at room temperature, but you may wonder what allows it to keep burning once it is lit. The reaction between candle wax and oxygen has a high activation energy, and very few molecules have enough energy at room temperature to react. The heat of a match is required if a large number of molecules are to overcome the activation energy barrier and cause the wax to burn. Now, the reaction between candle wax and oxygen is **exothermic,** meaning that energy is released as the reaction occurs. Exothermic reactions result when the products of the reaction have less potential energy than the reactants. Therefore, once these molecules begin to react, this reaction releases enough energy to boost additional molecules over the activation energy barrier. Hence the reaction becomes self-sustaining, and the candle continues to burn (Figure 10.2).

Some highly exothermic reactions can be explosive. Dynamite, for example, must be set off with a percussion cap. The very small explosion of the percussion cap releases enough energy for a small number of dynamite molecules to react, and the energy released by these molecules is enough to cause the remainder of the molecules to all react at once, releasing a tremendous amount of energy.

Other reactions, such as the decomposition of water into hydrogen and oxygen, or the formation of sugar molecules in the process of photosynthesis, require the addition of energy for the reaction to continue. Reactions requiring a constant input of energy are called **endothermic.** In such reactions the potential energy of the products is greater than that of the reactants. In effect, energy has been absorbed by the reaction. In endothermic reactions, reactants require an initial input of energy to get over the activation energy barrier, and then a continuing supply of energy to keep the reaction going. Stop the supply of electricity, and the decomposition of water will come to a halt; keep a green plant in the dark, and photosynthesis will stop, causing the plant to die (Figures 10.3 and 10.4).

Figure 10.5 contains potential energy diagrams for an exothermic and an endothermic reaction. The energy required or released by a reaction is

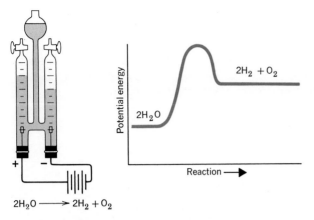

$$2H_2O \longrightarrow 2H_2 + O_2$$

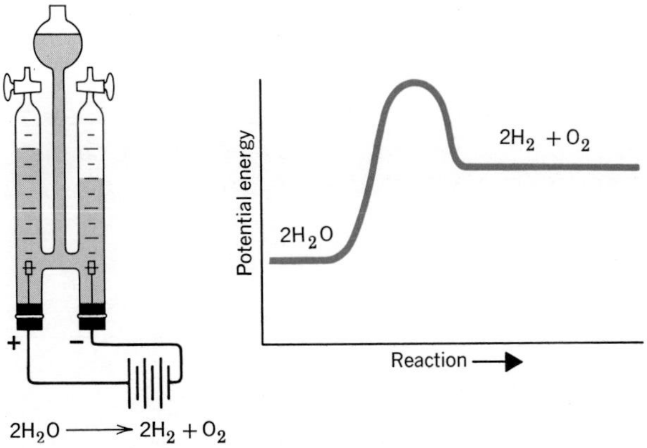

Figure 10.3 Electrolysis (the breaking apart of water molecules with electrical energy) is an endothermic reaction in which the products, hydrogen and oxygen, have more potential energy than the reactant, water.

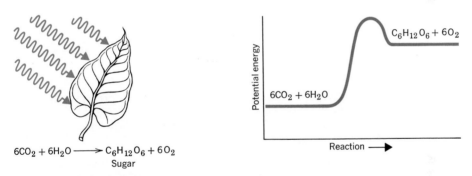

$$6CO_2 + 6H_2O \longrightarrow C_6H_{12}O_6 + 6O_2$$
Sugar

Figure 10.4 Photosynthesis is an endothermic process requiring a constant supply of energy from the sun.

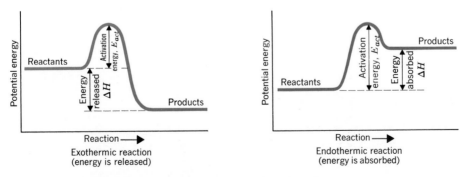

Figure 10.5 Potential energy diagrams for an exothermic and endothermic reaction. E_{act} = activation energy. ΔH = heat of reaction, the energy released or

called the **heat of reaction, ΔH.** This change in heat content (or enthalpy) is expressed in kilocalories per mole of reactant. The value of ΔH is the difference between the potential energy of the products and the potential energy of the reactants. The heat of reaction will have a positive value for an endothermic reaction, and a negative value for an exothermic reaction. Each reaction has its own characteristic activation energy and heat of reaction.

Example 10-1

1. Are the following reactions endothermic or exothermic?

(a) $2Na_2O_2 + 2H_2O \longrightarrow 4NaOH + O_2 + 30.2$ kcal
(b) $Si + 2H_2 + 8.2$ kcal $\longrightarrow SiH_4$

Reaction (a) is exothermic because energy is given off as a product by the reaction. Reaction (b) is endothermic because energy is required by the reaction; that is, energy is one of the reactants.

2. Draw the potential energy diagram for the following reaction, which has an activation energy of $E_{act} = 39$ kcal:

$$2ICl + H_2 \longrightarrow I_2 + 2HCl + 53 \text{ kcal}$$

To draw the potential energy diagram, we must first determine if the reaction is endothermic or exothermic. Because energy is given off by the reaction, it is exothermic. This means that the potential energy of the reactants is greater than the potential energy of the products: 53 kilocalories greater. There are several ways to draw the diagram, but to simplify things we can arbitrarily assign the substances with the least potential energy (the products) a relative potential energy of zero. The reactants, therefore, will have a potential energy of 53 kcal in relation to the products. The activation energy for this reaction is 39 kcal, so the top of the curve will be at 53 kcal (the potential energy of the reactants) + 39 kcal (the activation energy) = 92 kcal.

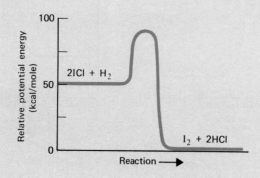

Factors Affecting Reaction Rates

10.3 The Nature of the Reactants

The nature of the reactants influences the rate of a chemical reaction. For example, when colorless nitric oxide escapes from a test tube into the air, reddish-brown nitrogen dioxide forms very quickly.

$$2NO + O_2 \longrightarrow 2NO_2, \qquad \text{Fast reaction at } 25°C$$

But when carbon monoxide from automobile exhaust fumes is released into the air, the reaction with oxygen to form carbon dioxide is quite slow (unfortunately for the residents of large urban areas).

$$2CO + O_2 \longrightarrow 2CO_2, \qquad \text{Very slow reaction at } 25°C$$

These two balanced equations look quite similar, but the large difference in their rates of reaction comes from the nature of the carbon monoxide and nitric oxide molecules themselves.

10.4 The Concentration of Reactants

The concentration of reactants can greatly influence the rate of a chemical reaction. We have seen that particles must collide in order for them to react, and it seems reasonable that the more reactant particles there are in a given space, the more collisions will occur. (We see the same thing on our highways; the more crowded the highways, the higher the probability of a fatal collision. This is especially apparent when you consider the high death rates on holiday weekends.) Increasing the concentration of the reactants will increase the number of collisions and, therefore, will increase the reaction rate (Figure 10.6).

Physicians make use of this principle when prescribing medicines to treat specific diseases. The appropriate treatment for tonsillitis or pneumonia may involve a dose of 250 mg of penicillin, but meningitis would require two to five times that dosage. This higher concentration of penicillin increases the rate of absorption into the bloodstream and, therefore, increases the effective concentration in the blood.

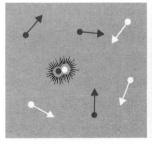

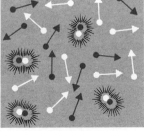

Figure 10.6 Increasing the concentration of the reactants will increase the frequency of collisions, and thereby increase the rate of the reaction.

Figure 10.7 The tops of these grain elevators were destroyed by the force of an explosion of grain dust. (Wide World Photos.)

10.5 The Surface Area of a Solid Reactant

Increasing the surface area of a solid reactant will increase the rate of reaction. An increase in the surface area increases the number of solid particles that are exposed, thereby increasing the number of collisions possible between the reactants. Increasing the surface area can sometimes increase the rate of reaction to explosive levels. For example, lumber mills seldom worry about their log piles spontaneously catching fire, but sawdust piles can burst into flame and, therefore, must be kept wet. You probably don't think of flour as being potentially dangerous, but finely divided flour dust can explode (Figure 10.7). As another example a crushed aspirin tablet, because of its increased surface area, will relieve your headache faster than a whole tablet.

10.6 The Temperature of the Reaction

Increasing the temperature of the reactants will increase the kinetic energy of the particles. This will not only increase the frequency of collisions, but will also increase the likelihood that the colliding particles will have enough energy to overcome the activation energy barrier. (This is similar to the idea that increasing speed limits on highways will not only increase the number of collisions, but will also increase the likelihood that any collision will be a fatal one.) (See Figure 10.8.)

Changes in temperature have important effects on living organisms. A fever increases the rate of chemical reactions in the body, as can be seen by the increased pulse rate, increased breathing rate, and abnormalities in digestive and nervous systems. When you run a fever, your basal metabolism rate goes up by 5% for each degree rise in body

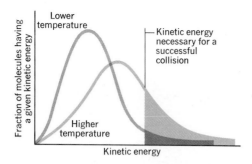

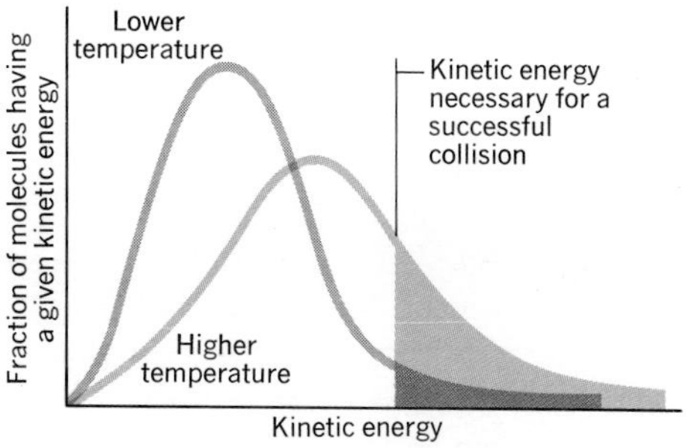

Figure 10.8 Increasing the temperature of a reaction increases the average kinetic energy of the molecules, thus increasing the number of molecules having enough energy to collide successfully and produce products.

temperature. The effects on living organisms from an increase in external temperatures is illustrated by what happens when warm water from power plants is released into streams and lakes, causing an increase in the metabolic rate in the aquatic life. In order to support their new higher rate of metabolism, these organisms require more oxygen. But the warming of the water also decreases the concentration of oxygen in the water, contributing to the death of fish in these streams and lakes. Another detrimental effect of these higher water temperatures is a resulting increase in the sensitivity of fish to pollutants in the water (Figure 10.9).

Figure 10.9 This fish kill resulted when warm waste water from a nuclear power plant was discharged into a nearby stream. (Wide World Photos)

Decreasing the body's temperature slows down bodily reaction rates. During open heart surgery, a patient's body temperature is usually lowered four to five degrees Fahrenheit to decrease the metabolism rate and oxygen requirements. At the beginning of this chapter we saw an extreme example of the decrease in bodily reactions from a lowering of body temperatures. As another example, an athletic trainer might spray a surface anesthetic called ethyl chloride on the injured knee of a basketball player to relieve the pain. The rapid evaporation of the ethyl chloride will lower the temperature of the tissue enough to block the reactions responsible for the transmission of nerve impulses.

Many animals hibernate during the cold winter months. Their body temperatures fall to a few degrees above freezing, and all of the chemical reactions in their bodies slow down. Their breathing and heart rates become very slow, and they require much less energy to sustain life. Therefore, the animals can live on stored body fat alone. To illustrate the extent of this slowdown, consider that a woodchuck's heart beats about 80 times a minute while he is active, but when hibernating his heart will beat only four times a minute (Figure 10.10). The interesting idea of keeping human beings in suspended animation by lowering their body temperatures has been the subject of many science fiction stories. In 1974 two scientists from the Darwin Research Institute discovered that bacteria

Figure 10.10 The body temperature of this hibernating woodchuck is only a few degrees above freezing; thus all chemical reactions in its body have slowed down. This means much less energy is used to sustain life, allowing the animal to survive the winter on stored body fat alone. (Leonard Lee Rue/Photo Researchers)

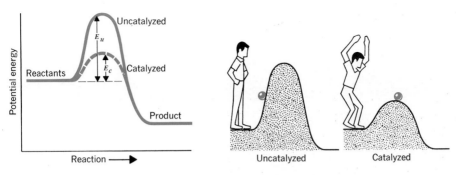

Figure 10.11 A catalyst lowers the activation energy, increasing the chances for a successful collision between reactant particles.

which had been frozen in the cold rock of Antarctica 10,000 to 1 million years ago could be revived in the warmth of the laboratory and, incredibly, could even reproduce.

10.7 Catalysts

Many reactions that proceed slowly can be made to occur at a more rapid rate by the introduction of substances called catalysts. A **catalyst** is a substance that increases the rate of a chemical reaction without being consumed in the reaction. The effect of the catalyst is to lower the activation energy required for the reaction. In our illustration of the bowler and the bowling ball, the action of a catalyst would resemble that of a bulldozer cutting a much lower path over the hill, allowing many more balls rolled by the bowler to reach the top and roll down the other side (Figure 10.11).

Industry makes wide use of catalysts, especially those catalysts which allow companies to produce large amounts of a product at a lower temperature, thereby saving on energy costs. An example of this occurs in the production of sulfuric acid, H_2SO_4, the so-called "king of chemicals." Over 20 million tons of sulfuric acid are used each year in the United States, mostly by the steel, fertilizer, and petroleum industries. One step in the production of sulfuric acid involves the production of sulfur trioxide (SO_3) from sulfur dioxide (SO_2) and oxygen (O_2). This reaction has a very high activation energy, and is quite slow even at high temperatures. But the reaction has been made economically worthwhile by the introduction of a catalyst, such as finely divided platinum, which greatly increases the reaction rate (Figure 10.12). Similarly, the cost of producing gasoline from much larger molecules of crude petroleum is kept low through the use of catalysts in the cracking process—the process by which large molecules are broken into small fractions.

Most of the chemical reactions that occur in our bodies would not ordinarily occur at a high enough rate at body temperature. But the body produces special compounds called **enzymes** which act as biological catalysts, permitting these reactions to readily occur at body temperature. The removal or damaging of body enzymes can have disastrous effects. For

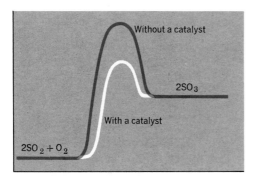

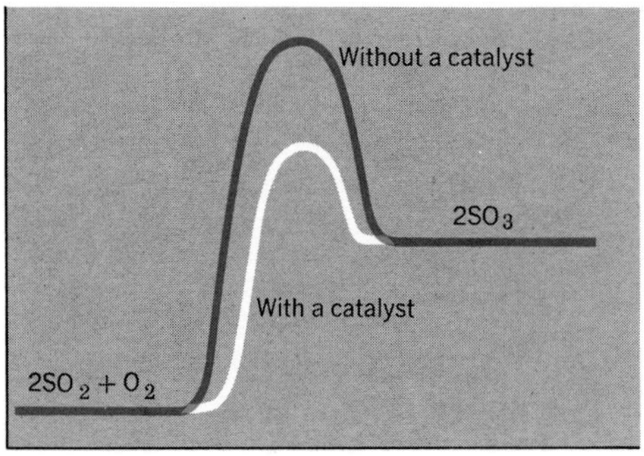

Figure 10.12 The production of sulfuric acid involves a reaction between sulfur dioxide and oxygen; this reaction has a very high activation energy. The addition of platinum as a catalyst lowers the activation energy, and thus makes the process economically feasible.

example, the poison cyanide prevents enzymes from catalyzing reactions essential to the life of the cell, thereby causing the death of the organism. "Everybody needs milk" does not hold for up to 90% of non-Caucasian adults, whose digestive tracts lack the enzyme lactase that catalyses the breakdown of lactose, the sugar found in milk. For these individuals, drinking milk results in nausea, gas, and diarrhea.

Chemical Equilibrium

10.8 What Is Chemical Equilibrium?

At one time or another you may have added sugar to a glass of iced tea, stirred, and watched the sugar dissolve. You may have noticed that additional sugar added to this solution continued to dissolve until you reached a certain point, after which any more sugar added just settled to the bottom of the glass. Although you may have continued stirring, or let the glass sit for a period of time, neither the amount of sugar on the

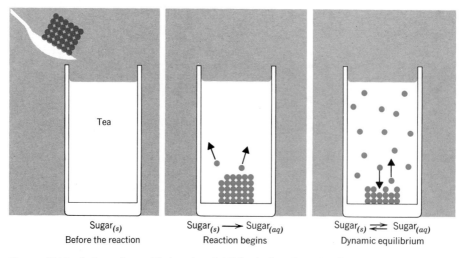

Figure 10.13 A dynamic equilibrium is established when the rate of the forward reaction equals the rate of the reverse reaction.

bottom of the glass nor the amount of sugar dissolved in the iced tea changed.

Under these circumstances the iced tea contained all the sugar molecules it could possibly hold, and we will see that the molecules of sugar dissolved in the iced tea and the molecules of sugar in the crystals on the bottom of the glass were in equilibrium. A molecular equilibrium such as this is not a static, motionless condition such as the equilibrium reached when two children are equally balanced on a seesaw. Rather, a chemical equilibrium is a dynamic state in which events are constantly occurring, but at exactly equal rates. In that glass of iced tea, for example, new sugar molecules were constantly dissolving and other sugar molecules were constantly crystallizing out of solution, but the rate of each of these processes was the same. Therefore, there was no overall change in the number of sugar molecules dissolved in the iced tea or lying on the bottom of the glass. In other words, for every sugar molecule that dissolved, one sugar molecule crystallized out of solution (Figure 10.13).

Let's examine this iced tea example in chemical terms. We said that when we first add sugar to the iced tea, it dissolves. We can represent this process by the following equation.

$$\text{Sugar}_{(s)} \longrightarrow \text{Sugar}_{(aq)}$$

(s) = solid $\qquad\qquad\qquad\qquad$ (aq) = aqueous, dissolved in water

As more and more sugar molecules are added to the iced tea, the number dissolved in the tea will increase to the point that the reverse reaction (that is, the crystallization of sugar from the tea) begins to occur.

$$\text{Sugar}_{(s)} \rightleftharpoons \text{Sugar}_{(aq)}$$

Notice that in this equation we use two arrows pointing in different directions to indicate that this reaction is reversible. As the number of sugar molecules in the tea continues to increase, the rate of the reverse reaction will increase until the rate of the reverse reaction equals the rate of the forward reaction. If we denote the rate of the forward reaction by rate_f and the rate of the reverse reaction by rate_r, we then have $\text{rate}_f = \text{rate}_r$.

$$\text{Sugar}_{(s)} \underset{\text{rate}_r}{\overset{\text{rate}_f}{\rightleftharpoons}} \text{Sugar}_{(aq)}$$

At this point, a state of equilibrium has been reached and no net change will occur in our glass of sugared iced tea. But it is again important to emphasize that the forward reaction and reverse reaction still continue to occur, only at exactly equal rates. We can define **chemical equilibrium,** then, as **a dynamic state in which the rate of the forward reaction equals the rate of the reverse reaction.**

*10.9 The Equilibrium Constant

We often need to discuss chemical equilibrium in precise terms and, therefore, need an indication of how far a reaction proceeds before

* This section is optional and may be skipped without loss of continuity.*

Table 10.1 Some Equilibrium Constants

Reaction	K_{eq}	Value of K_{eq} at Specific Temperatures
$HSO_4^- \rightleftharpoons H^+ + SO_4^{2-}$	$K = \frac{[H^+][SO_4^{2-}]}{[HSO_4^-]}$	1.3×10^{-2} at 25°C
$CH_3COOH \rightleftharpoons H^+ + CH_3COO^-$	$K = \frac{[H^+][CH_3COO^-]}{[CH_3COOH]}$	1.8×10^{-5} at 25°C
$N_2O_{4(g)} \rightleftharpoons 2NO_{2(g)}$	$K = \frac{[NO_2]^2}{[N_2O_4]}$	8.3×10^{-1} at 55°C
$Ag^+ + 2NH_3 \rightleftharpoons Ag(NH_3)_2^+$	$K = \frac{[Ag(NH_3)_2^+]}{[Ag^+][NH_3]^2}$	1.7×10^7 at 25°C

Note. In the equilibrium constant expression, the coefficient becomes the power to which the concentration is raised. In general, $aA \rightleftharpoons bB$

$$K_{eq} = \frac{[B]^b}{[A]^a}$$

equilibrium is established. Consider the following representation of a general reaction:

$$A + B \rightleftharpoons C + D$$

A principle of reaction kinetics called the Law of Mass Action states that the rate of the forward reaction is proportional to the product of the concentrations of the reactants, and can be determined by the following equation:

$$\text{rate}_f = k_f \times [A] \times [B]$$

The notation $[A]$ indicates the concentration of reactant A in moles per liter, $[B]$ indicates the concentration of reactant B in moles per liter, and k_f is the forward rate constant. The rate of the reverse reaction can be similarly expressed:

$$\text{rate}_r = k_r \times [C] \times [D]$$

At equilibrium we will have $\text{rate}_f = \text{rate}_r$, so at equilibrium we can write the equation,

$$k_f \times [A] \times [B] = k_r \times [C] \times [D]$$

If we solve this equation for the two rate constants, we get the following:

$$\frac{k_f}{k_r} = \frac{[C] \times [D]}{[A] \times [B]} = \frac{[C][D]}{[A][B]}$$

If we now define a new constant, $K_{\text{eq}} = k_f/k_r$, at equilibrium we will have the relationship,

$$K_{\text{eq}} = \frac{[C][D]}{[A][B]}$$

We call K_{eq} the **equilibrium constant** for that reaction. The value of K_{eq} for any reaction will be a constant at a given temperature. If we vary the temperature of the system, we will change the value of K_{eq} (Table 10.1).

The numerical value of K_{eq} gives us valuable information about the reaction. If the value of $K_{eq} > 1$, then $[C] \times [D]$ will be greater than $[A] \times [B]$ at equilibrium. If K_{eq} is a very large number, this indicates that the forward reaction goes almost to completion before the equilibrium is established. If $K_{eq} < 1$, then $[A] \times [B]$ is greater than $[C] \times [D]$ at equilibrium, so the forward reaction does not go very far before equilibrium is established.

10.10 Altering the Equilibrium

Altering a chemical equilibrium is an important technique in industrial or biological chemistry, especially when it is desirable for a reaction to go to completion even though the equilibrium state may be reached very early in the reaction. Various factors affecting the rate of a chemical reaction can alter a chemical equilibrium. Let's review each of the factors that we learned would cause a change in the rate of chemical reactions.

Changes in Concentration

Changing the concentration of a reactant or product in a reaction that is at equilibrium can affect the rate of the forward or reverse reaction. Increasing the concentration of a reactant increases the number of collisions between reactant molecules, thereby increasing the rate of the forward reaction. When this occurs, the system is no longer at equilibrium.

$$\text{rate}_f > \text{rate}_r$$

The forward reaction will proceed at a higher rate, producing more product molecules until the rate of the reverse reaction again increases enough to equal the rate of the forward reaction. At this point a new equilibrium is established.

One colorful example of the effect on a chemical equilibrium of changing reactant concentrations can be seen in the equilibrium between the yellow chromate ion, CrO_4^{2-}, and the orange dichromate ion, $Cr_2O_7^{2-}$.

$$\underset{\text{Yellow}}{2CrO_4^{2-}{}_{(aq)}} + 2H^+{}_{(aq)} \underset{\text{Rate}_r}{\overset{\text{Rate}_f}{\rightleftharpoons}} \underset{\text{Orange}}{Cr_2O_7^{2-}{}_{(aq)}} + H_2O$$

Increasing the concentration of hydrogen ions on the reactant side of the equation will increase the number of collisions between the chromate ions and the hydrogen ions, thus causing an increase in the rate of the forward reaction. As more dichromate forms, the rate of the reverse reaction also increases until it again equals the rate of the forward reaction. At that point a new equilibrium will be established. This new equilibrium will have a higher concentration of dichromate ions and a lower concentration of chromate ions than the previous equilibrium. Therefore, the solution will appear more orange (Figure 10.14). It is a common procedure for chemists to add more reactant, or to remove a product in a reaction, in order to upset a chemical equilibrium and to drive a reaction toward completion.

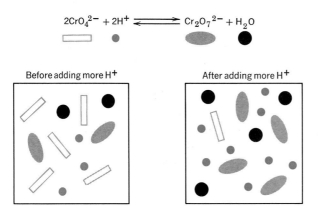

$$2CrO_4^{2-} + 2H^+ \rightleftharpoons Cr_2O_7^{2-} + H_2O$$

Before adding more H^+ After adding more H^+

Figure 10.14 A molecular view of the chromate-dichromate equilibrium.

In the plasma of the blood, an equilibrium exists between carbon dioxide and carbonic acid.

$$CO_2 \quad + \quad H_2O \quad \underset{Rate_r}{\overset{Rate_f}{\rightleftharpoons}} \quad H_2CO_3$$

Carbon dioxide Carbonic acid

In the tissues, carbon dioxide is a waste product that enters the blood. As the concentration of carbon dioxide in the blood increases, the above reaction is driven to the right (rate$_f$ > rate$_r$), and carbonic acid, which can be transported by the bloodstream, is formed. When the blood reaches the lungs, we exhale carbon dioxide. This lowers the concentration of carbon

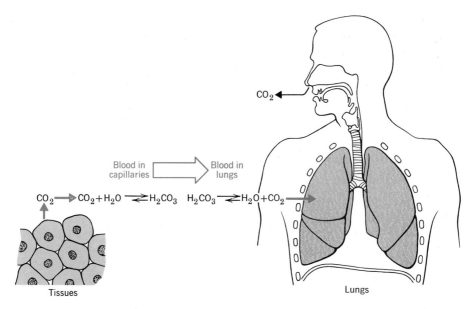

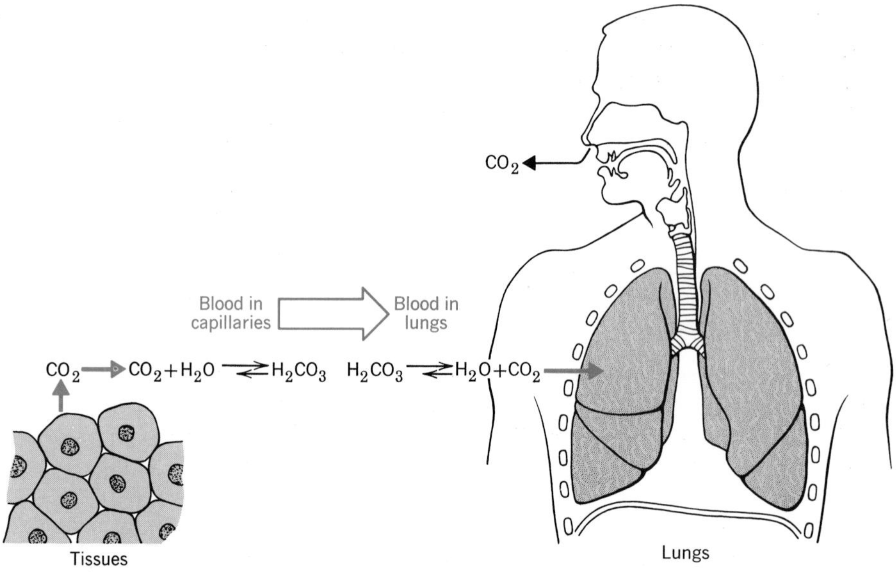

Figure 10.15 Changes in the levels of carbon dioxide in the body affect the carbon dioxide/carbonic acid equilibrium.

dioxide in the blood, driving the reaction to the left and allowing us to rid the bloodstream of the waste product carbonic acid (Figure 10.15).

Changes in Temperature

Changing the temperature of a system in equilibrium will alter the equilibrium. Increasing the temperature will favor the endothermic reaction (the reaction requiring energy), and lowering the temperature will favor the exothermic reaction (the reaction in which energy is released). For example, let's return to our iced tea example. The process of dissolving is an endothermic reaction. Therefore, increasing the temperature will increase the number of sugar molecules that will dissolve.

$$\text{Sugar}_{(s)} + \text{Energy} \underset{\text{rate}_r}{\overset{\text{rate}_f}{\rightleftharpoons}} \text{Sugar}_{(aq)}$$

Warming our tea will increase rate_f, because that is the endothermic reaction. Therefore, we will have $\text{rate}_f > \text{rate}_r$, and more sugar will be dissolving than is crystallizing. If we raise the temperature and hold it constant, rate_r will increase until it equals rate_f, thus establishing a new equilibrium with a greater number of sugar molecules dissolved in the tea (Figure 10.16). Obviously, if you like your tea sweet you should drink it warm!

Catalysts

We have discussed the fact that a catalyst will increase the rate of a chemical reaction, but you may be surprised to learn that a catalyst will have no effect on the equilibrium concentrations of the reactants and products in an equilibrium system. Why is this? Remember that the way a catalyst increases the rate of a chemical reaction is by lowering the activation energy of the reaction (Figure 10.11). At the same time, however, the activation energy for the reverse reaction will be lowered by an equal amount, so the rate of the reverse reaction will also be increased. A

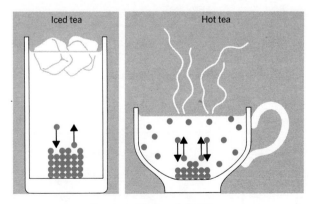

$$\text{Sugar}_{(s)} + \text{Heat} \rightleftharpoons \text{Sugar}_{(aq)}$$

Figure 10.16 Increasing the temperature of a system at equilibrium will increase the rate of the endothermic reaction. As the temperature of the iced tea is increased, more sugar will dissolve.

catalyst, then, increases the rate of both the forward and the reverse reactions by the same amount and, therefore, will have no effect on the equilibrium system.

10.11 Le Chatelier's Principle

We have seen that a chemical system is in equilibrium when the rates of the forward and reverse reactions are equal, and that changes made in the system can disrupt the equilibrium and thereby change the equilibrium concentrations of reactants and products. After studying a great number of equilibrium systems, the French chemist Le Chatelier was able to state a principle that helps to predict whether the reactants or products will be favored in a given change. **Le Chatelier's Principle states that when a stress is placed on a system in equilibrium, the system will change in a direction that will tend to remove the stress.**

When we talk of applying a stress to an equilibrium, we are referring to such actions as changing the concentrations of the reactants or products, changing the temperature of the system, or changing the pressure in a gaseous system. Le Chatelier's Principle tells us that if we increase the concentration of a reactant, the system will move in a direction to remove that increase; that is, $rate_f$ will be greater than $rate_r$. If we increase the temperature, the reaction which uses up that increase in energy will be favored; that is, the rate of the endothermic reaction will be greater than the rate of the exothermic reaction.

Example 10-2

Let's consider the effect of various stresses on the following reaction when it is at equilibrium.

$$2CO_{(g)} + O_{2(g)} \rightleftharpoons 2CO_{2(g)} + \text{Heat}$$

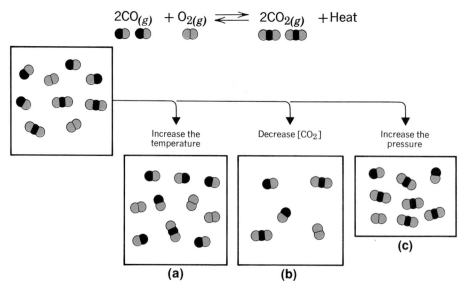

(a) *Increasing the temperature:* the endothermic reaction, which is the reverse reaction, would be favored. Therefore, $rate_r$ will be greater than $rate_f$.

(b) *Decreasing the concentration of CO_2:* Because the stress is a decrease in the amount of product CO_2, the forward reaction will be favored so as to counteract that change. Another way of reaching this conclusion is to realize that there will be fewer CO_2 molecules to react, so $rate_r$ will decrease.

(c) *Increasing the pressure:* The reaction that will be favored is the one that produces a system which exerts less pressure; that is, a system with fewer gas molecules. The forward reaction produces a system with two gas molecules, whereas the reverse reaction produces a system with three gas molecules. Therefore, to remove the stress, $rate_f$ will be greater than $rate_r$.

Le Chatelier's Principle can be very helpful in explaining the results of stresses on equilibrium systems in living organisms. For example, *calculus* is a word used in medicine to describe an abnormal build-up of ionic solids, called mineral salts, in a framework or matrix of other substances, such as proteins. Kidney stones, bladder stones, and gall stones are examples of calculi. The mineral salts found in kidney and bladder stones are formed from ions circulating in body fluids; the positive ions are calcium (Ca^{2+}) and magnesium (Mg^{2+}), and the negative ions are PO_4^{3-}, HPO_4^{2-}, or $H_2PO_4^-$. One of the equilibriums involved is

$$3Ca^{2+}{}_{(aq)} + 2PO_4^{3-}{}_{(aq)} \underset{rate_r}{\overset{rate_f}{\rightleftharpoons}} Ca_3(PO_4)_{2(s)}$$

When a person develops kidney stones, something has happened to disrupt the system that controls the amount of calcium and magnesium ions in body fluids. The concentration of calcium ions goes up, and this stress increases the rate of the forward reaction—the formation of the solid $Ca_3(PO_4)_2$. The stone will grow slowly as this solid, as well as other salts, are deposited in the matrix.

As another example, aspirin, which is taken by millions of people for relief of headaches, is a very safe drug. In some people, however, aspirin can disrupt the normal functioning of the stomach lining and can cause bleeding. The manner in which aspirin enters the stomach lining involves various stresses on the equilibrium which exists between the neutral aspirin molecule and the aspirin ion that is formed when the aspirin molecule is dissolved in water.

$$\underset{\text{(nonpolar)}}{\text{Aspirin}} \underset{Rate_r}{\overset{Rate_f}{\rightleftharpoons}} \underset{\text{(polar)}}{\text{Aspirin}^- + H^+}$$

When aspirin is dissolved in water, an equilibrium is formed between the polar and nonpolar forms of this compound. Therefore, when you swallow

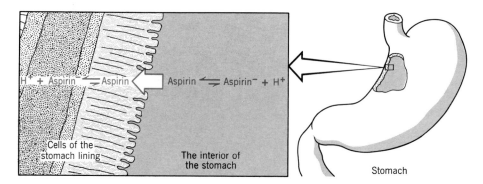

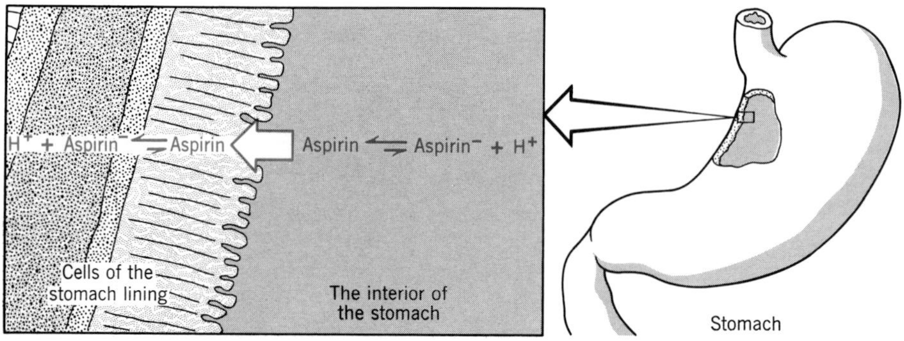

Figure 10.17 The passage of aspirin molecules through the stomach wall is accelerated by the stress that is applied to the aspirin/aspirin ion equilibrium by the hydrogen ions in stomach fluid.

an aspirin with some water, a significant amount of the aspirin will be found in the polar ion form, which cannot pass through the protective lining of the stomach. But the contents of the stomach are very high in hydrogen ions, H^+. This puts a stress on the equilibrium, increasing the rate of the reverse reaction and forming more nonpolar aspirin molecules, which can pass through the protective lining. Once these aspirin molecules have passed into the cells of the stomach, the concentration of hydrogen ions becomes very low, and the equilibrium shifts to the right. This again produces the polar aspirin ions, which cannot pass back into the stomach and, therefore, become trapped in the cells, where they produce the damage that causes the bleeding (Figure 10.17).

Chapter Summary

For a chemical reaction to occur, the reactants must collide with enough energy to overcome an energy barrier called the activation energy, E_{act}. When this occurs, the particles first form a reactive group called the activated complex, a transition state between reactants and products, and then go on to form the products. A reaction that gives off energy is called an exothermic reaction, and one that absorbs energy is called an endothermic reaction. The amount of energy given off or absorbed by a reaction is called the heat of reaction, ΔH. The following factors will all affect the rate at which a reaction occurs: the nature of the reactants, the concentration of the reactants, the temperature at which the reaction takes place, and the addition of a catalyst. A chemical reaction is in a state of equilibrium when the rate of the forward reaction equals the rate of the reverse reaction. According to Le Chatelier's Principle, when a system at equilibrium is put under stress (by changes in the concentration of reactants or products, or by changes in the temperature), the system will change in a direction that will tend to remove the stress.

Exercises and Problems

1. Using the concept of activation energy, explain why a dropped glass might remain intact on one occasion and shatter on another.
2. Draw potential energy diagrams for a general endothermic and exothermic reaction. Label the reactants, the products, the activation energy E_{act}, and the heat of reaction ΔH.
3. Which of the following reactions are endothermic and which are exothermic?

   (a) $2H_{2(g)} + O_{2(g)} \longrightarrow 2H_2O_{(g)} + 115.6$ kcal
   (b) $N_{2(g)} + 2O_{2(g)} + 16.2$ kcal $\longrightarrow 2NO_{2(g)}$
   (c) $2NH_{3(g)} + 22$ kcal $\longrightarrow N_{2(g)} + 3H_{2(g)}$
   (d) $3C_{(s)} + 2Fe_2O_{3(s)} + 110.8$ kcal $\longrightarrow 4Fe_{(s)} + 3CO_{2(g)}$
4. What is a catalyst?
5. Explain why special industrial procedures must be designed for handling large amounts of finely divided, dry, combustible materials.
6. Before dinner, you are cutting up peaches for dessert. What can you do to slow the browning of the peach slices?
7. Why do labels of certain antibiotic drugs state that the drugs must be kept under refrigeration?
8. Many animals, such as snakes, are cold-blooded; that is, their body temperature is the same as the temperature of their surroundings. Using your knowledge of reaction rates, explain why snakes are very sluggish on cold mornings, and often sun themselves on rocks before searching for food.
9. What is *equal* in a reaction at equilibrium?
10. Why is a chemical equilibrium described as dynamic rather than static?
11. Is an equilibrium established in a can of ether

    (a) with the lid off;
    (b) with the lid on? Give a reason for your answers.
12. What is the effect of a catalyst on a reaction at equilibrium?
13. State Le Chatelier's Principle in your own words.
14. State three ways to increase the concentration of NOCl in the following reaction

$$2NO_{(g)} + Cl_{2(g)} \rightleftharpoons 2NOCl_{(g)}$$

15. The heat of reaction for the conversion of graphite to diamond is quite small, and yet this reaction will take place only under extremely

high temperatures and pressure. Why is this? Draw a potential energy diagram for the reaction

$$0.45 \text{ kcal} + C_{(graphite)} \longrightarrow C_{(diamond)}$$

16. (a) Compare the difference in reaction rates for the following reaction if the carbon is supplied (1) in chunks and (2) as a finely divided powder.

$$C_{(s)} + O_{2(g)} \longrightarrow CO_{2(g)}$$

(b) Draw the potential energy diagrams for the reactions in (1) and (2).

17. The following is the potential energy curve for the reaction

$$CO + NO_2 \longrightarrow CO_2 + NO$$

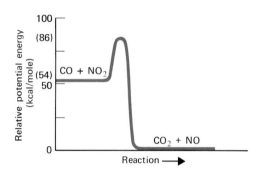

(a) Is this reaction endothermic or exothermic?
(b) What is the heat of reaction, ΔH?
(c) What is the activation energy E_{act} for the forward reaction?
(d) What is the activation energy E_{act} for the reverse reaction,

$$CO_2 + NO \longrightarrow CO + NO_2$$

18. An important industrial process is the production of ammonia through the following reaction.

$$N_{2(g)} + 3H_{2(g)} \rightleftharpoons 2NH_{3(g)} + 22 \text{ kcal}$$

This process uses a catalyst.

(a) Draw the potential energy diagram for the uncatalyzed and the catalyzed reaction.
(b) Describe three steps that manufacturers could take to maximize the yield of ammonia in this reaction.

19. What effect would the following changes have on the equilibrium concentration of $O_{2(g)}$ in the following reaction? Give a reason for each of your answers.

$$4HCl_{(g)} + O_{2(g)} \rightleftharpoons 2H_2O_{(g)} + 2Cl_{2(g)} + 27 \text{ kcal}$$

(a) Increasing the temperature of the reaction.
(b) Increasing the pressure.
(c) Decreasing the concentration of Cl_2.
(d) Increasing the concentration of HCl.
(e) Adding a catalyst.

20. Describe how each of the following changes would affect the equilibrium concentration of $H_{2(g)}$ in the following reaction:

$$H_{2(g)} + CO_{2(g)} + 9.9 \text{ kcal} \rightleftharpoons H_2O_{(g)} + CO_{(g)}$$

(a) Addition of H_2O
(b) Addition of CO_2
(c) Removal of CO
(d) Increasing the temperature of the reaction
(e) Decreasing the volume of the reaction container.

*21. Equilibrium constants are given for the following three reactions. In each case state whether there are more reactants or products at equilibrium.

(a) $CH_3COOH_{(aq)} \rightleftharpoons H^+_{(aq)} + CH_3COO^-_{(aq)}$ $\quad K_{eq} = 1.8 \times 10^{-5}$
(b) $CdS_{(s)} \rightleftharpoons Cd^{2+}_{(aq)} + S^{2-}_{(aq)}$ $\quad K_{eq} = 7.1 \times 10^{-28}$
(c) $H^+_{(aq)} + HS^-_{(aq)} \rightleftharpoons H_2S_{(aq)}$ $\quad K_{eq} = 1 \times 10^{7}$

* *Optional.*

chapter 11

Water and Solutions

Learning Objectives

By the time you have finished this chapter, you should be able to:

1. List three special properties of water and explain each of them in terms of the hydrogen bonding in water.

2. Describe in your own words the process by which sodium chloride dissolves in water.

3. State the difference between a strong and weak electrolyte.

4. Predict whether a precipitate will form when two ion-containing solutions are combined.

5. Explain how to prepare a solution of a given percentage or a given molarity of solute.

6. Given an ion, state its equivalent weight and calculate the number of milliequivalents of that ion in a given percentage solution.

7. Define *osmosis, hypertonic, isotonic,* and *hypotonic.*

8. Compare true solutions with colloids, and state three properties that are unique to colloids.

9. Describe a laboratory procedure for separating cellular colloids such as proteins from the crystalloid contents of the cell.

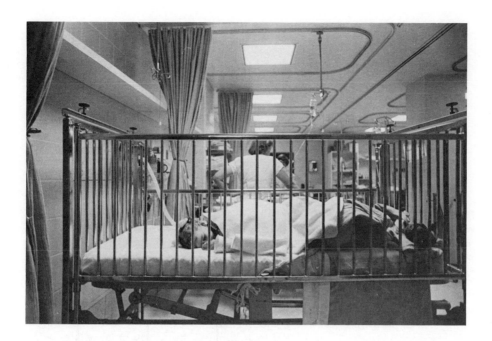

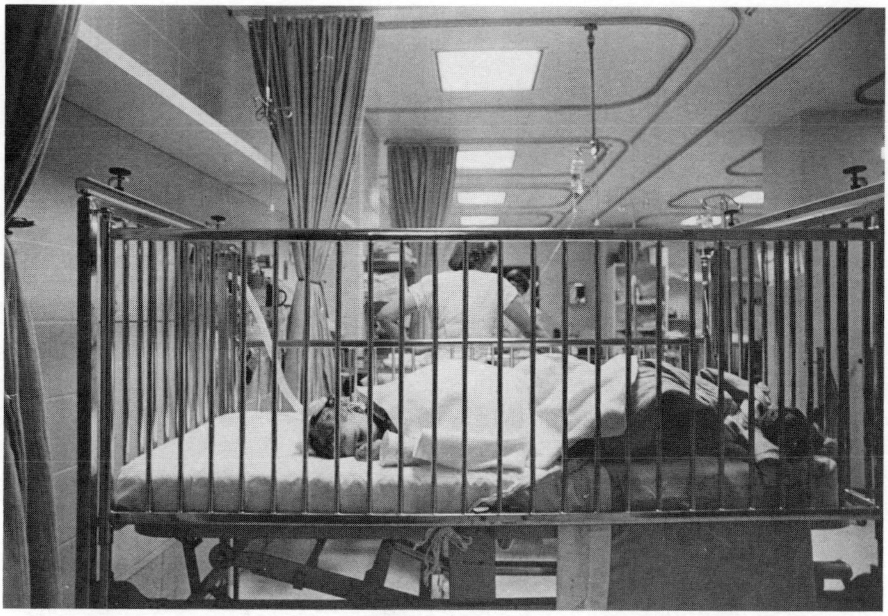

One afternoon in April, seven-year-old Betsy was sent home from school with the flu. She had vomited at school and she continued to vomit and have diarrhea for the next 12 hours, but then quickly improved. Her mother hoped that Betsy's 12-month-old sister Sarah wouldn't also catch the flu. However, on the third morning that Betsy was home, Sarah began vomiting and soon started showing signs of diarrhea. When the vomiting stopped, her mother gave Sarah one or two ounces of water every 15 minutes or so. The diarrhea continued for the next 24 hours and Sarah's mother became concerned as she noticed that Sarah continued to have a fever, that her mouth seemed very dry, and that her diapers had been dry for an entire day. She decided to call the pediatrician who, after hearing Sarah's symptoms, asked that Sarah be brought to the emergency room of the hospital. The pediatrician examined Sarah and found she had no saliva in her mouth, her eyeballs were soft and sunken, and her skin did not go back into place when stretched. Sarah's temperature was 101°F (38.3°C), and she had lost 2 pounds since her routine exam two weeks before. The pediatrician estimated that Sarah was suffering from 5 to 10% dehydration (or water loss). An immediate treatment of intravenous fluids was begun, and several laboratory tests were conducted to determine the degree of dehydration. The following chart shows the results of these tests.

Test	Value for Sarah	Normal infant value
1. Serum Na^+	128 mEq/*l*	138–142 mEq/*l*
2. Serum K^+	3.0 mEq/*l*	4.5–5.5 mEq/*l*
3. Serum Cl^-	94 mEq/*l*	100–105 mEq/*l*
4. Arterial pH	7.48	7.35–7.4
5. Serum HCO_3^-	42 mEq/*l*	20–30 mEq/*l*
6. Urine specific gravity	1.036	1.010–1.020
7. Blood urea nitrogen	32 mg%	10–20 mg%

Tests 1 and 2 indicate a low concentration of the important ions of potassium and sodium, which were lost from the body with the loss of water. These ions were restored by being added to the intravenous fluid that Sarah received. Tests 3 through 5 indicate that Sarah was suffering from alkalosis due to the loss of hydrochloric acid from vomiting. The correct balance of acids and bases will be restored as the body fluids again reach normal levels. Test 6 reveals a high specific gravity of the urine, showing that her urine was very concentrated from lack of water. The high level of urea nitrogen in the blood revealed by test 7 shows that there was not enough water for the kidneys to cleanse her body of its waste products.

Sarah remained in the hospital for three days receiving intravenous fluids until all seven tests were within the normal range, and until she was able to eat and drink enough on her own to prevent the recurrence of the dehydration.

The water loss or dehydration that Sarah suffered can, if untreated, have serious consequences. Water is critical to the survival of all living organisms and is second only to oxygen in importance for human survival. We can survive for several weeks without food, for a few days without water, and only minutes without oxygen. The water content of living organisms varies from less than 50% in some bacterial cells, to 96 and 97% in some marine invertebrates. Water is the most abundant chemical compound in the human body, making up 60 to 70% of the adult body weight. This water is found within the cells, in the extracellular fluid that bathes the cells, and in the blood plasma (Figure 11.1).

Water performs many biological functions; it is the fluid found throughout living organisms, transporting food and oxygen to the cells, and carrying away wastes. It is the fluid in which digestion takes place, and is a lubricant for the cells and tissues. Water plays a major role in the regulation of body temperature, and in the acid-base balance of body fluids and cells. Water is a reactant and product in many chemical reactions that take place in living organisms.

Every day you lose between 1500 and 3000 ml of water in the form of urine, perspiration, water vapor in your breath, and feces. This water must be replaced by the liquids you drink or foods that you eat. Diarrhea, high

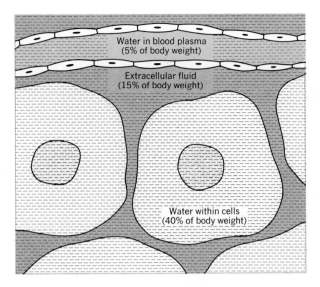

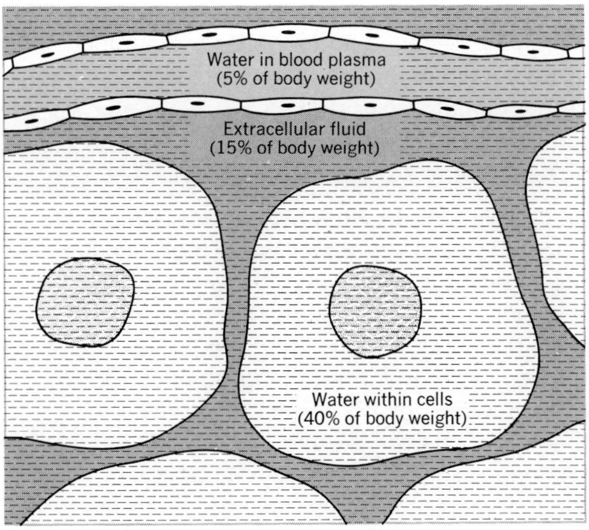

Figure 11.1 Water is the most abundant compound in the human body. It is distributed within the cells, in the extracellular fluid that bathes the cells, and in the blood plasma.

fever, bleeding, burns, or ulcers can disrupt the normal water balance in the body, causing dehydration. For adults, a 10% loss in total body fluids causes serious dehydration and disruption of normal body chemistry; a loss of 20% can be fatal. To understand the chemistry of living organisms, therefore, we need to examine the properties of water that make it so essential for life.

Properties of Water

11.1 Solvent Properties

Water is the most abundant liquid in the world, yet knowledge of its structure is far from complete. It is the special properties of water that make it so vital to life. Water, H_2O, is a molecule containing an oxygen atom covalently bonded to two hydrogen atoms. The shape of the molecule is bent.

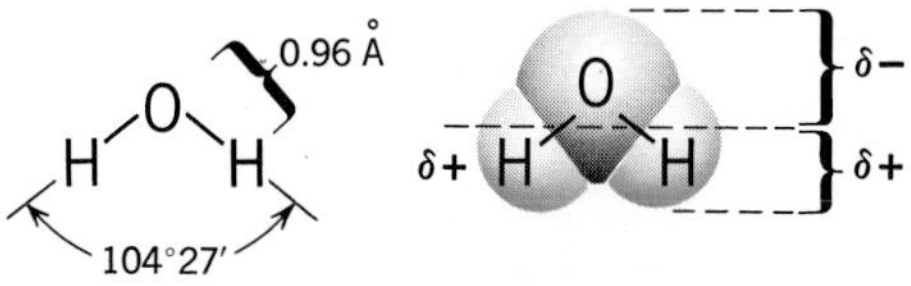

The oxygen atom is much more electronegative than the hydrogen atom, and each covalent bond is polar. In addition, the bent shape of the molecule makes the entire molecule polar, with the oxygen at the negative end and the hydrogens at the positive end. The polarity of water allows the formation of hydrogen bonds, and gives this molecule its solvent properties. Water is called the *universal solvent* because it is a better solvent than most other liquids. It will dissolve ionic compounds and molecular compounds that are polar or that contain polar groups.

11.2 High Melting and Boiling Points

Other properties of water can be explained by the hydrogen bonding that exists between water molecules (Figure 8.13). Water has a high melting point and a high boiling point when compared with other molecules of comparable formula weight. In fact, most compounds of comparable formula weight are gases at the temperature that water is a liquid. For example, water has a formula weight of 18 and a boiling point of 100°C, whereas ammonia (with a formula weight of 17) and methane (with a formula weight of 16) have boiling points of −33 and −164°C, respectively. The high melting point and boiling point result from the attraction of hydrogen bonding between the molecules of water, thus requiring more energy to pull these molecules apart.

11.3 Density of Ice

You will recall that the density of a substance is a measure of the substance's mass per unit volume (density = mass/volume). When heated, nearly all substances will increase in volume. Because the mass of the substance remains constant, its density will therefore decrease as the temperature increases (Figure 11.2a). As water cools, it contracts until it reaches its maximum density at 3.98°C. As it continues to cool, however, and then freeze at 0°C, it expands by 9% (Figure 11.2b). It is the open lattice structure of ice, in which each molecule of water is hydrogen bonded to four other water molecules, which makes ice less dense than the more compact liquid water. This is why pipes break when the water in them freezes. The fact that ice is less dense than water is critical to aquatic life.

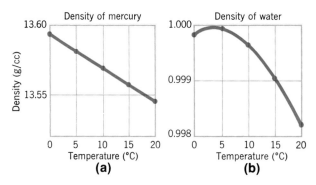

Figure 11.2 The density of most substances, such as mercury, gradually increases as the temperature decreases. The density of water, however, increases to a maximum at 4°C and then gradually decreases. As a result, ice is less dense than water.

When the surface of a lake freezes, the ice floats and the most dense water (at 4°C) sinks to the bottom, giving plants and animals a place to survive. If ice were more dense than water, lakes and oceans would freeze from the bottom up, and would probably never completely thaw in summer.

11.4 Surface Tension

We mentioned in Chapter 4 that water has a high surface tension. Water molecules beneath the surface are strongly attracted in all directions to other water molecules by hydrogen bonding. But the molecules on the surface are not attracted to the nonpolar air molecules. Therefore, rather than being attracted in all directions, these surface molecules are attracted only downward and inward toward the water. This surface tension pulls the surface molecules together, creating the effect of a thin, elastic membrane on the surface. This is why, for example, dust particles or water spiders can remain on the water's surface even though they are more dense than water. Also, you have probably noticed that water, when dropped onto waxed paper, forms small beads of liquid. This occurs because the surface water molecules are attracted neither to the nonpolar wax nor to the surrounding air, and thus the attraction between water molecules draws the liquid up to form a bead. Glass, on the other hand, is more polar than water. Thus, water is attracted to glass and, when placed on a glass surface, will spread out rather than form beads. Substances that reduce the surface tension of water are called **surface active agents,** or **surfactants.** In order for water to penetrate the dirt and oil in clothes in the laundry, or to aid in the digestion of fats, surfactants such as soaps, detergents, or bile must be present.

11.5 High Heat of Vaporization

Water has a high heat of vaporization (539 cal/g). The strong attractive forces between water molecules cause this liquid to boil away very slowly. It takes nearly seven times as much energy to boil away one liter of water at 100°C as it does to heat it from 21°C (room temperature) to 100°C. It is important to realize that evaporation is essentially the same process as boiling; it involves changing liquid water to gas, only at a slower rate. As with boiling, evaporation uses up a great deal of energy or heat. Animals make use of this principle to rid their bodies of excess heat through the evaporation of sweat.

11.6 High Heat of Fusion

Water also has a high heat of fusion (80 cal/g). That is, a great deal of heat must be released before ice can be formed. To convert one kilogram of water at 0°C to one kilogram of ice requires almost four times as much refrigeration as is needed to cool this much water from 21 to 0°C. This explains why ice forms slowly in the winter and melts slowly in the spring. Water stored as snow in the mountains will, therefore, run off relatively slowly, preventing disastrous floods in the spring.

Table 11.1 Specific Heat of Several Compounds (cal/g)

Compound	Specific Heat	Compound	Specific Heat
Water	1.0	Chloroform	0.23
Ethanol	0.58	Ethyl acetate	0.46
Methanol	0.60	Liquid ammonia	1.12
Acetone	0.53		

11.7 High Specific Heat

The specific heat of a liquid is the amount of heat required to raise the temperature of one gram of that liquid from 15 to 16°C. Water has a high specific heat compared with other liquids (Table 11.1). The higher the specific heat of a substance, the less its temperature will change when it absorbs a given amount of heat. The high specific heat of water enables this fluid to keep the temperature of an organism relatively constant in the face of fluctuating internal or external heat levels. On a larger scale, the water in lakes and oceans will absorb and store large quantities of solar energy, explaining why such large bodies of water have moderating effects on local climates.

Solutions

11.8 True Solutions

A **true solution** is a homogeneous, or uniform, mixture of two or more substances whose particles are of atomic or molecular size. Club soda (carbon dioxide in water), air (gas in gas), rubbing alcohol (alcohol in water), and steel (carbon in iron) are examples of solutions. We call the substance being dissolved the **solute,** whereas the **solvent** is the substance in which the solute is being dissolved. Our major interest in this text is **aqueous** solutions, which are solutions in which water is the solvent.

11.9 Aqueous Solutions

Water is often called the universal solvent because its highly polar nature allows a large number of substances to dissolve in it. When a sodium chloride crystal is placed in water, the surface of the crystal is bombarded by the water molecules, whose polar ends exert an attractive force on the surface ions of the crystal and pull them loose (Figure 11.3). As each ion is pulled loose, it is surrounded by water molecules. This action prevents the sodium and chloride ions from rejoining. Such an ion surrounded by water molecules is said to be **hydrated.** Note, however, that not all ionic substances dissolve in water. For some crystals, the attraction of the water molecules may not be strong enough to overcome the attraction between the oppositely charged ions within the crystal.

Highly polar molecular substances, such as sugar, are also soluble in

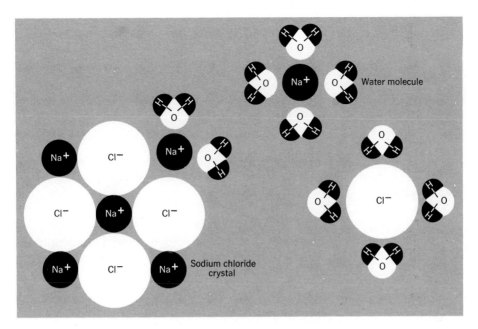

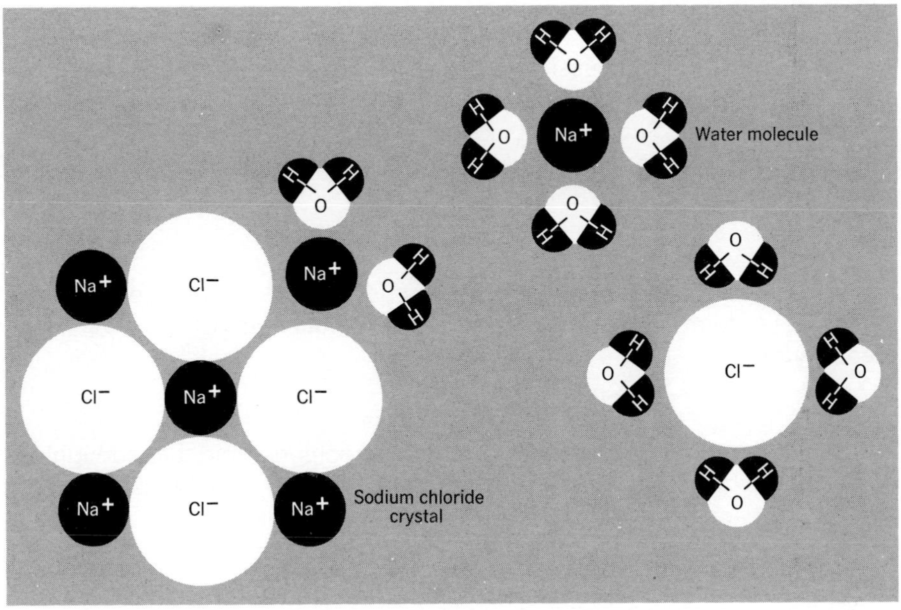

Figure 11.3 The ionic compound sodium chloride will dissolve in water because the attraction of the water molecules for the sodium and chloride ions is greater than the attraction between these ions in the crystal.

water. The hydration of sugar molecules occurs because of the attraction and hydrogen bonding between the oppositely charged polar regions on the sugar and water molecules (Figure 11.4).

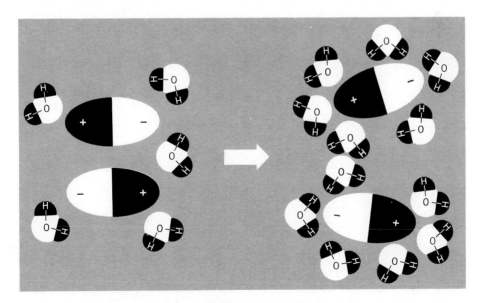

Figure 11.4 Highly polar molecular substances are soluble in water, especially if they contain atoms such as oxygen or nitrogen that can form hydrogen bonds with the water molecules.

11.10 Electrolytes and Nonelectrolytes

We can see from our discussion of the dissolving process that solute particles take two forms: they may be charged ionic particles, or uncharged molecular particles. Substances such as sugar, which form uncharged molecular particles when dissolved in a solvent, are called **nonelectrolytes.**

$$\text{Sugar}_{(s)} + H_2O \longrightarrow \text{Sugar}_{(aq)}$$

Ionic compounds such as sodium chloride, which dissolve (or dissociate) to form charged ionic particles in solution are called **electrolytes.**

$$NaCl_{(s)} + H_2O \longrightarrow Na^+{}_{(aq)} + Cl^-{}_{(aq)}$$

Some molecular substances, such as hydrochloric acid, are so highly polar that the molecule will be pulled apart or ionized by water molecules. These substances will also form electrolytes in solution.

$$HCl_{(g)} + H_2O \longrightarrow H_3O^+ + Cl^-{}_{(aq)}$$

The term *electrolyte* refers to the fact that solutions of electrolytes conduct electricity. Electrolytes may be classified as either strong or weak, depending upon the number of charged ionic particles formed when the substance dissolves. A **strong electrolyte** is a substance that completely ionizes, or dissociates into ions, in solution. A **weak electrolyte** is a substance that only partially ionizes in solution (Figure 11.5).

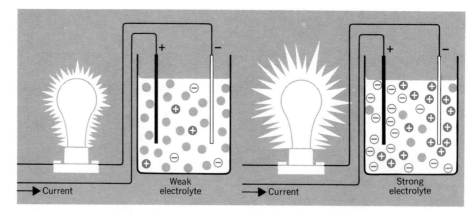

Figure 11.5 Solutions of electrolytes will conduct electricity. The stronger the electrolyte, the better the solution conducts electricity.

Electrolytes perform many important regulatory roles in our bodies, and are responsible for maintenance of the acid-base and water balance. The major positive ions, or **cations,** found in living tissue are Na^+, K^+, Ca^{2+}, and Mg^{2+}. The major negative ions, or **anions,** are HCO_3^-, Cl^-, HPO_4^{2-}, SO_4^{2-}, organic acids, and proteins (Figure 11.6).

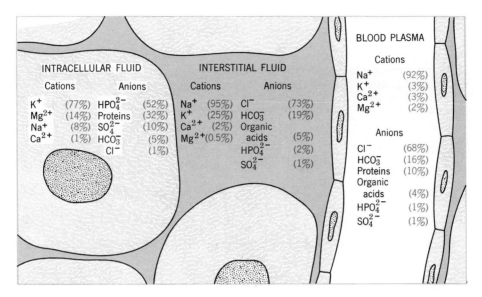

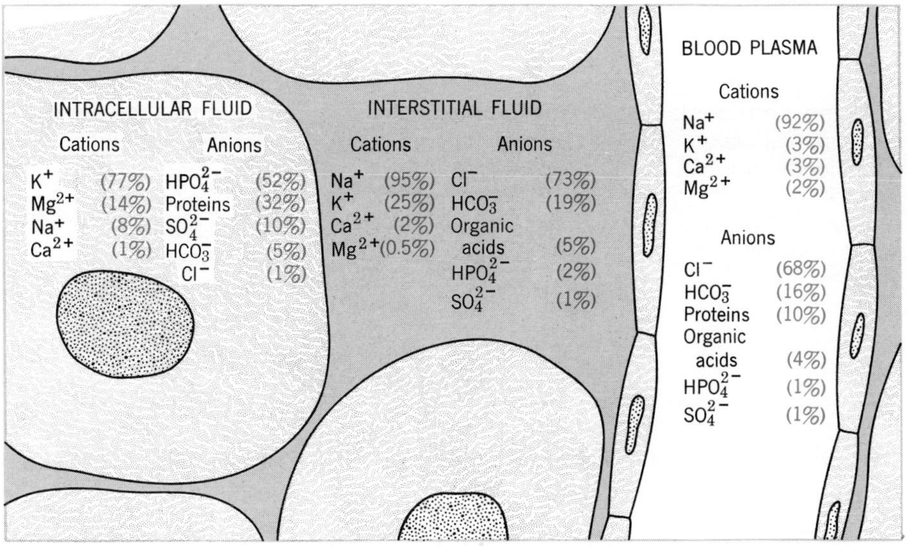

Figure 11.6 The principal electrolytes found in body fluids.

11.11 Factors Affecting the Solubility of a Solute

Many factors affect the solubility of a solute. A major factor is the nature of the solvent and the solute; polar solvents will dissolve polar solutes, and nonpolar solvents will dissolve nonpolar solutes. For most solid substances, increasing the temperature will increase the solubility of the solute. However, there are a few compounds, such as $CaCr_2O_7$, $CaSO_4$, and $Ca(OH)_2$, whose solubility decreases with an increase in temperature.

The solubility of a gas in a liquid decreases with an increase in temperature. For example, we have discussed how the warming of river water caused by heat discharged from a nuclear power plant decreases the supply of oxygen available for the river's fish. Finally, as mentioned in Chapter 4, the solubility of a gas in a liquid increases with an increase in pressure (Henry's law).

11.12 The Solubility of Ionic Solids

We have said that ionic solids vary in their solubility in water. Although the exact solubility of each compound must be determined by experiment, it is possible to make a few general statements about the solubility or insolubility of various solids.

Using Table 11.2 we can predict whether an insoluble solid, called a **precipitate,** will form when two solutions of soluble ionic compounds are mixed. For example, will a precipitate form when solutions of sodium chloride (NaCl) and silver nitrate ($AgNO_3$) are mixed? A solution of sodium chloride contains sodium ions (Na^+) and chloride ions (Cl^-), and a solution of silver nitrate contains silver ions (Ag^+) and nitrate ions (NO_3^-). The two new substances that could be formed from this combination are AgCl and $NaNO_3$. We see from Table 11.2 that $NaNO_3$ is soluble, but AgCl is not

Table 11.2 Some General Rules for the Solubility of Ionic Compounds

I	II	III
Compounds containing these cations are generally soluble:	Compounds containing these anions are generally soluble:	All ionic compounds formed from ions not listed in I or II are generally not soluble or, at most, only slightly soluble.
Li^+ Na^+ K^+ NH_4^+	NO_3^-, ClO_3^-, CH_3COO^-	
	Cl^-, Br^-, and I^- (except with Ag^+, Pb^{2+}, and Hg_2^{2+})	
	SO_4^{2-} (except with Pb^{2+}, Sr^{2+}, and Ba^{2+})	

soluble. Thus, we would expect a precipitate of AgCl to form. We can write these results in equation form:

$$\cancel{Na^+}_{(aq)} + Cl^-_{(aq)} + Ag^+_{(aq)} + \cancel{NO_3^-}_{(aq)} \longrightarrow AgCl_{(s)} + \cancel{Na^+}_{(aq)} + \cancel{NO_3^-}_{(aq)}$$

Cancelling the ions that appear on both sides of the equation, but that were not involved in the reaction, we get

$$Ag^+_{(aq)} + Cl^-_{(aq)} \longrightarrow AgCl_{(s)}$$

This equation is called the **net-ionic equation,** because it shows only those ions that actually react.

Concentration

11.13 Saturated Solutions

We may say that a solution is **dilute** if there are only a few solute particles dissolved in it, or **concentrated** if there are many solute particles dissolved. But the terms *dilute* and *concentrated* are not particularly precise, so various methods have been developed to more accurately describe the concentration of a solute in solution. A **saturated** solution is one that contains all the solute particles the solvent can normally hold at that temperature (if it contained fewer particles the solution would be **unsaturated**). A state of equilibrium will exist, in a saturated solution, between the dissolved solute and any undissolved solute in the container (Figure 11.7).

$$\text{Undissolved solute} \underset{\text{rate}_r}{\overset{\text{rate}_f}{\rightleftharpoons}} \text{Dissolved solute}$$

When a saturated solution is cooled, the solubility of the solute decreases. The rate of the crystallizing reaction (rate_r) will be greater than the rate of the dissolving reaction (rate_f). By Le Chatelier's Principle, the solute will continue to form crystals until a new equilibrium is established with fewer solute particles in solution. Occasionally, however, the solution will initially contain no crystals on which the dissolved particles can deposit. In such a case, when the temperature of the solution is lowered,

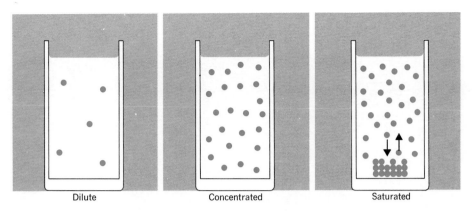

Figure 11.7 A dilute solution has very few solute particles. A concentrated solution has many solute particles. A saturated solution has all the solute particles the solvent can hold at that temperature.

no crystals will form and the solution is then said to be **supersaturated.** A supersaturated solution, however, is very unstable. If a small crystal is added, onto which the solute can deposit, a very rapid formation of crystals will often occur (Figure 11.8). Honey and jellies are often supersaturated solutions which, after long storage, may be found to contain sugar crystals.

11.14 Weight/Volume Percent

It is often necessary to be more precise about the concentration of a solution than to simply say that it is concentrated or dilute. One way to precisely define the concentration of a solute in a solution is by **weight/volume percent.** The concentration of a solution in weight/volume percent is defined to be the number of grams of solute in 100 milliliters of solution.

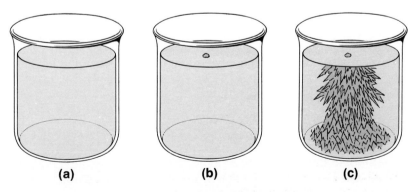

Figure 11.8 A supersaturated solution is very unstable (a). When a small crystal (called a seed crystal) is added to the solution (b), solute crystals will rapidly form on the seed crystal (c). (From Brady and Humisten, *General Chemistry*, copyright © 1976, John Wiley, New York.)

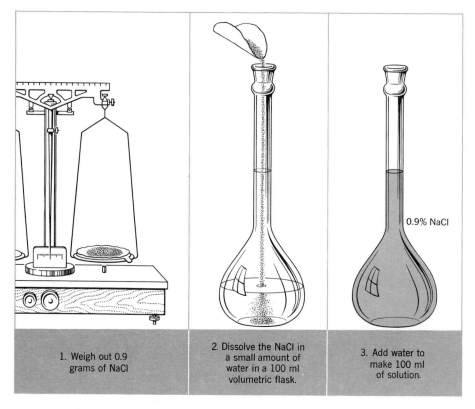

1. Weigh out 0.9 grams of NaCl

2. Dissolve the NaCl in a small amount of water in a 100 ml volumetric flask.

3. Add water to make 100 ml of solution.

0.9% NaCl

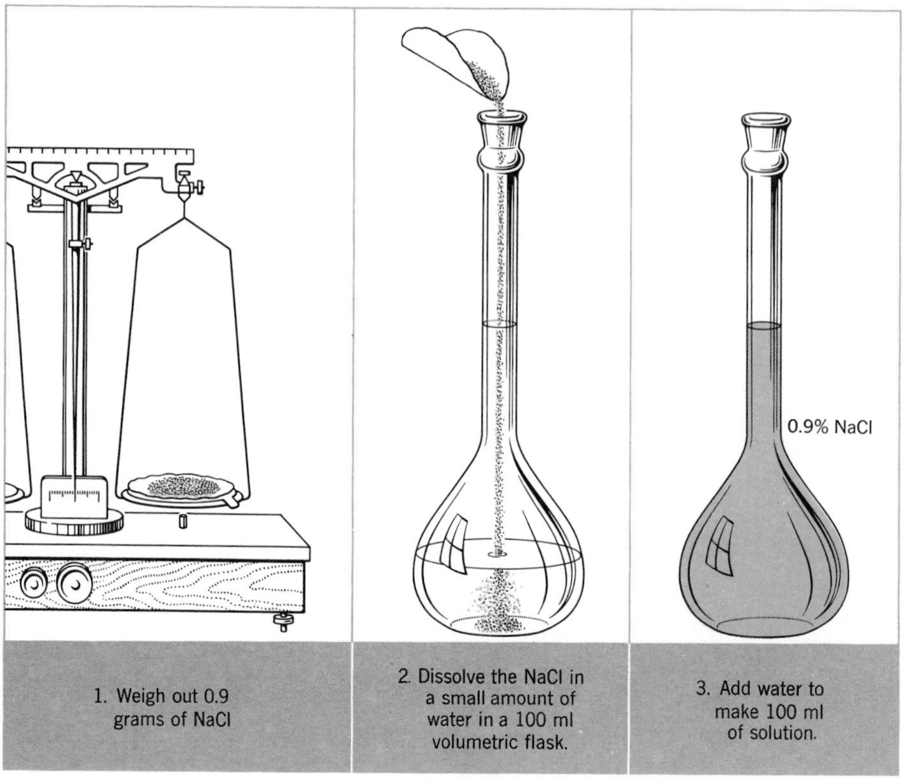

Figure 11.9 The procedure for making 100 ml of physiological saline (0.9% NaCl).

$$\text{Weight/Volume Percent} = \frac{\text{g of solute}}{100 \text{ ml of solution}} \times 100\%$$

Note that weight/volume percent solutions do not take into account the differences in formula weights of the solutes.

Example 11-1 _____

1. Intravenous solutions often contain normal (or physiological) saline, which is a 0.9% NaCl solution. How would you prepare 100 ml of normal saline? (See Figure 11.9.)

 0.9% NaCl will contain $\dfrac{0.9 \text{ g NaCl}}{100 \text{ ml}}$

 To prepare the solution, you would add 0.9 g of NaCl to enough water to make 100 ml of solution. (*Note:* it is the total volume of the solution that must equal 100 ml, not the volume of the water added.)

2. How many grams of glucose are there in 50 ml of a 10% glucose solution?

$$10\% \text{ glucose solution} = \frac{10 \text{ g glucose}}{100 \text{ ml}}$$

$$50 \text{ ml would contain } 50 \text{ ml} \times \frac{10 \text{ g glucose}}{100 \text{ ml}} = 5 \text{ g glucose.}$$

11.15 Molarity

A second way to describe the concentration of a solute—one that does take formula weights into account—is by means of molarity. **The molarity of a solution is defined to be the number of moles of solute in a liter of solution.** A one molar, or 1M solution, will contain one mole of solute in enough water to make a liter of solution.

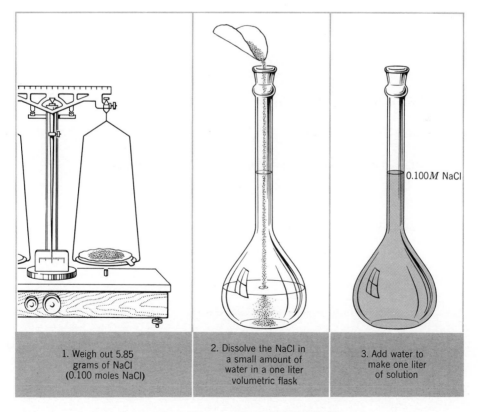

0.100M NaCl

1. Weigh out 5.85 grams of NaCl (0.100 moles NaCl)

2. Dissolve the NaCl in a small amount of water in a one liter volumetric flask

3. Add water to make one liter of solution

Figure 11.10 The procedure for making one liter of 0.100M NaCl.

Example 11-2 _____

1. How would you make a 0.100M NaCl solution?

 (a) First, you would weigh out 0.100 mole of NaCl.

 Formula weight of NaCl $= 23 + 35.5 = 58.5$
 1 mole of NaCl $= 58.5$ grams

 $$0.100 \text{ mole NaCl} \times \frac{58.5 \text{ grams}}{1 \text{ mole NaCl}} = 5.85 \text{ grams}$$

 (b) Second, you would dissolve 5.85 grams of NaCl in a small amount of water, and then add enough water to make one liter of solution (again note that it is the total volume of the solution that is measured, not the volume of water added) (Figure 11.10).

2. To carry out a particular reaction in the laboratory, you need 1 mole of hydrochloric acid dissolved in water. The only solution available to you is a 2M HCl solution. How many milliliters of this stock solution should you use?

 (a) 2M HCl means $\dfrac{2 \text{ moles HCl}}{1 \text{ liter}}$ or $\dfrac{2 \text{ moles HCl}}{1000 \text{ ml}}$.

 (b) The problem asks how many milliliters contain 1 mole HCl

 $$1 \text{ mole HCl} \times \frac{1000 \text{ ml}}{2 \text{ moles HCl}} = \tfrac{1}{2} \times 1000 \text{ ml} = 500 \text{ ml}$$

3. What is the molarity of a 5.0% NaOH solution?

 (a) 5.0% NaOH means $\dfrac{5.0 \text{ g NaOH}}{100 \text{ ml}}$

 (b) To determine the molarity, we need to determine
 $$\frac{\text{number of moles}}{1000 \text{ ml}}$$

 (c) 1000 ml of 5.0% NaOH will contain

 $$1000 \text{ ml} \times \frac{5.0 \text{ g NaOH}}{100 \text{ ml}} = 50 \text{ g NaOH}$$

 (d) How many moles is 50 g NaOH?

 The formula weight of NaOH is 40 g

 $$50 \text{ g} \times \frac{1 \text{ mole NaOH}}{40 \text{ g}} = 1.25 \text{ moles NaOH}$$

 (e) So we have $\dfrac{1.25 \text{ moles of NaOH}}{1000 \text{ ml}} = 1.25M \text{ NaOH}$

11.16 Parts per Million and Parts per Billion

When a solution is extremely dilute, the concentration units of parts per million (ppm) or parts per billion (ppb) may be used. One part per million is equivalent to 1 mg of solute per 1 liter of solution. One part per billion is 1 μg of solute per 1 liter of solution. It is hard to imagine concentrations this small, but the following comparison might help: If you were to earn a yearly salary of $100,000, then 1 ppm of your salary would be 10 cents, and 1 ppb would be only one one-hundreth of a penny! But it would be wrong to think that such concentrations are too small to even care about. Some of the industrial pollutants that are being released daily into the water we drink and the air we breathe can be extremely harmful in concentrations as small as 1 ppm.

11.17 Equivalents

Doctors often refer to the important ionic components of the blood in terms of their ionic charge. The unit that is used to describe the concentration of ionic components is the **equivalent (Eq).**

$$1 \text{ Eq} = 1 \text{ mole of charge (either + or −)}$$

For example, 1 mole of the sodium ion Na^+ contains one mole of positive charge, and is equal to one equivalent. For Mg^{2+}, each mole contains 2 moles of positive charge. Therefore, $\frac{1}{2}$ mole of Mg^{2+} is 1 Eq.

The **gram-equivalent weight** of a substance is the weight of the substance, in grams, that will contain one equivalent.

$$\text{Gram-equivalent weight} = \frac{\text{g of substance}}{1 \text{ Eq}}$$

For Na^+ we know that 1 Eq = 1 mole. Therefore, the gram-equivalent weight of Na^+ is 23 g. For Mg^{2+}, 1 Eq = $\frac{1}{2}$ mole. So the gram-equivalent weight is $\frac{24}{2} = 12$ g.

Because the concentration of ions in the blood is very dilute, medical reports will often state ion concentrations in **milliequivalents** (see Figure 11.6).

$$1000 \text{ milliequivalents (mEq)} = 1 \text{ Eq}$$

For example, it is important to regularly check the concentration of sodium ions (Na^+) and potassium ions (K^+) in the blood serum of a person with congestive heart failure who is also taking diuretics. A diuretic is a substance that increases the amount of water released into the urine by the kidneys. However, potassium and sodium ions will also pass into the urine with the water. In such a case, these ions will have to be added back into the person's diet if their concentrations in the blood fall below the normal range of 1.38 to 1.42 mEq/liter for Na^+ and 3.4 to 5.5 mEq/liter for K^+.

Example 11-3

The calcium ion Ca^{2+} serves many extremely important functions in the body.

(a) What is the gram-equivalent weight of the calcium ion?

The calcium ion is Ca^{2+}.
1 mole of Ca^{2+} contains 2 moles of + (positive charge).
Therefore, $\frac{1}{2}$ mole of Ca^{2+} contains 1 mole of +, or 1 Eq.
1 mole of $Ca^{2+} = 40$ g.
$\frac{1}{2}$ mole of $Ca^{2+} = \frac{40}{2} = 20$ g.
Therefore, the gram-equivalent weight of $Ca^{2+} = 20$ g.

(b) How many milliequivalents are present in 100 ml of a 0.1% Ca^{2+} solution?

From section 11.14 we know that

$$0.1\%\ Ca^{2+} = \frac{0.1\text{ g } Ca^{2+}}{100\text{ ml}}$$

Therefore, we want to know

$$0.1\text{ g} = ?\text{ milliequivalents}$$

Using the result of (a), 1 Eq of $Ca^{2+} = 20$ g

$$0.1\not{g}\ Ca^{2+} \times \frac{1\text{ Eq of } Ca^{2+}}{20\not{g}} = \frac{0.1}{20}\text{ Eq of } Ca^{2+}$$

$$= 0.005\text{ Eq of } Ca^{2+}$$

We know that 1000 mEq = 1 Eq

$$0.005\ \cancel{\text{Eq}}\text{ of } Ca^{2+} \times \frac{1000\text{ mEq}}{1\ \cancel{\text{Eq}}} = 5\text{ mEq of } Ca^{2+}$$

11.18 Dilutions of Solutions

A very common task in a chemical or medical laboratory is preparing a needed solution using a concentrated stock solution. You can use the following relationship to determine the amount of stock solution required for a desired solution, but be sure that you use the same units of concentration and units of volume on both sides of the equation.

(concentration desired) × (volume desired) =
(concentration of stock) × (volume of stock)

Example 11-4

How would you prepare 50 ml of 2.0*M* HCl from a 6.0*M* HCl stock solution?

$$(2.0M\ \text{HCl}) \times (50\ \text{ml}) = (6.0M\ \text{HCl}) \times V$$

$$\frac{2.0M\ \cancel{\text{HCl}} \times 50\ \text{ml}}{6.0M\ \cancel{\text{HCl}}} = V$$

$$16.7\ \text{ml} = V$$

Therefore, to make 50 ml of 2.0*M* HCl you would take 16.7 ml of the 6.0*M* HCl stock solution, and then add enough water to make 50 ml of solution.

Properties of Solutions

11.19 Freezing Point and Boiling Point of Water

Solutions have certain types of properties that depend only on the number of solute particles that are dissolved, regardless of the chemical nature of the solute. An example of such properties is the lowering of the freezing point and the raising of the boiling point of water by solute particles. The more solute particles in the solution, the lower the freezing point and the higher the boiling point. For example, automotive antifreeze lowers the freezing point of the water in a car's radiator to prevent it from freezing in the winter, and also raises the boiling point to keep it from boiling over in the summer.

11.20 Osmosis

Another such property of water is osmotic pressure. **Osmosis** is the flow of water molecules through a differentially permeable membrane (a barrier that will allow the passage of water, but not solute particles) from a region of lower solute concentration to a region of higher solute concentration. For example, if pure water and a sugar solution are separated by a differentially permeable membrane, the water molecules will flow through the membrane into the sugar solution at a higher rate than they return. Therefore, the volume of the sugar solution will increase (Figure 11.11).

Osmotic pressure is defined as the amount of pressure that must be applied to prevent the flow of water, or osmosis, through the membrane. Osmotic pressure depends only on the number of solute particles in the solution—the greater the concentration of solute, the greater the pressure that must be applied to prevent osmosis.

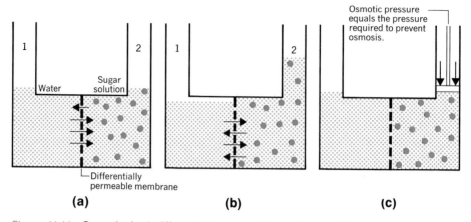

Figure 11.11 Osmosis. (a) A differentially permeable membrane will let water molecules pass through, but not sugar molecules. The water will move from the region of lower concentration to the region of higher concentration. (b) When equilibrium is established, the side with the sugar solution ends up with more material. (c) The osmotic pressure equals the pressure that must be applied to the sugar solution to prevent the movement of water, or osmosis, into the solution.

11.21 Osmolarity

The total concentration of particles in the blood or urine is often recorded in terms of osmolarity. **Osmolarity** describes the total number of moles of all particles in a solution. Because every particle in solution (whether an ion or molecule) contributes to the osmotic pressure of that solution, an **osmol** is defined as the unit of osmotic pressure created by one mole of any kind of particle in a liter of solution. For example, 1 mole of KCl will dissolve in a liter of solution and will produce 1 mole of K^+ and 1 mole of Cl^-, or 2 moles of particles—that is, 2 osmols. Therefore, a liter of 1*M* KCl will have an osmolarity of two osmols. A comparison of the osmolarity of the blood and urine is often a good indication of kidney function. Normal serum osmolarity is 280–290 osmols/liter, and normal urine osmolarity is 300–1000 osmols/liter.

11.22 Hemolysis

The passage of water in and out of living cells is an important biological process. If the concentration of solute is equal on both sides of the cell membrane, the osmotic pressures will be equal and the solutions are then called **isotonic.** The usual concentration of the salts in blood plasma is isotonic to a 0.9% NaCl solution. This concentration of NaCl is called **normal saline** or **physiological saline.**

A normal saline solution is isotonic to red blood cells. When red blood cells are placed in the saline, no change will be observed. If red blood cells are placed in a solution of lower solute concentration, called

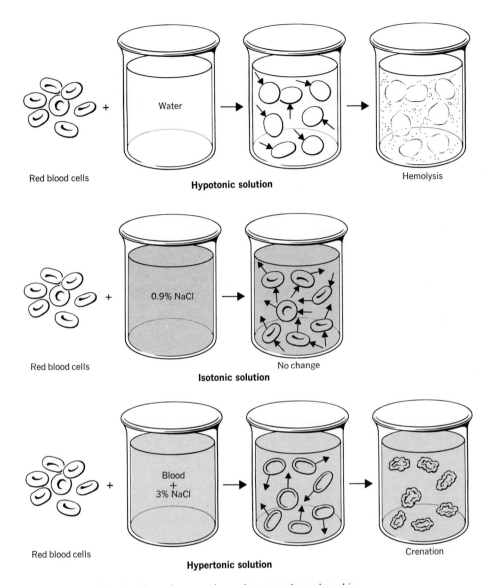

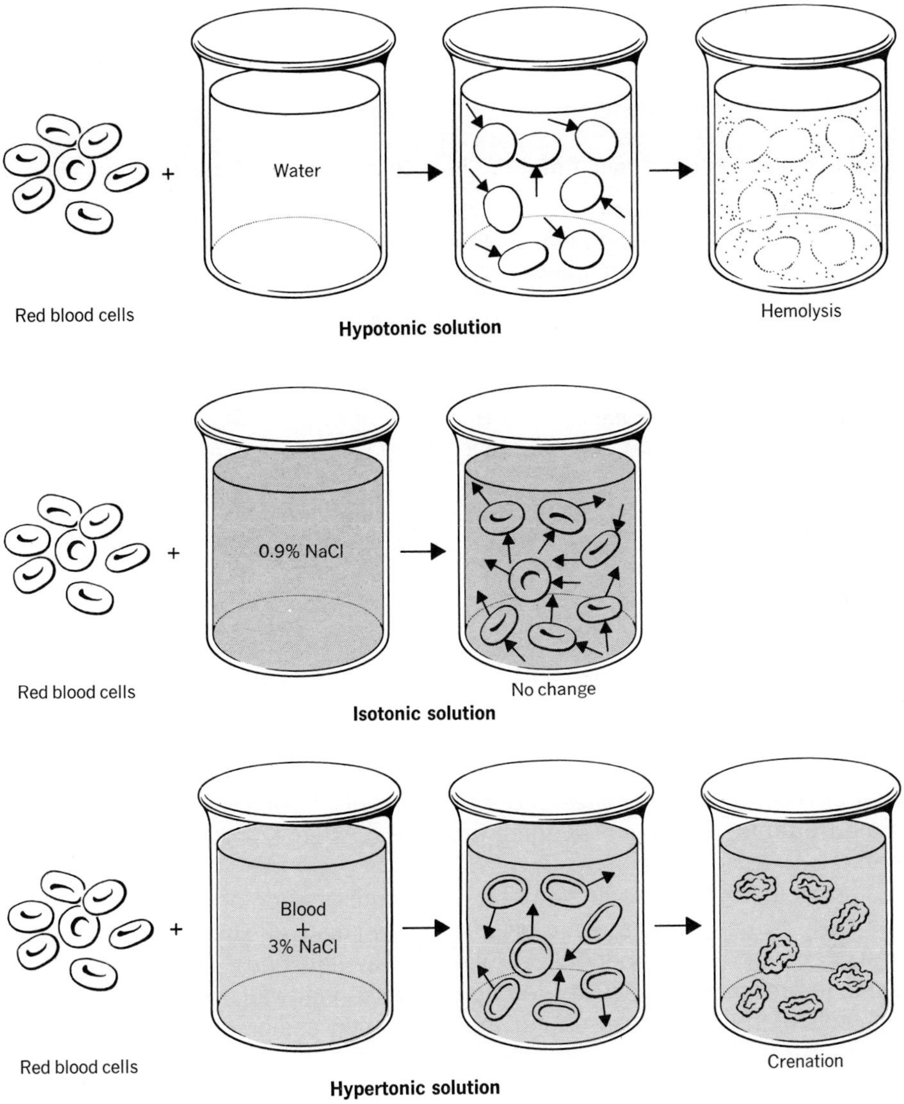

Figure 11.12 Blood cells undergo various changes when placed in solutions of different concentrations.

a **hypotonic solution** (for example, distilled water), water will move into the cells. The cells will then swell and rupture in a process called **hemolysis.** If red blood cells are placed in a solution of higher solute concentration, called a **hypertonic solution** (for example, blood plus 3% NaCl solution), the water will move from the cell to the solution. The cell will then shrink in a process called **crenation** (Figure 11.12). As you can see, solutions cannot be safely introduced into the bloodstream unless they are isotonic. All intravenous preparations are made with isotonic solutions.

Some of the solutes in the bloodstream are not capable of passing through the walls of the capillaries; this gives the blood a higher osmotic

Table 11.3 A Comparison of Some Properties of True Solutions, Colloids, and Suspensions

Property	Solution	Colloids	Suspension
Particle size	Less than 1 $m\mu$*	1 to 1000 $m\mu$	More than 1000 $m\mu$
Filtration	Will pass through filters and membranes	Will pass through filters but not membranes	Stopped by filters and membranes
Visibility	Invisible	Visible in an electron microscope	Visible to the eye or in a light microscope
Motion	Molecular motion	Brownian movement	Movement only by gravity
Passage of light	Transparent	Transparent, Tyndall effect	Translucent

* 1 $m\mu = 10^{-9}$ meters = 1 nanometer.

pressure than the surrounding extracellular fluid. As a result, water moves from the extracellular fluid into the capillaries, keeping the veins full and preventing the collapse of the blood vessels. In some diseases such as malnutrition or kidney failure, the concentration of particles in the blood goes down, decreasing the osmotic pressure and decreasing the flow of water from the tissues into the veins. This results in a condition called **edema,** which is a swelling of the tissues such as those of the hands or legs.

Sometimes the introduction of a hypertonic or hypotonic solution into the body can have therapeutic effects. Saline laxatives such as epsom salts act by forming hypertonic solutions in the intestines. The resulting osmotic pressure draws water through the intestinal wall, moistening the contents of the intestines and making evacuation easier.

Colloids

11.23 The Nature of Colloids

The aqueous systems of cells are not composed entirely of true solutions, but also contain colloidal dispersions, or colloids. **Colloids** are mixtures whose particles are larger than the molecules or ions forming true solutions, but are smaller than particles that will precipitate (or settle out of solution) after sitting for a long time. That is, particles larger than colloidal size can be seen by the eye and, when placed in water, will form suspensions that will settle out in time. Particles smaller than colloids (called **crystalloids**) are atomic or molecular in size and, when placed in water, form true

Table 11.4 Classes of Colloids

Class	Example
1. Solid in solid	Colored glass; certain alloys
2. Solid in liquid	Gelatin in water; protein in water
3. Solid in gas	Aerosols; dust in air; smoke
4. Liquid in solid	Water in gems, such as opals and pearls; jellies
5. Liquid in liquid	Emulsions; egg yolk; mayonnaise; protoplasm
6. Liquid in gas	Aerosols; fog, mist
7. Gas in solid	Activated charcoal; styrofoam
8. Gas in liquid	Foam; whipped cream
(Gas in gas)	(True solution)

solutions. Although there are not sharp boundaries, Table 11.3 will help define the nature of colloids.

Colloidal chemistry is quite important in the study of biological systems. Tissues and cells are colloidal in nature, and reactions occurring within them take place through colloidal chemistry. For example, food digestion involves the formation of colloids before the food can be digested; contraction of muscles can be explained by colloidal chemistry; and the body's proteins are of colloidal size. Colloids can be classified by the solvent (or dispersing medium), and by the colloidal matter (or the dispersed medium). Eight classes result, and are shown in Table 11.4.

11.24 Properties of Colloids

Brownian Movement

Have you ever watched particles of dust dancing in a sunbeam, or moving randomly about in the light from a movie projector? The random movement of the particles in a colloid is caused by the bombardment of these particles by the solvent molecules. The resulting movement is called **Brownian movement** after Robert Brown, who first observed this irregular movement of particles.

Tyndall Effect

When you watch dust dancing in a sunbeam, you are not seeing the actual dust particles—they are too small to be seen with the naked eye. What you are actually seeing is the reflection of light from these particles. A similar effect can be observed when a strong light is passed through a colloid. The path of the beam will be clearly visible because of the reflection of light from the dispersed colloidal particles (Figure 11.13).

Figure 11.13 The Tyndall effect. The container nearest the light source holds a true solution, and the second container holds a colloid. The light beam is visible in the second container, but it passes through the first container unscattered. (From C. W. Keenan, W. E. Bull and J. H. Wood. *Fundamentals of College Chemistry,* 3rd ed., p. 218, Harper & Row, New York, Copyright © 1972. Used by permission).

This property of colloids is known as the **Tyndall effect,** and can be used to distinguish true solutions from colloids.

Adsorption

Colloidal particles range in size from 1 to 1000 millimicrons (also called nanometers, 1×10^{-9} meter) in diameter. Particles of this size have very large surface areas compared with their weight. This large surface area gives colloidal particles the ability to take up, or **adsorb,** substances on their surface. For example, powdered charcoal, whose particles are of colloidal size, has many practical uses. It is put in gas masks to adsorb poisonous gases in the air, is used to remove gases and odors from city water supplies, is used to remove colored impurities from solutions in the laboratory and industry, and is used as an antidote for swallowed poisons. A newly developed device is saving lives by using a charcoal filter to remove toxic substances from the blood of persons who have swallowed poisons or overdoses of drugs.

11.25 Dialysis

Cellular membranes are not osmotic membranes. In order for a cell to live, the membrane must allow the passage of not only water molecules but also ions, nutrients, and waste products. Membranes that allow crystalloids to pass through them, but not large molecules or colloids,

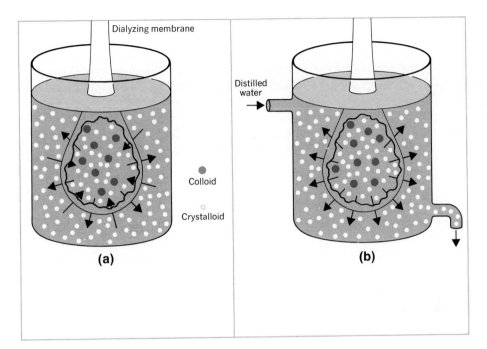

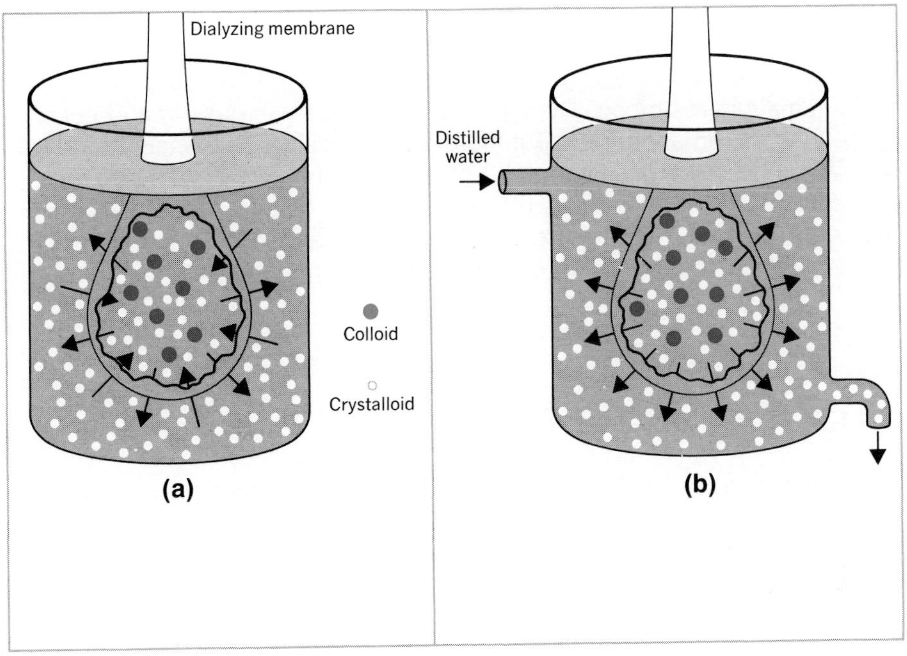

Figure 11.14 A dialysis system. (a) Crystalloid particles will readily pass through the membrane, while the colloidal particles cannot. An equilibrium will eventually be established between the crystalloid particles inside and outside the bag. (b) Complete separation of the crystalloid particles from the colloidal particles can be accomplished if distilled water is kept flowing slowly through the system. Because an equilibrium cannot be established, the crystalloid particles will continue to diffuse from the inside of the bag, leaving behind only colloidal particles.

are called **dialyzing membranes. Dialysis** is the movement of ions and small molecules through dialyzing membranes. Most animal membranes are dialyzing membranes.

Dialysis can be a useful laboratory tool for obtaining purified colloids from solutions of cell contents that may contain both crystalloids and colloids (Figure 11.14). For example, biochemists often use dialysis to separate protein molecules from aqueous ions. By controlling the nature of the dialyzing membrane, they can also separate large protein molecules from smaller ones.

11.26 Hemodialysis

The function of the kidneys is to remove waste products from the blood. If the kidneys are damaged and fail to function, waste products of crystalloid size will continue to build up in the blood, and the patient will die. However, a device called the artificial kidney is now saving the lives of patients with kidney failure. This machine passes blood from an artery in the arm or leg of such patients through long, coiled cellophane tubes that act as dialyzing membranes. Surrounding these tubes is a solution, called the dialysate, that is isotonic for all the components that are to remain in the blood. Therefore, these solutes will pass in and out of the blood at an equal rate. But the dialysate contains no waste products, so these wastes will pass out of the blood at a rate greater than they return. In this way the patient's blood is cleansed. Each week, dialysis patients require two to three treatments that can last 5 to 7 hours in order to remove a sufficient amount of waste products from their blood.

Chapter Summary

Water is critical to all living organisms. It is a good solvent for polar substances, has a high freezing and a high boiling point, is less dense as a solid than as a liquid at 0°C, and has a high surface tension, heat of vaporization, heat of fusion, and specific heat. All of these properties make water especially well-suited for performing many biological functions.

A solution is a homogeneous mixture of two or more substances whose particles, called crystalloids, are of atomic or molecular size. The dissolving medium is called the solvent, and the substance that is dissolved is the solute. Adding solute particles to a liquid solvent will raise the boiling point and lower the freezing point of the liquid. A differentially permeable membrane is one that will allow the passage of solvent, but not solute particles. Osmosis is the flow of solvent through such a membrane from a solution of lower solute concentration to a solution of higher solute concentration. Osmotic pressure is the pressure that must be applied to prevent such a flow of solvent.

A colloid is a mixture in which the particles of the dissolved substance are larger than crystalloids but smaller than particles that will precipitate from solution. Brownian movement, the Tyndall effect, and the ability to adsorb large amounts of other substances on the surface of the dissolved particles are all properties of colloids. Cellular membranes are dialyzing membranes. That is, they allow particles of crystalloid size to pass through, but not those of colloidal size.

An aqueous solution is one in which the solvent is water. The solute may be an electrolyte, which forms charged ions in solution, or a nonelectrolyte, which forms molecular particles in solution. The solubility of a substance in a solvent will depend upon the nature of the solute and the solvent, the temperature and, for a gas, the pressure.

The concentration of a solution may be expressed in many ways: dilute, concentrated, unsaturated, saturated, or supersaturated, which are all fairly general terms. To be more precise, solution concentrations can be stated in weight/volume percent, molarity, parts per million or parts per billion of a solute, or equivalents or milliequivalents of a particular ion.

Exercises and Problems

1. Explain the reason for each of the following properties of water in terms of the hydrogen bonding that exists in water.
   (a) High boiling point
   (b) Density of ice
   (c) High heat of vaporization
   (d) Solubility of molecular substances

2. Which of the following are cations and which are anions?
   (a) PO_4^{3-}
   (b) H^+
   (c) Br^-
   (d) Ba^{2+}

3. State in your own words the steps that occur when sodium chloride is dissolved in water.

4. Explain why a soft drink will go "flat" faster at room temperature than in the refrigerator.

5. State which of the following are soluble in water:
   (a) $MgCO_3$
   (b) NaOH
   (c) NH_4Br
   (d) $PbSO_4$
   (e) KNO_3
   (f) AgCl

6. Quite often celery stored for long periods in a refrigerator goes limp. It can be made crisp again by putting it in a container of water. Explain the principle behind the celery's return to its original crispness.

7. Referring to Figure 11.11(a), state which side, 1 or 2, would rise for each of the following:
   (a) 0.1*M* HCl in 1 and 0.01*M* HCl in 2
   (b) 1% dextrose in 1 and 5% dextrose in 2
   (c) 1*M* NaCl in 1 and 1% NaCl in 2

8. Potassium sulfate (K_2SO_4) is a strong electrolyte, and alcohol is a nonelectrolyte. Explain why the osmotic pressure of a 1*M* solution of K_2SO_4 is about three times that of a 1*M* alcohol solution.

9. What would happen to red blood cells if they were placed in the following solutions?
   (a) 1.5% NaCl solution
   (b) 0.154*M* NaCl solution
   (c) 0.15% NaCl solution

10. Identify each of the solutions in question 9 as isotonic, hypertonic, or hypotonic to red blood cells.

11. A pharmaceutical company recalled a batch of intravenous glucose solutions after a patient died from introduction of this solution into her bloodstream. It was discovered that the solution was 15% glucose. Explain how the introduction of this solution might lead to death.

12. Barium ions, when swallowed, are highly toxic. If a person swallows some barium chloride, the antidote that is given is sodium sulfate. Explain why this is an effective antidote, and write the net-ionic equation for the reaction that takes place.

13. Dextrose is often added to normal saline (0.9% NaCl) for intravenous feeding. How many grams of dextrose and NaCl would you need to prepare 500 ml of 5% dextrose in normal saline?

14. What volume of 0.1% sodium carbonate would contain 2.5 g of Na_2CO_3?

15. Describe how you would prepare each of the following solutions:
    (a) 50 ml of 0.20*M* NaOH
    (b) 250 ml of 0.10*M* Na_2SO_4
    (c) 0.50 liter of 5.0% KOH
    (d) 0.10 liter of 0.50*M* NH_4Cl
    (e) 125 ml of 0.90% NaCl
    (f) 55 ml of 6.0*M* HCl
    (g) 10 ml of 0.20*M* K_2HPO_4
    (h) 0.25 liter of 0.010*M* $(NH_4)_2SO_4$

16. If you dissolve 4.9 g of H_2SO_4 in 250 ml of solution, what is the molarity of the solution?

17. What is the molarity of NaCl in normal saline?

18. Give the gram-equivalent weight for the ions listed in question 2.

19. Lithium carbonate, Li_2CO_3, is being prescribed to control manic depression. The blood concentration of Li^+ must be carefully monitored so that it doesn't exceed 1.4 mEq/liter. If it exceeds this level, the drug is discontinued. After several weeks on lithium carbonate, a patient's blood was sampled and found to contain 1.4 mg of Li^+ in 100 ml. Should the drug dosage be stopped? Explain your answer.

20. Recently, drinking glasses given out by some fast food chains were found to have lead in their painted decorations. This lead could be leached out of the paint with mildly acidic drinks such as fruit juice. In one test, 100 ml of juice was found to contain 4 mg of lead. What is the concentration of lead in this solution in parts per million?

21. How would you prepare each of the following solutions from the given stock solution?
    (a) 250 ml of 8% NaCl from a stock of 10% NaCl
    (b) 150 ml of 0.03*M* NaOH from 0.10*M* NaOH

22. You have available three stock solutions to use in performing an experiment: 6*M* HCl, 0.1*M* NaOH, and 5% Na_2CO_3. State the amount of each stock solution required to supply the following amounts of reactants needed for the experiment.
    (a) 0.3 mole HCl
    (b) 0.15 mole of NaOH
    (c) 2.5 g Na_2CO_3
    (d) 40 mg of NaOH
    (e) 21.9 g of HCl
    (f) 0.25 mole Na_2CO_3

23. Dextran is a substance that is administered in saline solution to help maintain the blood pressure of persons in shock. The recommended

dosage for the first 24 hours is 2 g of dextran per kilogram of body weight. How many milliliters of 10% dextran in normal saline should be administered to a 165 lb man during the first 24 hours?

24. Draw a schematic diagram of an artificial kidney, label its parts, and describe what takes place at the molecular level along the cellophane tube.

chapter 12

Acids and Bases

Learning Objectives

By the time you have finished this chapter, you should be able to:

1. Give the Brønsted-Lowry definition of an acid and base, and identify the conjugate acid-base pairs in a chemical reaction.
2. Write the equation for the ionization of water.
3. State in your own words the difference between a strong and weak acid, and a strong and weak base.
4. Write a balanced equation for a neutralization reaction between an acid and base.
5. Given the pH of a solution, calculate the hydrogen ion concentration and hydroxide ion concentration, and state whether the solution is acidic, basic, or neutral.
6. Perform the equivalence point calculations for a titration.
7. Describe conditions in the body that will cause acidosis to occur.
8. Give an example of the way in which the blood plasma is protected against large changes in pH.

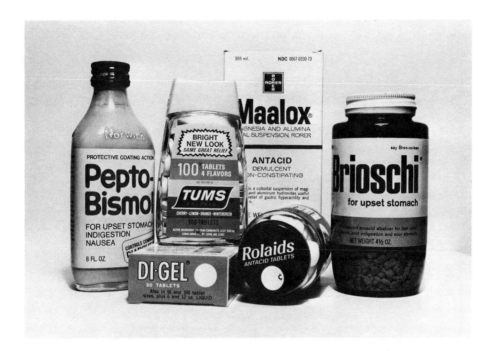

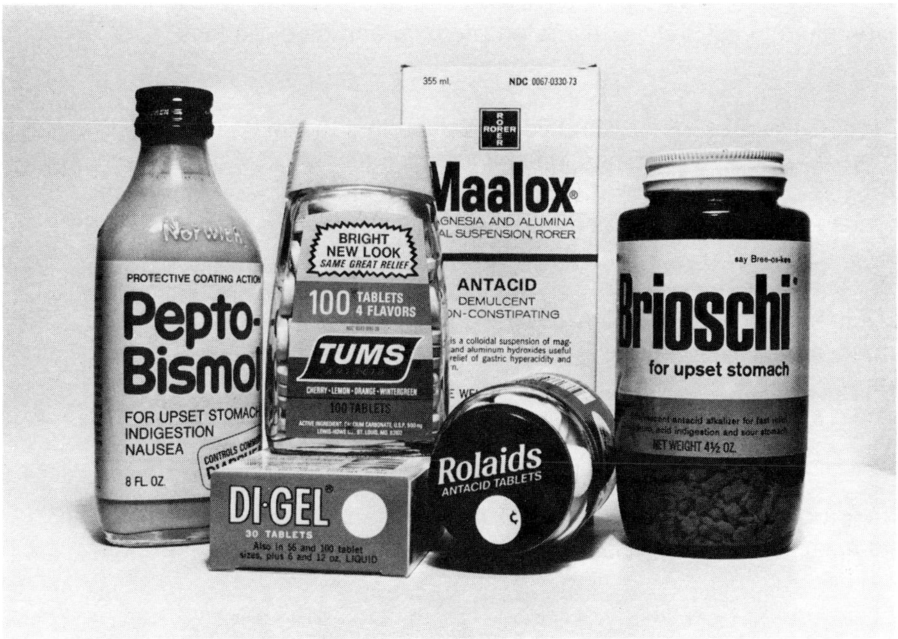

Fred Masterson was a middle-aged insurance agent with a very sensitive stomach. He kept a roll of antacids handy in his pocket to control the acid indigestion from which he constantly suffered, and also drank large amounts of milk in hopes of preventing an ulcer from forming. One afternoon Fred began to suffer some very unusual symptoms. He was extremely dizzy, had high and irregular blood pressure, felt nauseous, and had an acid stomach that was worse than usual. His doctor put Fred into the hospital for a series of tests, but was unable to determine what was wrong. The symptoms soon disappeared, and Fred was released. But not long after he returned to work, the symptoms reappeared. Again he was hospitalized, and again the doctors could find nothing wrong. On Fred's third visit to the hospital, his doctor even called in psychiatrists for consultation on Fred's case. Finally, after long discussions with Fred about his daily habits, a very determined intern found a reference to the "milk-alkali" syndrome in a medical textbook. This syndrome, which included the symptoms suffered by Fred, is caused by a high intake of calcium-containing antacids and large quantities of milk.

To understand what was happening to Fred we must take a closer look at the work done by the stomach. Cells in our stomach lining produce hydrochloric acid (HCl), which is very corrosive. In fact, at the concentrations secreted by the stomach (0.1 moles/liter), hydrochloric acid is able to dissolve a piece of zinc, and is deadly to cells. In the stomach, this acid works to kill bacteria in foods, to soften foods, and to convert the inactive enzyme pepsinogen into its active form, pepsin, to begin the digestion of protein.

Hydrochloric acid reacts with water in the stomach fluid to form hydronium ions (H_3O^+, or hydrated protons) and chloride ions.

$$HCl + H_2O \longrightarrow H_3O^+ + Cl^-$$

It is the presence of the hydronium ion that makes the stomach contents acidic. Sometimes the stomach produces excess acid, resulting in indigestion and heartburn. The various products sold as antacids all contain compounds that can react with hydronium ions to neutralize the excess acid.

The "active ingredient" in the antacid tablets Fred was taking is calcium carbonate ($CaCO_3$). Dissolved in the water mixture in the stomach, calcium carbonate breaks apart to form the calcium ion and the carbonate ion.

$$CaCO_{3(s)} \overset{H_2O}{\rightleftharpoons} Ca^{2+}{}_{(aq)} + CO_3^{2-}{}_{(aq)}$$

The carbonate ions then react with and neutralize hydronium ions in the stomach, relieving the indigestion and heartburn.

$$CO_3^{2-}{}_{(aq)} + 2H_3O^+ \longrightarrow 3H_2O + CO_{2(g)}$$

Calcium carbonate is an excellent antacid for many reasons: it acts rapidly, can neutralize a large amount of acid, and has a long-lasting effect at low cost. As Fred discovered, however, it can also have serious side effects when used over long periods of time. Recent research has shown that calcium carbonate can stimulate increased secretion of stomach acid and, in effect, become self-defeating. Calcium carbonate also tends to cause constipation.

A more important side effect, however, results from the high levels of calcium ions in the blood that can occur when a person takes calcium-based antacids repeatedly. This possibility of high blood calcium is greatly increased in persons who drink large amounts of milk, as was the case with Fred, or with persons who have kidney problems. Although calcium ions are essential for normal body functioning, excessively high blood calcium levels will disrupt reactions in the body, and will cause the unusual symptoms suffered by Fred. Such elevated calcium levels can also cause impaired kidney function, and lead to the formation of kidney stones. Fred's symptoms cleared up about a week after doctors limited his intake of milk and had him switch to an aluminum-magnesium antacid to relieve his indigestion. Most medical authorities now recommend antacids that contain aluminum or magnesium compounds, because they produce less severe side effects.

A critical balance between acids and bases must be maintained in all parts of the living organism to permit normal physiological reactions to occur. In this chapter we will discuss the nature of acids and bases and the ways in which their concentrations are regulated in the human body.

Acids and Bases

12.1 What Are Acids and Bases?

Acids and bases play important roles in living organisms. Strong acids and bases can be quite harmful; they will destroy tissue by dissolving protein material and drawing out water. For example, concentrated sulfuric acid is a strong dehydrating (water-removing) agent, and will rapidly injure tissues on contact. Concentrated bases will react with the fats that make up the protective membranes of cells, destroying such membranes and causing even more widespread destruction to tissues than acids. Strong laundry soaps and detergents contain bases. Clothes containing wool and silk (which are animal proteins) cannot be washed in such soaps because the base in the soap will cause the fibers of these materials to shrink and partially dissolve.

In general, **acids** are compounds which, when dissolved in water, produce solutions that conduct electricity, react with metals such as zinc or magnesium to produce hydrogen gas, taste sour, and turn paper containing litmus dye from blue to red. **Bases** also form solutions that conduct electricity; however, they taste bitter, feel slippery to the touch, and turn litmus paper from red to blue. Bases will react with acids to neutralize each others' properties. For example, medicines for the relief of pain, indigestion, and constipation often contain the base bicarbonate (HCO_3^-), which will neutralize the hydrochloric acid (HCl) in the stomach (Figure 12.1).

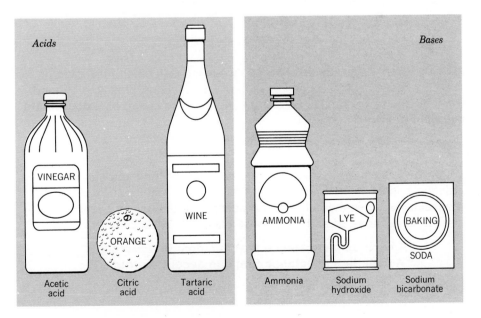

Figure 12.1 Some household acids and bases.

12.2 Ionization of Water

In the previous chapter we discussed water's ability to dissolve ionic compounds and to ionize some highly polar molecular compounds. But to a very small extent water molecules are also able to ionize other water molecules. One water molecule out of every 550 million will be pulled apart to form a hydronium ion, H_3O^+, and hydroxide ion, OH^-. In a liter of water, then, the concentration of H_3O^+ and OH^- is 1×10^{-7} moles.

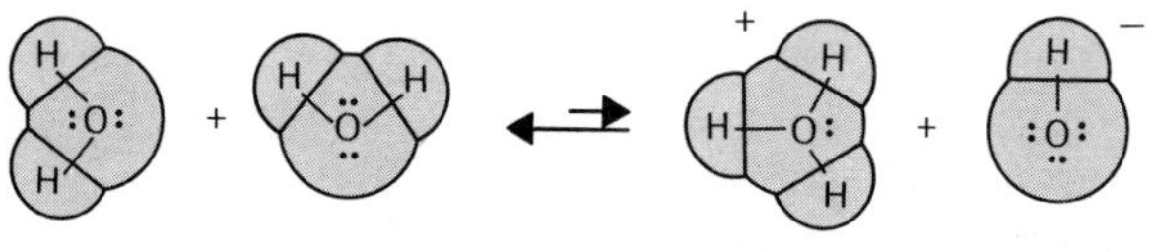

$$H_2O + H_2O \rightleftharpoons H_3O^+ + OH^-$$

For convenience, we often write the equation for the ionization of water as follows:

$$H_2O \rightleftharpoons \underset{\substack{\text{Hydrogen ion} \\ \text{(proton)*}}}{H^+} + \underset{\text{Hydroxide ion}}{OH^-}$$

But it is important to understand that although we often use the terms *protons* or *hydrogen ions* when talking about aqueous solutions, single hydrogen ions never exist in water. They are always associated with a water molecule in the form of a hydronium ion (hydrated proton).

$$H^+ + :\overset{H}{\underset{..}{\ddot{O}}}:H \longrightarrow H:\overset{H+}{\underset{..}{\ddot{O}}}:H$$

In the covalent bond that is formed between the oxygen and the hydrogen ion, the oxygen molecule donates both of the electrons. This type of bond, in which one atom donates both of the electrons, is known as a **coordinate covalent bond.**

12.3 Acids and Bases: The Brønsted-Lowry Definition

What makes a compound an acid or a base? Although there are several definitions that can be used, we will find it convenient for our study of biochemistry to use the definitions proposed in 1923 by the Danish chemist J. N. Brønsted and the English chemist T. M. Lowry:

An acid is a substance that can donate a proton, H^+.
A base is a substance that can accept a proton, H^+.

Let's look at some examples. Consider the following reaction:

---

* *A hydrogen atom has just one proton in its nucleus, and one electron. Therefore, the hydrogen ion (H^+, a hydrogen atom that has lost one electron) is simply a proton.*

$$HNO_2 \quad + \quad H_2O \quad \rightleftharpoons \quad NO_2^- \quad + \quad H_3O^+ \qquad (1)$$

Nitrous Nitrite
acid ion

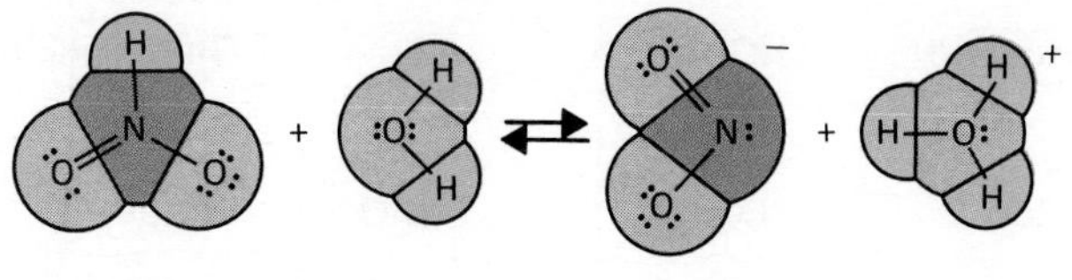

In this reaction, nitrous acid is donating a proton to water. Therefore, nitrous acid is the acid (the proton donor), and water acts as the base (the proton acceptor).

$$:NH_3 \quad + \quad H_2O \quad \rightleftharpoons \quad NH_4^+ \quad + \quad OH^- \qquad (2)$$

Ammonia Ammonium Hydroxide
 ion ion

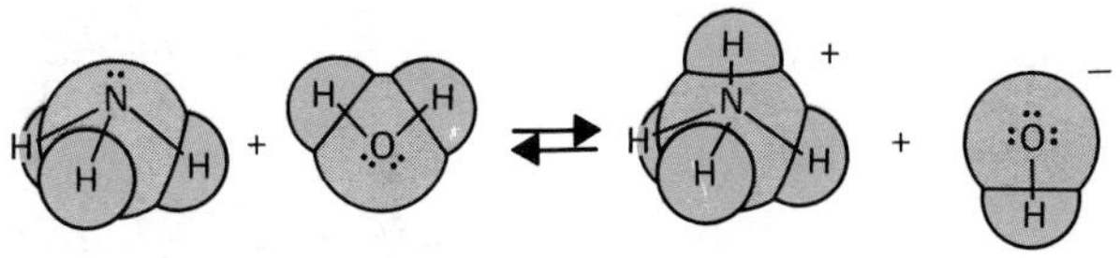

In this reaction, ammonia (NH_3) accepts a proton from a water molecule. Therefore, ammonia is the base, and water (as the proton donor) is the acid. You can see that water may function as either an acid or a base depending upon the substance with which it is reacting.

Consider for a moment the reverse reaction in equation (1). Here, the hydronium ion (H_3O^+) is the acid and the nitrite ion (NO_2^-) is the base. Similarly, in the reverse of equation (2), the ammonium ion (NH_4^+) is the acid and the hydroxide ion (OH^-) is the base.

$$HNO_2 + H_2O \rightleftharpoons NO_2^- + H_3O^+$$

Acid$_1$ Base$_2$ Base$_1$ Acid$_2$

$$NH_3 + H_2O \rightleftharpoons NH_4^+ + OH^-$$

Base$_1$ Acid$_2$ Acid$_1$ Base$_2$

You might notice that HNO_2 and NO_2^-, and H_2O and H_3O^+, differ only by a proton. They are known as conjugate acid-base pairs. The nitrite ion is the conjugate base of nitrous acid, and the hydronium ion is the conjugate acid of water.

12.4 The Strength of Acids and Bases

We will define strong and weak acids in much the same way that we defined strong and weak electrolytes. A strong acid is one that completely, or almost completely, ionizes to donate all of its protons. The result of adding a strong acid to water is a large increase in the concentration of

Table 12.1 Relative Strengths of Some Conjugate Acid-Base Pairs

ACID		Increasing Acid Strength ↑	Increasing Base Strength ↓	BASE	
Sulfuric acid	H_2SO_4			HSO_4^-	Hydrogen sulfate ion
Hydrochloric acid	HCl			Cl^-	Chloride ion
Nitric acid	HNO_3			NO_3^-	Nitrate ion
Hydronium ion	H_3O^+			H_2O	Water
Sulfurous acid	H_2SO_3			HSO_3^-	Hydrogen sulfite ion
Hydrogen sulfate ion	HSO_4^-			SO_4^{2-}	Sulfate ion
Phosphoric acid	H_3PO_4			$H_2PO_4^-$	Dihydrogen phosphate ion
Nitrous acid	HNO_2			NO_2^-	Nitrite ion
Acetic acid	$HC_2H_3O_2$			$C_2H_3O_2^-$	Acetate ion
Carbonic acid	H_2CO_3			HCO_3^-	Hydrogen carbonate ion
Hydrogen sulfite ion	HSO_3^-			SO_3^-	Sulfite ion
Dihydrogen phosphate ion	$H_2PO_4^-$			HPO_4^{2-}	Hydrogen phosphate ion
Ammonium ion	NH_4^+			NH_3	Ammonia
Hydrogen carbonate ion	HCO_3^-			CO_3^{2-}	Carbonate ion
Hydrogen phosphate ion	HPO_4^{2-}			PO_4^{3-}	Phosphate ion
Water	H_2O			OH^-	Hydroxide ion
Ammonia	NH_3			NH_2^-	Amide ion

hydronium ions. For example, nitric acid is a strong acid. In 0.1*M* HNO_3, 92% of the nitric acid molecules are ionized to hydronium ions and nitrate ions, and only 8% remain as whole molecules.

$$HNO_3 + H_2O \rightleftharpoons H_3O^+ + NO_3^-$$

Other strong acids are hydrochloric acid (HCl), hydrobromic acid (HBr), hydroiodic acid (HI), and sulfuric acid (H_2SO_4). A weak acid only partly ionizes in water to donate protons. Therefore, the addition of a weak acid to water results in only a small increase in the concentration of hydronium ions. Acetic acid is an example of a weak acid. In a 0.1*M* CH_3COOH solution, only 1.3% of the molecules will be ionized.

$$CH_3COOH + H_2O \rightleftharpoons H_3O^+ + CH_3COO^-$$

Other weak acids are nitrous acid (HNO_2), carbonic acid (H_2CO_3), and boric acid (H_3BO_3).

Similarly, a strong base will have a very large attraction for protons. A weak base will have a weak attraction, and only a small percentage of its molecules will accept protons. The hydroxide ion is a strong base, whereas ammonia is a weak base.

If an acid is strong and, therefore, has a strong tendency to donate protons, its conjugate base will be weak and will have a low attraction for protons. The reverse will be true for a weak acid; a weak acid will have a conjugate base with a strong attraction for protons. Table 12.1 lists the relative strengths of some conjugate acid-base pairs. (Note that the number of hydrogens in the formula of an acid is not an indication of its strength as an acid.)

Example (optional)

Mathematically, the strength of an acid can be determined from its ionization constant, K_a. Consider a general acid HA, with

$$HA \rightleftharpoons H^+ + A^-$$

Then, K_a is given by the equation

$$K_a = \frac{[H^+] \times [A^-]}{[HA]}$$

where $[H^+]$, $[A^-]$, and [HA] represent concentrations in moles/liter. The weaker the acid, the smaller the concentration of H^+ and A^- in comparison with the concentration of HA; therefore, the smaller the value of K_a.

1. Given the K_a values for the following acids, list them in order of increasing acid strength.

CH_3COOH	$K_a = 1.75 \times 10^{-5}$	H_2CO_3	$K_a = 4.3 \times 10^{-7}$
HNO_2	$K_a = 4.5 \times 10^{-4}$	H_3BO_3	$K_a = 7.3 \times 10^{-10}$

   Weakest: H_3BO_3, H_2CO_3, CH_3COOH, HNO_2 :Strongest

2. Phosphoric acid, H_3PO_4, has a K_a value for the ionization of each of its hydrogens.

   $$H_3PO_4 \rightleftharpoons H^+ + H_2PO_4^- \qquad K_{a_1} = 7.5 \times 10^{-3}$$
   $$H_2PO_4^- \rightleftharpoons H^+ + HPO_4^{2-} \qquad K_{a_2} = 6.2 \times 10^{-8}$$
   $$HPO_4^{2-} \rightleftharpoons H^+ + PO_4^{3-} \qquad K_{a_3} = 2.2 \times 10^{-12}$$

   (a) Which is the strongest acid? H_3PO_4, $H_2PO_4^-$, or HPO_4^{2-}

   K_{a_1} has the highest value, therefore H_3PO_4 is the strongest acid.

   (b) Which conjugate base is the strongest? $H_2PO_4^-$, HPO_4^{2-}, or PO_4^{3-}

   The weakest acid will have the strongest conjugate base. Because K_{a_3} has the smallest value, HPO_4^{2-} is the weakest acid and, therefore, PO_4^{3-} is the strongest base.

12.5 Naming Acids and Bases

When dissolved in water, certain binary compounds containing hydrogen form solutions that have acidic properties. In Chapter 8 we learned that HCl is called hydrogen chloride, but that its aqueous solution is called hydrochloric acid. Such compounds, called binary acids, are composed of hydrogen and one other nonmetallic element. However, not all binary hydrogen compounds are acids. When the formula of a binary hydrogen compound has the hydrogen listed first (HCl, HI, H_2S), it is an acid; but compounds such as CH_4 and NH_3 are not acidic. To name a binary acid, (1) use the prefix *hydro-*, (2) add an *-ic* ending on the name of the other element, and (3) add the word *acid*. For example, HBr is hydrobromic acid.

Some acids contain three different elements: hydrogen, oxygen, and another element which can be either a metal or a nonmetal. In Chapter 8 we learned that H_2SO_4 is called dihydrogen sulfate, and H_2SO_3 is dihydrogen sulfite. These names refer to the compounds in their pure state. In water these substances become acids. Their acid names are formed by using the ending of *-ic* or *-ous* on the name of the third element, and then adding the word *acid*. The *-ic* ending is used for the compound containing the greater number of oxygens. Therefore, H_2SO_4 is called sulfuric acid and H_2SO_3 is sulfurous acid. If there are more than two acids formed with the same third element, the prefix *hypo-* is used for the acid having fewer oxygens than the *-ous* acid. The prefix *per-* is used when the acid has more oxygens than the *-ic* acid. For example, there are four acids formed from hydrogen, oxygen, and chlorine:

HClO	Hypochlorous acid
$HClO_2$	Chlorous acid
$HClO_3$	Chloric acid
$HClO_4$	Perchloric acid

You can see from Table 8.5 that the names of the negative ions formed when all the hydrogens are removed from an *-ic* acid (such as SO_4^{2-} from H_2SO_4) end in *-ate*. SO_4^{2-} is the sulfate ion: sulfuric acid $\longrightarrow$ sulfate ion. The name of the negative ions formed from an *-ous* acid (such as SO_3^{2-} from H_2SO_3) end in *-ite*. SO_3^{2-} is the sulfite ion: sulfurous acid $\longrightarrow$ sulfite ion.

12.6 Neutralization Reaction

Acids will react with bases to form water and the dissolved ions of an ionic compound, commonly called a **salt.** For example,

$$\underset{\text{Acid}}{\text{HCl}} \; + \; \underset{\text{Base}}{\text{NaOH}} \; \longrightarrow \; \underset{\text{Water}}{\text{H}_2\text{O}} \; + \; \underset{\text{Salt}}{\text{NaCl}}$$

If equal amounts of hydronium ions and hydroxide ions react, the resulting solution will have neither acidic nor basic properties, and is said to be **neutral.** A **neutralization reaction,** therefore, is one in which either an acid or basic solution is converted to a neutral solution.

The reaction between, for example, hydrochloric acid and sodium hydroxide actually occurs between aqueous ions. The complete ionic equation for the reaction is

$$H^+ + \cancel{Cl^-} + \cancel{Na^+} + OH^- \longrightarrow H_2O + \cancel{Na^+} + \cancel{Cl^-}$$

The net-ionic equation for this reaction is

$$H^+ + OH^- \longrightarrow H_2O$$

Example 12-1

1. Write the balanced equation for the neutralization of sulfuric acid by potassium hydroxide.

   The reactants are H_2SO_4 and KOH. The products are water (the H^+ from H_2SO_4 and the OH^- from NaOH) and the salt potassium sulfate (formed by K^+ and SO_4^{2-}). The correct formula for potassium sulfate is K_2SO_4. The unbalanced equation is:

   $$H_2SO_4 + KOH \longrightarrow H_2O + K_2SO_4$$

   If we first balance the potassium atom, we have

   $$H_2SO_4 + 2KOH \longrightarrow H_2O + K_2SO_4$$

   The sulfur atoms already balance, so next we balance the hydrogen atoms.

   $$H_2SO_4 + 2KOH \longrightarrow 2H_2O + K_2SO_4$$

   Finally, check the total number of oxygen atoms on both sides. The equation is balanced.

12.7 Ion Product of Water, K_w

We stated earlier in this chapter that only a very small number of water molecules ionize to form hydrogen ions* and hydroxide ions. In pure water at room temperature the concentration of hydrogen ions equals only 0.0000001*M*, or $1 \times 10^{-7}M$. Because a hydroxide ion is formed for every hydrogen ion that is formed, the concentration of OH^- is also $1 \times 10^{-7}M$. Therefore,

$$[H^+] \times [OH^-] = [1 \times 10^{-7}][1 \times 10^{-7}] = 1 \times 10^{-14} = K_w$$

This product, known as the **ion product of water, K_w,** always equals 1×10^{-14} at room temperature. From this equation we can calculate the concentration of either H^+ or OH^- if we know the concentration of the other ion.

$$[H^+] = \frac{1 \times 10^{-14}}{[OH^-]} \qquad [OH^-] = \frac{1 \times 10^{-14}}{[H^+]}$$

* The terms *hydrogen ions* and *hydronium ions* will be used interchangeably in this discussion.

Example 12-2

The concentration of hydrogen ions in a slightly acid solution is 1×10^{-5}. What is the hydroxide ion concentration in this solution? Substituting in the equation above,

$$[OH^-] = \frac{1 \times 10^{-14}}{[H^+]} = \frac{1 \times 10^{-14}}{1 \times 10^{-5}} = 1 \times 10^{-9}$$

(If you are unsure about how to divide numbers in exponential form, see Appendix 1).

Measuring the Concentrations of Acids and Bases

12.8 The pH Scale

Small changes in hydrogen ion concentration can be of great importance to living cells, and are critical in many fields of scientific investigation. As a result, scientists are constantly measuring hydrogen ion concentrations. The pH scale was developed in 1909 by a chemist named Sorenson in order to provide a way of indicating the hydrogen ion concentration more conveniently than by using a negative exponent (10^{-7}) or a decimal fraction (0.0000001). pH is the negative power to which the number 10 must be raised to express the concentration of hydrogen ions in moles/liter. Mathematically,

$$[H^+] = 1 \times 10^{-pH} \qquad \text{or} \qquad pH = -\log [H^+]$$

At room temperature, the $[H^+]$ in pure water is 1×10^{-7}. Therefore, the pH of pure water is 7. Because in pure water we have $[H^+] = [OH^-]$, the

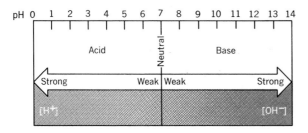

Figure 12.2 The pH scale. Acidic solutions have a range of pH from 0 to 7, and basic solutions a pH range from 7 to 14. A solution with a pH of exactly 7 is neutral, and will have equal concentrations of hydrogen ions and hydroxide ions.

Table 12.2 The pH Scale and the Corresponding Concentrations of Hydrogen and Hydroxide Ions

	$[H^+]$	pH	$[OH^-] = \frac{1 \times 10^{-14}}{[H^+]}$
Acidic	$10^0 = 1$	0	10^{-14}
	$10^{-1} = 0.1$	1	10^{-13}
	$10^{-2} = 0.01$	2	10^{-12}
	$10^{-3} = 0.001$	3	10^{-11}
	$10^{-4} = 0.0001$	4	10^{-10}
	$10^{-5} = 0.00001$	5	10^{-9}
	$10^{-6} = 0.000001$	6	10^{-8}
Neutral	$10^{-7} = 0.0000001$	7	10^{-7}
Basic	$10^{-8} = 0.00000001$	8	10^{-6}
	$10^{-9} = 0.000000001$	9	10^{-5}
	$10^{-10} = 0.0000000001$	10	10^{-4}
	$10^{-11} = 0.00000000001$	11	10^{-3}
	$10^{-12} = 0.000000000001$	12	10^{-2}
	$10^{-13} = 0.0000000000001$	13	10^{-1}
	$10^{-14} = 0.00000000000001$	14	10^0

water is neutral. This means that the pH of a neutral solution is seven. Acidic solutions are ones in which the hydrogen ion concentration is greater than the hydroxide ion concentration. From Table 12.2 we see that for an acidic solution the exponent of ten will be less than −7. Thus, the pH of an acidic solution will be less than seven. A basic solution is one in which the hydrogen ion concentration is less than the hydroxide ion concentration, and the pH will be greater than seven (Figure 12.2).

Example 12-3

A sample of rain water was found to have a pH of 3. (a) Is the solution acidic or basic? (b) State the $[H^+]$ and $[OH^-]$.

(a) The pH is less than 7 and, therefore, the solution is acidic.

(b) $[H^+] = 1 \times 10^{-pH} = 1 \times 10^{-3}$

$$[OH^-] = \frac{1 \times 10^{-14}}{[H^+]} = \frac{1 \times 10^{-14}}{1 \times 10^{-3}} = 1 \times 10^{-11}$$

Table 12.3 The Normal pH Range of Some Body Fluids

Fluid	pH	Fluid	pH
Gastric juice	1.0–3.0	Blood	7.35–7.45
Vaginal secretion	3.8	Intestinal secretions	7.7
Urine	5.5–7.0	Bile	7.8–8.8
Saliva	6.5–7.5	Pancreatic juice	8

12.9 Measuring pH

The measurement of pH is an important laboratory procedure because the pH of a solution affects the activity of biological molecules. This means that it can influence the behavior of cells and even entire organisms. For example, bacteria will grow best in a very narrow range of pH. Therefore, the pH of culture media must be carefully controlled. Enzymes, the biological catalysts, work best in a very narrow range of pH which can vary from an optimum pH range of 1 to 4 for pepsin (an enzyme in the stomach), to an optimum pH range of 8 to 9 for trypsin (an enzyme in the small intestine). Most body fluids remain in a very narrow range of pH which, if changed, can be toxic to the organism (Table 12.3).

The pH of a solution is best measured using a pH meter (Figure 12.3). These instruments make use of the fact that the voltage of an electric current passing through a solution will change according to the pH of the solution. A second, but less accurate, method of measuring pH is by a colorimetric indicator. Such methods use chemical dyes, called acid-base indicators, which will change color at certain hydrogen ion concentrations

Figure 12.3 A pH meter. (Courtesy Beckman Instruments)

Table 12.4 Colors of Some Acid-Base Indicators at Various pH Levels

Indicator	pH 0–1	1–2	2–3	3–4	4–5	5–6	6–7	7–8	8–9	9–10	10–11	11–12	12–13	13–14
Thymol blue*	Red	Transition		Yellow										
Methyl orange			Red	Transition	Yellow									
Methyl red					Red	Transition	Yellow							
Litmus					Red		Transition	Blue						
Bromothymol blue						Yellow	Transition	Blue						
Metacresol purple							Yellow		Transition	Purple				
Thymol blue*								Yellow	Transition		Blue			
Phenolphthalein								Colorless		Transition	Red			

*Thymol blue indicator undergoes two color changes—one in the acid range and one in the base range.

(Table 12.4). For example, paper on which the dye nitrazine has been placed is yellow at a pH of 4.5, and blue at a pH of 7.5. Such paper is used in hospitals to test the pH of urine. Acidic urine—urine with a pH of 4.5 or lower—will turn the paper yellow, and is often an indication of a serious disorder.

12.10 Titrations

We can use a neutralization reaction to measure the amount of acid or base in a solution by means of a procedure called **titration** (Figure 12.4). In a titration, a solution of known concentration of acid or base is added from a buret to a solution of unknown concentration of base or acid. The neutralization reaction can be monitored by a pH meter or an acid-base indicator. The indicator is chosen so that it changes color at the **equivalence point,** the pH at which all the hydroxide (or hydrogen) ions have been neutralized. Titrations can be used to determine the alkali (or basic) constituents of the blood, the acidity of the stomach, or the acidity of the urine.

Example 12-4

1. What is (a) the concentration of hydroxide and (b) the pH of a solution, if 5.00 ml of 0.0200*M* HCl completely neutralizes 100 ml of the solution?

   (a) A 0.0200*M* HCl solution contains 0.0200 moles of H^+ in 1000 ml. Therefore, 5.00 ml will contain

$$5.00\ \cancel{\text{ml}} \times \frac{0.0200\ \text{moles}}{1000\ \cancel{\text{ml}}} = \frac{0.100\ \text{moles}}{1000} = 1.00 \times 10^{-4}\ \text{moles of H}^+$$

   Because this many moles of H^+ completely neutralize the base, the 100 ml of solution contains 1.00×10^{-4} moles of OH^-.

(b) To calculate the pH we must know the concentration of H^+, in moles per liter, of the original solution. In part (a) we calculated that the concentration of OH^- was 1.00×10^{-4} moles in 100 ml. Therefore, 1 liter of this solution would contain

$$\frac{1.00 \times 10^{-4}\ \text{moles}}{100\ \text{ml}} \times \frac{1000\ \text{ml}}{1\ \text{liter}} = \frac{1.00 \times 10^{-3}\ \text{moles}}{1\ \text{liter}} = 1.00 \times 10^{-3}\ \text{moles of}\ OH^-$$

Using the relationship between $[OH^-]$ and $[H^+]$ which we discussed in Section 12.7, we can calculate $[H^+]$ as follows:

$$[H^+] = \frac{1 \times 10^{-14}}{[OH^-]}$$

$$= \frac{1 \times 10^{-14}}{1 \times 10^{-3}}$$

$$= 1 \times 10^{-11}$$

Therefore, the pH of this solution is 11.

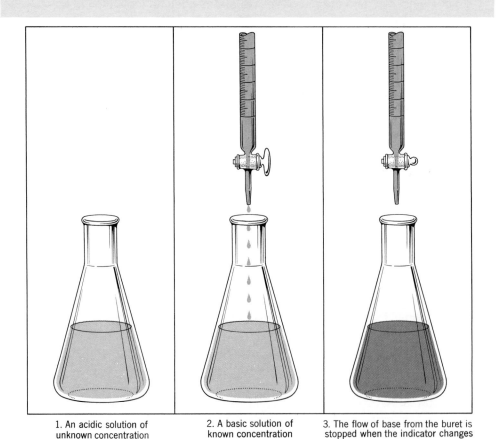

1. An acidic solution of unknown concentration is placed in a flask with an indicator.

2. A basic solution of known concentration is added slowly from the buret.

3. The flow of base from the buret is stopped when the indicator changes color, showing that just enough base has been added to react with all the acid in the flask.

Figure 12.4 A titration.

2. Titration of the contents of the stomach can be very useful in medical diagnosis. What is the molar concentration of acid in 100 ml of stomach fluid if complete neutralization occurs when 27.0 ml of 0.10M NaOH has been added?

0.10M NaOH will contain 0.10 moles of OH$^-$ in 1000 ml. Therefore, 27.0 ml will contain:

$$27.0 \text{ ml} \times \frac{0.10 \text{ moles}}{1000 \text{ ml}} = 0.0027 \text{ moles of OH}^-$$

This amount of base completely neutralized the acid, so the 100 ml of stomach fluid contained 0.0027 moles of H$^+$. The concentration of acid in moles per liter is then

$$\frac{0.0027 \text{ mole}}{100 \text{ ml}} \times \frac{1000 \text{ ml}}{1 \text{ liter}} = \frac{0.027 \text{ moles}}{1 \text{ liter}} = 0.027M$$

In general, a person is considered to be suffering from hyperacidity when 44 ml or more of 0.10M base are required to neutralize 100 ml of stomach fluid (pH < 1.4). The person is suffering from hypoacidity when 10 ml or less is required (pH > 2).

Buffer Systems

12.11 What Are Buffers?

Buffers are substances which, when present in solution, resist sudden changes in pH. In particular, they protect against large changes in pH when acids or bases are added to the solution. Living cells are extremely sensitive to even very slight changes in pH. As we stated earlier, the reason for this sensitivity is that the enzymes that catalyze metabolic reactions operate in only a small range of pH. Altering the pH will slow down or stop the action of the enzyme. Fortunately, the contents of cells, the extracellular fluid, and the blood have all developed buffer systems that protect against pH changes.

The best buffer systems consist of a weak acid and its conjugate base, or a weak base and its conjugate acid. Such systems have their highest buffering capacity at a pH where the [concentration of acid] = [concentration of conjugate base], or [concentration of base] = [concentration of conjugate acid]. The following are some acid-base pairs that can be used in buffer systems.

$$\underset{\text{Carbonic acid}}{H_2CO_3} \quad + \quad H_2O \quad \rightleftharpoons \quad \underset{\text{Bicarbonate ion}}{HCO_3^-} \quad + \quad H_3O^+$$

$$\underset{\text{Acetic acid}}{CH_3COOH} \quad + \quad H_2O \quad \rightleftharpoons \quad \underset{\text{Acetate ion}}{CH_3COO^-} \quad + \quad H_3O^+$$

$$H_2PO_4^- \quad + \quad H_2O \quad \rightleftharpoons \quad HPO_4^{2-} \quad + \quad H_3O^+$$

Dihydrogen phosphate ion $\qquad$ Monohydrogen phosphate ion

$$NH_3 \quad + \quad H_2O \quad \rightleftharpoons \quad NH_4^+ \quad + \quad OH^-$$

Ammonia $\qquad$ Ammonium ion

12.12 Control of pH in Body Fluids

In the human body the blood plasma has a normal pH of 7.4. If the pH should fall below 7.0 or rise above 7.8, the results would be fatal. The buffer systems in the blood are very effective in protecting this fluid from large changes in pH. For example, if 1 ml of 10.0M HCl were added to 1 liter of unbuffered physiological saline (0.15M NaCl) at a pH of 7, the pH would fall to 2. But if 1 ml of 10.0M HCl is added to 1 liter of blood plasma at pH 7.4, the pH will drop to only 7.2 (Figure 12.5).

How do such buffer systems protect the blood? The major buffer system in the blood is the carbonic acid-bicarbonate system. Consider the following equilibrium equation:

$$H_2CO_3 \rightleftharpoons HCO_3^- + H^+$$

Adding a strong acid to the system will increase the concentration of H$^+$, driving the reaction to the left and forming more carbonic acid.

$$H_2CO_3 \rightleftharpoons HCO_3^- + H^+$$

But carbonic acid is unstable, and will decompose to form carbon dioxide and water.

$$H_2CO_3 \longrightarrow CO_{2(g)} + H_2O$$

The carbon dioxide so formed can be removed from the blood and exhaled by the lungs. This buffer system will continue to protect against the pH change until all the carbonate has reacted.

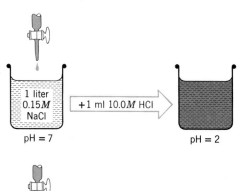

Figure 12.5 The buffer systems in the blood are very effective in preventing large changes in pH, which could be fatal.

1 liter 0.15M NaCl

+1 ml 10.0M HCl

pH = 7

pH = 2

1 liter blood plasma

+1 ml 10.0M HCl

pH = 7.4

pH = 7.2

Various factors can cause abnormal increase in acid levels in the blood. Such factors are hypoventilation caused by emphysema, by congestive heart failure or bronchopneumonia; an increase in the production of metabolic acids, such as occurs in diabetes mellitus or some low-carbohydrate/high-fat diets; ingestion of excess acids; excess loss of bicarbonate in severe diarrhea; or decreased excretion of hydrogen ions through kidney failure. Each of these conditions will cause an increase in the hydrogen ion level in the blood, and a decrease in the concentration of basic components (such as bicarbonate), known as the alkaline reserves. The pH of the blood can drop to 7.1 or 7.2, resulting in a condition known as **acidosis** (called respiratory acidosis if its origin is in the respiratory system, and metabolic acidosis if the origin is other than respiratory) (Table 12.5). However, the body has ways to restore the blood pH to normal. First, it can expel the excess carbon dioxide, formed from the carbonic acid, through an increase in the rate of breathing. Second, it can increase the excretion of H^+ and the retention of HCO_3^- by the kidneys, resulting in acidic urine (pH about 4).

The bicarbonate buffer system also protects against an addition of strong base to the system. A base will react with the hydrogen ions to produce water, decreasing the concentration of hydrogen ions in the system. This will drive the reaction to the right.

$$H_2CO_3 \rightleftharpoons HCO_3^- + H^+$$

Such an increase in base in the blood can occur in cases of hyperventilation during extreme fevers or hysteria, from excessive ingestion of basic substances such as antacids, and in severe vomiting. The pH of the blood can increase to a pH of 7.5, resulting in a condition known as **alkalosis.** Alkalosis is not as common as acidosis. The body's means for returning the pH to normal are a decrease in expulsion of carbon dioxide by the lungs and an increase in excretion of HCO_3^- by the kidneys, resulting in an alkaline urine ($pH > 7$) (Figure 12.6).

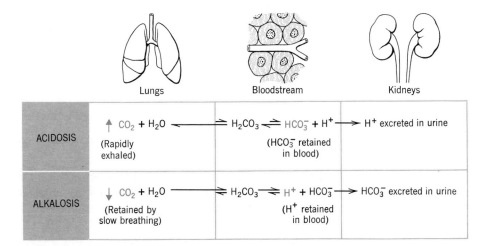

	Lungs	Bloodstream	Kidneys
ACIDOSIS	↑ $CO_2 + H_2O$ ⟵ (Rapidly exhaled)	$\rightleftharpoons H_2CO_3 \rightleftharpoons HCO_3^- + H^+$ ⟶ (HCO_3^- retained in blood)	H^+ excreted in urine
ALKALOSIS	↓ $CO_2 + H_2O$ ⟶ (Retained by slow breathing)	$\rightleftharpoons H_2CO_3 \rightleftharpoons H^+ + HCO_3^-$ ⟶ (H^+ retained in blood)	HCO_3^- excreted in urine

Figure 12.6 The carbonic acid/bicarbonate buffer system acts in the blood to prevent the conditions of both acidosis and alkalosis.

Table 12.5 Acidosis and Alkalosis

ACIDOSIS			
Type	Condition Results from	Causes	Compensatory Reactions
Respiratory	Retention of CO_2 Blood pH decreases	Hypoventilation Emphysema Congestive heart failure Bronchopneumonia Hyaline membrane disease Drugs that depress brain respiratory center	Increased rate of breathing Kidneys excrete acidic urine
Metabolic	Increase in H^+ Blood pH decreases	Diabetes mellitus Kidney failure Ingestion of acidic drugs such as aspirin Loss of HCO_3^-	Increased rate of breathing Kidneys excrete acidic urine

ALKALOSIS			
Type	Condition Results from	Causes	Compensatory Reactions
Respiratory	Rapid expulsion of CO_2 Blood pH increases	Hyperventilation High fevers Trauma Hysteria	Slower rate of breathing Kidneys excrete less acid
Metabolic	Increase in basic components in the blood Blood pH increases	Severe vomiting causing loss of stomach acid Excess ingestion of basic substances Kidney disease	Slower rate of breathing Kidneys excrete less acid

Another buffer system, active mainly within the cells, is the phosphate buffer system, which has a maximum buffering action at a pH of 7.2.

$$H_2PO_4^- \rightleftharpoons HPO_4^{2-} + H^+$$

Adding strong acid to this system will drive the reaction to the left, increasing the concentration of $H_2PO_4^-$, which is only weakly acidic. Large amounts of $H_2PO_4^-$ will result in acidosis, but the body will eliminate the excess in the urine. Adding strong base to the system will drive the reaction to the right, as the hydrogen ions react with the base to form water. Large amounts of HPO_4^{2-} would be found in alkalosis, but under normal kidney function the HPO_4^{2-} is also excreted in the urine (Figure 12.7).

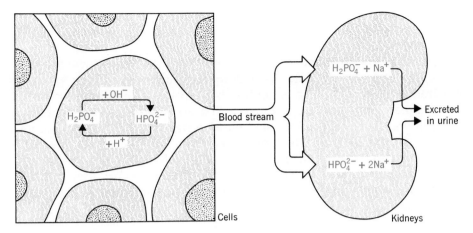

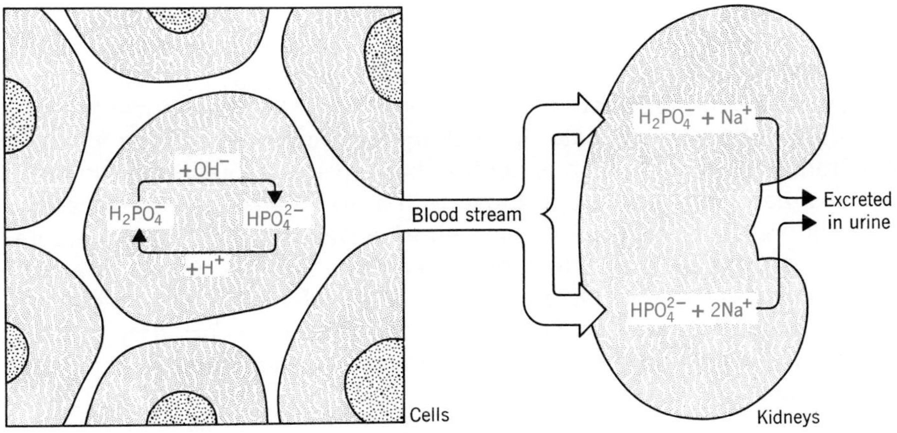

Figure 12.7 The phosphate buffer is active mainly in the cells to prevent changes in pH. The kidneys remove any excess HPO_4^{2-} or $H_2PO_4^-$ from the blood and excrete it in the urine.

Normal metabolic reactions in the body result in the continuous production of acids. Cells produce an average of about 10 to 20 moles of carbonic acid each day, which is equivalent to 1 or 2 liters of concentrated HCl. This acid must be removed from the cells and carried to the organs of excretion without disrupting the pH of the blood. It is through the action of the buffer systems in the cells and extracellular fluid that our bodies are protected from changes in pH that would otherwise be caused by these acids.

Chapter Summary

Acids are compounds which, when dissolved in water, produce solutions that conduct electricity, react with metals to produce hydrogen, taste sour, and turn litmus paper red. Bases are compounds that form solutions which conduct electricity, taste bitter, feel slippery, and turn litmus paper blue. The Brønsted-Lowry definition of acids and bases states that an acid is a substance that can donate a proton (H^+), and a base is a substance that can accept protons. A strong acid is one that has a large tendency to donate protons and, when dissolved in water, is almost completely ionized or dissociated. Only some of the molecules of a weak acid will donate their protons, so a weak acid added to water will cause a much smaller increase in the hydrogen ion concentration than the same amount of strong acid. A strong base will have a very large attraction for protons, whereas a weak base will have a weak attraction.

Acids will neutralize bases. If equal amounts of acid and base are reacted, a neutralization reaction occurs. A titration is a procedure by which a neutralization reaction is used to determine an unknown concentration of acid or base.

A few molecules of pure water will ionize to form hydroxide ions and hydrogen ions. In water the hydrogen ion concentration $[H^+]$ times the hydroxide ion concentration $[OH^-]$ will always equal 1×10^{-14}, or K_w, called the ion product of water. The pH scale has been invented as a convenient way to indicate the hydrogen ion concentration of a solution. Pure water, which is neutral, has a pH of 7. Solutions with a pH less than 7 are acidic, and those with a pH greater than 7 are basic.

Buffers are systems containing either a weak acid and its conjugate base, or a weak base and its conjugate acid, which will protect against sudden changes in pH caused by the addition of acid or base. Because living organisms are so sensitive to sudden changes in pH, they contain buffer systems within the cells, the extracellular fluid, and the blood to protect against such changes.

Exercises and Problems

1. In your own words, state the Brønsted-Lowry definition of (a) an acid and (b) a base.

2. Explain the difference between a coordinate-covalent bond and a covalent bond.

3. Label the conjugate acid-base pairs in the following reactions:

   (a) $HI + H_2O \rightleftharpoons H_3O^+ + I^-$
   (b) $CO_3^{2-} + H_2O \rightleftharpoons OH^- + HCO_3^-$
   (c) $CH_3COOH + H_2O \rightleftharpoons H_3O^+ + CH_3COO^-$
   (d) $HF + NH_3 \rightleftharpoons NH_4^+ + F^-$
   (e) $O^{2-} + H_2O \rightleftharpoons OH^- + OH^-$
   (f) $NO_2^- + N_2H_5^+ \rightleftharpoons HNO_2 + N_2H_4$
   (g) $HCl + NH_2OH \rightleftharpoons NH_3OH^+ + Cl^-$

4. Using Table 12.1, state which is the stronger acid in each of the following pairs:

   (a) NH_4^+ or H_3O^+
   (b) HCO_3^- or H_2CO_3
   (c) $H_2PO_4^-$ or HPO_4^{2-}
   (d) HCl or H_2SO_3

5. Using Table 12.1, state which is the stronger base in each of the following pairs.

   (a) Cl^- or OH^-
   (b) PO_4^{3-} or HPO_4^{2-}
   (c) NO_3^- or NH_3
   (d) HCO_3^- or CO_3^{2-}

6. Hydrogen sulfide, H_2S, is a stronger acid than phosphine, PH_3. Compare the relative strengths of their conjugate bases, HS^- and PH_2^-.

7. One of the uses of baking soda ($NaHCO_3$) described on its label is to relieve acid indigestion. Explain how baking soda acts in the relief of indigestion (use words or an equation).

8. You may have noticed that when you add lemon juice to your tea, the tea changes color. Suggest a reason for this color change.

9. Arrange the following pH values in order of increasing acid strength (from the least acidic to the most), and indicate which values represent basic, neutral, and acidic solutions.

   4, 6.3, 9.5, 1.4, 7, 5.5, 8.4, 12, 7.4

10. What is the pH of a solution having an $[H^+]$ of

    (a) 0.01 (b) 1×10^{-8} (c) $[OH^-] = 1 \times 10^{-4}$

11. What is the $[H^+]$ and the $[OH^-]$ of a solution having a pH of

    (a) 1 (b) 6 (c) 12

12. Indicate whether each of the solutions in questions 10 and 11 is acidic or basic.

13. Solution A has a pH of 3 and solution B a pH of 5. Which solution is more acidic, A or B? What is the hydrogen ion concentration in each solution, and by what factor do the two hydrogen ion concentrations differ?

14. For the neutralization reaction between ammonium chloride and sodium hydroxide, write the following:

    (a) a balanced equation for the reaction
    (b) the net-ionic equation for the reaction

    Would the odor of ammonia be greater before or after the sodium hydroxide was added to a solution of ammonium chloride? Give the reason for your answer.

15. How many milliliters of a 0.1*M* NaOH solution are required to completely neutralize 30 ml of a solution of strong acid that has a pH of 2?

16. Four aqueous solutions were prepared as follows:

    1. 36.5 mg HCl in 100 ml
    2. 2.45 g of H_2SO_4 in 500 ml
    3. 126 mg HNO_3 in 2 liters
    4. 0.4 g NaOH in 100 ml

    For each of the above solutions, calculate:

    (a) The molarity of the solution.
    (b) The hydrogen ion concentration in moles per liter (assume that each compound is 100% ionized).
    (c) The pH of the solution.

17. A small private laboratory specializes in analyzing samples of waste water to see if they meet federal pollution standards. The analyst uses titrations to determine the molarity of acid or base in each sample. For each of the following samples, (a) calculate the molarity of the sample and (b) indicate whether the analyst was testing for acid or base in the sample.

(a) Sample 1: 100.0 ml required 52.0 ml of 0.100*M* NaOH
(b) Sample 2: 1.00 liter required 150.0 ml of 0.200*M* HCl
(c) Sample 3: 25.0 cc required 30.0 ml of 0.150*M* KOH
(d) Sample 4: 50.0 ml required 15.0 ml of 0.090*M* HNO_3

18. Use words and equations to explain how an acetic acid-acetate ion buffer system can protect against pH changes when (a) acid is added to the system and (b) when base is added.

19. A large amount of lactic acid is produced in the muscles during very hard exercise. This lactic acid must be transported by the blood to the liver, where it is broken down. Why doesn't the pH of the blood change drastically after very hard exercise?

20. What means does the body use to correct the conditions of (a) acidosis and (b) alkalosis?

21. Would you expect a person in a diabetic coma to be breathing faster or slower than normal? Upon analysis would the pH of her urine be higher (more basic) or lower (more acidic) than normal? Give reasons to support both of your answers.

22. The following equilibrium is formed between ammonia and the ammonium ion:

$$NH_3 + H^+ \rightleftharpoons NH_4^+$$

Use the following statements and your knowledge of Le Chatelier's Principle to explain how ammonia is removed from the blood and excreted into the kidney tubules that carry urine to the bladder.

1. NH_3 is soluble in, and can pass through, cell membranes, but NH_4^+ cannot.
2. The normal pH of the urine in the kidney tubules is between 5.5 and 6.5.

section III

the elements necessary for life

chapter 13

Introduction to Organic Chemistry

Learning Objectives

By the time you have finished this chapter you should be able to:

1. Define *organic chemistry.*

2. Describe the difference between molecular and structural formulas of a compound.

3. Given the molecular formula of a compound, draw the structural formula of its isomers.

4. Given two structural formulas, state whether they represent the same compound, isomers, or two unrelated compounds.

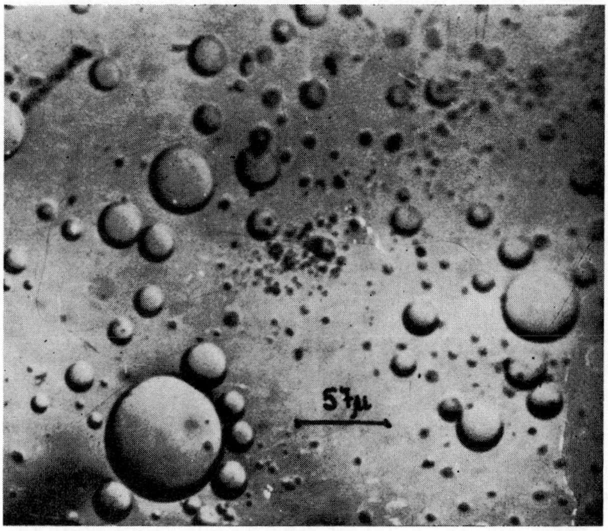

Four and one-half billion years ago the planet earth was newborn, having formed as a ball of hot gases from huge masses of gas in outer space. Over endless centuries the surface of the earth slowly cooled and became solid, forming the rocks known as the earth's mantle. As this cooling continued, steam condensed to form water which filled the low places in the mantle, creating the oceans of the earth. This infant earth was a lifeless landscape of rock and water, constantly pounded by violent storms and heavy rains. The first atmosphere that may have formed, enriched by gases pouring from the interior of the earth in volcanic eruptions, consisted mostly of methane (CH_4) and nitrogen (N_2), with hydrogen (H_2), ammonia (NH_3), carbon monoxide (CO), and water vapor (H_2O) in smaller quantities. There was no free oxygen (O_2).

In this very earliest, or primordial, atmosphere, chemical reactions were started by ultraviolet radiations, ionizing radiations, and lightning. Many of the products of these reactions were simple carbon compounds. Such molecules, formed in the atmosphere, were brought to earth with the rains. They settled in the pools and oceans, forming a warm, dilute, aqueous solution of dissolved minerals and carbon compounds—a "primordial soup" rich in the building blocks for basic life processes. Many theories have been proposed describing how life began and developed; these theories range from the Biblical story of creation in seven days to recent scientific explanations covering billions of years of slow change. The following is a theory proposed by the Russian chemist A. I. Oparin, describing how life may have first begun early in the earth's history.

Oparin suggests that, as the centuries passed, the carbon compounds floating in the primordial soup began to join together to form larger and larger molecules, forming a thin layer on the surface of the water. Once these molecules reached a certain critical size, they began to clump together to form droplets that floated through the surrounding medium. These droplets, or coacervates, had the extraordinary ability to absorb only selected materials from the surrounding environment and to form a surface layer, or boundary, having properties different from the droplet itself. Various molecules could be held very close to one another within the coacervate, allowing their concentrations to be increased to levels much higher than the surrounding environment. This represented the first major step toward the beginning of life.

The coacervates that survived over time were those in which certain types of chemical reactions that would help with their maintenance (such as repairing or replacing the surface molecules) took place at a faster rate than in the outside primordial soup. As the ages passed, coacervate systems developed which were capable of growth as well as simple maintenance. This capability required the development of complicated chemical pathways to trap chemical energy that could be obtained from other molecules found in the primordial soup.

These coacervate systems multiplied by fragmentation due to wave action or to internal stress. However, the droplets that best survived such fragmentation were mainly those with improved chemical systems having specific chemical pathways for the types of reactions most important to the maintenance of life. The coacervate droplets having such improved internal organization are called protobionts, and may be thought of as a link between the original droplets and the most primitive of living organisms. Oparin suggests that the evolution of the protobiont into a primitive living organism occurred about 3.5 billion years ago, when the protobionts were able to develop methods of controlling the chemical reactions taking place within themselves. This led to a more efficient use of nutrients and a much more rapid growth, and allowed the development of reproductive processes that insured the survival of the organism. With the development of mechanisms to control the organism's own reproduction, we can truly say life began.

Elements Necessary for Life

13.1 Elements Abundant in Living Organisms

Living organisms, like all other matter on earth, are composed of atoms of the 90 naturally occurring elements. However, not all 90 of these elements are found in such organisms. The periodic table in Figure 13.1 shows the 24 elements that have thus far been shown to be essential to life. Hydrogen,

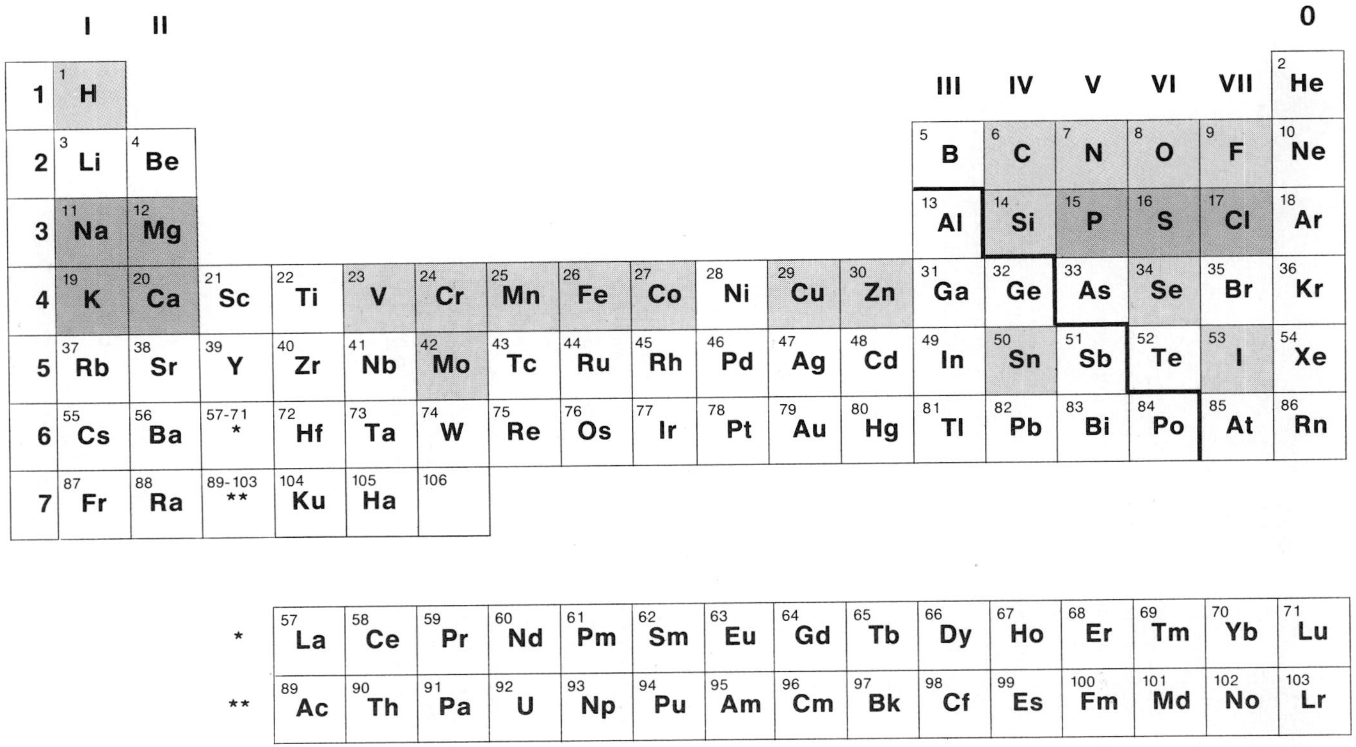

Figure 13.1 The elements that are essential to life. The elements most abundant in the living organism, making up 99.3% of all the atoms in the human body, are shown in the brightest shading. The seven elements shown in medium shading make up only about 0.7% of the atoms in the body. The remaining elements that are essential to life are called the trace elements and make up less than 0.01% of the atoms in the body.

Figure 13.2 The tetrahedral carbon atom. The most stable arrangement for the four covalent bonds formed by carbon occurs when these bonds are directed toward the four corners of a tetrahedron.

carbon, nitrogen, and oxygen are the most plentiful, or abundant, elements in the living organism. They make up 99.3% of all the atoms in your body, whereas the remaining 20 elements account for only 0.7%.

13.2 The Role of Carbon

At the beginning of this chapter we described a primordial soup rich in molecules containing carbon atoms. What is so special about carbon-containing molecules that they should be the building blocks of life?

Look at the periodic table in Figure 13.1. Carbon is the first member of group IV, a family also containing silicon (Si), germanium (Ge), tin (Sn), and lead (Pb). Each member of this family has four valence electrons. To attain a stable octet of electrons these elements can lose four electrons, or share four electrons. Under ordinary conditions carbon has a strong tendency to share four electrons, and thus to form four covalent bonds. Because the negatively charged electron pairs repel each other, the most stable arrangement for these four electron pairs occurs when they are as far as possible from the other pairs. Therefore, the bonds formed by these electron pairs will be found directed toward the corners of a tetrahedron, with the carbon atom in the center (Figure 13.2).

A carbon atom can form four single bonds with four different atoms or with other carbon atoms. In this second case the other carbon atoms can, in turn, be bonded to up to three more carbon atoms, and so on. The very stable molecule that results when every carbon atom is bonded to four other carbon atoms is diamond (Figure 13.3). A major feature, then, making carbon very special among the elements is its ability to form strong stable bonds with up to four other atoms identical to itself.

We have mentioned that diamond is the molecule formed when every carbon is bonded to four other carbon atoms. But all four bonds need not be to other carbon atoms, and the molecules found in living organisms are rarely of this type. Rather, in such molecules we find carbon atoms bound to other carbon atoms in the form of long chains, branched chains, or rings (Figure 13.4). As the number of carbon atoms in a molecule increases, the number of ways in which they can be arranged increases,

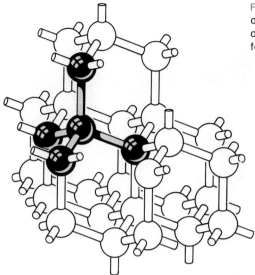

Figure 13.3 A model of the bonding in diamond. Each carbon atom sits at the center of a tetrahedron, and is bonded to four other carbon atoms.

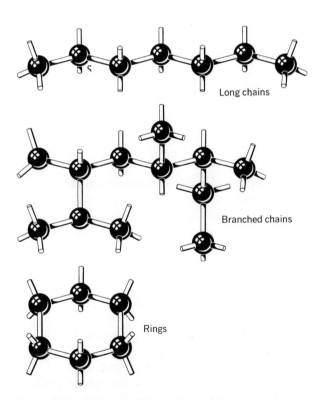

Figure 13.4 Some possible arrangements of carbon atoms found in organic compounds. (Only the carbon atoms are shown.)

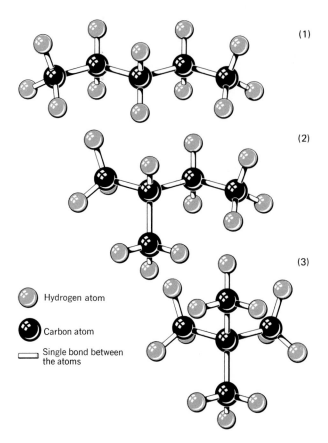

(1)

(2)

Hydrogen atom

Carbon atom

Single bond between
the atoms

(3)

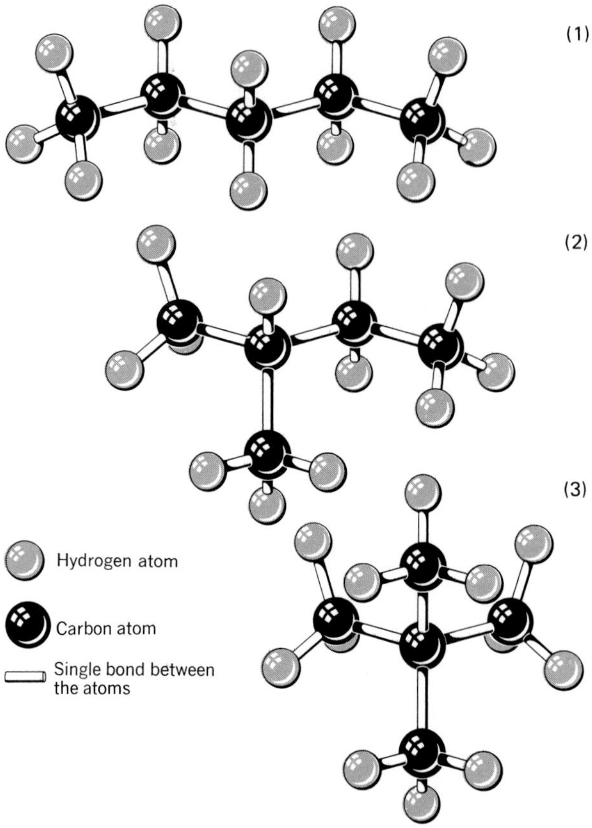

Figure 13.5 These molecules all have the molecular formula C_5H_{12}. They have different structural
formulas and, therefore, are isomers.

thus making it possible to form compounds with the same chemical
composition (the same number of atoms of each element), but with
different structures and with correspondingly different chemical and
physical properties. Compounds having the same molecular formulas but
with different geometric structures are called **isomers** (Figure 13.5).

13.3 Organic Chemistry

Carbon's special properties make it unique among the elements. First,
it is the only element whose atoms can bond together to form very long
straight and branched chains, as well as intricate ring structures. (The
atoms of silicon, directly below carbon on the periodic table, can form
only short chains.) Second, carbon can also form strong bonds with a
number of other elements: hydrogen, oxygen, nitrogen, sulfur, phosphorus,
and the halogens. Third, the resulting large compounds can have more
than one arrangement of atoms (or many isomers). The resulting large
number of different compounds has given rise to a field of chemistry
devoted entirely to the study of just such carbon compounds; this is the

field of **organic chemistry.** The number of known organic compounds, both natural and synthetic, is greater than two million, whereas the number of known inorganic compounds (those formed from combinations of all the other known elements) is only about 500,000. The reason for this comparatively small number of inorganic compounds is that molecules of covalent inorganic compounds are composed of only a few atoms. Two or three atoms will form a very stable inorganic molecule, but as more atoms are added the molecule becomes unstable and is more likely to fall apart. Inorganic molecules containing more than 12 atoms are quite rare. However, it is not uncommon for a large organic molecule such as a complex protein to contain more than a million atoms!

You might not look forward to studying a field of chemistry covering more than two million compounds, but fortunately the chemical properties of many of these compounds are similar, allowing them to be conveniently grouped into several classes or series of compounds. In that way, by looking at one or two examples of the chemistry of organic compounds belonging to a given class, we can obtain an understanding of the chemistry of all compounds belonging to that class.

Formulas and Nomenclature

Before beginning our study of organic chemistry, we will pause to take a brief look at two concepts that often cause some confusion. These are the topics of formula writing and nomenclature, and a basic understanding of these areas should give you a good headstart on your studies of this section.

13.4 Writing Structural Formulas

First of all, let's take a look at formula writing. Previously, we mentioned that in organic chemistry it is common to find several different compounds having the same molecular formula. Therefore, we will often find it important to present the structural formula of the molecule. Structural formulas are no more than chemical diagrams. They make organic chemistry easier to follow in much the same way that diagrams can help in assembling a bicycle or in sewing a dress. Although chemical compounds actually have three-dimensional structures, it is often very difficult to draw them in that way. Instead, therefore, we will use two-dimensional diagrams to represent these three-dimensional structures. For example,

instead of [structure] we will write

$$H-\underset{\underset{H}{|}}{\overset{\overset{H}{|}}{C}}-\underset{\underset{H}{|}}{\overset{\overset{H}{|}}{C}}-\underset{\underset{H}{|}}{\overset{\overset{H}{|}}{C}}-H$$

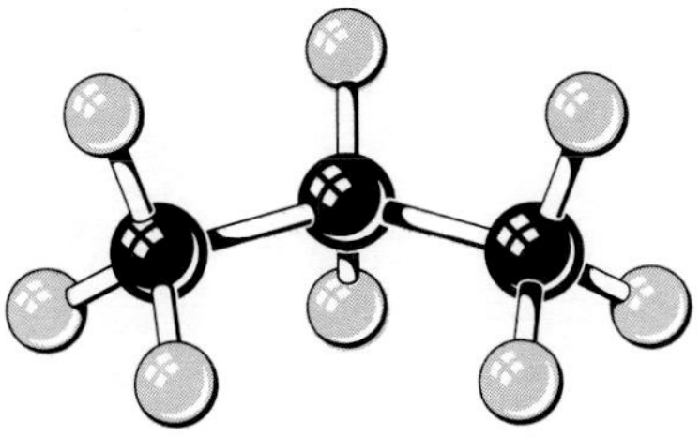

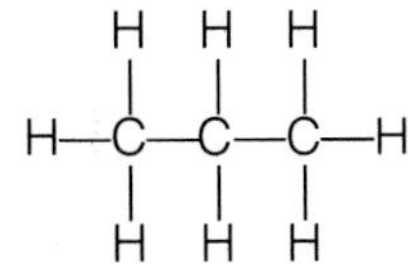

In the structural formula of a compound, the shared pair of electrons in a

covalent bond is indicated by a line drawn between the two atoms connected by that bond. Each atom is represented by the symbol of its element. For example, the three isomers of C_5H_{12} shown in Figure 13.5 can be written as follows:

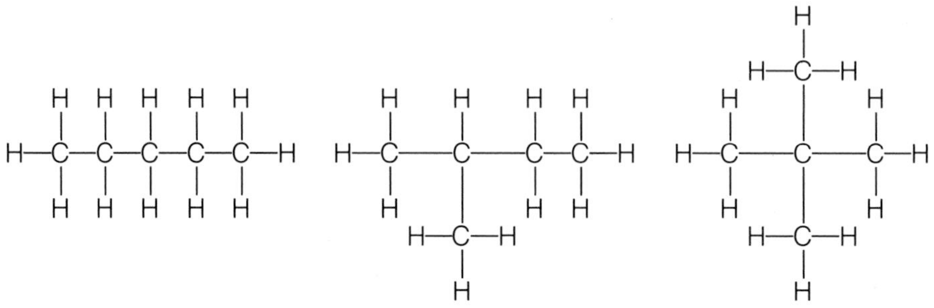

(1) (2) (3)

This subject would be fairly straightforward if we could stop our discussion right here, but drawing the structural formulas of large organic molecules can be quite time consuming. Therefore, organic chemists have devised several ways of shortening the procedure. One commonly used method is to indicate the atoms that are bonded to the carbon by writing their symbol after the carbon symbol, but without bothering to use a dash for the bonds. For example,

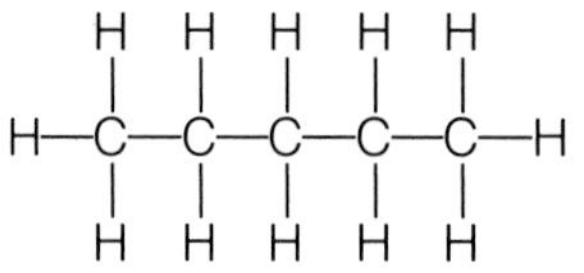

becomes $CH_3CH_2CH_2CH_2CH_3$

or $CH_3(CH_2)_3CH_3$

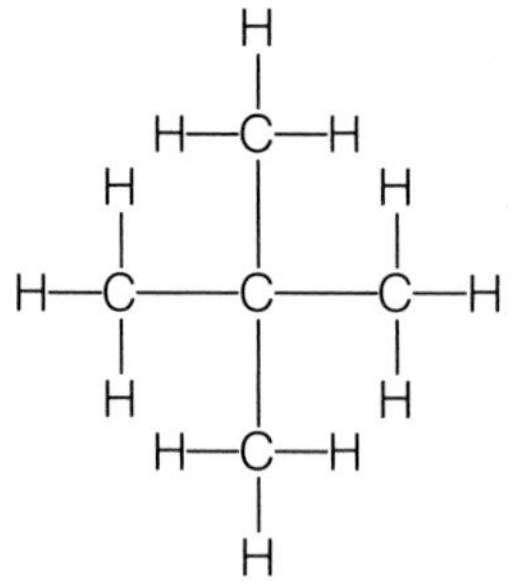

becomes $CH_3{-}\underset{\displaystyle CH_3}{\overset{\displaystyle CH_3}{C}}{-}CH_3$

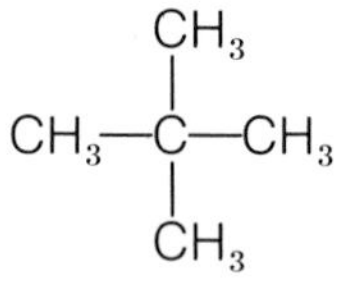

or $CH_3C(CH_3)_2CH_3$

Or, quite often the part of the molecule in which we are interested is drawn out in detail, while the rest of the molecule is represented in shorthand.

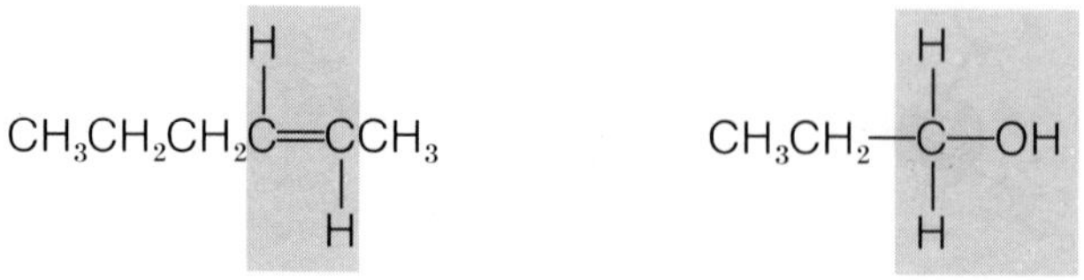

As you read the following chapters and see more structural formulas, you will find that they soon become quite easy to read, and help make chemical discussions much easier to follow.

13.5 IUPAC Nomenclature

Nomenclature is an equally important topic for the beginner to understand, because it is almost impossible to discuss any of the molecules of life without a basic knowledge of the way in which these compounds are named. Before a standard procedure for naming compounds was finally established, most organic compounds were known by common names, usually indicating the source of the compound rather than its chemical structure. For example, lactic acid is the common name of a compound found in sour milk, getting its name from the Latin word for milk, *lactis*. The name *lactic acid,* however, does not give us any clue as to the structure of this compound. We would have to have memorized the structure of the lactic acid molecule in order to recognize its chemical properties. Given that there are more than two million known organic compounds, the study of organic chemistry would clearly be impossible without some system of nomenclature which could be standardized and which would indicate the structure of the molecule being named. For this reason, the International Union of Pure and Applied Chemistry (IUPAC) began meeting in Geneva in 1892 to establish the rules for such a naming system.

A complete discussion of the IUPAC rules for naming compounds must be left for other texts, but we will briefly describe some of the basic rules.

Rule 1. Prefixes are used to indicate the number of carbon atoms in the main carbon chain.

Number of Carbon Atoms	Prefix	Number of Carbon Atoms	Prefix
1	meth-	6	hex-
2	eth-	7	hept-
3	prop-	8	oct-
4	but-	9	non-
5	pent-	10	dec-

Rule 2. Organic compounds can be categorized into groups or classes. Each group is given a suffix that identifies the group.

Suffix	Class of Compounds	Example
-ane	alkanes	prop*ane*
-ene	alkenes	prop*ene*
-yne	alkynes	prop*yne*
-ol	alcohols	propan*ol*
-one	ketones	propan*one*
-oic acid	carboxylic acids	propan*oic acid*

Rule 3. Atoms or groups of atoms (other than hydrogen) attached to the carbon chain are listed before the name of the compound. For example, bromomethane is a compound in which a bromine atom has replaced one of the hydrogen atoms on a molecule of methane, and chloroethane is a compound in which a chlorine atom has replaced one of the hydrogens on a molecule of ethane.

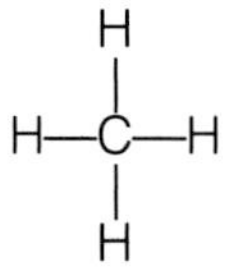

Methane

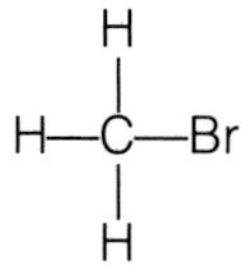

Bromomethane

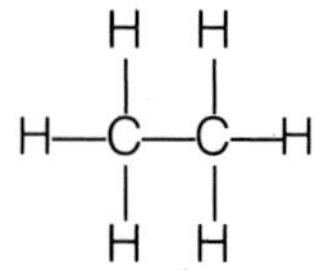

Ethane

Chloroethane

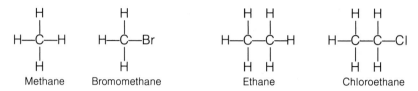

The names of several important groups often found attached to the carbon chain are:

Group	Name	Group	Name
−F	fluoro-	−I	iodo-
−Cl	chloro-	$-NH_2$	amino-
−Br	bromo-	$-NO_2$	nitro-

Rule 4. Groups of atoms that are attached to the main carbon chain and that contain only carbon and hydrogen arranged in straight or branched chains are called **alkyl groups.** They are named using the prefixes shown in Rule 1.

$-CH_3$ is the methyl group
$-CH_2CH_3$ is the ethyl group
$-CH_2CH_2CH_3$ is the propyl group, etc.

Appendix 3 gives the names of the most commonly used alkyl groups.

Rule 5. If the organic compound contains more than one of the same type of these "other" atoms, the number is indicated by a prefix: di- indicates two, tri- indicates three, and tetra- indicates four. For example, dibromomethane is a methane molecule containing two atoms of bromine, and tetrachloromethane is a methane molecule containing four atoms of chlorine.

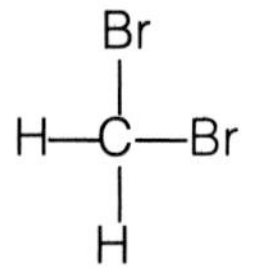

Dibromomethane

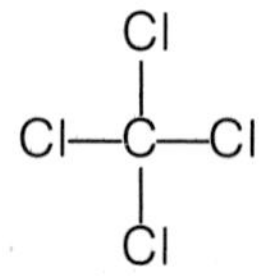

Tetrachloromethane
(Carbon tetrachloride)

Rule 6. If two or more different groups are attached to the carbon chain, they are named either in order of increasing number of carbons in the attached chain (for example, methylethylamine), or simply in alphabetical order (for example, ethylmethylamine).

Rule 7. To indicate the carbon atoms to which these "other" atoms are attached, numbers corresponding to the carbon atoms precede the name of the attached group. The carbon atoms are numbered from the end of the chain nearest the attached group. For example, 1, 2-dichloroethane is an ethane molecule containing two atoms of chlorine—one attached to carbon number 1 and one attached to carbon number 2.

$$\begin{array}{cc} Cl & Cl \\ | & | \\ H-C-C-H \\ | & | \\ H & H \end{array} \qquad \begin{array}{cc} Cl & Br \\ | & | \\ H-C-C-Br \\ | & | \\ H & H \end{array}$$

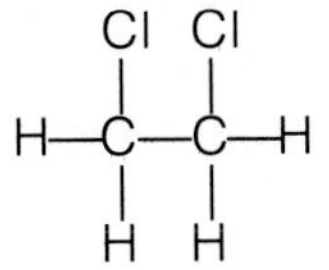

1, 2-Dichloroethane 1, 1-Dibromo-2-chloroethane

To see how these rules work in actual practice, let's return to our example of lactic acid. The IUPAC name for lactic acid is 2-hydroxypropanoic acid. The structure of lactic acid is as follows:

$$\begin{array}{ccc} H & H & O \\ | & | & \| \\ H-C_3-C_2-C_1-OH \\ | & | \\ H & OH \end{array}$$

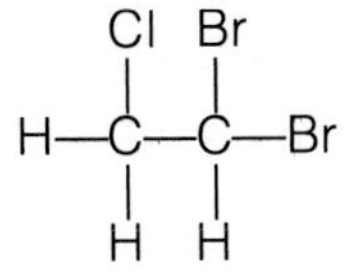

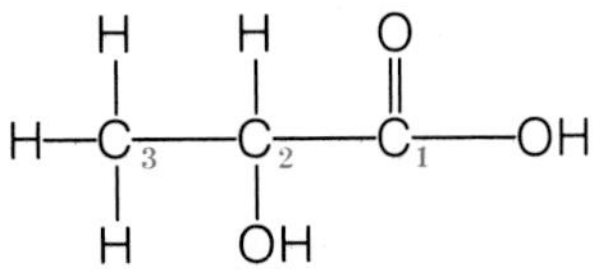

In the IUPAC name, the prefix "prop-" tells us this compound has three carbon atoms in its main carbon chain. The suffix "-oic acid" tells us that this compound belongs to the class of carboxylic acids, which are molecules containing the following group of atoms:

$$\begin{array}{c} O \\ \| \\ -C-OH \end{array}$$

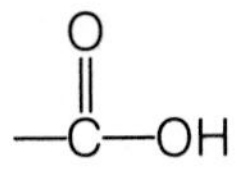

And finally, "2-hydroxy" tells us that this compound contains a hydroxyl group of atoms (the combination —OH), and that this group is attached to carbon number 2. To complete the structure of the molecule, any other possible carbon bonds not already described are assumed to lead to hydrogen atoms. Thus, this IUPAC name allows us to determine the entire structure of the lactic acid molecule. Let's face it, though—the name lactic acid is certainly easier to say and to remember than 2-hydroxypropanoic acid. This is the reason that common names are still used, and we will occasionally use common names in this book (especially when the structure is very complex).

In the next three chapters we will discuss some major classes of organic compounds. We will explain the specific rules used for naming compounds in these classes, but you will often need to refer back to the general rules we have just discussed when doing the exercises in those chapters.

Chapter Summary

Living organisms are made up of compounds containing only 24 of the 90 naturally occurring elements. Of these elements, the four most abundant are carbon, hydrogen, oxygen, and nitrogen. Carbon-containing compounds are the building blocks of life. The carbon atom is especially suited for this purpose because it can form strong stable bonds with up to four other carbon atoms. Molecules can have carbon atoms connected in long chains, branched chains, or rings—resulting in a great variety of molecules. Compounds that have the same molecular formula but different structural formulas are called isomers. The field of chemistry that studies carbon compounds is called organic chemistry. The study of this large number of compounds is simplified by using structural formulas instead of molecular formulas, and by following rules of nomenclature established by the International Union of Pure and Applied Chemistry (IUPAC).

Exercises and Problems

1. Carbon is often said to form the "backbone" of life molecules. What does this mean, and what special properties of carbon make this possible?

2. Define *organic chemistry* in your own words.

3. What is the difference between a molecular and a structural formula of a compound?

4. Write the structural formulas for the five isomers of C_6H_{14}.

5. State whether the following pairs of structures are (a) the same compound, (b) isomers, or (c) unrelated compounds. (Note that to be the same compound, the structural formula must have the identical sequence (or order) of atoms, no matter how they are arranged on the paper. If the molecular formulas are the same, but the sequence of atoms is different, then the compounds are isomers.)

(a) CH_3—CH_2—CH_3

(b) CH_3—CH_2—CH_2—CH_3

(c) CH_3—$\overset{\overset{\displaystyle CH_3}{|}}{C}H$—$CH_2$—$CH_3$

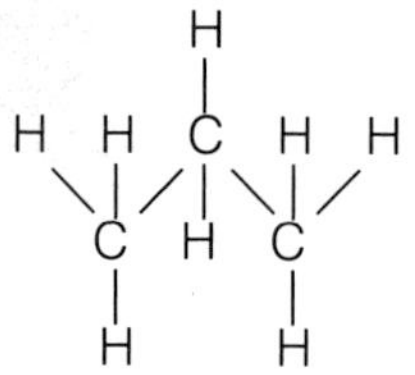

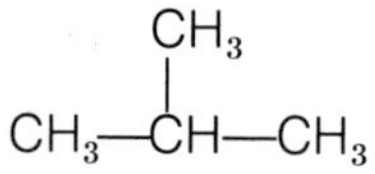

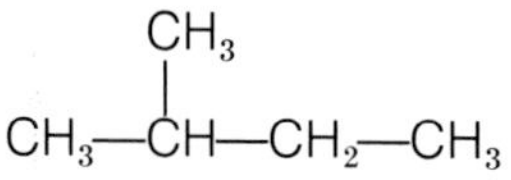

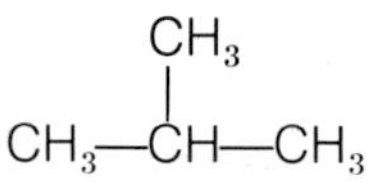

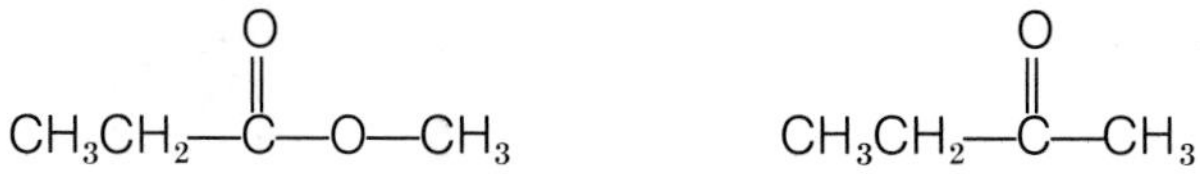

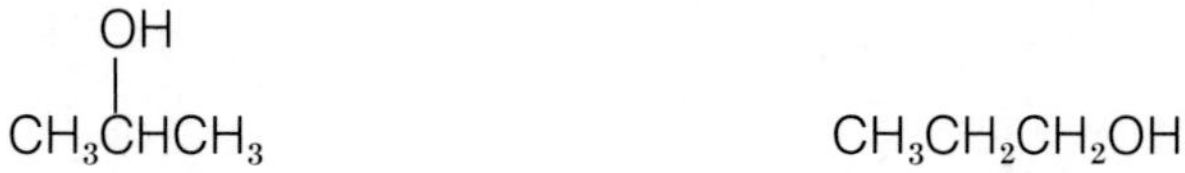

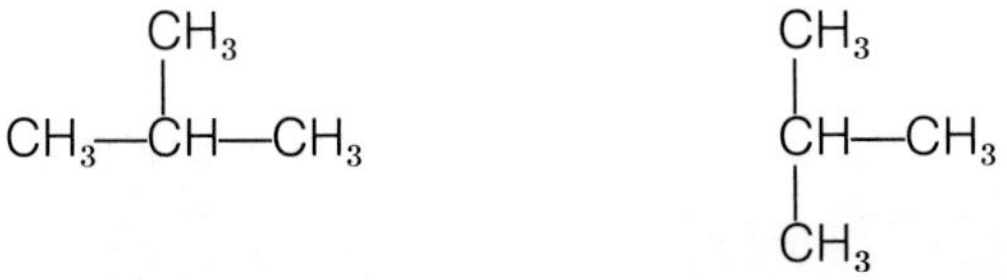

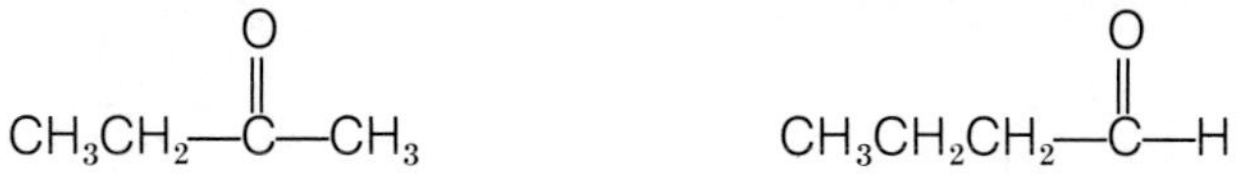

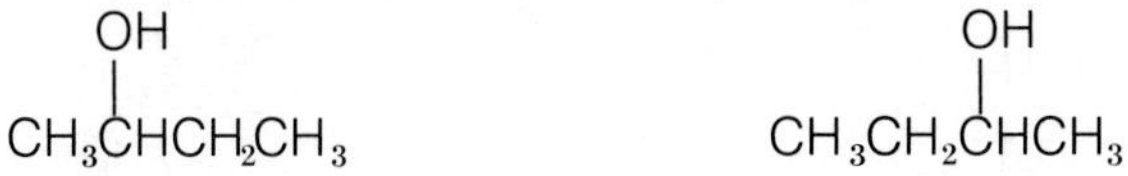

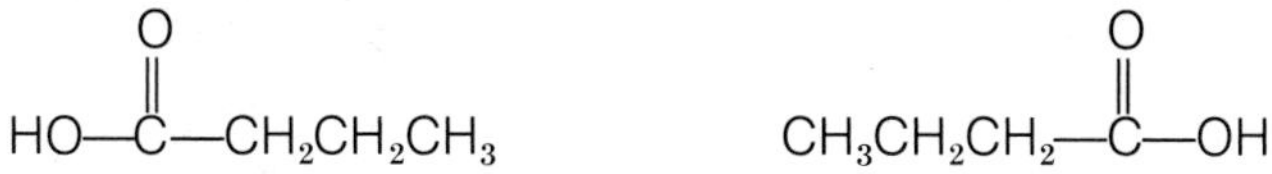

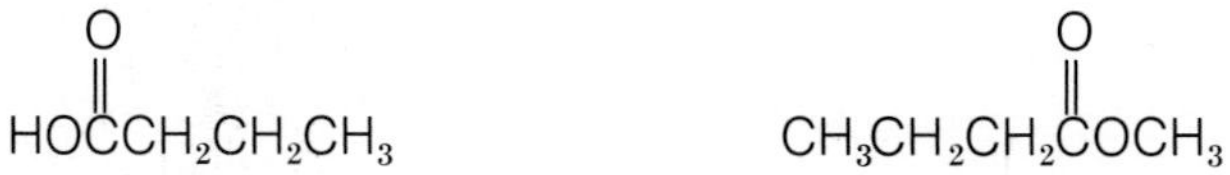

(d) $CH_3-\overset{\overset{\displaystyle CH_3}{|}}{C}H-CH_3$

$\overset{\overset{\displaystyle CH_3}{|}}{\underset{\underset{\displaystyle CH_3}{|}}{C}H}-CH_3$

(e) $CH_3\overset{\overset{\displaystyle OH}{|}}{C}HCH_2CH_3$

$CH_3CH_2\overset{\overset{\displaystyle OH}{|}}{C}HCH_3$

(f) $CH_3\overset{\overset{\displaystyle OH}{|}}{C}HCH_3$

$CH_3CH_2CH_2OH$

(g) CH_3-O-CH_3

CH_3CH_2-OH

(h) $CH_3CH_2-\overset{\overset{\displaystyle O}{\|}}{C}-CH_3$

$CH_3CH_2CH_2-\overset{\overset{\displaystyle O}{\|}}{C}-H$

(i) $HO-\overset{\overset{\displaystyle O}{\|}}{C}-CH_2CH_2CH_3$

$CH_3CH_2CH_2-\overset{\overset{\displaystyle O}{\|}}{C}-OH$

(j) $CH_3CH_2-\overset{\overset{\displaystyle O}{\|}}{C}-O-CH_3$

$CH_3CH_2-\overset{\overset{\displaystyle O}{\|}}{C}-CH_3$

(k) $HO\overset{\overset{\displaystyle O}{\|}}{C}CH_2CH_2CH_3$

$CH_3CH_2CH_2\overset{\overset{\displaystyle O}{\|}}{C}OCH_3$

(l) $CH_3CH_2NH_2$

CH_3NHCH_3

(m) $H-\overset{\overset{\displaystyle H}{|}}{\underset{\underset{\displaystyle H}{|}}{C}}-\overset{\overset{\displaystyle OH}{|}}{\underset{\underset{\displaystyle H}{|}}{C}}-\overset{\overset{\displaystyle H}{|}}{\underset{\underset{\displaystyle CH_3}{|}}{C}}-\overset{\overset{\displaystyle H}{|}}{\underset{\underset{\displaystyle H}{|}}{C}}-\overset{\overset{\displaystyle O}{\|}}{C}-H$

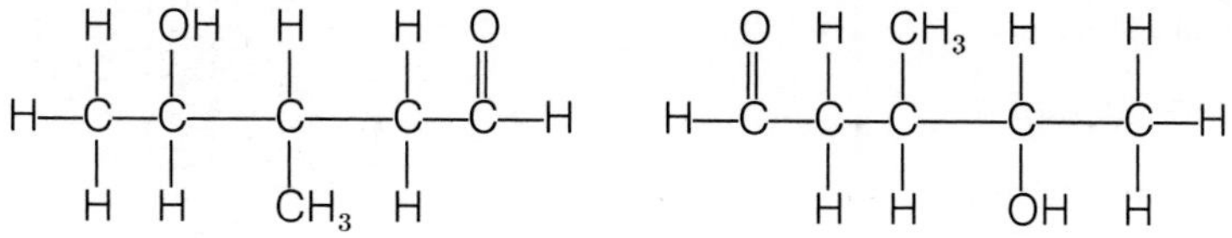

$H-\overset{\overset{\displaystyle O}{\|}}{C}-\overset{\overset{\displaystyle H}{|}}{\underset{\underset{\displaystyle H}{|}}{C}}-\overset{\overset{\displaystyle CH_3}{|}}{\underset{\underset{\displaystyle H}{|}}{C}}-\overset{\overset{\displaystyle H}{|}}{\underset{\underset{\displaystyle OH}{|}}{C}}-\overset{\overset{\displaystyle H}{|}}{\underset{\underset{\displaystyle H}{|}}{C}}-H$

(n) $CH_3-\overset{\overset{\displaystyle CH_3}{|}}{C}H-CH_2-\overset{\overset{\displaystyle O}{\|}}{C}-CH_3$

$CH_3-\overset{\overset{\displaystyle CH_3}{|}}{C}H-CH_2-\overset{\overset{\displaystyle OH}{|}}{C}H-CH_3$

(o) $CH_3-\overset{\underset{\underset{\displaystyle CH_3}{|}}{C}H}{}-\overset{\overset{\displaystyle O}{\|}}{C}-NH_2$

$CH_3-NH-\overset{\overset{\displaystyle O}{\|}}{C}-\overset{\underset{\underset{\displaystyle CH_3}{|}}{C}H}{}-CH_3$

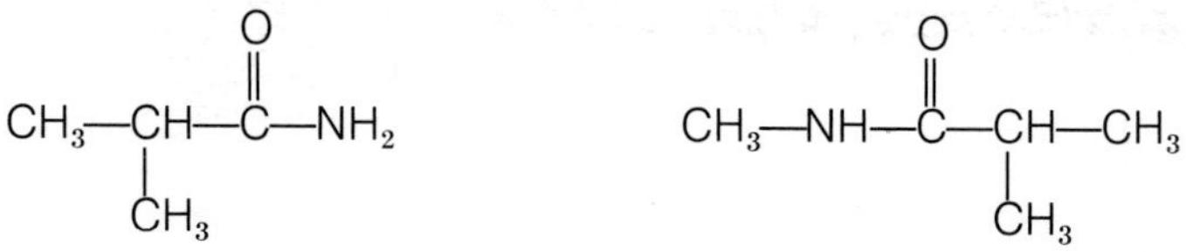

(p) $CH_3-\overset{\overset{\displaystyle CH_3}{|}}{C}H-CH_2-CH-\overset{\overset{\displaystyle CH_3}{|}}{\underset{\underset{\displaystyle CH_2CH_3}{|}}{C}H}-CH_3$

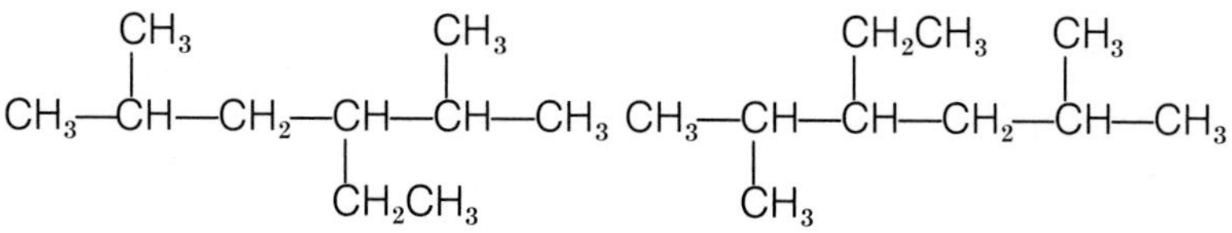

$CH_3-\overset{\overset{\displaystyle CH_2CH_3}{|}}{\underset{\underset{\displaystyle CH_3}{|}}{C}H}-CH-CH_2-\overset{\overset{\displaystyle CH_3}{|}}{C}H-CH_3$

chapter 14

Carbon and Hydrogen

Learning Objectives

By the time you have finished this chapter, you should be able to:

1. Describe the difference between the alkanes, alkenes, and alkynes, and give examples of molecules found in each class.

2. Draw the structural formula of a hydrocarbon when given its IUPAC name, and state its IUPAC name when shown its structural formula.

3. Given the structural formula of a hydrocarbon, identify whether it is saturated or unsaturated.

4. Define *structural isomer* and give two examples.

5. Define *geometric isomer* and give an example of the *cis* and *trans* forms of an alkene and a cyclic hydrocarbon.

6. Describe the polymerization reaction of an alkene.

7. Compare the chemical reactivity of the alkanes, alkenes, and alkynes.

8. Describe the structure of the benzene molecule, and explain why the molecule is extremely stable.

9. Define *carcinogen.*

10. Give several examples of compounds belonging to the halogenated hydrocarbons, and explain their uses.

In the summer of 1973 Bill Hamilton drove to Battle Creek, Michigan, to pick up a load of feed for his dairy herd. The feed mill in Battle Creek supplied many of the dairy farmers of Michigan, who bought a special fodder enriched with magnesium to improve their herds' milk production. Shortly before Bill's visit, the mill had prepared a new batch of fodder, adding bags of magnesium oxide (MgO) to the feed to provide the needed Mg^{2+} ion. However, this time there had been a disastrous mistake: Among the bags of MgO delivered to the feed mill by Michigan Chemical Corporation there were also four bags of a fire-fighting chemical made by the same company. The bags of this chemical—polybrominated biphenyls, or PBB—were never noticed, and were added to the feed together with the MgO.

Two years later Bill Hamilton's farm still looked much the same, with its neat white farmhouse and large red dairy barn. But it was a lifeless place. All that now remained of Bill's once proud dairy herd were three scrawny cows that could not produce milk. The chicken coops were empty, and the barnyard cats had long since died. Bill Hamilton, his wife, and his two children all suffered from aching joints, muscle weakness, and fatigue. Chores that used to take only a few minutes were now taking hours to finish, and Bill would often have trouble remembering even simple details of his farmwork.

Hundreds of other Michigan farm families also lost their dairy herds and suffered similar illnesses. Although animals had started showing symptoms soon after the contaminated feed was distributed, it wasn't until eight months later that PBB was identified as the cause of the problem, and then only because of the determined efforts of a dairy farmer who was also a chemist by training. Meanwhile, farm families had been eating eggs, meat, and milk contaminated with PBB. By 1975 thousands of cows and other animals found to be contaminated with even extremely small amounts of PBB were killed and buried, completely destroying many of the dairy herds in Michigan.

What makes this story different from many other cases of livestock poisoning that have occurred throughout history are the special properties of PBB. Much like a similar environmental pollutant, the industrial PCB's (polychlorinated biphenyls), PBB is an extremely stable molecule that is practically indestructible. It is not broken down by reactions with moisture or the air, and is not metabolized by microorganisms. When taken up by living organisms it is not excreted, but rather is stored in the fatty (adipose) tissue. There it is passed on when eaten by other creatures, or passed from mother to child through the placenta and the milk. In fact, in a test of the milk of new mothers in Michigan, PBB was detected in every case!

The great stability of PBB and other related compounds, therefore, is a very important property. In this chapter we will see how this stability is related to the properties of many molecules in the large class of compounds called hydrocarbons.

Hydrocarbons

14.1 Petroleum

The **hydrocarbons** are a large class of organic compounds, each of which contains atoms of only carbon and hydrogen. These compounds are the basic building blocks from which all other organic compounds can be formed. A major source of hydrocarbons in commercial quantities is petroleum, which is a mixture of hundreds of different hydrocarbons. Petroleum was formed over a period of millions of years as large regions of plant life died, decayed, and were covered over as the earth slowly changed. After crude petroleum is pumped out of the ground, it is transported to refineries where the hydrocarbons are separated by a process called **fractional distillation.** This process makes use of the fact that compounds composed of long hydrocarbon chains (high molecular weight) require more heat and higher temperatures to vaporize than do compounds composed of shorter chains (lower molecular weight). In other words, a hydrocarbon with a long carbon chain has a higher boiling point than one with a shorter carbon chain. In fractional distillation, the petroleum is heated slightly; this causes the molecules with short carbon chains to vaporize, and they are drawn off. As the heat is increased, molecules with longer and longer carbon chains vaporize, are drawn off, and are collected. The actual contents of each fraction that is drawn off may vary with the source of the petroleum. Each fraction may then be further processed to produce the particular hydrocarbons that are desired (Figure 14.1).

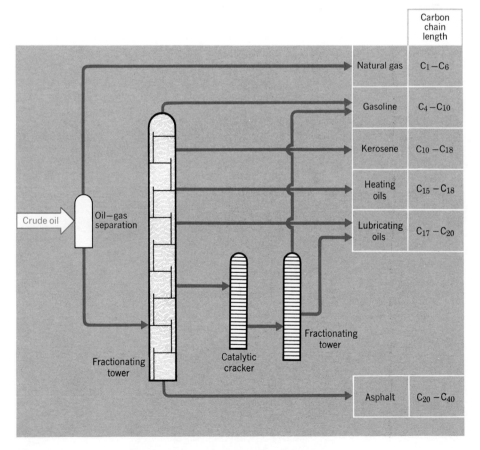

	Carbon chain length
Natural gas	$C_1 - C_6$
Gasoline	$C_4 - C_{10}$
Kerosene	$C_{10} - C_{18}$
Heating oils	$C_{15} - C_{18}$
Lubricating oils	$C_{17} - C_{20}$
Asphalt	$C_{20} - C_{40}$

Figure 14.1 Fractional distillation of petroleum. The molecules found in petroleum (crude oil) are separated on the basis of their boiling points. The molecules that remain are called heavy bottoms; they have very long carbon chains and are used to make asphalt and coke.

Saturated Hydrocarbons

14.2 Alkanes

The simplest of all hydrocarbons is the gas methane, CH_4. In this molecule the carbon is bonded to four hydrogen atoms that are located at the four corners of a tetrahedron. This makes the methane molecule symmetrical and nonpolar. It is not soluble in water (a highly polar solvent), and has a very low boiling point, −161.5°C (Figure 14.2).

Methane is a major component of the natural gas used to heat homes and cook food. It is occasionally found concentrated in pockets in fields of coal, and is one of the causes of explosions in coal mines. Methane can be formed by the action of bacteria on decaying matter, and is found among the gases that bubble out of marshes and swamps (and for that reason was once known as marsh gas). In the Middle Ages, the fires that resulted when these marsh gases were ignited were thought to be the spirits of the dead.

The second member of this group is ethane, CH_3CH_3 or C_2H_6, which is formed when a hydrogen on methane is replaced by a methyl group, $—CH_3$ (Figure 14.3). Methane and ethane are members of the hydrocarbon class of compounds called the **alkanes,** whose formulas all fit the general pattern of C_nH_{2n+2}. The identifying characteristic of the alkanes is that in each molecule the carbons are bonded singly to four other atoms. That is, each carbon atom forms four single bonds, so additional atoms cannot be

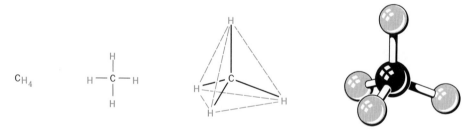

CH_4

Figure 14.2 Methane.

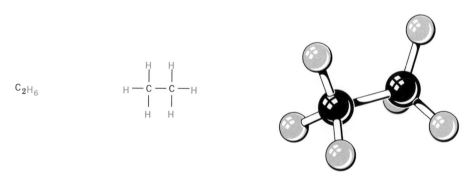

C_2H_6

Figure 14.3 Ethane.

Table 14.1 Some Alkanes

Number of Carbons	Molecular Formula	IUPAC Prefix	Name	Structural Formula	Boiling Point in °C
1	CH_4	meth-	methane	CH_4	−162
2	C_2H_6	eth-	ethane	CH_3CH_3	−89
3	C_3H_8	prop-	propane	$CH_3CH_2CH_3$	−42
4	C_4H_{10}	but-	butane	$CH_3CH_2CH_2CH_3$	0
5	C_5H_{12}	pent-	pentane	$CH_3CH_2CH_2CH_2CH_3$	36
6	C_6H_{14}	hex-	hexane	$CH_3CH_2CH_2CH_2CH_2CH_3$	69
7	C_7H_{16}	hept-	heptane	$CH_3CH_2CH_2CH_2CH_2CH_2CH_3$	98
8	C_8H_{18}	oct-	octane	$CH_3CH_2CH_2CH_2CH_2CH_2CH_2CH_3$	126
9	C_9H_{20}	non-	nonane	$CH_3CH_2CH_2CH_2CH_2CH_2CH_2CH_2CH_3$	151
10	$C_{10}H_{22}$	dec-	decane	$CH_3CH_2CH_2CH_2CH_2CH_2CH_2CH_2CH_2CH_3$	174

added to the molecule. Molecules having this property are said to be **saturated.** Single bonds between carbon atoms are strong and stable, making the alkanes the least reactive class of hydrocarbons.

Propane ($CH_3CH_2CH_3$ or C_3H_8) and butane ($CH_3CH_2CH_2CH_3$ or C_4H_{10}) are the next two members of the alkane series. They can be liquified in tanks under pressure, allowing them to be stored and transported easily. This gives them wide use as fuels for lighters, torches, and furnaces in rural homes.

14.3 Naming the Alkanes

The names of all molecules belonging to the alkane group of hydrocarbons end in **-ane.** The prefix used indicates the number of carbons in the main carbon chain (Table 14.1). The naming of groups attached to the main carbon chain follows the rules given in Section 13.5.

Example 14-1

Name the following compound.

$$_5CH_3 \!-\!\! _4CH_2 \!-\!\! \underset{\underset{H}{|}}{\overset{\overset{CH_3}{|}}{_3C}} \!-\!\! \underset{\underset{H}{|}}{\overset{\overset{CH_3}{|}}{_2C}} \!-\!\! _1CH_3$$

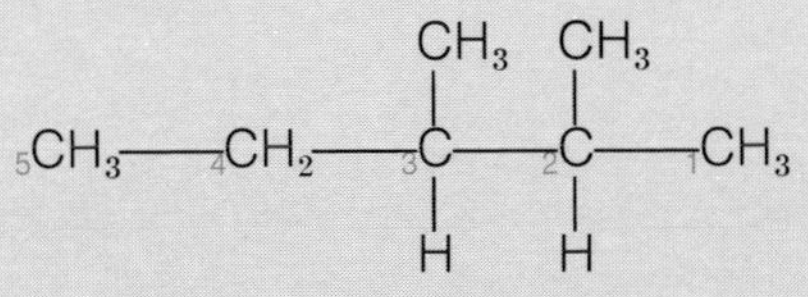

(a) All the bonds in this compound are single bonds, so it is an alkane. Therefore, the name of the compound will end in -ane.

(b) The number of carbons in the main, or longest, carbon chain is five (prefix pent-), so the compound is a pentane.

(c) There are two methyl groups attached to the carbon chain. Therefore, the prefix dimethyl- is added, making the compound a dimethylpentane.

(d) To indicate the position of the methyl groups we number the carbons in the main carbon chain, starting with the end that is nearest the attached groups. Thus, there is a methyl group on carbon 2 and carbon 3.

(e) The name of the compound, therefore, is 2,3-dimethylpentane.

14.4 Structural Isomers

There are two different compounds that have the molecular formula C_4H_{10}. These two compounds have their carbon and hydrogen atoms arranged in a different order in the three-dimensional structure of the molecule, and are known as **structural isomers.** The two isomers having this molecular formula C_4H_{10} are butane (common name, *n*-butane), whose carbons all lie in a straight chain, and 2-methylpropane (common name, isobutane), whose carbons are arranged in a branched chain (Figure 14.4).

Although there are only two structural isomers of butane, the number of possible isomers increases as the number of carbon atoms in a molecule increases. Octane, which is the eight-carbon alkane, has 18 different structural isomers. Each octane isomer behaves in a slightly

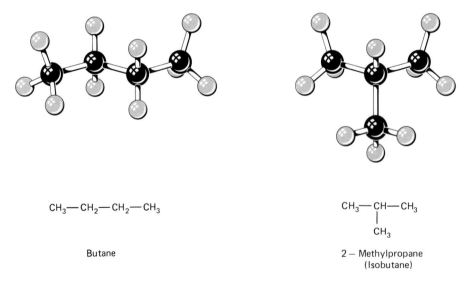

$$CH_3-CH_2-CH_2-CH_3$$

Butane

$$CH_3-CH-CH_3$$
$$\qquad\ \ |$$
$$\qquad\ CH_3$$

2 — Methylpropane
(Isobutane)

Figure 14.4 Butane and 2-methylpropane (isobutane) are structural isomers having the same molecular formula, C_4H_{10}.

different way. One of them, "isooctane," burns very well in car engines, and is used as a standard in determining the octane rating of a gasoline.

$$CH_3\text{—}\underset{\underset{CH_3}{|}}{\overset{\overset{CH_3}{|}}{C}}\text{—}CH_2\text{—}\underset{\underset{H}{|}}{\overset{\overset{CH_3}{|}}{C}}\text{—}CH_3$$

"Isooctane" (2,2,4-Trimethylpentane)

Someone once calculated that a 40-carbon compound would have more than 60 trillion possible isomers!

14.5 Optical Isomerism*

A second type of isomerism results only from the arrangement of atoms in space (not from the order in which they are arranged), and is called optical isomerism. We have seen that a carbon atom can have four groups attached to it, each directed toward one of the corners of a tetrahedron. If these four groups are all different, the carbon atom will possess a property called **asymmetry** and is called a **chiral carbon.** A chiral molecule (one containing a chiral carbon), as with other asymmetric objects, has a mirror image that cannot be superimposed on itself. That is, if you take a three-dimensional model of a chiral molecule and a model of its mirror image, there is no way that you can superimpose the two models so that all of the groups match up (Figure 14.5). Consider a common example: Your hand is asymmetric. Your right and left hands are not superimposable, and there is no way that a right-hand glove will fit on your left hand (except by turning the glove inside out) (Figure 14.6).

What difference will a chiral carbon atom make in a molecule? If molecules are nonsuperimposable mirror images, they will have identical

* This section is optional and may be omitted without loss of continuity.

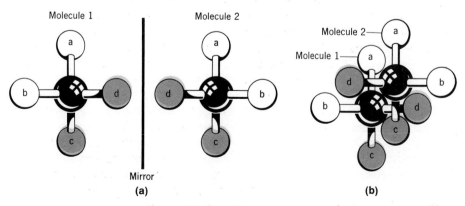

Figure 14.5 Molecules 1 and 2 are optical isomers. (a) Molecule 1 contains a chiral carbon, and molecule 2 is the mirror image of molecule 1. (b) The two molecules cannot be superimposed.

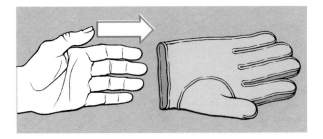

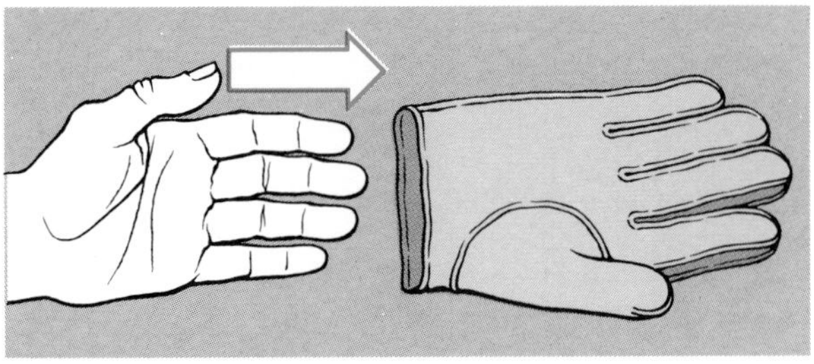

Figure 14.6 Your hand is asymmetric; there is no way that you can put a right-hand glove on your left hand (without turning it inside out).

physical properties (such as boiling point, density, and vapor pressure), except for one. They will differ in the way in which they interact with polarized light—light in which the waves vibrate in only one plane. When placed in solutions, chiral molecules will interact with polarized light by rotating the direction of the plane of vibration in either a clockwise or counterclockwise direction. Molecules having this property are said to be optically active. **Optical isomers** are optically active compounds that share the same molecular formula, but which have the ability to rotate the plane of polarized light in opposite directions. This one difference between optical isomers may seem quite unimportant, but as we will see in Chapters 18 through 20, it is critical to the well-being of living organisms.

14.6 Reactions of Alkanes

Oxidation

You are probably most familiar with alkanes as the fuels that heat your home, run your car, and power your camping stove and lantern. Although we mentioned earlier that the saturated hydrocarbons are fairly unreactive, one of the reactions which they will easily undergo is the reaction with oxygen known as burning. In this reaction, oxygen atoms from the air combine with the carbon and hydrogen atoms in the hydrocarbon until the carbon and hydrogen atoms are all bonded to oxygens. The reaction is exothermic, and the end products of the reaction are carbon dioxide (CO_2) and water (H_2O). This reaction with oxygen, by the way, is occurring all the time, even at room temperature. However, it usually occurs at a rate too slow to be noticed. If the compound is heated in the presence of air, this process will speed up until a point is reached, called the ignition point, where the heat that is generated can be seen and felt. It is when we actually see and feel the reaction occurring that we say the substance is burning, and refer to the reaction as *combustion.* The general name for any reaction with oxygen, no matter what the rate, is **oxidation.** Although the process of oxidation occurring in a forest fire is easily seen and is impressive in its force, even the results of very slow oxidation can be observed by noticing the yellowing of pages in old books (Figure 14.7).

Figure 14.7 Both of these photographs illustrate oxidation reactions, but these reactions are occurring at greatly different rates. (Top, Courtesy U.S. Forest Service; bottom, Ron Nelson).

The equation for the oxidation reaction that occurs when methane is burned is

$$CH_4 + 2O_2 \longrightarrow CO_2 + 2H_2O + \text{Energy}$$

Oxidation reactions are occurring continuously in our bodies and, in fact, are the source of energy for our cells. If this oxidation occurred in the same way as methane burns, our cells would be destroyed by heat. Oxidation occurs in the body at a very controlled rate, however, allowing cells to trap and use a portion of the energy that is released. Actually, our bodies do not use alkanes as a fuel supply, but rather use derivatives of alkanes—carbohydrates, fats, and proteins. The oxygen needed for oxidation in the body is supplied to the tissues by the red blood cells, which pick up oxygen in the lungs and distribute it throughout the body, returning the waste product carbon dioxide to the lungs to be exhaled. The water produced in the oxidation reaction either remains in the cells and tissues, or is excreted through sweat and urine. Just as a fire can be smothered by a blanket which cuts off the oxygen supply, we can be smothered by anything which cuts off our supply of oxygen for as short a period as five minutes.

If there is an insufficient amount of oxygen present when alkanes are burned, incomplete combustion (or incomplete oxidation) occurs, producing end products of carbon monoxide (CO) and water. For example, if methane is burned in an oxygen-poor environment, the following reaction occurs.

$$2CH_4 + 3O_2 \longrightarrow 2CO + 4H_2O + \text{Energy}$$

The carbon monoxide produced by incomplete combustion in factories and cars is the largest source of carbon monoxide pollution in our society. It was computed that in 1965, 91 million tons of carbon monoxide entered the

Table 14.2 The Effects of Carboxyhemoglobin* Blood Levels on the Human Body

Blood Levels of Carboxyhemoglobin	Effects
2 to 5%	Impairment of the central nervous system.
5%	Impairment of perception and psychomotor performance.
10%	Oxygen transport significantly impaired.
15%	Headaches, dizziness, and lassitude.
15 to 40%	Ringing ears, nausea, vomiting, heart palpitations, difficulty breathing, muscular weakness, apathy.
40% and above	Collapse, coma, and death.

*Carboxyhemoglobin is a hemoglobin–carbon monoxide complex that is formed when carbon monoxide enters the blood.

Table 14.3 Median Carboxyhemoglobin Levels in Smokers and
Nonsmokers*

Location	Cigarette smokers	Nonsmokers
Anchorage	4.7%	1.5%
Chicago	5.8%	1.7%
Denver	5.5%	2.0%
Houston	3.2%	1.2%
Los Angeles	6.2%	1.8%
Vermont and New Hampshire	4.8%	1.2%
Washington, D.C.	4.9%	1.2%

*The maximum level recommended by the Clean Air Act of 1971 is 1.5%.
Reprinted from the *Journal of the American Medical Association,* August 26, 1974, Volume 229.
Copyright © 1974. American Medical Association.

atmosphere from automobile exhaust alone. Recent studies have shown that
the carbon monoxide from automobile exhaust can speed up the formation
of photochemical smog.

Although carbon dioxide is a normal part of the environment of living
cells, carbon monoxide is toxic. When inhaled, carbon monoxide greatly
reduces the oxygen-carrying ability of the blood by binding very strongly
with the hemoglobin molecules in red blood cells. At high levels, this
reduction can be so great as to produce coma and death; but even at low
levels, disruption of the central nervous system can be measured.
Cigarette smokers inhale carbon monoxide into their lungs together with
the cigarette smoke, producing elevated levels of carbon monoxide in their
bloodstreams. These levels are often sufficient to produce measurable
effects on the central nervous system. City drivers who smoke can easily
elevate their carbon monoxide level enough to produce headaches,
dizziness, and fatigue—the first symptoms of carbon monoxide poisoning
(Tables 14.2 and 14.3).

Our bodies have several different ways of responding to elevated
carbon monoxide levels, all of which result in increased strain on the heart
and increased risk of heart disease. Short-term high levels cause the heart
to pump faster, and long-term low levels cause the body to increase the
number of red blood cells that carry oxygen, thereby thickening the blood
and increasing the work load on the heart.

Substitution Reactions

Under the proper chemical conditions, alkanes can also react with nitric
acid (HNO_3) and with the elements of group VII (the halogens: F_2, Cl_2,
Br_2, I_2) in reactions called **substitution reactions.** These reactions are so
named because another atom or group of atoms is substituted for one or
more of the hydrogens on the alkane. For example, when methane reacts
with nitric acid, nitromethane is formed.

$$H-\underset{\underset{H}{|}}{\overset{\overset{H}{|}}{C}}-H \quad + \quad HONO_2 \quad \xrightarrow{\text{> 400°C}} \quad H-\underset{\underset{H}{|}}{\overset{\overset{H}{|}}{C}}-NO_2 \quad + \quad HOH$$

Methane Nitric acid Nitromethane Water

In the above chemical equation you will notice that > 400°C has been written above the arrow. Whenever special conditions are required for a reaction to occur, they are often indicated in this way. In this case, then, the temperature must be greater than 400°C for the reaction to occur. Nitromethane, the product of this reaction, is used as a solvent, as an important chemical in the production of other organic compounds, and as a high energy fuel for racing cars.

Unsaturated Hydrocarbons

Our study of the alkanes, which have carbon-to-carbon single bonds, gives us just a hint of the wide variety of compounds making up the chemistry of carbon. In other compounds carbon can share four electrons with another carbon atom, forming a double bond, or can share six electrons with another carbon atom, forming a triple bond (Figure 14.8). A double bond is less stable, or more reactive than a single bond, and a triple bond is even more reactive. Such multiple bonds, therefore, form an unstable, or reactive,

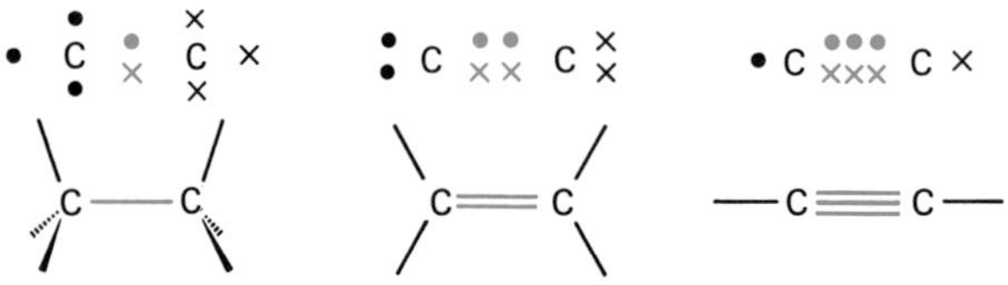

Figure 14.8 A carbon atom can form a single, double, or triple bond with another carbon atom.

Saturated

Unsaturated Polyunsaturated

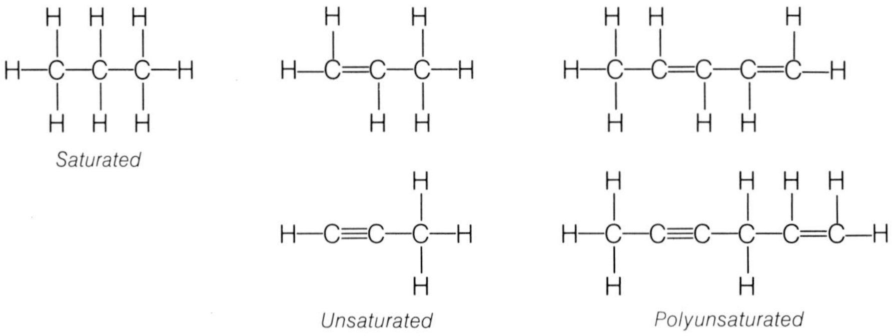

Figure 14.9 The degree of unsaturation of an organic molecule is determined by the number of carbon-to-carbon double or triple bonds in the molecule.

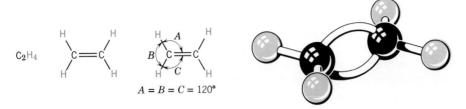

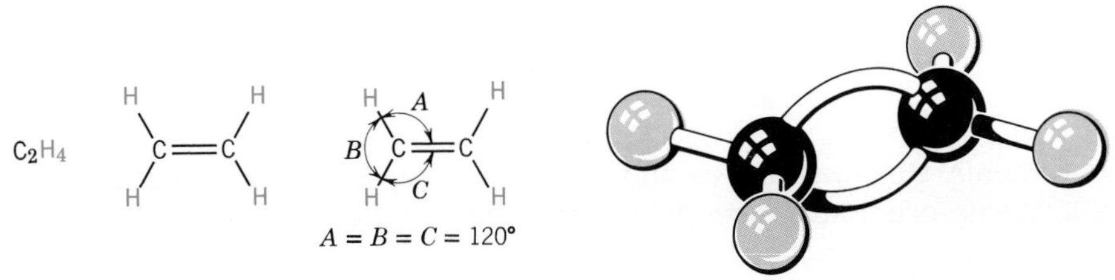

Figure 14.10 Ethene (ethylene).

spot in the organic molecule. Compounds having double or triple bonds can add more atoms to their molecules, and are said to be **unsaturated.** If a compound has several double or triple bonds, then it is **polyunsaturated** (Figure 14.9). You probably have seen that word before in margarine advertisements, and we will discuss polyunsaturated fats and oils in the chapter on lipids.

14.7 Alkenes

The class of compounds containing double bonds between carbon atoms is known as the **alkenes.** The simplest alkene is ethene, or ethylene, CH_2CH_2 or C_2H_4 (Figure 14.10). Ethylene is a flammable, anesthetic gas that is nontoxic to tissues even in high concentrations. Its anesthetic effects are rapid; a patient is ready for surgery two to four minutes after administration. A commercially useful property of ethylene gas is its ability to shorten the ripening time of citrus fruits.

Notice in Figure 14.10 that the structure of ethylene is no longer tetrahedral. The most stable arrangement for this molecule results when the bonds are as far apart as possible, and this occurs when all the atoms in the ethylene molecule are located in the same plane. Alkene molecules can also have two or more double bonds, which can be located anywhere in the molecule. Large alkene molecules can have dozens of double bonds, leading to millions of possible isomers (Table 14.4).

14.8 Naming the Alkenes

The IUPAC rules for naming alkenes are much the same as the rules for naming alkanes, but with the following changes:

1. All the names of alkenes end in **-ene.**
2. The carbon chain chosen for the main carbon chain is the longest carbon chain containing the double bond.
3. The carbons in the carbon chain are numbered starting at the end closest to the first double bond. A number prefix is used to indicate the carbon on each double bond that is nearest the beginning of the chain. For example, $CH_3CH{=}CHCH_2CH_3$ is 2-pentene, not 3-pentene or 2,3-pentene.

4. If two or more double bonds are found in the molecule, the prefixes di-, tri-, and so on are added to the name of the alkene.

Carefully study the names and structures given in Table 14.4 and in the following examples to see how these rules are used in the naming of alkenes. (Be sure to notice the way in which commas and hyphens are used in the names.)

Table 14.4 Some Alkenes

Number of Carbons	Number of Double Bonds	Molecular Formula	Name	Structural Formula
3	1	C_3H_6	Propene	CH_2=$CHCH_3$
4	1	C_4H_8	1-Butene	CH_2=$CHCH_2CH_3$
4	1	C_4H_8	2-Butene	CH_3CH=$CHCH_3$
4	1	C_4H_8	2-Methyl-1-propene	CH_3 \| CH_2=C—CH_3
5	2	C_5H_8	1,3-Pentadiene	CH_2=$CHCH$=$CHCH_3$
5	2	C_5H_8	2-Methyl-1,3-butadiene	CH_3 \| CH_2=C—CH=CH_2

Example 14-2 _____

1. Name the following compound.

$$CH_3CHCH=CCH_3$$
with a CH_3 group on the second carbon and a CH_3 group on the fourth carbon

(a) The longest chain containing the double bond has 5 carbons. If you number the carbons from the end closest to the double bond, you will see that the last part of the name is 2-pentene.

(b) There are two methyl groups attached to the chain, one on carbon 2 and the other on carbon 4.

(c) The name is 2,4-dimethyl-2-pentene.

2. Draw the structural formula for 2-methyl-1,3-hexadiene.

(a) 1,3-hexadiene tells us that the main carbon chain has 6 carbons with 2 double bonds: one starting on carbon 1 and the other starting on carbon 3.

$$C=C—C=C—C—C$$

(b) 2-methyl tells us that there is one group attached to the chain: a methyl group on carbon 2. The rest of the bonds on the carbon atoms will be attached to hydrogen atoms. (But make sure that each carbon atom has only 4 bonds connected to it.)

$$H-\underset{\underset{H}{|}}{\overset{\underset{}{|}}{C}}=\underset{\underset{}{|}}{\overset{\underset{CH_3}{|}}{C}}-\underset{\underset{H}{|}}{\overset{\underset{H}{|}}{C}}=\underset{\underset{H}{|}}{\overset{\underset{H}{|}}{C}}-\underset{\underset{H}{|}}{\overset{\underset{H}{|}}{C}}-H \quad \text{or} \quad CH_2{=}\overset{\overset{CH_3}{|}}{C}CH{=}CHCH_2CH_3$$

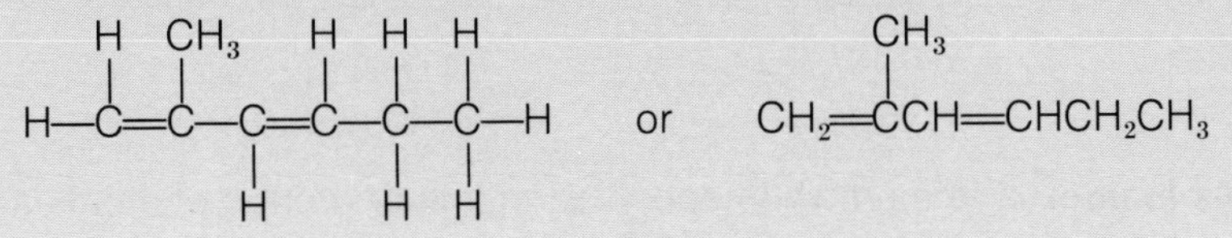

14.9 Geometric Isomers

A carbon-to-carbon double bond makes another type of isomerism possible, **geometric isomerism.** A single bond between carbon atoms, as in the alkanes, does not restrict the rotation of atoms around that bond. The carbon atoms can twist freely around their single bonds just as two balls can twist freely when connected by a string (Figure 14.11). A double bond between two carbon atoms is structurally rigid, however, preventing free rotation of the carbon atoms. This results in two possible arrangements of atoms on either side of the unsaturated bond (Figure 14.12). In terms of our comparison, replacing a single bond between two carbon atoms by a double bond is much like replacing the string between the two balls with a rigid pole.

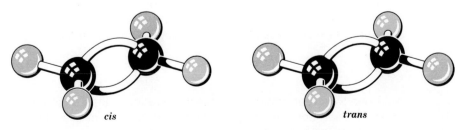

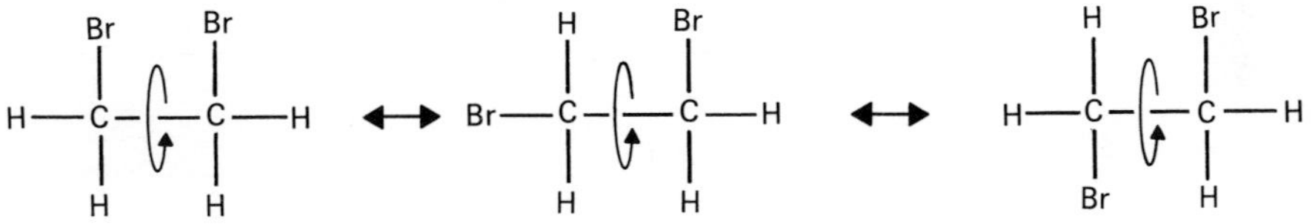

Figure 14.11 *Cis-trans* isomerism does not exist in 1,2-dibromoethane because the carbon-to-carbon single bond is free to rotate; as a result, the atoms bonded to the carbons can be in any position.

cis *trans*

Figure 14.12 *Cis-trans* isomerism does exist in 1,2-dibromoethene. The double bond is structurally rigid, preventing free rotation of the carbon atoms.

When some specified atoms or groups of atoms attached to the doubly bonded carbons appear on the same side of the bond, the molecule is called a ***cis* isomer.**

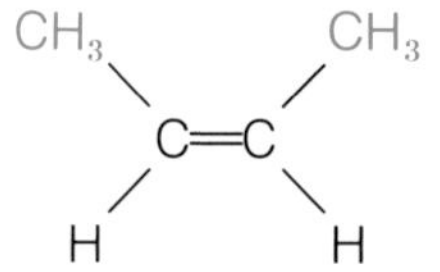

cis-2-Butene

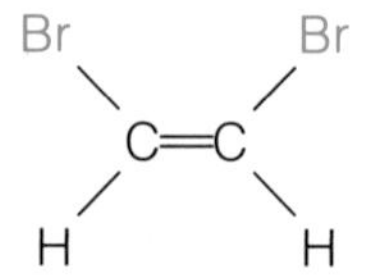

cis-Dibromoethene

A ***trans* isomer** is formed when specified atoms or groups of atoms appear on opposite sides of the double bond.

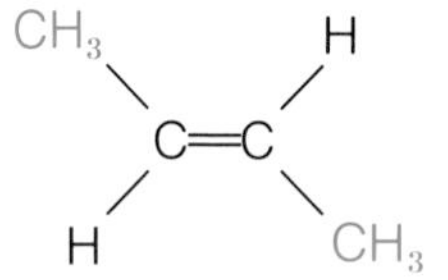

trans-2-Butene

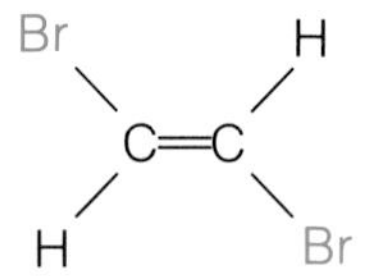

trans-Dibromoethene

You may wonder why we are bothering to point out these different types of isomers. A major reason is that isomers are critically important to the chemical reactions of living cells. For example, although you may find it difficult to tell the difference between the *cis* and *trans* isomer of the following complicated compound, the cells in the retina of your eye certainly can.

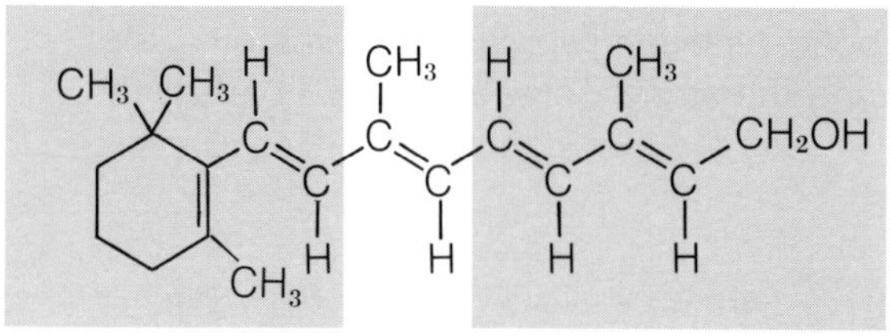

Vitamin A.
(all *trans*—100% activity)

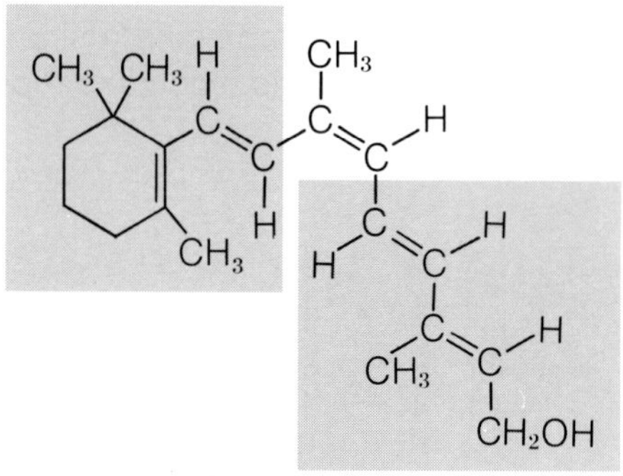

9-*cis* isomer
(22% activity)

The *trans* isomer is Vitamin A, a compound necessary for us to see in dim light. The corresponding *cis* isomers can also be used by the retina cells. These isomers cannot be used as efficiently as the all *trans* isomer, however, and they are much less effective in reducing night blindness.

14.10 Reactions of Alkenes

Because a carbon-to-carbon double bond is more reactive than a single bond, alkenes are used commercially in a wide variety of reactions. They are especially important in the chemical industry, where they are used in the production of many other compounds. Three important types of reactions for which alkenes are well suited are oxidation reactions, addition reactions, and polymerization reactions.

Oxidation Reactions

Earlier in the chapter we discussed the oxidation of alkanes by describing the combustion reaction. But compounds do not have to be burned in order to be oxidized. There is actually a wide range of reactions which are oxidation reactions, or more precisely, oxidation-reduction (redox) reactions. Because an oxidation reaction involves the loss of electrons by a reactant, and a reduction reaction involves the gain of electrons by another reactant, redox reactions always occur together. For the purposes of our study of organic chemistry, we can define an **oxidation reaction** as a reaction in which one of the reactant molecules gains oxygen atoms or loses hydrogen atoms (called dehydrogenation). Similarly, a **reduction reaction** can be defined as a chemical reaction in which a molecule loses oxygen atoms or gains hydrogen atoms (called hydrogenation). In organic chemistry and biochemistry, oxidation reactions involving dehydrogenation are much more common than those involving the addition of oxygen.

As was the case with the alkanes, alkenes can be oxidized completely to produce carbon dioxide (CO_2) and water.

$$CH_2{=}CH_2 + 3O_2 \longrightarrow 2CO_2 + 2H_2O$$

Ethene Oxygen Carbon dioxide Water

Notice that in this reaction the molecule of ethene undergoes oxidation, and the oxygen gas undergoes reduction. Under certain conditions the oxidation of ethene may be incomplete, resulting in the formation of glycol (a type of alcohol) rather than carbon dioxide.

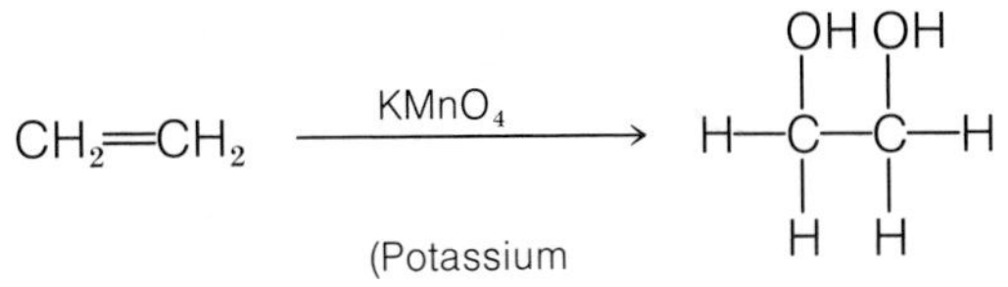

Ethene (Potassium permanganate) Glycol

Addition Reactions

An **addition reaction** takes place when atoms react with a double bond, causing the double bond to become a single bond.

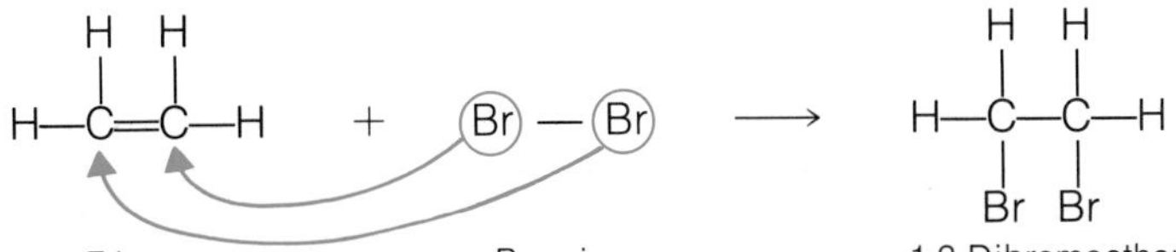

Ethene Bromine (Br_2) 1,2-Dibromoethane

This bromine reaction is often used to test for the presence of unsaturated bonds (that is, double or triple bonds) in a molecule. Bromine in water or in carbon tetrachloride forms a reddish-brown solution, but the dibromides formed from the addition reaction are colorless. Therefore, if the reaction of bromine water with a hydrocarbon results in a colorless solution, the hydrocarbon may have contained unsaturated bonds.

Compounds other than the halogens can also be used in addition reactions. Here are a few examples:

Hydrogenation reaction	H—C=C—H (with H, H above; Ethene)	+ H₂ Hydrogen	⟶	H—C—C—H (with H, H above and H, H below; Ethane)

Hydrohalogenation reaction	H—C=C—H (with H, H above; Ethene)	+ HCl Hydrogen chloride	⟶	H—C—C—Cl (with H, H above and H, H below; Ethyl chloride)

Hydration reaction	H—C=C—H (with H, H above; Ethene)	+ HOH Water	⟶	H—C—C—OH (with H, H above and H, H below; Ethyl alcohol)

Polymerization

You are probably most familiar with the word *ethylene* from seeing it in the name of a type of plastic, polyethylene. Polyethylene is formed by an addition reaction involving thousands of ethylene units. This is how three ethylene units would bond together:

$$\cdots \quad C{=}C \quad + \quad C{=}C \quad + \quad C{=}C \quad \cdots$$

yields

Polyethylene

The process of joining many simple units, called **monomers,** together to form very large molecules, called **polymers,** is **polymerization.** Thus, polyethylene is a polymer of the monomer ethylene. Synthetic (that is, artificial) polymers appear constantly in our daily lives, from the plastic containers, bags, and wrapping we use, to the synthetic fibers such as orlon, rayon, and acrylics that we wear, to the synthetic rubber that we use for tires and for parts in appliances (Table 14.5). However, don't think that polymers are the result of human ingenuity alone; the synthetic polymer industry resulted from the attempt to imitate natural polymers. For example, tropical plants produce natural rubber in the form of a milky sap called latex, which is a large molecule composed of 4500 isoprene units. Isoprene is a five-carbon compound with two double bonds.

Table 14.5 Some Commonly Used Polymers

Monomer	Polymer	Uses
Propylene $H_2C=CHCH_3$	Polypropylene $-CH_2CHCH_2CHCH_2CH-$ 　　　$\|$　　　$\|$　　$\|$ 　　　CH_3　CH_3　CH_3	Film and molded parts
Styrene $H_2C=CHC_6H_5$	Polystyrene $-CH_2CH-CH_2CH-CH_2CH-$ 　　　$\|$　　　　$\|$　　　　$\|$ 　　　C_6H_5　　C_6H_5　　C_6H_5	Molded objects, insulation, and foam plastics
Vinyl chloride $H_2C=CHCl$	Polyvinyl chloride (PVC) $-CH_2CHCH_2CHCH_2CH-$ 　　　$\|$　　　$\|$　　$\|$ 　　　Cl　　Cl　　Cl	Plastic bottles and containers, plastic pipe, and insulation
Acrylonitrile $H_2C=CHCN$	Polyacrilonitrile $-CH_2CHCH_2CHCH_2CH-$ 　　　$\|$　　　$\|$　　$\|$ 　　　CN　　CN　　CN	Orlon and clothing fibers
Tetrafluoroethylene $F_2C=CF_2$	Polytetrafluoroethylene $-CF_2CF_2CF_2CF_2CF_2CF_2-$	Teflon and lubricating films

A molecule of isoprene has carbon-to-carbon double bonds alternating with single bonds. This alternating arrangement is given the name **conjugated double bonds.** Any compound having conjugated double bonds will be more stable than a similar compound having double bonds arranged in other patterns.

$$
\begin{array}{ccc}
 & CH_3 & \\
 & | & \\
H-C=C-C=C-H \\
 | & | & | \\
 H & H & H
\end{array}
$$

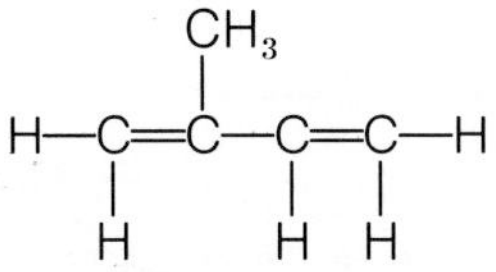

Isoprene (2-Methyl-1,3-butadiene)

14.11 Alkynes

Alkynes are the class of hydrocarbons that contains triple bonds between carbon atoms. These triple bonds, having three pairs of electrons shared between the two carbon atoms, put quite a strain on the molecule, making such bonds very reactive. Nevertheless, it is possible for a molecule to have more than one triple bond, and there are some molecules that contain both double and triple bonds.

Alkynes are named following the same rules given for alkenes in Section 14.7, except that the names of alkynes end in **-yne.** For example, $CH_3CH_2C\equiv CCH_3$ is 2-pentyne.

The simplest alkyne is ethyne, or acetylene, CHCH or C_2H_2 (Figure 14.13). Acetylene is a flammable and explosive gas that burns with a bright flame, making it useful for lighting. When acetylene is burned

C_2H_2 H — C ≡ C — H

$A = B = 180°$

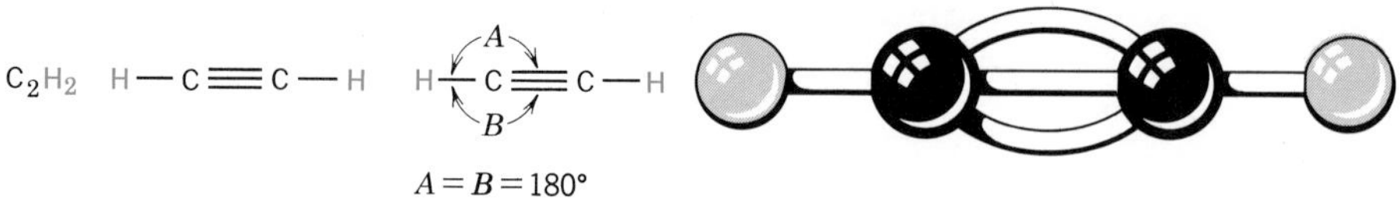

Figure 14.13 Ethyne (acetylene).

together with oxygen in an oxyacetylene welding torch, the resulting reaction is highly exothermic, providing enough energy to cut and weld metals. Acetylene is so reactive that it forms a convenient starting material for the industrial production of almost every simple organic compound. However, compounds with triple bonds do not play a part in the chemistry of living organisms. In fact, the human body does not naturally contain any molecules with triple bonds in their structures.

Carbon in Rings

14.12 Cyclic Hydrocarbons

The great variety and complexity of carbon chemistry becomes even more obvious when we realize that carbon atoms not only form straight chain molecules, but also form stable rings. These compounds, called **cyclic hydrocarbons,** may be found in various sizes and often contain both single and double bonds (Table 14.6). The most common cyclic hydrocarbons have rings composed of five or six carbon atoms, but both larger and smaller rings are possible. The cyclic hydrocarbons are named following the same rules given for the straight chain hydrocarbons, except that the unsubstituted name begins with **cyclo-.**

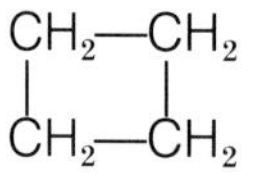

Cyclobutane

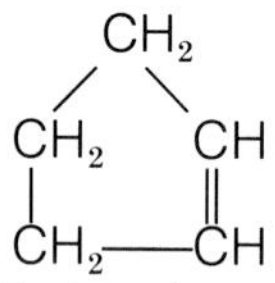

Cyclopentene

The saturated cyclic hydrocarbon with the smallest number of carbon atoms is cyclopropane. Cyclopropane is a highly potent anesthetic, but care must be taken in its use because it is both inflammable and explosive. Many hydrocarbons, in addition to cyclopropane, act as anesthetics when inhaled. (Among the others are methane, acetylene, ethylene, and cyclobutane). An anesthetic is a compound that decreases a person's sensitivity to pain and, in most cases, also causes the person to lose consciousness. The anesthetic property of the hydrocarbons comes from the nonpolar nature of the molecules, and the absence of a polar hydrogen atom for hydrogen bonding.

An anesthetic compound acts on the nerves, preventing nerve impulses from moving along the nerve fibers. This happens because each nerve is surrounded, in part, by a protective coating composed of molecules that

Table 14.6 Some Cyclic Hydrocarbons

Molecular Formula	Name	Structural Formula		
C_3H_6	Cyclopropane	$\begin{array}{c}\text{CH}_2\\ \diagup\quad\diagdown\\ \text{CH}_2\text{—CH}_2\end{array}$		
C_4H_8	Cyclobutane	$\begin{array}{c}\text{CH}_2\text{—CH}_2\\	\qquad	\\ \text{CH}_2\text{—CH}_2\end{array}$
C_5H_{10}	Cyclopentane	$\begin{array}{c}\text{CH}_2\\ \diagup\quad\diagdown\\ \text{CH}_2\quad\text{CH}_2\\	\qquad	\\ \text{CH}_2\text{—CH}_2\end{array}$
C_5H_6	1, 3-Cyclopentadiene	$\begin{array}{c}\text{CH}\\ \diagup\!\!\!\!\diagup\quad\diagdown\\ \text{CH}\quad\text{CH}_2\\	\qquad	\\ \text{CH}\text{==}\text{CH}\end{array}$
C_6H_{12}	Cyclohexane	$\begin{array}{c}\text{CH}_2\\ \diagup\quad\diagdown\\ \text{CH}_2\quad\text{CH}_2\\	\qquad	\\ \text{CH}_2\quad\text{CH}_2\\ \diagdown\quad\diagup\\ \text{CH}_2\end{array}$

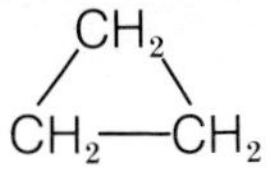

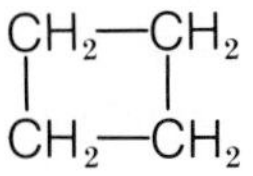

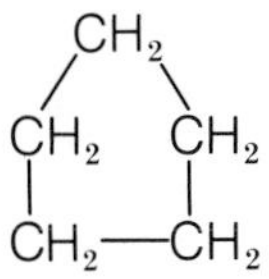

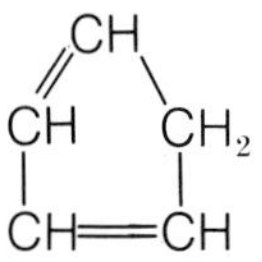

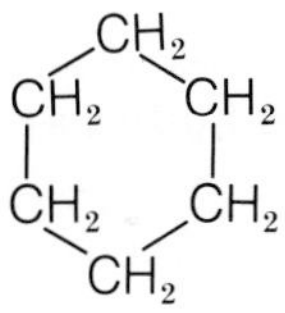

are nonpolar. When an anesthetic is inhaled, it enters the bloodstream from the lungs and travels to these tissues. There it dissolves in the nonpolar protective coating around the nerves. As the anesthetic accumulates in this coating, it causes a "short circuit" in the transmission of nerve impulses. Because the nerve impulses no longer reach the brain, there is no sensation of pain. The many deaths caused by inhaling methane, or sniffing the solvents in glue, have resulted from these nonpolar hydrocarbon compounds dissolving in the nonpolar protective coatings of the nerves and short-circuiting the critical nerve impulses from the brain to the lungs and heart.

14.13 *Cis* and *Trans* Isomers

When carbon atoms are arranged in a ring, free rotation around even the carbon-to-carbon single bond is restricted, similar to the restricted rotation we have seen in the alkenes. This results in *cis* and *trans* isomers in ring compounds. A *cis* isomer of a ring compound will have specified atoms or groups of atoms lying on the same side of the carbon ring.

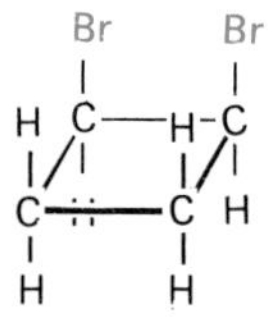

cis–1, 2–Dibromocyclobutane *cis*–1, 2–Dichlorocyclohexane

Similarly, a *trans* isomer will have these atoms or groups of atoms located on opposite sides of the carbon ring.

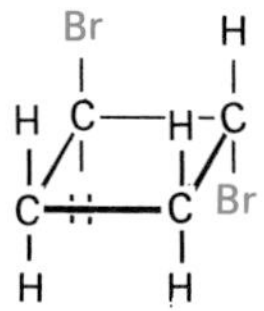

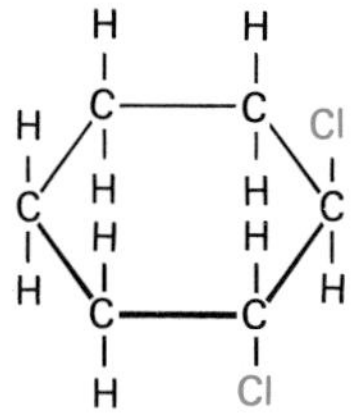

trans–1, 2–Dibromocyclobutane *trans*–1, 2–Dichlorocyclohexane

14.14 Benzene and Its Derivatives: the Aromatic Hydrocarbons

The most common and important of the cyclic hydrocarbons is benzene. Benzene and its derivatives make up the class of compounds called the **aromatic hydrocarbons.** In spite of their name, the aromatic hydrocarbons are no more "smelly" than other hydrocarbons. However, the first natural compounds identified as part of this class had a definite and fairly pleasant odor—from which came the name *aromatic.* Benzene has the molecular formula C_6H_6, but its structural formula puzzled scientists for many years. The six carbon atoms in a molecule of benzene are arranged in a ring, and are often drawn with alternating single and double bonds.

Benzene

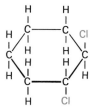

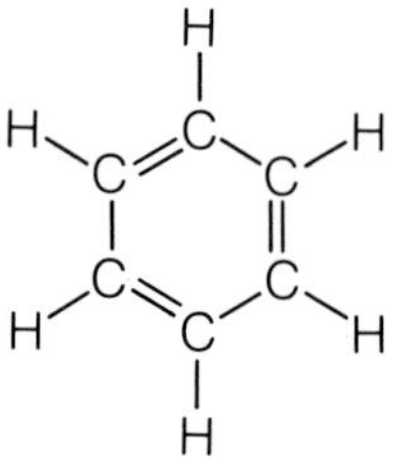

However, this structure does not accurately explain the properties of benzene. The actual benzene ring is much more stable than would be expected from this conjugated alkene structure, and does not break open during reactions. Moreover, each carbon-to-carbon bond is the same strength, and the distance between each of the carbon atoms is equal.

Two-dimensional structural formulas do not adequately represent benzene's structure. It is

neither nor

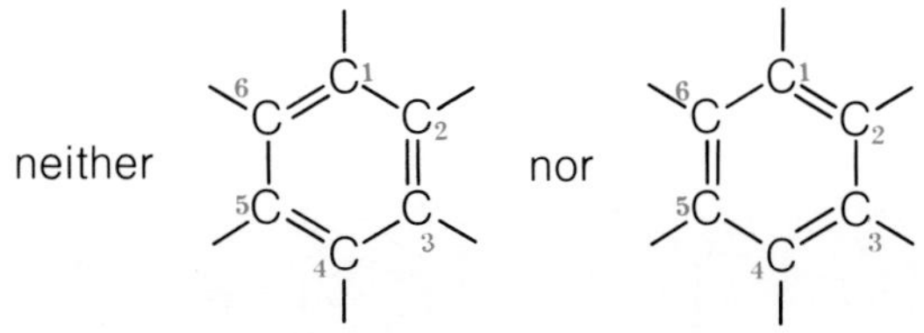

but halfway between these structures, as if the electrons from the three double bonds were delocalized, or spread out, among all six of the carbon atoms. The name *resonance hybrid* is often used to describe this combined structure (*resonance* refers to the ability of a compound to exist in more than one arrangement of electrons). Symbolically, we can represent this resonance structure by drawing a circle in the center of the benzene ring.

Benzene

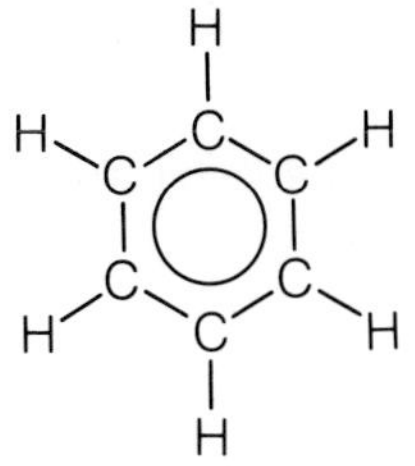

All of the atoms in a molecule of benzene are located in one plane (with bond angles of 120°). As a result, *cis* and *trans* isomers do not exist. Its resonance structure makes the benzene molecule extremely stable, and this molecule forms part of the structure of many complex natural compounds.

To simplify the drawing of the structural formulas of such complex molecules, organic chemists have devised a simplified representation of a hydrocarbon or benzene ring.

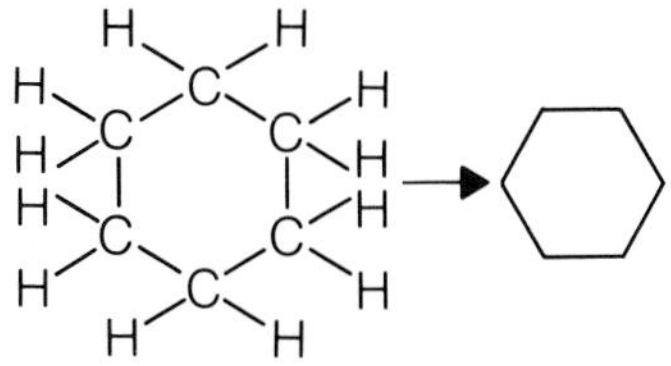

Cyclohexane Benzene

To translate such a diagram back to the full structural formula, place a carbon atom at each angle of the figure and then draw hydrogen atoms on the bonds not involved in the ring. If any elements other than carbon or hydrogen are involved in the compound, they will be separately shown. Here are some examples:

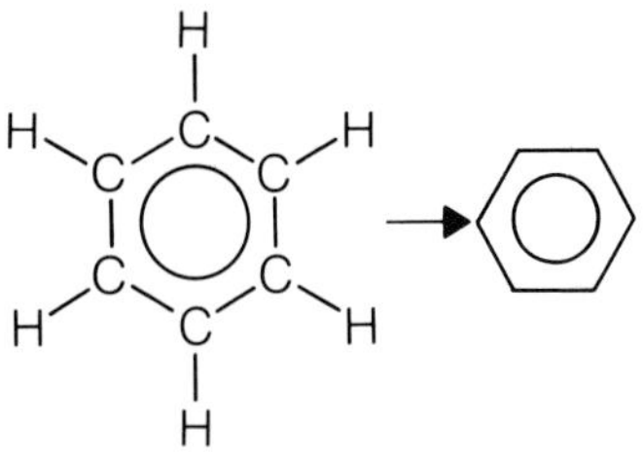

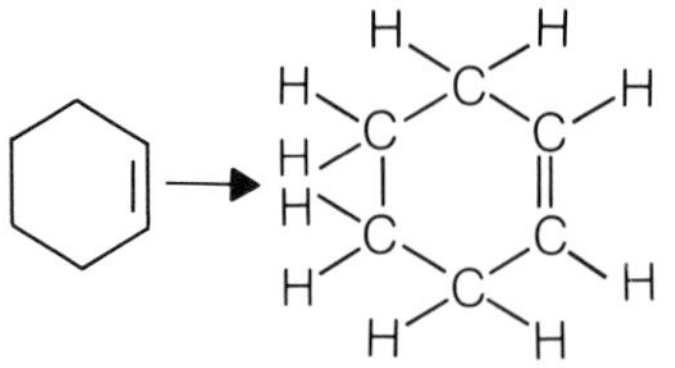

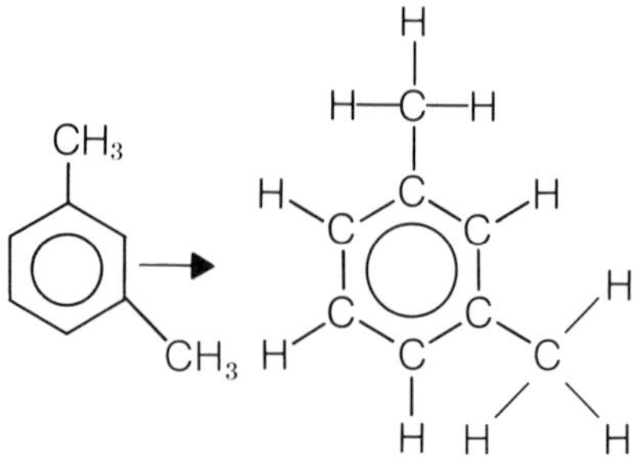

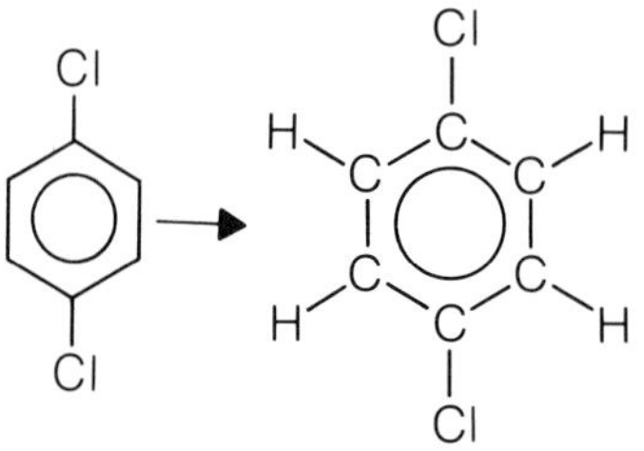

Benzene and other aromatic hydrocarbons can be recovered from coal tar, which is produced by heating bituminous (or soft) coal in the absence of air. Benzene is used by the chemical industry as a solvent, and as a starting material in the production of a variety of compounds. Its fumes are toxic and, when inhaled, can cause nausea or death from respiratory and heart failure. Again let us stress that the danger of these nonpolar compounds, many of which are used as solvents in such common products as paints, glues, and cleaners, is their ability to act on the central nervous system and to interfere with nerve action. Some important aromatic hydrocarbons are listed in Table 14.7. Appendix 3 explains the names of some of the common benzene derivatives.

The benzene ring is found in many compounds which are important to the life processes of the living organism. Phenylalanine, an amino acid whose importance was illustrated in Chapter 1, is just one example.

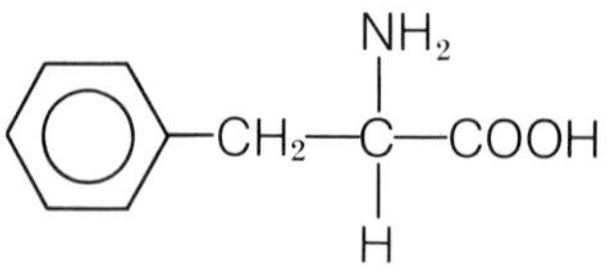

Phenylalanine

Plants are able to synthesize benzene rings from carbon dioxide, water, and inorganic materials. Animals, however, cannot synthesize these aromatic rings, and their survival depends upon obtaining the essential aromatic compounds through their diets.

14.15 Reactions of Benzene

The ring structure of benzene is very stable, and remains intact during most chemical reactions. Many aromatic compounds are produced both naturally and synthetically by **substitution reactions,** in which one or more hydrogens on the aromatic ring are replaced with other atoms or groups of atoms.

Table 14.7 Some Aromatic Compounds

Name	Strucual Formula	Name	Structural Formula
Benzene		Toluene	CH_3
Phenol	OH	Nitrobenzene	NO_2
Benzoic acid	O∥C—OH	Aniline	NH_2

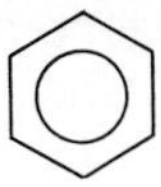

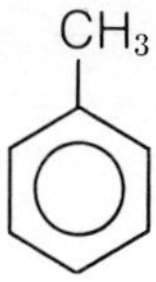

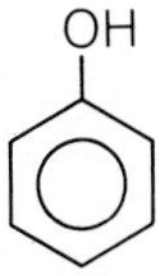

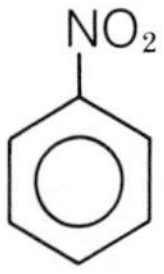

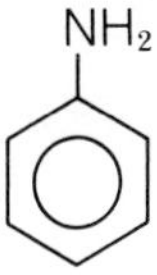

We saw in Section 14.10 that alkenes react readily with bromine, adding bromine atoms across the double bond. However, very special conditions are required for benzene to react with bromine and, when this occurs, the benzene ring remains intact and one bromine atom is substituted for a hydrogen atom.

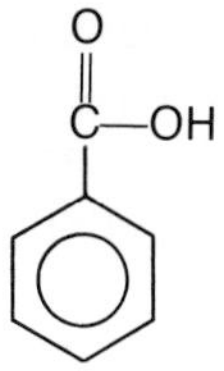

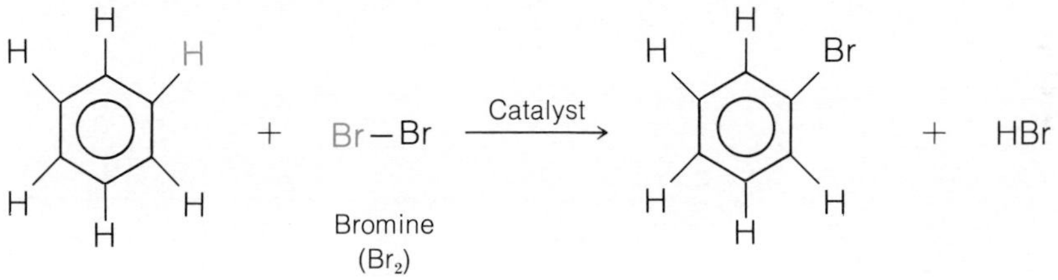

$$\text{Benzene} \quad + \quad Br-Br \quad \xrightarrow{\text{Catalyst}} \quad \text{Product} \quad + \quad HBr$$

Bromine
(Br_2)

Under certain conditions, benzene can be forced to undergo a substitution reaction with concentrated acids.

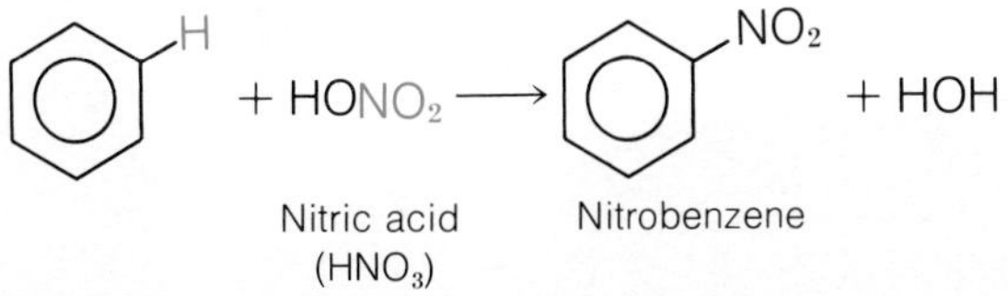

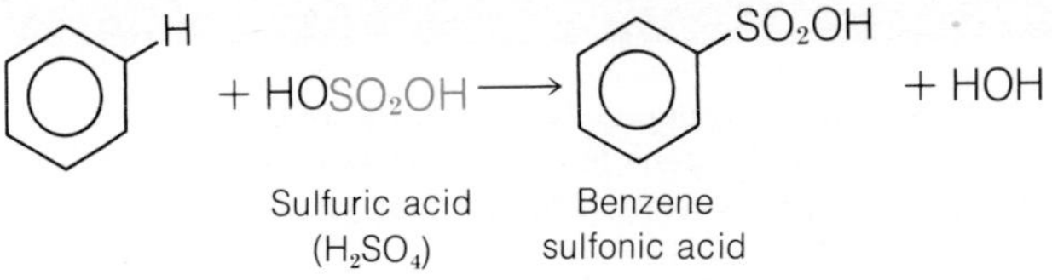

$$+ \; HONO_2 \longrightarrow \text{(}NO_2\text{)} + HOH$$

Nitric acid Nitrobenzene
(HNO_3)

$$+ \; HOSO_2OH \longrightarrow \text{(}SO_2OH\text{)} + HOH$$

Sulfuric acid Benzene
(H_2SO_4) sulfonic acid

14.16 Other Aromatic Compounds

Benzene rings may be found in large molecules whose chemical diagrams resemble honeycombs. As a simple example, the compound naphthalene, which gives moth balls their characteristic odor, is composed of two benzene rings joined or fused together. Three benzene rings joined together form anthracene, an important starting material in the production of dyes.

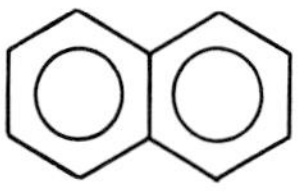

Naphthalene

Anthracene

Compounds containing multiple benzene rings are obtained in the production of coal tar. Their harmful effects to humans became evident when workers in European coal tar factories developed skin cancer. A later study of the chemical components of coal tar determined that several aromatic fused-ring compounds were capable of causing cancer in mice (Figure 14.14). Chemicals that cause cancer in animals are known as **carcinogens.** It was found that the carcinogenic hydrocarbons in coal tar all had a similar arrangement of fused benzene rings, as shown in Figure 14.15. These compounds are formed in the partial combustion of many large organic molecules. One of the most active carcinogens, 3,4-benzpyrene, is discharged each year, in very large quantities, into the

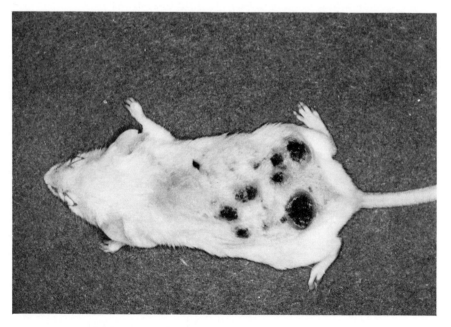

Figure 14.14 Skin tumors were produced by applying a solution containing 7,12-dimethylbenzanthracene to the shaved skin on the back of this mouse. (Courtesy Kanematsu Suguira, Sloan Kettering Institute for Cancer Research)

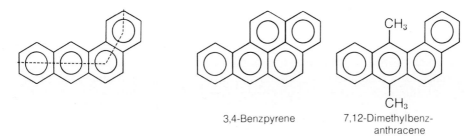

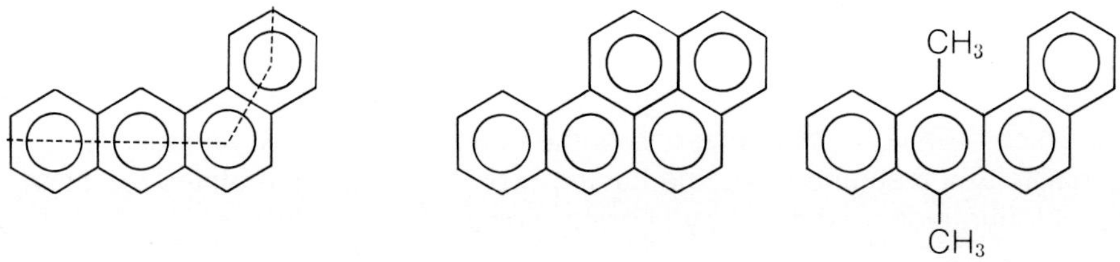

3,4-Benzpyrene 7,12-Dimethylbenz-
 anthracene

Figure 14.15 These carcinogenic hydrocarbons found in coal tar all
have a similar arrangement of fused benzene rings.

atmosphere of industrial nations. It is also one of the major carcinogens
found in cigarette smoke. Only a few milligrams (0.001 grams) of
3,4-benzpyrene are enough to produce cancer in experimental animals.
 The way in which these rather inert compounds produce cancer in
humans and animals has puzzled scientists, but recent studies are
beginning to shed light on this process. Our bodies contain a certain set
of enzymes located mostly in the liver and kidneys, but also found in the
lungs and other tissues, whose function is to detoxify foreign chemicals that
enter the body. One way in which this is done is by making nonpolar
compounds (such as 3,4-benzpyrene) more polar. This makes these
compounds more soluble in water (through hydrogen bonding) and more
easily excreted by the kidneys. Most of the products of such detoxifying
reactions are less harmful to the cells than were the original reactants, but
in some cases the products turn out to be very carcinogenic. This is the
way in which these enzymes (some of which are located in the lungs) can
contribute to the synthesis of carcinogenic hydrocarbons from cigarette
smoke. The way that such carcinogens turn a normal cell into a cancer cell
is still an area of active research.

Hydrocarbon Derivatives

14.17 Halogenated Hydrocarbons

Hydrocarbon derivatives, such as those hydrocarbons containing one or
more atoms of the halogen family, have a variety of uses in our society.
For example, ethylene bromide is a compound that is added to leaded
gasoline to prevent the lead from settling out and ruining an automobile's
engine.

$$\begin{array}{cc} Br & Br \\ | & | \\ H-C-C-H \\ | & | \\ H & H \end{array}$$

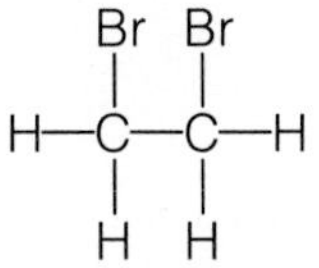

1,2-Dibromoethane
(ethylene bromide)

The bromine in this compound combines with the lead atoms from the gasoline to form lead bromide ($PbBr_2$), which is a vapor at engine temperatures and leaves the engine as part of the exhaust gases.

Dichlorodifluoromethane, one of the compounds known as Freon®, consists of a methane molecule having two of its hydrogen atoms replaced by chlorine atoms, and two by fluorine atoms.

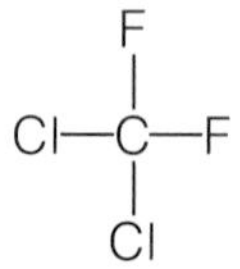

$$\begin{array}{c} F \\ | \\ Cl-C-F \\ | \\ Cl \end{array}$$

Dichlorodifluoromethane

Freon is odorless, nonpoisonous, noncorrosive, and nonflammable, thus making Freon far better than other compounds for use as a refrigerant in refrigerators and other cold-storage equipment.

Fluorocarbons, compounds with carbon and fluorine atoms, are much more stable than hydrocarbons and are less affected by heat and other chemicals. Fluorocarbons are used as artificial rubber compounds, as lubricants, and as chemicals for fire extinguishers. An important polymer made from a fluorocarbon is teflon, which is constructed from the monomer tetrafluoroethene (see Table 14.5). Teflon is chemically inert in the presence of almost all compounds, and is widely used as a nonstick coating for pots and pans, and in electric insulation materials.

Widespread controversy surrounds the use of another halogenated hydrocarbon, dichlorodiphenyltrichloroethane—better known as DDT. DDT was first synthesized by a German chemist in 1874, but not until 60 years

Figure 14.16 The higher up the food chain, the greater the concentration of hologenated hydrocarbons such as DDT, PBB, or PCB.

later was its effectiveness as an insecticide proven. DDT is a broad spectrum insecticide; that is, it indiscriminately kills both good and bad insects. It is still not known how DDT produces death in insects, but this compound has a high toxicity for insects and no apparent toxic effects to humans on brief exposure. The widespread and successful use of DDT in the 1940s and 1950s to combat the malaria mosquito has led to the development of resistance to DDT in many insect species. As a result, there has been a whole series of chlorinated hydrocarbons synthesized for use as pesticides (Table 14.8).

Table 14.8 Chlorinated Hydrocarbons Used in Pesticides

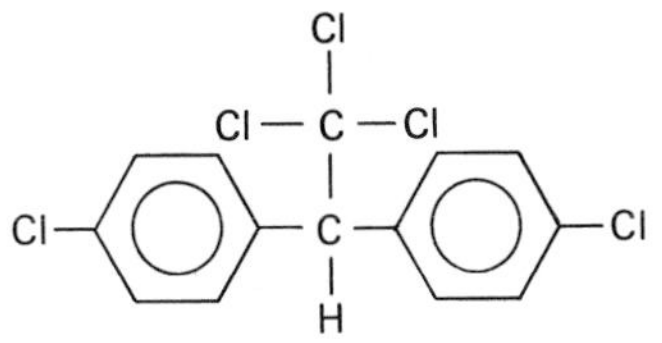

DDT

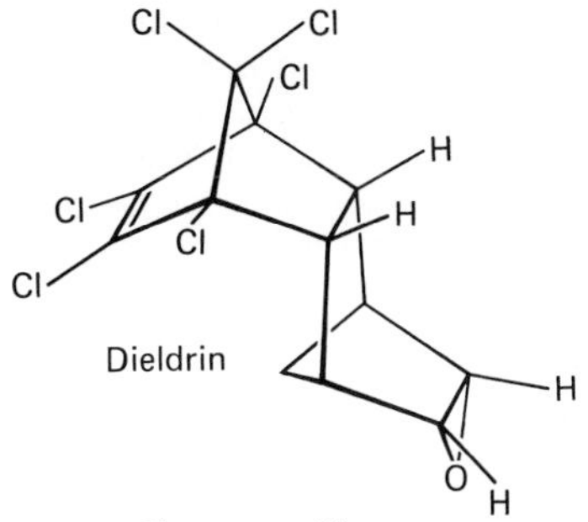

Dieldrin

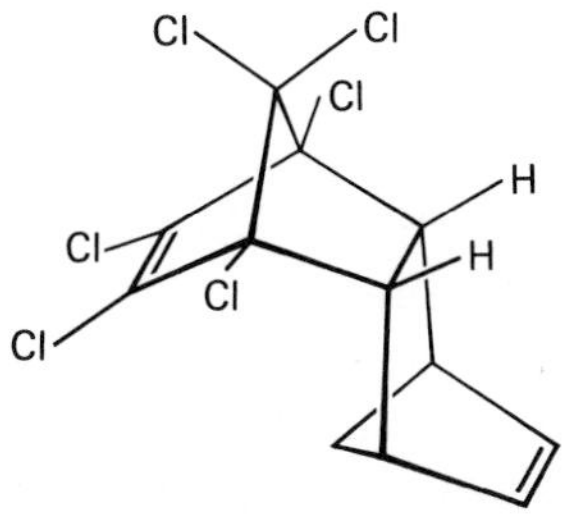

Aldrin

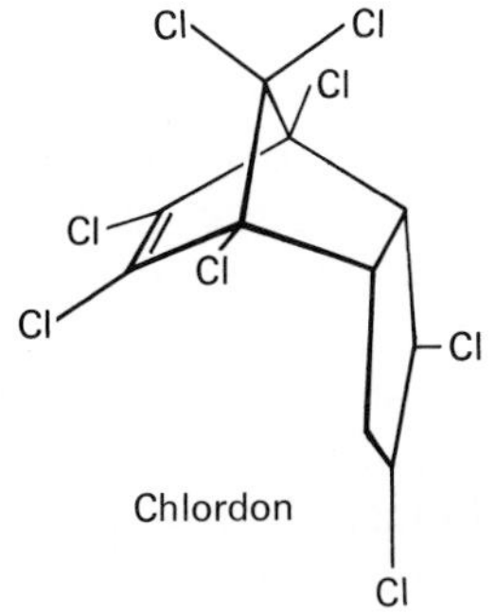

Chlordon

The benzene rings in the structure of DDT give the molecule great stability and, as is the case with PBB and PCB, it breaks down very slowly in the environment. Because these molecules are nonpolar, they tend to accumulate in the fatty (adipose) tissues of living organisms. The concentrations of these compounds in the fatty tissues increases as one moves up the food chain (Figure 14.16). Although the exact process is not known, DDT is suspected of playing a role in the extinction of several species of birds (Figure 14.17). However, it was not until the publication of Rachel Carson's book *Silent Spring* that the harmful side effects of these extremely stable chlorinated hydrocarbons were brought to the attention of the public. The Environmental Protection Agency has since banned the use of DDT, and has placed restrictions on the production, use, and disposal of other halogenated hydrocarbons.

Figure 14.17 An egg found in the nest of a brown pelican off the coast of California. The weight of the adult bird sitting on the egg was enough to crush the thin shell that resulted from high concentrations of DDT in the female bird. The concentration of DDT in the egg was measured at 2500 parts per million. None of the eggs laid by the 300-pair colony of birds hatched. (Courtesy Joseph R. Jehl, Jr.)

Chapter Summary

Hydrocarbons are a large group of organic compounds composed only of hydrogen and carbon. Hydrocarbon molecules are nonpolar, and will dissolve only in other nonpolar substances. Hydrocarbons may be saturated, containing only carbon-to-carbon single bonds, or unsaturated, containing carbon-to-carbon double or triple bonds. These molecules can have carbon atoms in straight chains, branched chains, or rings. Saturated hydrocarbons belong to the class of compounds called alkanes. Alkenes are hydrocarbons containing carbon-to-carbon double bonds, and alkynes contain triple bonds. Alkanes are relatively unreactive, but will enter into oxidation and substitution reactions. Alkenes and alkynes are increasingly more reactive and will undergo oxidation, addition, and polymerization reactions.

Several types of isomerism result from the various possible arrangements of carbon atoms. Structural isomers are molecules with the same molecular formula, but with a different ordering of carbon and hydrogen atoms within the molecule. Geometric isomers are molecules with the same ordering of carbon atoms, but with a different arrangement of those atoms in space.

The compound benzene and its derivatives make up a large class of compounds called aromatic hydrocarbons. The ring structure of benzene has unusual stability, and remains intact through most chemical reactions. Derivatives of benzene are formed through substitution reactions, in which other atoms or groups of atoms are substituted for a hydrogen on a benzene ring.

Exercises and Problems

1. Identify each of the following as (1) an alkane, alkene, or alkyne, and (2) as saturated or unsaturated.

 (a) $CH_3CH_2CH_2CH_3$

 (b)

 (c) $CH_3\overset{\displaystyle CH_3}{\underset{\displaystyle CH_2CH_3}{\overset{|}{\underset{|}{C}}}HCHCH_2CH_2CH_3}$

 (d) $CH_3C\!\equiv\!CH$

 (e) $CH_3\overset{\displaystyle CH_3}{\overset{|}{C}}\!=\!CHCH_3$

 (f) $CH_3(CH_2)_6CH_3$

 (g)

 (h) $CH_3\overset{\displaystyle CH_3}{\overset{|}{C}}\!=\!CHCH\!=\!C\underset{\displaystyle CH_3}{\overset{|}{}}CH_2CH_3$

 (i) $CH_3\overset{\displaystyle CH_3}{\underset{\displaystyle CH_3}{\overset{|}{\underset{|}{C}}}C\!\equiv\!CCH_3}$

 (j)

 (k) $CH_3\overset{\displaystyle CH_3}{\underset{\displaystyle CH_2CH_2CH_2CH_3}{\overset{|}{\underset{|}{C}}}CH_2CH_3}$

 (l) $CH_3\overset{\displaystyle CH_2CH_3}{\overset{|}{C}}\!=\!CCH_3$
 $CH_3C\!=\!CHCH_3$

 (m)

2. Give the IUPAC name for the compounds listed in question 1. (In addition to reviewing the rules given in this chapter, you might also refer to Section 13.5. *Hint:* For (k) and (l) be sure that you identify the longest carbon chain.)

3. Write the structural formulas for the following compounds:

 (a) cyclopentane
 (b) 2-hexyne
 (c) 2-methyl-3-heptene
 (d) 2,3-dichlorobutane
 (e) 2,3-dimethyl-2-butene
 (f) benzene
 (g) 1,3,5-octatriene
 (h) 3,5-diethylheptane
 (i) 3,3,4-trimethyl-1-pentyne
 (j) 2,2,3,3-tetrabromohexane

4. Which of the following compounds are structural isomers of each other?

 (a) $CH_3CH_2CH_2CH_2CH_3$

 (d) $CH_3-\overset{\overset{\displaystyle CH_3}{|}}{\underset{\underset{\displaystyle H}{|}}{C}}-CH_3$

 (b) $CH_3-\overset{\overset{\displaystyle CH_3}{|}}{\underset{\underset{\displaystyle CH_3}{|}}{C}}-CH_3$

 (e) $CH_3\overset{\overset{\displaystyle CH_3}{|}}{C}HCH_2CH_3$

 (c) cyclopentane ring: CH_2-CH_2 / CH_2 \ CH_2 / CH_2

 (f) $CH_3\overset{\overset{\displaystyle CH_3}{|}}{C}HCH\underset{\underset{\displaystyle CH_3}{|}}{C}H_3$

5. Draw the geometric isomers for the following compounds.

 (a) 2-pentene (b) 1,2-dichloroethylene (c) 1,2-dichlorocyclopentane

6. Can dibromoacetylene exist as *cis* and *trans* isomers? Why or why not?

7. For each of the following pairs of molecules, identify the more reactive compound. Give the reason for your choices.

 (a) $CH_3CH_2CH_3$ or $CH_3CH=CH_2$

 (b) $CH_3C\equiv CCH_3$ or $CH_3-\overset{\overset{\displaystyle CH_3}{|}}{\underset{\underset{\displaystyle H}{|}}{C}}-CH_3$

8. Which of the following compounds are aromatic hydrocarbons?

 (a) cyclopropane ring: CH_2 / \ CH_2—CH_2

 (c) benzene ring with CH_2CH_3

 (b) benzene ring with NO_2

 (d) cyclohexene ring with NO_2

9. Electric switches in a hospital operating room may not be turned on or off while gaseous anesthetic is being administered. Suggest a reason for this regulation.

10. Write the equation for the following reactions:

 (a) The complete oxidation of butene.
 (b) The substitution reaction between propane and nitric acid.
 (c) The substitution reaction between benzene and chlorine.
 (d) The addition reaction between 2-pentene and hydrogen.

11. Propylene (propene) can undergo a reaction to form the polymer called polypropylene. Show the polymerization of three propylene molecules.

12. Compounds in several commercially available products have been found to be carcinogenic in mice in laboratory experiments. What does it mean for a compound to be carcinogenic? Should humans continue to use such products? Why?

13. Ozone (O_3) is a compound, found in the atmosphere, which has both good and bad effects. In the lower atmosphere ozone, in high concentrations, can cause respiratory problems in humans and can kill plant life. In the earth's upper atmosphere, however, ozone acts as a protective layer that screens out the harmful ultraviolet rays of the sun. Consider the following statements:

    Hydrocarbon pollutants released into the atmosphere increase the ozone at lower levels.

    Fluorocarbons released into the air can reach the upper atmosphere unreacted, and may be removing the protective layer of ozone.

    (a) Why are fluorocarbons able to pass through the earth's lower atmosphere unreacted?
    (b) Discuss the possible future effects on the earth's population of continued production and release of hydrocarbons and fluorocarbons into the atmosphere.

14. A sample group of mothers who smoke were found to have delivered twice as many low birth-weight infants as a similar group of mothers who did not smoke. Suggest a possible physiological reason for this finding.

chapter 15

Oxygen

Learning Objectives

By the time you have finished this chapter, you should be able to:

1. Classify the organic molecules having oxygen-containing functional groups.

2. Explain why alcohols have comparatively high melting and boiling points.

3. Explain why short carbon chain alcohols are soluble in water, whereas longer carbon chain alcohols are insoluble.

4. State the difference between a primary, secondary, and tertiary alcohol.

5. Describe two methods for preparing alcohols in the laboratory.

6. Write the dehydration and oxidation reactions of an alcohol.

7. Compare the polarity and water solubility of ethers and alcohols.

8. Identify the difference in structure between aldehydes and ketones.

9. Describe the effects of ethanol and methanol on the human body.

10. Describe one method of preparing aldehydes and ketones.

11. Compare the ease with which aldehydes and ketones can be oxidized and reduced.

12. Define an *acid* and compare the strength of carboxylic acids and inorganic acids.

13. Describe one method of preparing carboxylic acids.

14. Define *condensation reaction, hydrolysis reaction,* and *saponification reaction,* and give an example of each.

Joyce was thoroughly enjoying the wedding reception for her friend's son. She was feeling relaxed and happy, and was having a great time. The next morning, however, she awoke with a splitting headache and a general feeling of nausea. The substance responsible for these effects on her body was ethanol, the liquid commonly known as alcohol. Ethanol, or ethyl alcohol, is actually just one of many alcohols, all of which have molecules containing a special group of atoms. This group is called the hydroxyl group, and it contains an oxygen atom and a hydrogen atom connected by a single bond, —O—H. When this hydroxyl group replaces one of the hydrogen atoms on a hydrocarbon, an alcohol is formed.

$$\begin{array}{ccc} \text{H} & \text{H} \\ | & | \\ \text{H}-\text{C}-\text{C}-\text{H} \\ | & | \\ \text{H} & \text{H} \end{array} \qquad \begin{array}{ccc} \text{H} & \text{H} \\ | & | \\ \text{H}-\text{C}-\text{C}-\text{OH} \\ | & | \\ \text{H} & \text{H} \end{array}$$

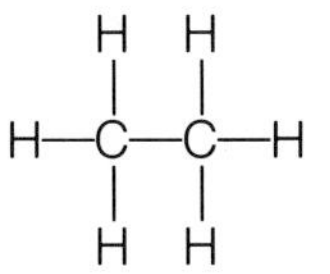

Ethane Ethanol

You are probably quite familiar with the many effects of ethanol on the body. It can create either a relaxed, easy feeling or a nauseating, aching one. As mentioned above, it is often responsible for a "great" evening, followed by a totally miserable morning. Ethanol, in the form of alcoholic beverages, is one of the most widely used of all drugs. It can be used to provide food energy, to either stimulate or inhibit the appetite, to aid

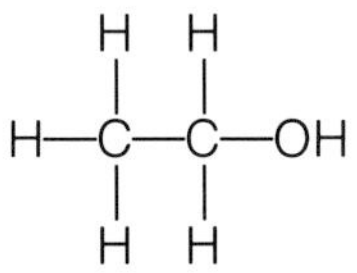

Table 15.1 Blood Alcohol Levels and Stages of Intoxication

Percent Alcohol in Blood*	Stage of Intoxication	Clinical Effects
0.05 or less	Subclinical	Normal by ordinary observation.
0.05 to 0.09	Subclinical	Normal by observation; slight change by special tests.
0.09 to 0.21	Stimulation	Decreased inhibition; emotional instability; slight uncoordination; slowing of response to stimuli.
0.18 to 0.30	Confusion	Disturbance of sensation; decreased sense of pain; staggering gait; slurred speech.
0.27 to 0.39	Stupor	Marked stimuli response decrease; approaching paralysis.
0.36 to 0.48	Coma	Complete unconsciousness; depressed reflexes; subnormal temperature; anesthesia; impairment of circulation; heavy snoring; possible death.

*Overlapping stages of intoxication have been shown above; this was done to take into account the variation observed in blood alcohol levels and the corresponding physical and mental impairment in different individuals.

From E. C. Hoff, *Aspects of Alcoholism,* J. B. Lippincott Co., Philadelphia, 1963, p. 54. Used by permission.

digestion, to bring relief from tension, anxiety, or pain, and it can function as a sedative.

As ethanol is swallowed, it is rapidly absorbed by the mucous membranes in the nose and throat, and its vapors are absorbed through the lungs. Absorption of ethanol into the body begins immediately after it is swallowed, with about 20% of the alcohol absorbed through the stomach, and the remainder through the small intestines. It can take up to six hours for the alcohol in a single drink to be absorbed and distributed throughout the body when the stomach is full, but only about one hour when the stomach is empty. Therefore, you certainly would feel the effects of a drink faster on an empty stomach than on a full one.

Ethanol dissolves completely in water. Once in the bloodstream it moves rapidly into the tissues, especially into organs with large blood supplies, such as the brain. This movement into the tissues continues until the concentration of ethanol in the tissues equals the concentration of ethanol in the blood. Because the ethanol will be equally distributed throughout the tissues of the body, the concentration of ethanol in one's breath or urine can be used as an accurate indicator of the level of ethanol in the blood. Police make use of this fact when they give a breath test to determine the state of intoxication of a driver.

A small amount of ethanol in the blood will act as a stimulant to most organs and body systems. But as the level increases, the ethanol becomes a depressant. This is especially true in the brain, where ethanol has a disruptive effect on the nerve cell membranes, making the brain less responsive to stimuli. When the blood alcohol level lies below 0.10%, a person is generally in a mood of pleasant relaxation; tensions and anxieties are eased. As the alcohol level rises, inhibitions become decreased as the control center in the brain becomes less active. At higher levels, the increased disruption of the nervous system results in a lack of muscular coordination, slurred speech, and difficulty in understanding what is seen and heard. A concentration of alcohol in the blood above 0.36% can result in delirium, anesthesia, coma, and even death (Table 15.1). Death from an overdose of ethanol occurs when the activity of the respiratory center in the brain becomes so disrupted that breathing stops. There have been reported cases of persons betting that they could drink a pint of whiskey without removing their lips from the bottle, doing just that, collecting their bets, and collapsing and dying as they walk out the door.

Compounds Containing Oxygen

You are probably most familiar with oxygen as the component of air that we breathe to sustain life. Without oxygen, we would die. But oxygen plays a further role in living systems, not just as free oxygen (O_2) or as part of water (H_2O), but as an element in the structure of many molecules that are extremely important in life processes. We have seen that alcohols, for example, are a class of compounds identified by the presence of a hydroxyl group, —OH. Such special groups of atoms, only some of which contain oxygen, are called **functional groups.** Functional groups create a reactive area on an organic molecule, giving the molecule specific chemical properties. In Chapter 13 we mentioned that the study of organic chemistry is simplified by categorizing compounds having similar properties into classes, and then studying several examples of each class.

Table 15.2 Functional Groups Containing Oxygen

Functional group	Class of Compound	Typical Compound	
R—OH	Alcohol	CH_3CH_2—OH	Ethanol
R—O—R′	Ether	CH_3—O—CH_2CH_3	Ethyl methyl ether
R—C(=O)—H	Aldehyde	CH_3CH_2—C(=O)—H	Propanal
R—C(=O)—R′	Ketone	CH_3—C(=O)—CH_3	Propanone (Acetone)
R—C(=O)—OH	Carboxylic acid	CH_3CH_2—C(=O)—OH	Propanoic acid
R—C(=O)—O—R′	Ester	CH_3CH_2—C(=O)—O—CH_3	Methyl propanoate

In this chapter we will examine classes of organic compounds that have functional groups containing oxygen (Table 15.2).

Before beginning this discussion, it is important to explain the symbol R that is often used when writing general formulas for organic compounds (see the first column in Table 15.2). General formulas are useful in describing properties and reactions of a whole class of compounds. The letter R is used to represent any alkyl group. That is, it could be a methyl group, ethyl group, or propyl group, and so on. In these general formulas, R' and R" are used to represent alkyl groups that are different from R.

15.1 Alcohols

To repeat, alcohols are organic compounds whose molecules contain the hydroxyl functional group, —OH. Thus, the general formula for any alcohol is ROH. (The hydroxyl group on an alcohol has no basic (alkaline) properties, and should not be confused with the basic hydroxide ion, OH^-, which we discussed in Chapter 12.) This hydroxyl group forms a reactive spot on the molecule, and alcohols are an important class of organic compounds because they are easily formed, are quite reactive, and are good starting materials for the synthesis of many other compounds.

Alcohols are named by changing the -e name ending of the parent alkane to **-ol.** The position of the hydroxyl group is indicated by adding a number in front of the name.

Example 15-1 _____

1. Name the following compounds:

 (a)
 $$CH_3—CH_2—\overset{\displaystyle OH}{\underset{\displaystyle H}{\overset{|}{\underset{|}{C}}}}—CH_3 \qquad \text{2-Butanol}$$

 (b)
 $$CH_3—\overset{\displaystyle CH_3}{\underset{\displaystyle H}{\overset{|}{\underset{|}{C}}}}—\overset{\displaystyle H}{\underset{\displaystyle H}{\overset{|}{\underset{|}{C}}}}—OH \qquad \text{2-Methyl-1-propanol}$$

2. Write the structural formula for 2-ethyl-1-pentanol.

 This compound is an alcohol with 5 carbons in its main chain, the chain containing the —OH functional group. The —OH group is on carbon 1, and an ethyl group is on carbon 2.

 $$CH_3—CH_2—CH_2—\overset{\displaystyle CH_3—CH_2}{\underset{\displaystyle H}{\overset{|}{\underset{|}{C}}}}—\overset{\displaystyle H}{\underset{\displaystyle H}{\overset{|}{\underset{|}{C}}}}—OH$$

Alcohols can be divided into three categories according to the placement of the hydroxyl group on the molecule. **Primary alcohols** have the hydroxyl group attached to a carbon atom which is bonded at most to one other carbon atom.

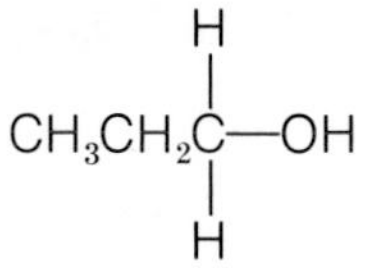

$$CH_3CH_2\overset{\displaystyle H}{\underset{\displaystyle H}{\overset{|}{\underset{|}{C}}}}-OH$$

1-Propanol

$$CH_3-\overset{\displaystyle CH_3}{\underset{\displaystyle H}{\overset{|}{\underset{|}{C}}}}\overset{\displaystyle H}{\underset{\displaystyle H}{\overset{|}{\underset{|}{C}}}}-OH$$

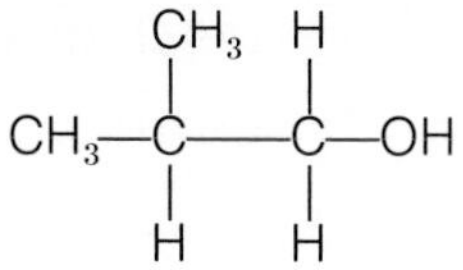

2-Methyl-1-propanol

Secondary alcohols have the hydroxyl group attached to a carbon atom which is bonded to two other carbon atoms.

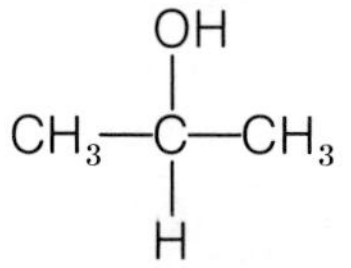

$$CH_3-\overset{\displaystyle OH}{\underset{\displaystyle H}{\overset{|}{\underset{|}{C}}}}-CH_3$$

2-Propanol

$$CH_3CH_2-\overset{\displaystyle OH}{\underset{\displaystyle H}{\overset{|}{\underset{|}{C}}}}-CH_3$$

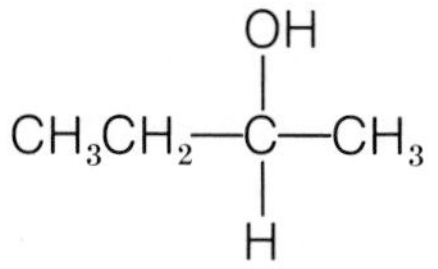

2-Butanol

Tertiary alcohols have the hydroxyl group attached to a carbon atom which is bonded to three other carbon atoms.

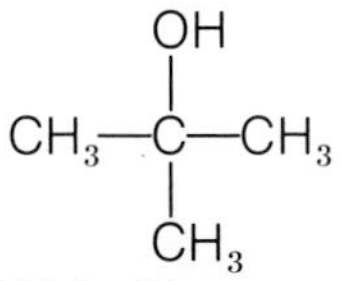

$$CH_3-\overset{\displaystyle OH}{\underset{\displaystyle CH_3}{\overset{|}{\underset{|}{C}}}}-CH_3$$

2-Methyl-2-propanol

The hydroxyl group on an alcohol molecule forms a polar area on the otherwise nonpolar carbon chain. The oxygen atom in the hydroxyl group has a stronger attraction for additional electrons than does the hydrogen, and thus will form an unequal, or polar, covalent bond with the hydrogen. This hydrogen, then, is capable of hydrogen bonding with other alcohol molecules or with water molecules (Figure 15.1).

The hydroxyl group on an alcohol molecule forms a polar area on the otherwise nonpolar carbon chain. The oxygen atom in the hydroxyl group has a stronger attraction for additional electrons than does the hydrogen, and thus will form an unequal, or polar, covalent bond with the hydrogen. This hydrogen, then, is capable of hydrogen bonding with other alcohol molecules or with water molecules (Figure 15.1).

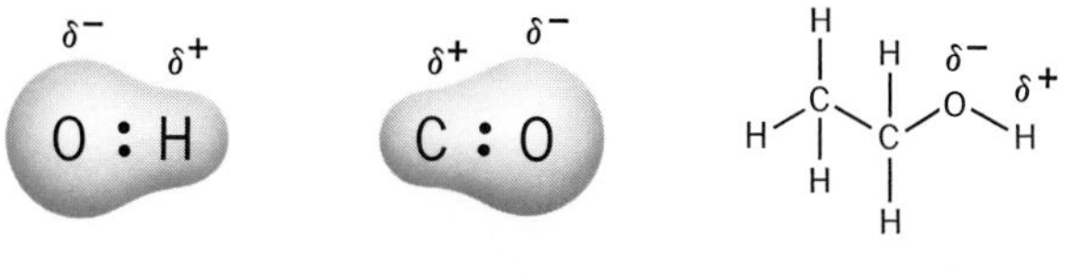

Because of this hydrogen bonding, alcohols with short carbon chains are soluble in polar solvents such as water. As the length of the carbon chain increases, however, the nonpolar nature of the carbon chain becomes more important than the attraction of the hydroxyl group for the water. This makes larger molecules less and less soluble in polar solvents, and more soluble in nonpolar solvents such as fats, benzene, and carbon tetrachloride. Increasing the number of hydroxyl groups on a long alcohol molecule increases the number of areas available for hydrogen bonding, making such molecules again more soluble in water (Table 15.3).

Table 15.3 Solubility of Some Alcohols (in Water)

Compound	Formula	Solubility
Ethanol	CH_3CH_2OH	Completely soluble
1-Pentanol	$CH_3CH_2CH_2CH_2CH_2OH$	Slightly soluble
1,2-Pentanol	$CH_3CH_2CH_2CHOHCH_2OH$	Soluble
2-Hexanol	$CH_3(CH_2)_3CHOHCH_3$	Very slightly soluble
2,3-Hexanediol	$CH_3(CH_2)_2CHOHCHOHCH_3$	Soluble
1-Decanol	$CH_3(CH_2)_8CH_2OH$	Insoluble

The polar hydroxyl group also affects the melting point and the boiling point of the alcohol molecule. The hydrogen bonding that occurs between alcohol molecules increases the amount of energy necessary to pull the molecules apart, thereby increasing the melting and boiling points. Alcohols have much higher boiling points than alkanes with comparable molecular weights. For example, ethanol has a molecular weight of 46 and propane a molecular weight of 44. Yet ethanol boils at 78°C, whereas propane boils at −45°C. Similarly, 1-butanol (molecular weight 74) boils at 117°C, whereas pentane (molecular weight 72) boils at 36°C.

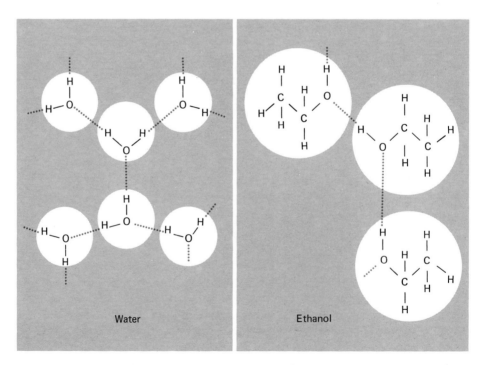

Figure 15.1 Hydrogen bonding in water and ethanol. The polar hydroxyl group on alcohols can enter into hydrogen bonding in much the same way as the —OH group in water. The hydrogen bond is indicated by the colored dotted line.

15.2 Straight Chain Alcohols

Methanol

The alcohol that is derived from methane is methanol, CH_3OH. Its common name, wood alcohol, comes from the fact that it was first obtained by heating wood in the absence of air. Methanol is very poisonous; less than 10 cc (2 teaspoons) can cause blindness, and 30 cc (2 tablespoons) can cause death.

Ethanol

Ethanol has been known for thousands of years. It was originally produced by mixing honey, fruits, berries, cereals, or other plant materials with water, and leaving them in the sun. This created a liquid prized as food, as a ceremonial or religious potion, and as medicine. The production of ethanol through this process known as fermentation became quite an art long before the actual chemistry was understood (Figure 15.2). In fermentation, yeast cells in the fruit or cereal mixture use the nutrients found there to supply themselves with energy. The waste products of this energy-producing reaction are ethanol and carbon dioxide.

$$\underset{\substack{\text{Glucose}\\\text{(a sugar)}}}{C_6H_{12}O_6} \xrightarrow{\text{Yeast}} \underset{\text{Ethanol}}{2CH_3CH_2OH} + 2CO_2 + \text{Energy}$$

The ethanol produced by the yeast cells dissolves in the surrounding liquid. When the concentration of ethanol rises above 12 to 18% (depending upon the type of yeast), however, it becomes toxic to the yeast. The cells die and settle to the bottom of the container, leaving behind an alcoholic beverage. Beverages with an alcoholic content higher than 18% can be produced by first distilling the ethanol to obtain it in pure form, and then mixing this alcohol in various proportions with other ingredients.

Polyhydric Alcohols

Alcohols can have more than one hydroxyl group on their molecules. We have already mentioned that the addition of more polar hydroxyl groups will make the molecule more water soluble and will also raise the boiling point (remember, this is due to more hydrogen bonding). Multiple hydroxyl groups, for a reason still unknown, also make a compound taste sweet.

Ethylene glycol, a dihydric alcohol (that is, an alcohol containing two hydroxyl groups) is a colorless, sweet liquid that is water soluble. It is just as toxic as methanol when taken internally. However, ethylene glycol has great commercial value as the basic ingredient in permanent antifreeze for automobiles, and also in the production of synthetic fabrics.

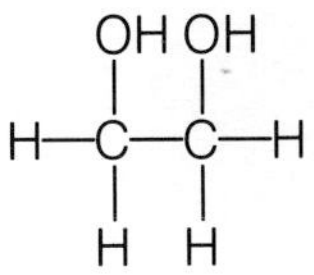

Ethylene glycol

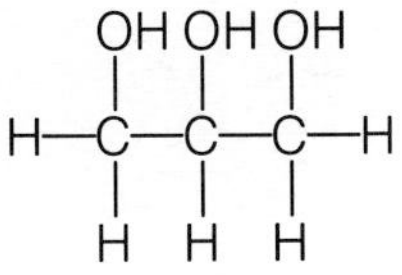

Glycerol

Glycerol or glycerin, which has three hydroxyl groups, is a thick,

Figure 15.2 Wine making—old and new. (Top, New York Public Library Picture Collection; bottom, Courtesy The Taylor Wine Company)

sweet-tasting liquid which is not toxic and which is a component of all natural fats and oils. These properties make it suitable for wide commercial use. It protects the skin and, therefore, is used in hand lotions and cosmetics. Glycerol is also used in inks, tobaccos, cream-filled candies, and plastic clays to prevent loss of water (or dehydration). It is a sweetening agent and a solvent for medicines, a lubricant used in chemical laboratories, and a component of plastics, surface coatings, and synthetic fabrics.

15.3 Cyclic and Aromatic Alcohols

Cyclic Alcohols

Hydroxyl groups can be found on all types of carbon chains. For example, menthol, which is found in peppermint oil, is a cyclic alcohol containing a cyclohexane ring.

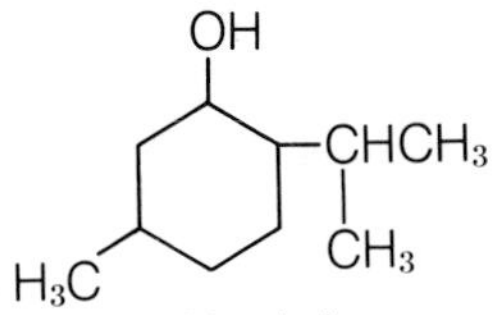

Menthol

The properties of menthol have been known for thousands of years. It causes an unusual cooling and refreshing sensation when rubbed on the skin, leading to its use in aftershave lotions and cosmetics. Menthol soothes inflamed mucous tissues, as in the mucous membranes of the nose and throat, leading to its use in nose and throat sprays, in cough drops, and in cigarettes.

Aromatic Alcohols

Aromatic alcohols are produced when one or more of the hydrogen atoms on a benzene ring is replaced by a hydroxyl group. When one hydrogen is replaced, the compound phenol is produced. Phenol is a powerful germicide, as are most of its derivatives. In 1865, the English surgeon Lister was the first to apply chemicals to a wound to prevent infection. For this purpose he used a dilute solution of phenol. As other antiseptic chemicals were found, the use of phenol was discontinued because of its damaging effects on the tissues to which it was applied, its absorption through the skin, and its extreme toxicity. However, it is still used by the drug industry as a standard for measuring the germicidal activity of other antiseptics.

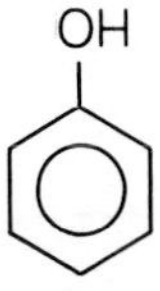

Phenol

Substituted phenols (phenols having other groups on the benzene ring) are also used as antiseptics. Hexachlorophene, a strong antiseptic compound used in soaps and deodorants, was taken off the market when it was shown to be absorbed through the skin and was found to cause brain damage in monkeys. Another complicated phenol that everyone would like to avoid is urushiol, which is one of the irritants in poison ivy.

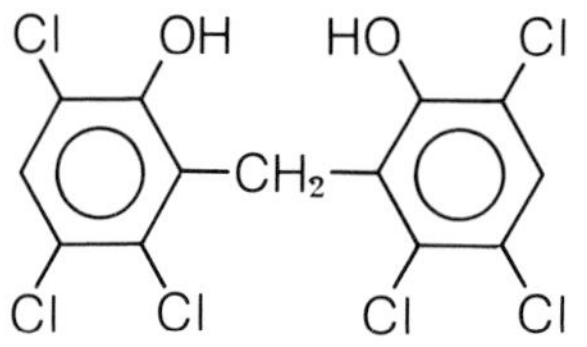

Hexachlorophene

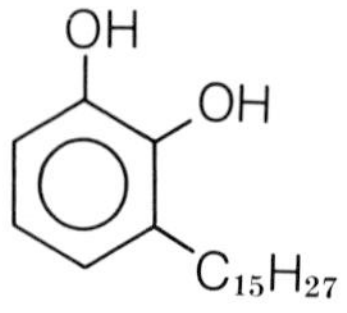

Urushiol

15.4 Preparation of Alcohols

Alcohols can be prepared through the addition of water to the double bond of an alkene in the presence of an acid (hydration reaction).

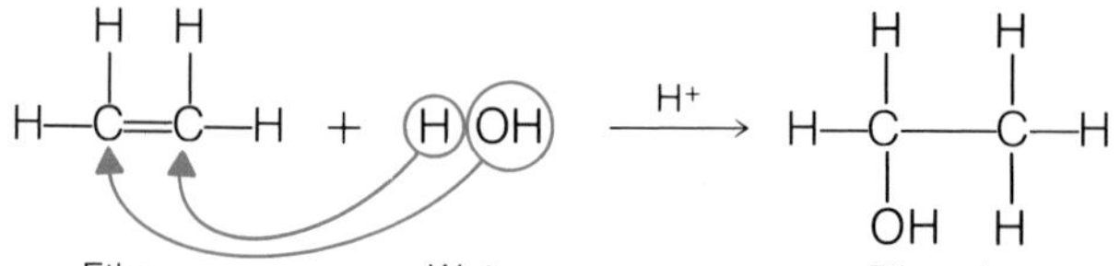

Ethene Water Ethanol

Approximately half of the ethanol used in the United States is produced in this way. Alcohols can also be produced by the addition of hydrogen atoms to aldehydes or ketones in the presence of a catalyst (which we will discuss shortly).

15.5 Reactions of Alcohols

Dehydration

A dehydration reaction of an alcohol is the removal of a water molecule from an alcohol, thereby producing an unsaturated hydrocarbon. Some alcohols undergo dehydration more easily than others. Tertiary alcohols are the easiest to dehydrate and primary alcohols are the most difficult. The dehydration process can be accomplished by heating, but in most cases a dehydrating agent such as sulfuric acid is required.

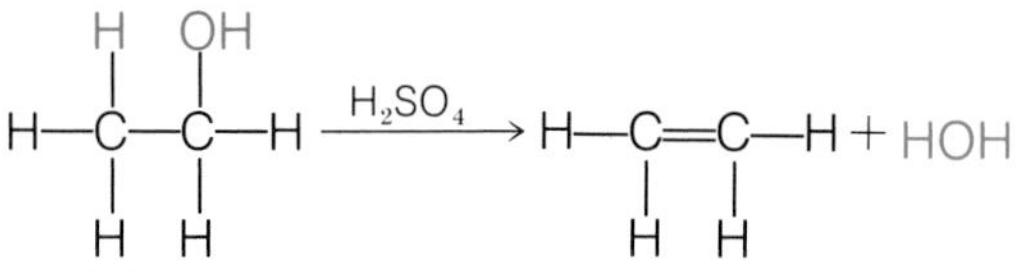

Ethanol Ethene

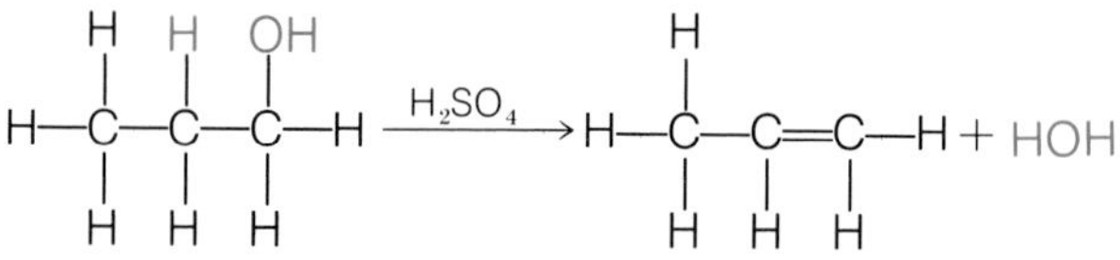

1-Propanol Propene

Oxidation

We have defined an oxidation reaction to be one in which a molecule gains oxygen atoms or loses hydrogen atoms (dehydrogenation reaction). The oxidation of an alcohol involves dehydrogenation, the loss of two hydrogen atoms—one from the hydroxyl group and one from the carbon to which the hydroxyl group is attached. These hydrogen fragments form water by a hydrogenation (reduction) reaction with the oxygen from the oxidizing agent. Potassium permanganate ($KMnO_4$) or potassium dichromate ($K_2Cr_2O_7$) are commonly used as oxidizing agents for alcohols.

Primary Alcohols The oxidation (dehydrogenation) of primary alcohols produces members of the class of organic compounds called aldehydes.

$$CH_3-\overset{\overset{\displaystyle H}{|}}{\underset{\underset{\displaystyle H}{|}}{C}}-OH \xrightarrow{-2H} CH_3-\overset{\overset{\displaystyle H}{|}}{C}=O$$

| Ethanol | Ethanal (Acetaldehyde) |

$$CH_3CH_2CH_2-\overset{\overset{\displaystyle H}{|}}{\underset{\underset{\displaystyle H}{|}}{C}}-OH \xrightarrow{-2H} CH_3CH_2CH_2-\overset{\overset{\displaystyle H}{|}}{C}=O$$

| 1-Butanol | Butanal |

Secondary Alcohols The oxidation (dehydrogenation) of secondary alcohols produces members of the class of organic compounds called ketones.

$$CH_3-\overset{\overset{\displaystyle OH}{|}}{\underset{\underset{\displaystyle H}{|}}{C}}-CH_3 \xrightarrow{-2H} CH_3\overset{\overset{\displaystyle O}{||}}{C}CH_3$$

| 2-Propanol | 2-Propanone (Acetone) |

$$CH_3CH_2-\overset{\overset{\displaystyle OH}{|}}{\underset{\underset{\displaystyle H}{|}}{C}}-CH_3 \xrightarrow{-2H} CH_3CH_2\overset{\overset{\displaystyle O}{||}}{C}CH_3$$

| 2-Butanol | 2-Butanone |

Tertiary Alcohols Tertiary alcohols cannot be easily dehydrogenated because the carbon attached to the hydroxyl group does not have a hydrogen atom attached to it.

15.6 Ethers

An organic molecule having an oxygen atom bonded to two carbon atoms belongs to the class of compounds called **ethers**, $-\overset{|}{C}-O-\overset{|}{\underset{|}{C}}-$ (general formulas ROR or ROR′). The placement of the oxygen atom between the carbon atoms eliminates the possibility of hydrogen bonding. Therefore, ethers are only slightly more soluble in water than are alkanes, and are generally soluble in nonpolar solvents. Because there is no hydrogen bonding, ethers boil at much lower temperatures than do alcohols with comparable molecular weights. For example, dimethyl ether (molecular weight 46) boils at −23°C, but ethanol (molecular weight 46) boils at 78°C. Similarly, ethyl methyl ether (molecular weight 60) boils at 11°C, but 1-propanol (molecular weight 60) boils at 97°C. Ethers are extremely flammable, and great care must be taken in their use.

Simple ethers are frequently given common names that just list both alkyl groups attached to the oxygen. For example,

CH_3OCH_3 is dimethyl ether, and

$CH_3OCH_2CH_3$ is ethyl methyl ether.

IUPAC nomenclature is used mostly for complicated ethers which we will not discuss in this book.

Most ethers have some anesthetic properties. Diethyl ether is the anesthetic compound most commonly referred to as "ether" (Figure 15.3). Although it takes a long time for diethyl ether to take full effect on a patient (10 to 15 minutes), it is quite useful in long operations because there is a large safety margin in its use. That is, there is a large difference between the concentration that causes anesthesia and the concentration that will

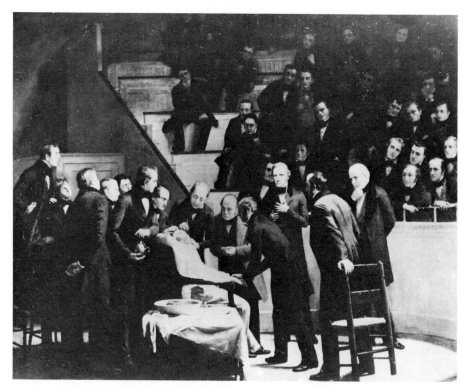

Figure 15.3 The first public demonstration of the use of ether as a surgical anesthetic took place at Massachusetts General Hospital in Boston on October 16, 1846. (Courtesy Massachusetts General Hospital)

kill the patient. Often a chemical having a more rapid anesthetic effect is used first, and then the patient is switched to diethyl ether.

$CH_3CH_2OCH_2CH_3$

$CH_2{=}\overset{\overset{\displaystyle H}{|}}{C}{-}O{-}\overset{\overset{\displaystyle H}{|}}{C}{=}CH_2$

OH — ether group
—OCH₃
$CH_2CH{=}CH_2$
Eugenol

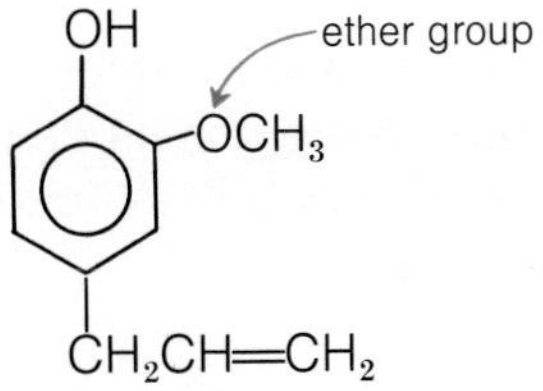

Diethyl ether Divinyl ether

Divinyl ether is a fast-acting anesthetic that can be used first, before diethyl ether. But the double bonds in its structure make divinyl ether highly reactive, and this compound must be handled with care to prevent decomposition. Eugenol, which is obtained from cloves, is a mild local anesthetic used by dentists to lessen pain when filling cavities in teeth.

15.7 Aldehydes and Ketones

Two classes of compounds, **aldehydes** and **ketones,** contain the **carbonyl**

group: an oxygen doubly bonded to a carbon atom, $-\overset{\overset{\displaystyle O}{\|}}{C}-$. The difference in properties between the aldehydes and the ketones is due to the position of the carbonyl group. Aldehydes have a terminal carbonyl group—that is, a carbonyl group which is bonded to at most one other carbon atom,

$-\overset{|}{\underset{|}{C}}-\overset{\overset{\displaystyle O}{\|}}{C}-H$ (general formula RCH or RCHO). The carbonyl group in a

ketone, by contrast, will appear somewhere in the middle of the molecule. It will be bonded to two other carbon atoms, $-\overset{|}{\underset{|}{C}}-\overset{\overset{\displaystyle O}{\|}}{C}-\overset{|}{\underset{|}{C}}-$ (general

formula RCR′ or RCOR′). Aldehydes and ketones are both highly reactive compounds; aldehydes are even more reactive than ketones. The large class of foods called carbohydrates contains aldehyde and ketone functional groups, and will be the subject of a later chapter.

Aldehydes are named by adding the suffix **-al** to the name of the longest carbon chain containing the carbonyl group. For example,

$CH_3\overset{\overset{\displaystyle O}{\|}}{C}H$ is ethanal, and

$CH_3CH_2\overset{\overset{\displaystyle CH_3}{|}}{\underset{\overset{|}{H}}{C}}{-}\overset{\overset{\displaystyle O}{\|}}{C}{-}H$ is 2-methylbutanal.

However, we will use the common names for methanal (formaldehyde) and

ethanal (acetaldehyde) because they are so often referred to by these names. When the carbonyl group is attached to a benzene ring, the compound is called benzaldehyde (see Section 15.8).

Ketones are named by adding the suffix **-one** to the name of the longest carbon chain containing the carbonyl group. The position of the carbonyl group is indicated by a number appearing before the name. For example,

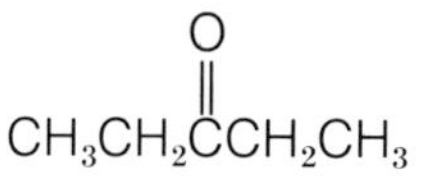

$$CH_3CH_2\overset{\overset{\displaystyle O}{\|}}{C}CH_2CH_3 \text{ is 3-pentanone, and}$$

$$CH_3\overset{\overset{\displaystyle O}{\|}}{C}CH_2\underset{\underset{\displaystyle CH_3}{|}}{C}HCH_2CH_3 \text{ is 4-methyl-2-hexanone.}$$

15.8 Important Aldehydes and Ketones

Formaldehyde is the simplest of the aldehydes.

$$H-\overset{\overset{\displaystyle H}{|}}{C}=O$$

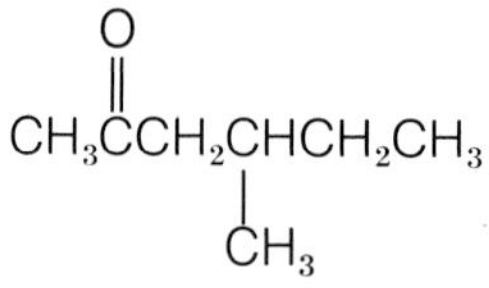

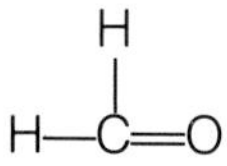

Formaldehyde

You may be aware of the use of formaldehyde as a tissue preservative. It combines easily with proteins, killing microorganisms and hardening tissues. If you have ever come in contact with formaldehyde, you know how it can irritate the membranes of your eyes, nose, and throat. In our earlier discussion of alcohols we mentioned that even very small amounts of methanol in the body can cause blindness. This happens because the liver, in attempting to rid the body of methanol, changes this compound into formaldehyde. But formaldehyde in the bloodstream will build up in the retina of the eye, destroying the cells and causing blindness. When formaldehyde is polymerized with the alcohol phenol, substances called resins, which are the gummy saps of evergreens, are formed.

Many aldehydes have pleasant odors or tastes, and are used in perfumes and artificial flavorings.

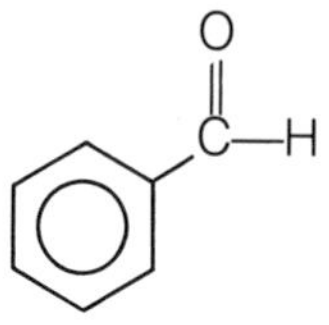

Benzaldehyde
(Almond flavoring)

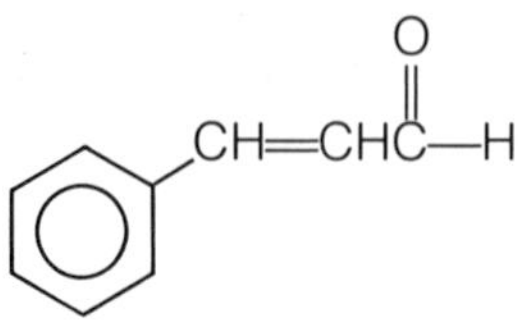

Cinnamaldehyde
(Cinnamon flavoring)

The aldehydes used in perfumes and flavorings are relatively nonpolar and hence only slightly soluble in water, but will readily dissolve in ethanol. As a result, many perfumes and flavorings use ethanol as their solvent.

Ketones are widely used in the chemical industry as solvents. One of the more common ketone compounds is acetone (propanone).

$$CH_3\overset{\overset{\displaystyle O}{\|}}{C}CH_3$$

Acetone

$$CH_3\overset{\overset{\displaystyle O}{\|}}{C}CH_2CH_3$$

2-Butanone

Acetone can be produced in the body as the result of a metabolic side reaction, not normally occurring except when there is some metabolic disorder. In the disease diabetes mellitus, for example, some normal metabolic reactions do not occur and acetone is produced in large quantities. This compound builds up in the tissues and appears in the urine and breath of untreated diabetics. 2-Butanone (methyl ethyl ketone) is a solvent with a characteristic odor, and is commonly used in nail polish remover.

Some unusual cyclic ketones, isolated from animals, have recently become very popular in the cosmetic industry. Are you familiar with musk oil? Civetone, isolated from the African civet cat, and muscone, isolated from the male musk deer, have very unpleasant odors in high concentrations. When used in extremely small quantities in perfumes, however, they intensify and give long-lasting qualities to the scent.

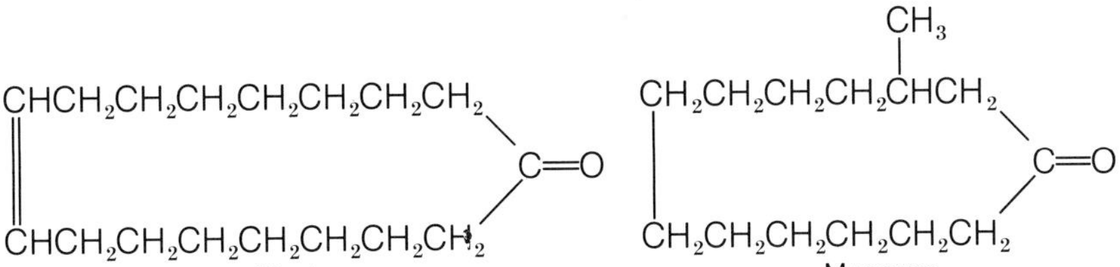

Civetone

Muscone

15.9 Preparation of Aldehydes and Ketones

Aldehydes and ketones are produced by the oxidation (dehydrogenation) of primary and secondary alcohols in the presence of oxidizing agents such as potassium permanganate or potassium dichromate.

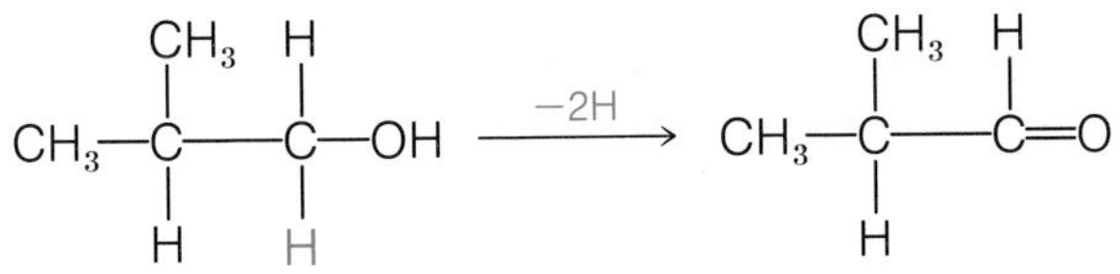

2-Methylpropanol $\xrightarrow{-2H}$ 2-Methylpropanal

3-Pentanol $\xrightarrow{-2H}$ 3-Pentanone

15.10 Reactions of Aldehydes and Ketones

Reduction

Primary alcohols can be produced from aldehydes, and secondary alcohols can be produced from ketones, by the reduction (hydrogenation) of these compounds in the presence of a proper catalyst such as platinum (Pt).

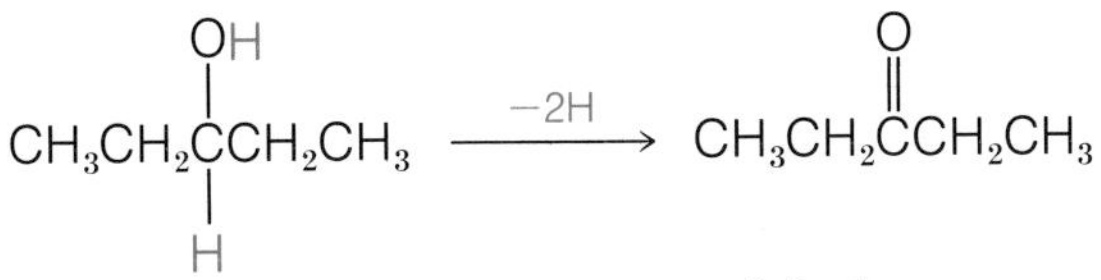

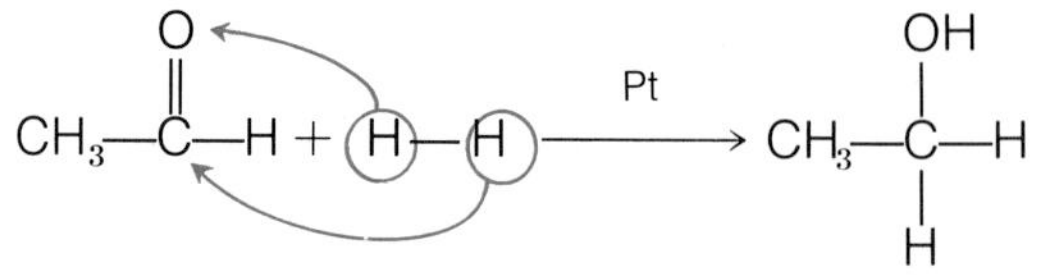

Acetaldehyde Ethanol

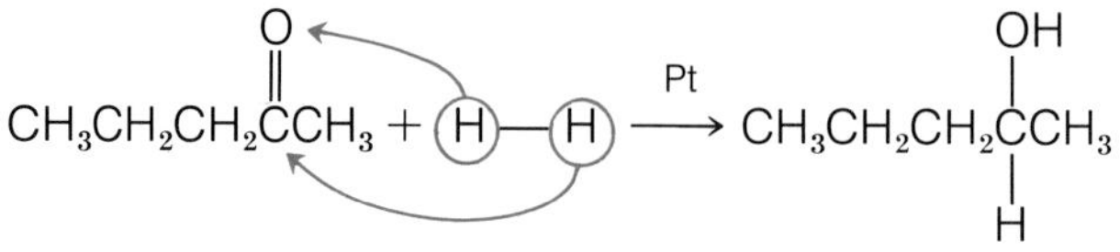

2-Pentanone 2-Pentanol

Oxidation

Aldehydes are very easy to oxidize, and such oxidation produces carboxylic acids (which we will discuss shortly). Ketones, however, can be oxidized only under extreme conditions. This, therefore, gives us a convenient way to distinguish between aldehydes and ketones in the laboratory. Notice that in this case the oxidation reaction results in the gain of oxygen atoms by the aldehyde molecule.

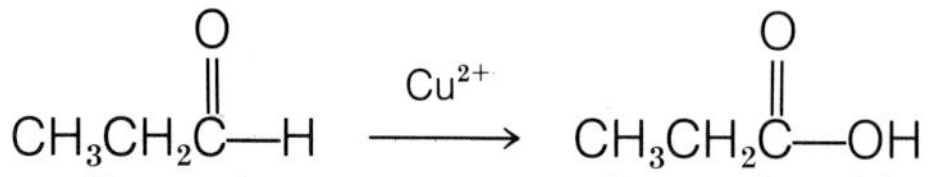

Propanal Propanoic acid

Formation of Hemiacetals and Hemiketals

Dissolving an aldehyde in an alcohol results in the formation of an equilibrium between the aldehyde and a product called a **hemiacetal.**

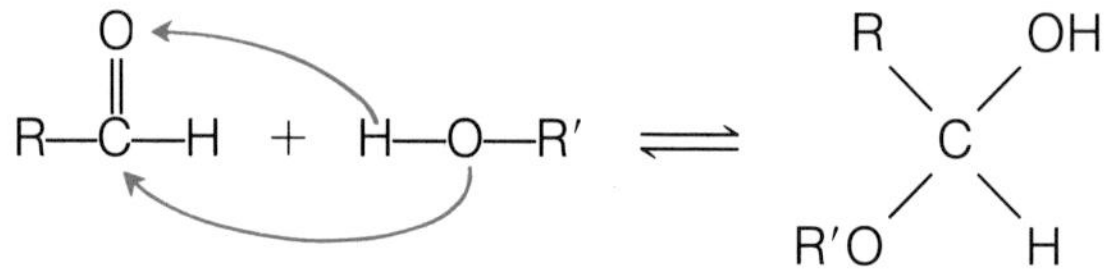

A hemiacetal

Hemiacetals can form between an aldehyde group and an alcohol group on the same molecule, resulting in the formation of a stable ring structure.

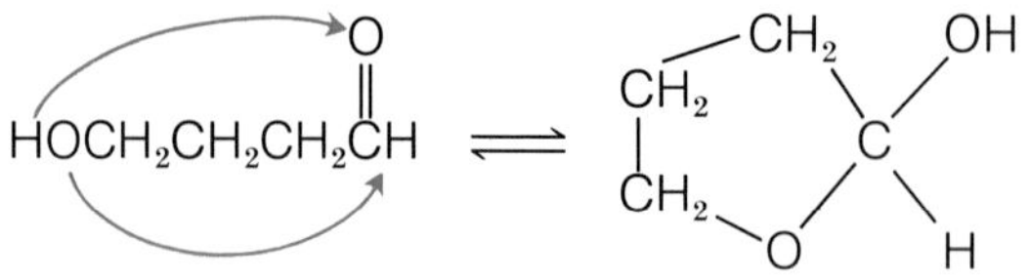

A similar reaction results when a ketone is dissolved in an alcohol, forming a **hemiketal.**

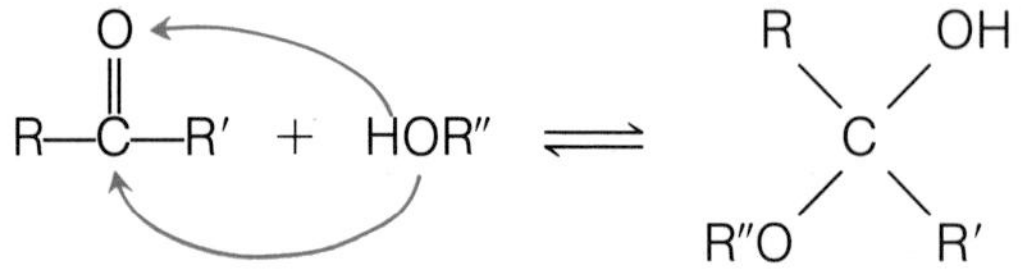

A hemiketal

15.11 Carboxylic Acids

Carboxylic acids (general formula $R\overset{O}{\overset{\|}{C}}OH$, or RCOOH) are organic acids that contain a hydroxyl group attached to a carbonyl group. This combined group is called the **carboxyl** or **carboxylic acid** functional group, $-\overset{O}{\overset{\|}{C}}-OH$. The bond between the oxygen atom and the hydrogen atom in the carboxyl group is extremely polar, so much so that when placed in water some of the molecules ionize as follows:

$$\underset{\text{Acetic acid}}{CH_3\overset{O}{\overset{\|}{C}}-OH} + H_2O \longrightarrow \underset{\text{Acetate ion}}{CH_3\overset{O}{\overset{\|}{C}}-O^-} + \underset{\text{Hydronium ion}}{H_3O^+}$$

The stronger the acid, the more easily it gives up its hydrogen to form the hydronium ion. But organic acids are very weak compared with inorganic acids such as nitric or sulfuric acid. Because some carboxylic acids are found in fats, these acids are also referred to as fatty acids.

Carboxylic acids are named by removing the ending letter *-e* from the alkane with the same length carbon chain, and adding **-oic acid.** For example,

$$H\overset{O}{\overset{\|}{C}}OH \text{ is methanoic acid,}$$

$$CH_3\overset{O}{\overset{\|}{C}}OH \text{ is ethanoic acid, and}$$

$$CH_3\underset{\underset{CH_3}{|}}{C}H\overset{O}{\overset{\|}{C}}OH \text{ is 2-methylpropanoic acid.}$$

The common names for acids often come from the Latin or Greek words indicating their source, and are still very much in use. We will always use the common names formic acid (for methanoic acid) and acetic acid (for ethanoic acid).

15.12 Important Carboxylic Acids

Formic acid is very irritating to tissues; this is the compound released in bee and ant stings, which causes skin inflammation. (Table 15.4 shows the structure of each of the carboxylic acids that we will discuss.) Formic acid, like formaldehyde, is formed by the liver in the oxidation of methanol. When it enters the bloodstream it can cause severe acidosis, a condition that disrupts the blood's ability to carry oxygen.

Table 15.4 Some Common Carboxylic Acids

Common Name	IUPAC Name	Formula
Formic acid	Methanoic acid	$H-\overset{\overset{\displaystyle O}{\|\|}}{C}-OH$
Acetic acid	Ethanoic acid	$CH_3-\overset{\overset{\displaystyle O}{\|\|}}{C}-OH$
Butyric acid	Butanoic acid	$CH_3CH_2CH_2-\overset{\overset{\displaystyle O}{\|\|}}{C}-OH$
Capric acid	Decanoic acid	$CH_3(CH_2)_8-\overset{\overset{\displaystyle O}{\|\|}}{C}-OH$
Benzoic acid	Benzoic acid	⬡$-\overset{\overset{\displaystyle O}{\|\|}}{C}-OH$
Salicylic acid	o-Hydroxybenzoic acid*	⬡$-\overset{\overset{\displaystyle O}{\|\|}}{C}-OH$, OH

* See Appendix 3 for this nomenclature.

Acetic acid has been known as long as alcohol, for it is the substance that is formed when wine turns sour. This souring occurs when certain bacteria oxidize ethanol to form acetic acid, producing the familiar taste and smell of vinegar. Acetic acid is the compound used by living cells to produce the longer-chain fatty acids, all of which have an unmistakably strong smell. For example, butyric acid is the unpleasant odor of rancid butter, and is one of the compounds that causes "body odor." Capric acid gives limburger cheese its smell.

Benzoic acid and its sodium salt, sodium benzoate, are used when making bread and cheese to prevent spoilage from the growth of mold or bacteria (Figure 15.4). Salicylic acid, whose structure is similar to that of benzoic acid, cannot be used in foods because it is so irritating to tissues. It can destroy horny growths such as corns and warts, and is used in products that treat such conditions.

Oxalic acid, a dicarboxylic acid, ionizes to form the oxalate ion, which is a normal product of the body's metabolism.

$$HO-\overset{\overset{\displaystyle O}{\|\|}}{C}-\overset{\overset{\displaystyle O}{\|\|}}{C}-OH + 2H_2O \longrightarrow {}^-O-\overset{\overset{\displaystyle O}{\|\|}}{C}-\overset{\overset{\displaystyle O}{\|\|}}{C}-O^- + 2H_3O^+$$

Oxalic acid Oxalate ion ($C_2O_4{}^{2-}$) Hydronium ion

Oxalate ions bond strongly with calcium ions to form an insoluble salt. This reaction occurs in the urine of humans, but large crystals of calcium

Figure 15.4 These are some of the many foods containing sodium benzoate as a preservative. (Ron Nelson)

oxalate normally are kept from forming. In some cases, however, such crystals can build up to form small stones known as kidney or bladder stones (urinary calculi).

$$Ca^{2+}{}_{(aq)} \; + \; C_2O_4{}^{2-}{}_{(aq)} \longrightarrow CaC_2O_{4(s)}$$

Calcium ion Oxalate ion Calcium oxalate

15.13 Preparation of Carboxylic Acids

Carboxylic acids are prepared by the oxidation of primary alcohols or aldehydes in the presence of a strong oxidizing agent.

$$CH_3CH_2CH_2CH_2OH \xrightarrow{KMnO_4} CH_3CH_2CH_2\overset{\displaystyle O}{\overset{\|}{C}}-OH$$

1-Butanol Butanoic acid

$$CH_3\overset{\displaystyle O}{\overset{\|}{C}}-H \xrightarrow{KMnO_4} CH_3\overset{\displaystyle O}{\overset{\|}{C}}-OH$$

Acetaldehyde Acetic acid

Aromatic acids can be prepared by oxidizing a carbon side-chain on a benzene compound with a strong oxidizing agent. The carbon atom attached to the ring becomes the carbon atom in the carboxylic acid group, and the remaining carbons on the side-chain go to form carbon dioxide.

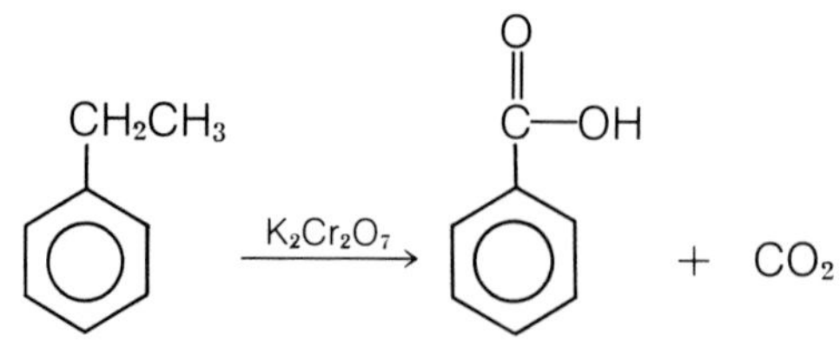

Ethylbenzene Benzoic acid

15.14 Esters

An important class of compounds derived from carboxylic acids is the esters, whose molecules contain the ester functional group.

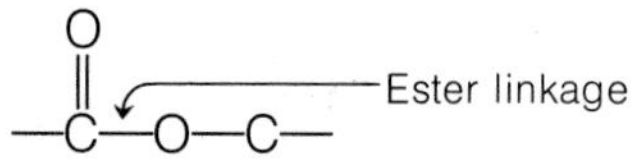

Ester functional group

Esters are noted for their pleasant aromas, and are the source of the familiar smells of flowers and the tastes of fruits (Table 15.5).

Esters can be produced through a reaction that removes a molecule of water from a molecule of a carboxylic acid and a molecule of an alcohol.

$$\underset{\text{Carboxylic acid}}{RC(=O)OH} + \underset{\text{Alcohol}}{HOR'} \longrightarrow \underset{\text{Ester}}{RC(=O)OR'} + \underset{\text{Water}}{HOH}$$

$$\underset{\text{Formic acid}}{H{-}C(=O){-}OH} + \underset{\text{Ethanol}}{HOCH_2CH_3} \longrightarrow \underset{\text{Ethyl formate}}{H{-}C(=O){-}O{-}CH_2CH_3} + H_2O$$

Such a reaction, in which a molecule of water is removed from two reactant molecules with the resulting formation of one product molecule, is known as a **condensation reaction.** This type of reaction accounts for many of the chemical reactions occurring in biological systems.

The names of esters are formed by first listing the alcohol portion with a *-yl* ending, and then the acid portion with an *-ate* ending. If the alcohol portion contains a branched chain, the name used is one of the alkyl groups listed in Appendix 3. For example,

$\underbrace{CH_3{-}O}_{\text{Alcohol portion: methanol}}{-}\underbrace{C(=O){-}CH_2CH_2CH_3}_{\text{Acid portion: butanoic acid}}$ is methyl butanoate, and

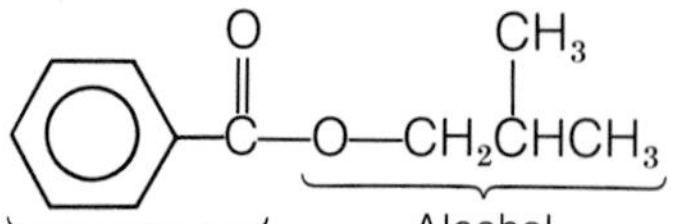

is isobutyl benzoate

Acid portion Alcohol portion

Table 15.5 Some Esters Used as Flavorings

IUPAC Name (Common Name)	Formula	Flavor
Ethyl methanoate (Ethyl formate)	$CH_3CH_2-O-\overset{\overset{O}{\|}}{C}-H$	Rum
Pentyl ethanoate (Amyl acetate)	$CH_3(CH_2)_4-O-\overset{\overset{O}{\|}}{C}-CH_3$	Banana
Octyl ethanoate (Octyl acetate)	$CH_3(CH_2)_7-O-\overset{\overset{O}{\|}}{C}-CH_3$	Orange
Pentyl butanoate (Amyl butyrate)	$CH_3(CH_2)_4-O-\overset{\overset{O}{\|}}{C}-CH_2CH_2CH_3$	Apricot

15.15 Important Esters

Nitroglycerin

Nitroglycerin, which is the active substance in dynamite, is an ester formed by the condensation reaction between an inorganic acid (nitric acid) and an alcohol (glycerol). In addition to its use in dynamite, nitroglycerin is used to relieve pain in the heart disorder angina pectoris. This compound acts to enlarge (or dilate) smaller blood vessels and to relax smooth muscles found in the arteries, thereby lowering high blood pressure and the accompanying pain.

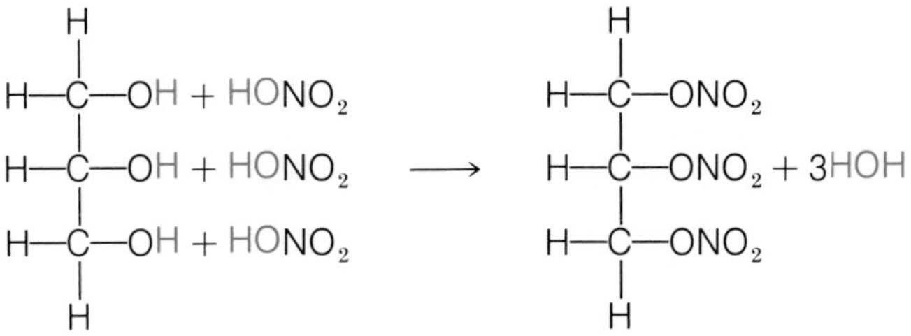

Glycerol   Nitric acid   Nitroglycerin

Salicylic Acid Esters

Salicylic acid is found in the bark of willow trees. It is a molecule containing a carboxylic acid group and an alcohol group, both of which can undergo condensation reactions to form esters. We have already mentioned that salicylic acid is a very irritating substance with a disagreeable taste, and is not to be taken internally. But the products of condensation reactions with this compound are not as irritating and have a wide range of uses. For example, if the acid group on salicylic acid is condensed with methanol, methyl salicylate is formed. This compound is more commonly known as oil of wintergreen, and is used in perfumes, candies, and in ointments which cause a mild burning sensation on your skin in an attempt to take your mind off your sore muscles.

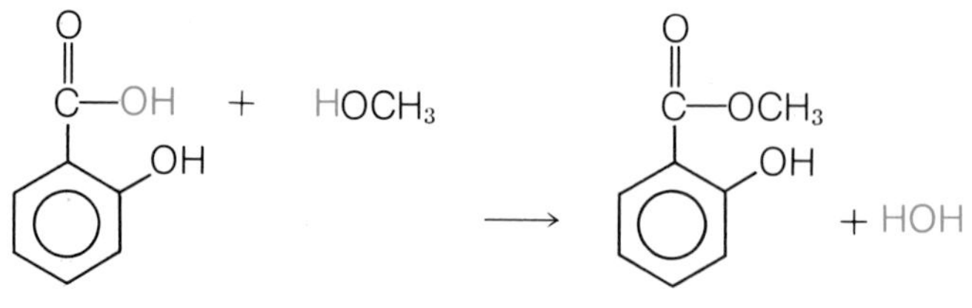

Salicylic acid Methanol Methyl salicylate

If acetic acid reacts with the alcohol group on salicylic acid, acetylsalicylic acid, one of the most widely used drugs in the world, is formed.

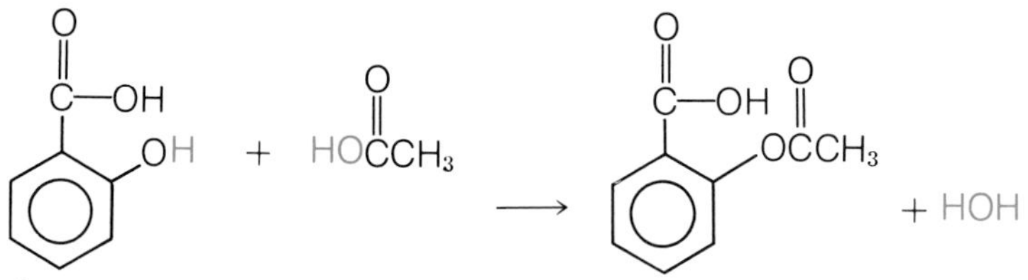

Salicylic acid Acetic acid Acetylsalicylic acid (Aspirin)

Aspirin, as acetylsalicylic acid is more commonly known, has the properties of an analgesic (pain reliever) and antipyretic (fever reducer). Despite the claims of various aspirin producers, repeated testing has shown that there is no difference among competing brands of aspirin except for their price. Adult aspirin tablets contain only 5 grains (0.33 g) of acetylsalicylic acid. The rest of the tablet is mostly a binder which the manufacturer uses to put the aspirin in tablet form. The difference in size between various brands of aspirin tablets is most often due to the amount of binder used, and not the aspirin itself.

Aspirin is by far the most widely used medicine in the United States, with an annual production of 30 million pounds a year. (This averages out to approximately 200 tablets per year for every man, woman, and child!) Although the exact reason for aspirin's therapeutic effects is not known, its use is considered relatively safe. However, the constant use of aspirin can produce several unpleasant side effects. Aspirin can irritate the lining of the stomach, and has been shown to produce stomach ulcers in rats. As molecules of aspirin pass through the stomach wall they injure the cells of this tissue, causing small hemorrhages. In most people, the common dosage of two 5-grain aspirin tablets will cause only small blood loss in the stomach—from 0.2 ml to 5 ml. Some people, however, are very sensitive to this drug, and experience stomach bleeding at potentially dangerous levels.

Polyesters

Esters have become extremely important in the production of synthetic fibers. Ester polymers called **polyesters,** formed through condensation reactions, give synthetic fabrics many desirable properties. These advantages include the ability to set the fabric into permanent creases and pleats, and the resistance of the fabric to turning gray or yellow with long use. The familiar material Dacron is a polyester made of esters of ethylene glycol and terephthalic acid. Notice that both ethylene glycol and terephthalic acid have two positions where ester bonds can be formed.

Ethylene glycol

Terephthalic acid

Dacron + Water

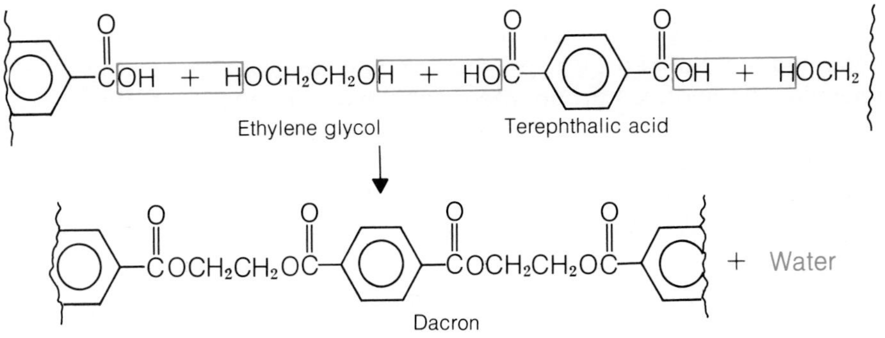

15.16 Chemistry of Esters

As we have mentioned, esters can be produced by a condensation reaction between a carboxylic acid and an alcohol. This reaction can be made to occur in the laboratory in the presence of a strong acid.

Hydrolysis

Esters can be broken apart by an addition reaction with water to form the carboxylic acid and the alcohol. This reaction, called **hydrolysis,** is the exact reverse of the condensation reaction by which the ester may have been formed. It can be made to occur by heating the ester in a water solution of a strong inorganic acid.

$CH_3CH_2OCCH_3 + H_2O \xrightarrow{H^+} CH_3CH_2OH + HOCCH_3$

Ethyl acetate Ethanol Acetic acid

Ethyl benzoate Benzoic acid Ethanol

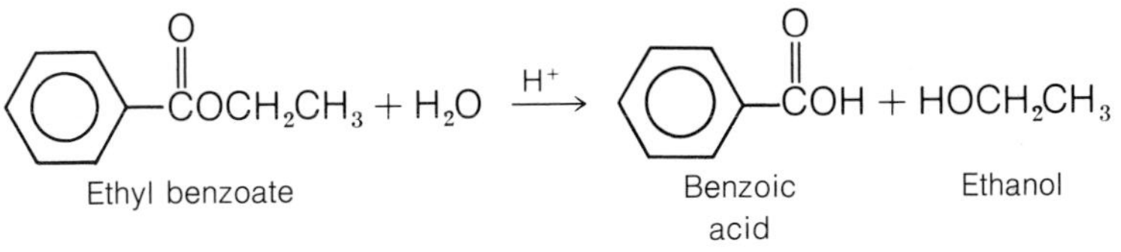

Saponification

Esters can also be broken apart when heated in an aqueous solution of strong base, such as NaOH or KOH. This reaction, called **saponification,** is just like hydrolysis except that the carboxylic acid formed exists not as the free acid but as the sodium or potassium salt.

$CH_3CH_2OCCH_3 + NaOH_{(aq)} \longrightarrow CH_3CH_2OH + NaOCCH_3$

Ethyl acetate Ethanol Sodium acetate

Ethyl benzoate Potassium benzoate Ethanol

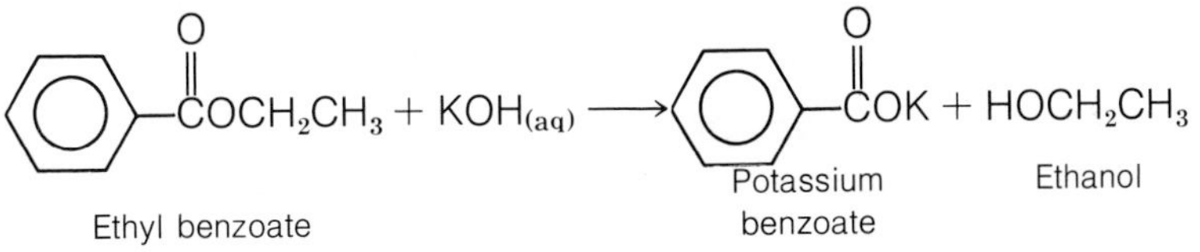

Chapter Summary

When substituted on a hydrocarbon molecule, functional groups containing oxygen add a reactive spot to the molecule. Each oxygen-containing functional group adds its own special properties to the molecule. The hydroxyl group, the aldehyde group, and the carboxylic acid group are all polar, and make molecules with short carbon chains water soluble. The following are the general formulas for the different classes of compounds that we have studied:

Alcohol	ROH	Ketone	$\overset{\displaystyle O}{\overset{\displaystyle \|}{\text{RCR}'}}$	or RCOR'
Ether	ROR or ROR'	Carboxylic Acid	$\overset{\displaystyle O}{\overset{\displaystyle \|}{\text{RCOH}}}$	or RCOOH
Aldehyde	RCHO	Ester	$\overset{\displaystyle O}{\overset{\displaystyle \|}{\text{RCOR}'}}$	or RCOOR'

The chemical reactions that we have discussed for these compounds can be summarized as follows:

I. OXIDATION

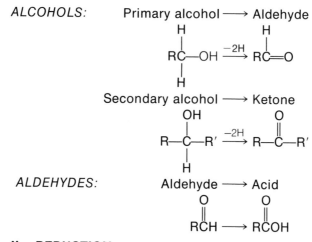

ALCOHOLS: Primary alcohol ⟶ Aldehyde

$$\overset{H}{\underset{H}{\text{RC}}}\!\!-\!\text{OH} \xrightarrow{-2H} \overset{H}{\underset{}{\text{RC}}}\!\!=\!\text{O}$$

Secondary alcohol ⟶ Ketone

$$\overset{OH}{\underset{H}{\text{R}\!-\!\text{C}\!-\!\text{R}'}} \xrightarrow{-2H} \overset{O}{\overset{\|}{\text{R}\!-\!\text{C}\!-\!\text{R}'}}$$

ALDEHYDES: Aldehyde ⟶ Acid

$$\overset{O}{\overset{\|}{\text{RCH}}} \longrightarrow \overset{O}{\overset{\|}{\text{RCOH}}}$$

II. REDUCTION

ALDEHYDE: Aldehyde ⟶ Primary alcohol

$$\overset{H}{\underset{}{\text{RC}}}\!\!=\!\text{O} + H_2 \longrightarrow \overset{H}{\underset{H}{\text{RC}}}\!\!-\!\text{OH}$$

KETONE: Ketone ⟶ Secondary alcohol

$$\overset{O}{\overset{\|}{\text{R}\!-\!\text{C}\!-\!\text{R}'}} + H_2 \longrightarrow \overset{OH}{\underset{H}{\text{R}\!-\!\text{C}\!-\!\text{R}'}}$$

III. DEHYDRATION

$$\text{Alcohol} \longrightarrow \text{Alkene} + H_2O$$
$$RCH_2CH_2OH \longrightarrow RCH{=}CH_2 + H_2O$$

IV. CONDENSATION

$$\text{Alcohol} + \text{Acid} \longrightarrow \text{Ester} + \text{Water}$$

$$
\overset{\quad\quad\quad\quad\quad O}{\underset{\quad\quad\quad\quad\quad \|}{ROH + HOCR'}} \longrightarrow \overset{O}{\underset{\|}{ROCR'}} + H_2O
$$

V. HYDROLYSIS

$$\text{Ester} + H_2O \longrightarrow \text{Alcohol} + \text{Acid}$$

$$
\overset{O}{\underset{\|}{ROCR'}} + H_2O \longrightarrow ROH + \overset{O}{\underset{\|}{HOCR'}}
$$

VI. SAPONIFICATION

$$\text{Ester} + \text{Strong Base} \longrightarrow \text{Alcohol} + \text{Acid Salt}$$

$$
\overset{O}{\underset{\|}{ROCR'}} + NaOH \longrightarrow ROH + \overset{O}{\underset{\|}{NaOCR'}}
$$

Exercises and Problems

1. Identify each of the following as either an alcohol, ether, carboxylic acid, ester, aldehyde, or ketone:

(a)
$$
CH_3\overset{\overset{\textstyle CH_3}{|}}{CH}-\overset{\overset{\textstyle O}{\|}}{C}-OH
$$

(b) $CH_3CH_2CH_2OCH_2CH_3$

(c)
$$
CH_3CH_2O\overset{\overset{\textstyle O}{\|}}{C}CH_2CH_2CH_3
$$

(d)
$$
CH_3\overset{\overset{\textstyle CH_3}{|}}{CH}-\overset{\overset{\textstyle O}{\|}}{C}H
$$

(e)
$$
CH_3CH_2CH_2\overset{\overset{\textstyle O}{\|}}{C}CH_2CH_3
$$

(f)
$$
CH_3\overset{\overset{\textstyle OH}{|}}{C}HCH_2\overset{\overset{\textstyle O}{\|}}{C}OH
$$

(g)

(h)
$$
CH_3CH_2\overset{\overset{\textstyle OH}{|}}{\underset{\underset{\textstyle CH_2CH_3}{|}}{C}}CH_2CH_2CH_3
$$

(i)
$$
CH_3CH_2CH_2CH_2\overset{\overset{\textstyle O}{\|}}{C}H
$$

(j)
$$
CH_3\overset{\overset{\textstyle}{}}{\underset{\underset{\textstyle CH_3}{|}}{C}}H-\overset{\overset{\textstyle O}{\|}}{C}-\underset{\underset{\textstyle CH_3}{|}}{C}HCH_3
$$

(k)
$$
CH_3O\overset{\overset{\textstyle O}{\|}}{C}CH_2CH_2CH_2CH_3
$$

2. Name each of the compounds listed in question 1.

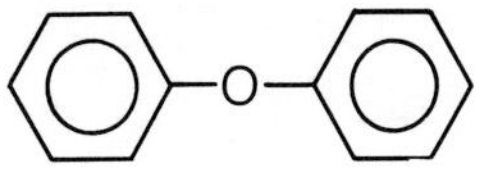

3. Why is it that propane boils at $-45°C$ and methyl ether at $-23°C$, but ethanol boils at $78°C$?

4. Explain why 2-hexanol is only very slightly soluble in water, whereas 2,3-hexanediol is soluble in water.

5. Identify each of the following compounds as either a primary, secondary, or tertiary alcohol.

(a) $CH_3\overset{\overset{\displaystyle OH}{|}}{\underset{\underset{\displaystyle H}{|}}{C}}CH_2CH_3$

(d) $CH_3CH_2\overset{\overset{\displaystyle OH}{|}}{\underset{\underset{\displaystyle H}{|}}{C}}\!\!-\!\!H$

(b) ⟨O⟩—CH_2OH

(e) $CH_3\!\!-\!\!\overset{\overset{\displaystyle CH_3}{|}}{\underset{\underset{\displaystyle CH_3}{|}}{C}}\!\!-\!\!\overset{\overset{\displaystyle OH}{|}}{\underset{\underset{\displaystyle H}{|}}{C}}\!\!-\!\!CH_3$

(c) $CH_3CH_2\!\!-\!\!\overset{\overset{\displaystyle CH_3}{|}}{\underset{\underset{\displaystyle CH_3}{|}}{C}}\!\!-\!\!OH$

(f) ⟨O⟩—$\overset{\overset{\displaystyle OH}{|}}{\underset{\underset{\displaystyle CH_3}{|}}{C}}\!\!-\!\!CH_3$

6. Assume you are making chocolate-covered cherries as a fund-raising project. Why might you want to add some glycerol to the cherry filling?

7. Why might a drink of methanol cause blindness and death?

8. Which products used in your home might contain aldehydes and ketones?

9. A solution containing equal amounts of acetic acid ($CH_3\overset{\overset{\displaystyle O}{\|}}{C}OH$) and acetate ion ($CH_3\overset{\overset{\displaystyle O}{\|}}{C}O^-$) has a pH that remains quite constant when small amounts of strong acid or base are added. Explain why this happens.

10. Name the products of the following reactions and write the general equation for the reaction:

(a) oxidation of a primary alcohol
(b) oxidation of a secondary alcohol
(c) oxidation of an aldehyde
(d) reduction of an aldehyde
(e) reduction of a ketone
(f) dehydration of an alcohol
(g) condensation reaction between a carboxylic acid and an alcohol
(h) hydrolysis of an ester
(i) saponification of an ester

11. Write the structure of the hemiacetal or hemiketal that will form between the following pairs of compounds:

    (a) acetaldehyde and methanol
    (b) acetone and ethanol
    (c) 2-butanone and methanol
    (d) formaldehyde and ethanol
    (e) propanal and methanol
    (f) acetone and 1-propanol

12. Write the structure of the aldehyde or ketone that will form from the oxidation of the following alcohols:

    (a) 1-propanol
    (b) 2-propanol
    (c) 3-methyl-1-butanol
    (d) 3-methyl-2-butanol
    (e) cyclopentanol
    (f) ethanol
    (g) 1-pentanol
    (h) 2-hexanol

13. Write the structure of the carboxylic acid that will form from the oxidation of the following aldehydes:

    (a) acetaldehyde
    (b) butanal
    (c) hexanal
    (d) benzaldehyde

14. Write the structural formula of the alcohol that will result from the reduction of the following aldehydes and ketones:

    (a) butanone
    (b) butanal
    (c) benzaldehyde
    (d) 2-pentanone
    (e) acetone
    (f) hexanal

15. Write the structure of the product of the dehydration of

    (a) 1-butanol, and (b) 3-hexanol.

16. Write the equation for the condensation reaction between:

    (a) butanoic acid and methanol
    (b) benzoic acid and ethanol
    (c) formic acid and 1-propanol
    (d) acetic acid and 2-butanol
    (e) salicylic acid and ethanol
    (f) salicylic acid and propanoic acid

17. Write the equation for the hydrolysis of the following esters:

    (a) ethyl benzoate
    (b) isopropyl butanoate
    (c) octyl acetate
    (d) t-butyl propanoate

18. Write the equation for the saponification of the following esters by sodium hydroxide:

    (a) methyl acetate
    (b) ethyl propanoate
    (c) *sec*-butyl benzoate
    (d) diethyl oxalate

19. How would you prepare, in a laboratory, the flavoring for wintergreen candies?

chapter 16

Nitrogen

Learning Objectives

By the time you have finished this chapter, you should be able to:

1. Identify the nitrogen-containing functional groups.

2. Given the structural formula of an amine, identify it as a primary, secondary, or tertiary amine.

3. Show how a nitrogen atom is able to form a coordinate covalent bond.

4. Define a *base,* and write an equation showing the reaction of an amine and water.

5. Write the structural formula of an amine or amide when given its IUPAC name, and state the IUPAC name when given the structural formula.

6. Write an equation for the hydrolysis of an amide.

7. Define *heterocyclic rings.*

8. Describe three uses of alkaloids in drug therapy.

"Well, this is it—they've just called for the 100-meter freestyle. Now, the team's hopes for the state championship rest on me, Ed Scott, the kid who barely made the team his senior year! Two minutes to go—I must get control of myself. I was fine until they called the race, now my heart is pounding (can't everyone hear it?). My hands are sweating and I have a sick feeling in the pit of my stomach. Shake it off, Ed—try to relax those muscles! You'll need everything you've got to beat that guy from Crescent Tech. Slow down that breathing, man, you're panting like the race was over! TAKE YOUR MARK! Concentrate, Ed. GET SET! Well, it's now or never—come on, body, give it all you've got. GO!!"

Faced with a situation full of stress, we've all had reactions much like Ed's. A sudden, loud squeal of brakes close by, or a frightening, suspenseful movie can make your heart pound, your body perspire, and your stomach feel as though it were climbing into your throat. These various body reactions are all caused by a chemical compound which is produced in a pair of hat-shaped organs called the adrenals, located on top of the kidneys. The compound is one of many chemical messengers, called hormones, that travel through the bloodstream and cause changes in body cells. This particular compound is called epinephrine, but you are probably more familiar with its common name, adrenalin. Under normal conditions, small amounts of epinephrine are released into the blood to help control blood pressure and to maintain the level of sugar in the blood. But this compound is also the body's way of meeting emergencies and dealing with stress such as emotional excitement, exercise, extreme temperature changes, severe hemorrhaging, and the administration of certain anesthetics.

Epinephrine affects many parts of the body in reacting to stress. It increases the rate and strength of the heart beat, which increases the cardiac output. It raises the blood pressure by causing constriction of blood vessels in all parts of the body, except for the blood vessels in such vital organs as the skeletal muscles, heart, brain, and liver. It relaxes the smooth muscles in the lungs, making epinephrine a very effective drug in the treatment of severe bronchial asthma attacks. It also increases the rate and depth of breathing, enabling more oxygen to get into the lungs. Epinephrine causes an increase in blood sugar, and a general increase in the metabolic activity of the cells. It slows down the action of the digestive tract, and accelerates blood clotting. It delays the fatigue of skeletal muscles, and increases the strength of contractions of these muscles. This last property has allowed people to show amazing strength under great stress; for example, after an auto accident, a man lifted his car to free his son who was trapped underneath. Each of these changes in normal body function caused by epinephrine enables the body to meet the initial challenge of the stress. The effects last only a short time, because the epinephrine released into the bloodstream is inactivated by the liver in about 3 minutes.

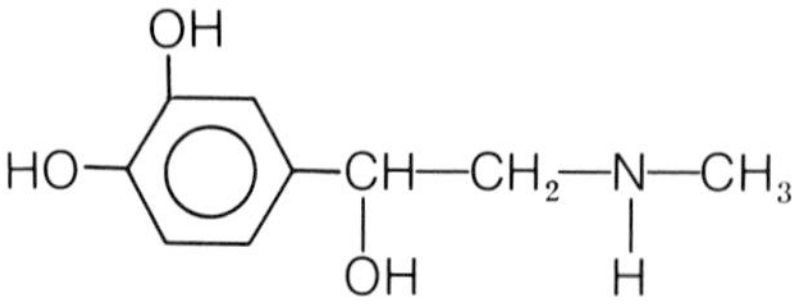

Epinephrine (adrenalin)

Epinephrine is a fairly complex molecule containing a benzene ring, hydroxyl groups, and a functional group that contains nitrogen. In the last few chapters we have discussed important biological compounds composed of carbon, hydrogen, and oxygen. Nitrogen is the fourth most abundant element in the human body, making up 1.4% of the total number of atoms. Nitrogen can form strong bonds with carbon, oxygen, and hydrogen, thus allowing this element to be found in many different

Table 16.1 Some Nitrogen-Containing Functional Groups

Functional Group	Class of Compound	Typical Compound	
R—NH$_2$	Amine	CH$_3$NH$_2$	Methylamine
R—$\overset{\overset{\textstyle O}{\|\|}}{C}$—NH$_2$	Amide	CH$_3$$\overset{\overset{\textstyle O}{\|\|}}{C}$—NH$_2$	Acetamide
R—$\overset{\overset{\textstyle R'}{\|}}{C}$=NH	Imine	H$_2$N—$\overset{\overset{\textstyle NH_2}{\|}}{C}$=NH	Guanidine
R—C≡N	Nitrile	H$_2$C=CHC≡N	Acrylonitrile

arrangements in organic molecules. As in the previous chapters, we can categorize these nitrogen-containing molecules by means of similarities in chemical properties resulting from the presence of important functional groups (Table 16.1).

Compounds Containing Nitrogen

16.1 Nitrogen

Nitrogen is the first member of group V on the periodic table. An atom of nitrogen has five valence electrons, and will form three covalent bonds to become stable.

$$\cdot \overset{\cdot\cdot}{\underset{\cdot}{N}} \cdot$$

Nitrogen

Such bonding can occur in several ways: by the formation of three single bonds, a double and a single bond, or one triple bond (Figure 16.1).

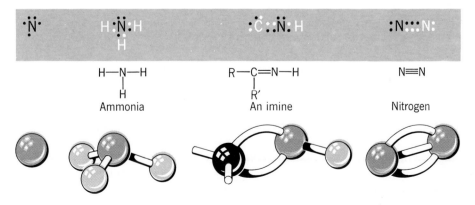

Figure 16.1 Nitrogen can form single, double, and triple covalent bonds.

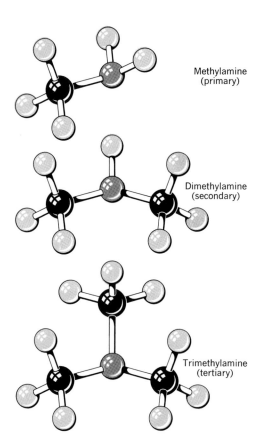

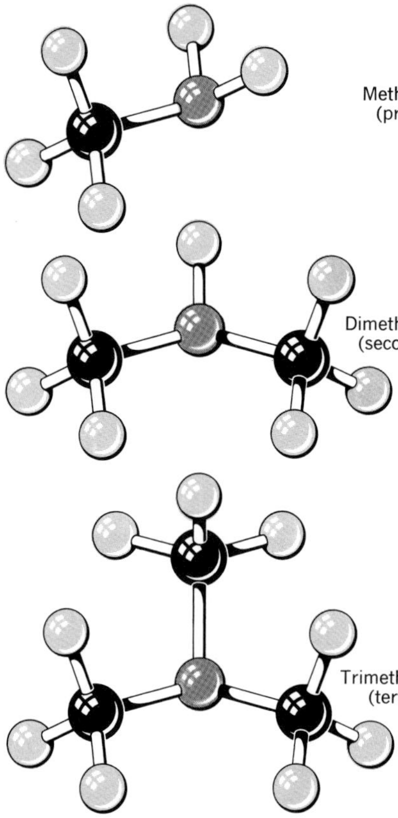

Figure 16.2 Amines can be either primary, secondary, or tertiary.

Nitrogen is found in the atmosphere as the molecule N_2, two nitrogen atoms connected by a triple bond. This molecule accounts for 80% of the gases in the atmosphere; it is stable and relatively unreactive. Nitrogen in the form of N_2 is useless to most forms of life. We inhale it with every breath we take, but then exhale it because our bodies cannot force this molecule to react with other atoms. There is only one type of organism that is able to force atmospheric nitrogen to form chemical combinations with other atoms, which then makes it usable by other forms of life. These organisms are bacteria which live in the soil or in the roots of plants such as peas and alfalfa. The bacteria convert gaseous nitrogen into compounds that can be used by the plants. Animals then get the nitrogen they need by eating such plants, or by eating other animals that have eaten such plants.

16.2 Amines

Amines are organic compounds which contain the amino functional group, —NH_2 (general formula, RNH_2). They are known for their very strong, pungent odors. For example, methylamine (CH_3NH_2) smells like spoiled fish. Amines are also produced during the natural decay of living organisms, and such amines are called **ptomaines.**

$H_2NCH_2CH_2CH_2CH_2NH_2$
Putrescine

$H_2NCH_2CH_2CH_2CH_2CH_2NH_2$
Cadaverine

Table 16.2 Some Amines

Name	Formula	Type
Propylamine	$CH_3CH_2CH_2NH_2$	Primary
Isopropylamine	CH_3 over CH_3—CH—NH_2	Primary
Methylethylamine	CH_3CH_2—N—H with CH_3 below N	Secondary
Dimethylethylamine	CH_3CH_2—N—CH_3 with CH_3 below N	Tertiary
Aniline	⬡—NH_2	Primary
N-Methylaniline	⬡—N—CH_3 with H below N	Secondary
N,N-Dimethylaniline	⬡—N—CH_3 with CH_3 below N	Tertiary

The common names of these two ptomaines might give you a hint about their smells. It was once thought that such amines were responsible for the vomiting and diarrhea that results from eating spoiled food, so these symptoms were referred to as ptomaine poisoning. However, it is now known that such body reactions result from more complicated causes.

Amines can be classified as primary, secondary, or tertiary amines depending upon the number of carbon atoms bonded directly to the nitrogen atom. In a primary amine, the nitrogen atom will be bonded to one carbon atom and to two hydrogen atoms.

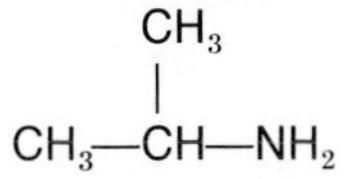

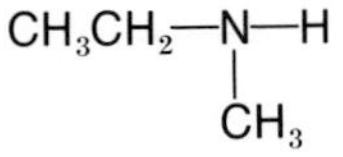

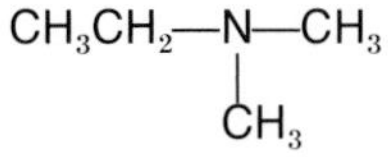

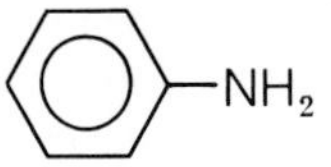

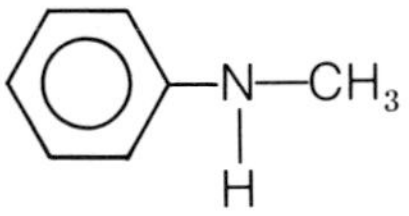

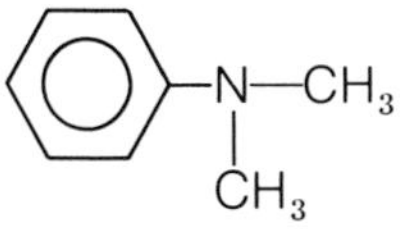

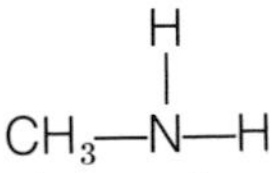

$$CH_3—N—H \text{ with } H \text{ above N}$$
Methylamine, a primary amine

In a secondary amine, the nitrogen will be bonded to two carbon atoms and one hydrogen atom.

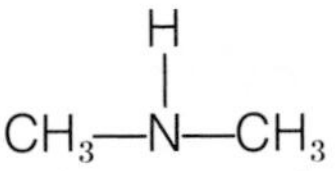

$$CH_3—N—CH_3 \text{ with } H \text{ above N}$$
Dimethylamine, a secondary amine

In a tertiary amine, the nitrogen will be bonded to three carbon atoms (Figure 16.2 and Table 16.2).

$$CH_3$$
$$|$$
$$CH_3-N-CH_3$$

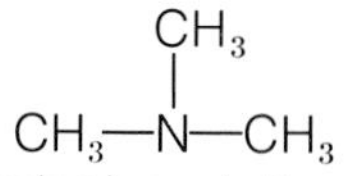

Trimethylamine, a tertiary amine

Amines have been used successfully to treat patients who suffer from severe depression. Interestingly, it has been found that tertiary amine antidepressants help to calm depressed patients who are highly agitated, whereas secondary amine antidepressants stimulate those who suffer from depression and total immobility. The reason for this difference is not entirely clear, but these two classes of amines seem to interact differently with nerve cells in the brain.

In certain instances, nitrogen can form a strong fourth bond. After forming three covalent bonds, an atom of nitrogen has a pair of electrons which it can share with another atom, forming a **coordinate covalent,** or **donor-acceptor, bond.** If a nitrogen atom in an amine forms a fourth bond, or coordinate bond, an organic ammonium ion is formed.

$$CH_3 \qquad\qquad CH_3$$
$$| \qquad\qquad\qquad |+$$
$$CH_3-N-CH_3 \qquad CH_3-N-CH_3$$
$$.. \qquad\qquad\qquad |$$
$$\qquad\qquad\qquad\qquad CH_3$$

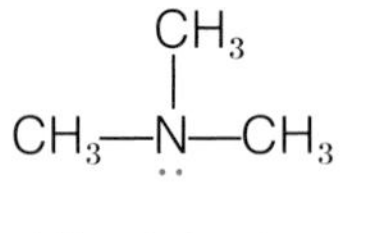

Trimethylamine Tetramethylammonium ion

Choline is a compound which contains an organic ammonium ion, and derivatives of the choline molecule are extremely important to the structure of the cell and to the transmission of nerve impulses.

$$\left[HOCH_2CH_2-\overset{\overset{CH_3}{|}}{\underset{\underset{CH_3}{|}}{N}}-CH_3 \right]^+ OH^- \qquad \left[CH_3\overset{O}{\overset{||}{C}}OCH_2CH_2-\overset{\overset{CH_3}{|}}{\underset{\underset{CH_3}{|}}{N}}-CH_3 \right]^+ OH^-$$

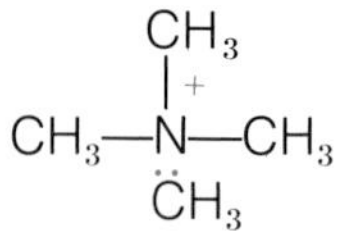

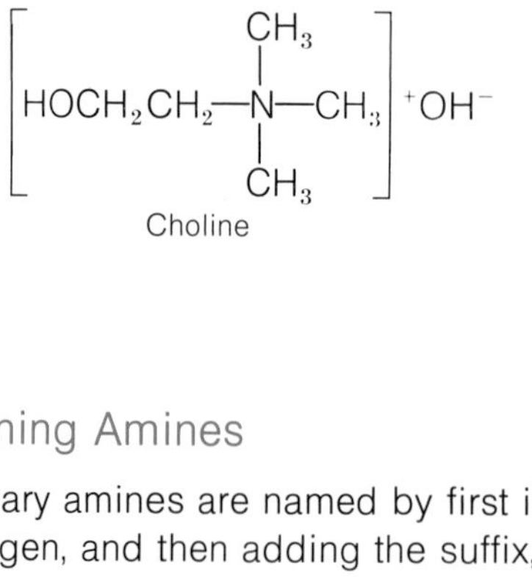

Choline Acetylcholine
(neurochemical transmitter)

16.3 Naming Amines

Primary amines are named by first identifying the group attached to the nitrogen, and then adding the suffix, **-amine.** For example, $CH_3CH_2NH_2$ is ethylamine, and $CH_3CH_2CH_2CH_2NH_2$ is butylamine. Secondary and tertiary amines are named in the same way: Identify all groups attached to the nitrogen if they are different, or use the prefixes di- and tri-. For example,

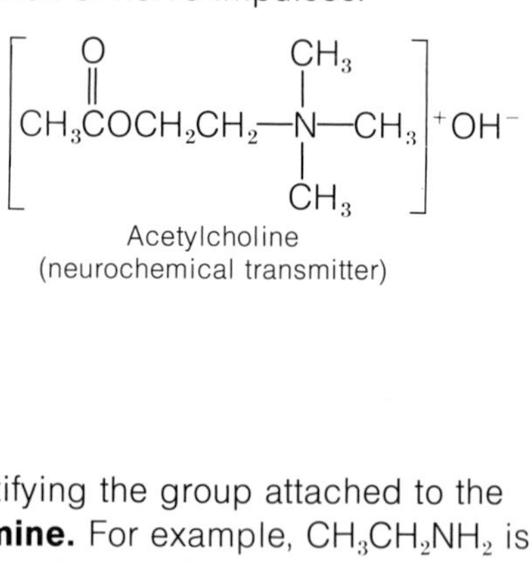

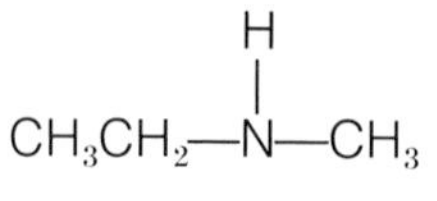

$$H$$
$$|$$
$$CH_3CH_2-N-CH_3 \text{ is methylethylamine, and}$$

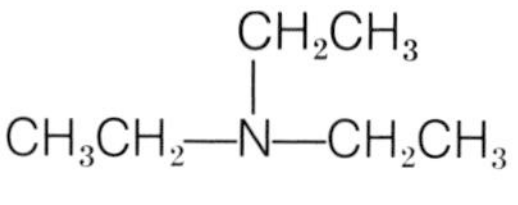

$$CH_2CH_3$$
$$|$$
$$CH_3CH_2-N-CH_2CH_3 \text{ is triethylamine}$$

The more complicated amines are often named by using the word *amino-* to identify the —NH_2 group in the molecule. For example,

$$\overset{\displaystyle O}{\overset{\displaystyle \|}{H_2NCH_2CH_2COH}} \text{ is 3-aminopropanoic acid, and}$$

$$H_2NCH_2CH_2CH_2CH_2NH_2 \text{ is 1,4-diaminobutane.}$$

And finally, an *N* appearing before the name of a substituted group indicates that the group named after the *N* is attached to the nitrogen of the amino group (and not somewhere else in the molecule) (see Table 16.2).

Example 16-1 _____

1. Name the following compound:

$$\begin{array}{c} CH_3 \\ | \\ CH_3CH{-}N{-}H \\ | \\ CH_2CH_3 \end{array}$$

 This is a secondary amine with an ethyl group and an isopropyl group attached to the nitrogen. Therefore, the name is ethylisopropylamine.

2. Draw the structural formula for 3-(*N,N*-dimethylamino)-butanoic acid.

 First draw the carbons for butanoic acid: 4 carbon atoms.

$$\overset{\displaystyle O}{\overset{\displaystyle \|}{C{-}C{-}C{-}C{-}OH}}$$

 The "3" indicates that on the third carbon we will find the group *N,N*-dimethylamino-, which is an amino group with two methyl groups attached to the nitrogen.

$$\begin{array}{c} \overset{\displaystyle O}{\overset{\displaystyle \|}{C{-}C{-}C{-}C{-}OH}} \\ | \\ N{-}CH_3 \\ | \\ CH_3 \end{array}$$

 Now all that remains is to add hydrogens so that each carbon has 4 bonds.

$$\begin{array}{c} \begin{array}{cccc} H & H & H & O \\ | & | & | & \| \end{array} \\ H{-}C{-}C{-}C{-}C{-}OH \quad \text{or} \quad CH_3CHCH_2COH \\ \begin{array}{cccc} | & | & & \\ H & H & & \end{array} \qquad\qquad\quad | \\ \;\; N{-}CH_3 \qquad\qquad\qquad N(CH_3)_2 \\ \;\; | \\ \;\; CH_3 \end{array}$$

16.4 Basic Properties of Amines

The ability to form a fourth, or coordinate, bond gives the amines the properties of a base. As you will remember, we defined an acid as a substance that donates or gives off hydrogen ions (H^+), and a base as a substance that accepts or picks up hydrogen ions. The stronger the base, the greater its attraction for hydrogen ions. All amines are derivatives of the weak inorganic base ammonia, NH_3. When ammonia is placed in water, some of the ammonia molecules will form the ammonium ion.

$$:NH_3 + H_2O \longrightarrow NH_4^+ + OH^-$$

Ammonia Ammonium ion

Similarly, the organic amines are weak bases. For example, when methylamine is placed in water, some of its molecules will attract hydrogens and will form the methylammonium ion.

$$CH_3-\overset{\overset{\displaystyle H}{|}}{\underset{\underset{\displaystyle H}{|}}{N}}: \quad +.H_2O \longrightarrow CH_3-\overset{\overset{\displaystyle H}{|}}{\underset{\underset{\displaystyle H}{|}}{\overset{+}{N}}}-H \quad + OH^-$$

Methylamine Methylammonium ion

The strength of an amine as a base depends upon the organic groups attached to the nitrogen atom. Carbon chains attached to the nitrogen atom increase the basic properties of the molecule, whereas benzene rings attached to the nitrogen atom greatly decrease the basic properties of the molecule.

$CH_3CH_2-\overset{..}{\underset{\underset{CH_2CH_3}{|}}{N}}-H$	$CH_3-\overset{..}{\underset{\underset{H}{|}}{N}}-H$	$H-\overset{..}{\underset{\underset{H}{|}}{N}}-H$	$\underset{}{\bigcirc}-\overset{..}{\underset{\underset{H}{|}}{N}}-H$
Diethylamine >	Methylamine >	Ammonia >	Aniline
100× stronger base than ammonia	10× stronger base than ammonia	weak inorganic base	100,000× weaker base than ammonia

16.5 Reactions of Amines

Amines form salts when they react with acids.

$$CH_3NH_2 + HCl \longrightarrow CH_3NH_3^+Cl^-$$

Methylamine Hydrochloric acid Methylammonium chloride

$$CH_3-\overset{\overset{\displaystyle CH_3}{|}}{\underset{\underset{\displaystyle CH_3}{|}}{N}}: \quad + \quad HBr \longrightarrow CH_3-\overset{\overset{\displaystyle CH_3}{|}}{\underset{\underset{\displaystyle CH_3}{|}}{N}}-H^+Br^-$$

Trimethylamine Hydrobromic acid Trimethylammonium bromide

When heated with alkyl halides, amines form compounds known as quarternary ammonium salts.

$$(CH_3)_3N + CH_3I \longrightarrow (CH_3)_4N^+I^-$$

Trimethylamine Methyl iodide Tetramethylammonium
iodide

Secondary amines can react with nitrites (general formula XNO_2) to produce nitrosamines (*N*-nitrosodialkylamines), a class of highly potent carcinogens. These compounds have been shown to cause cancer in all types of laboratory animals, and are suspected of causing human cancers. In general,

$$\begin{array}{c} R' \\ \diagdown \\ NH + HNO_2 \longrightarrow \\ \diagup \\ R'' \end{array} \qquad \begin{array}{c} R' \\ \diagdown \\ N-NO + H_2O \\ \diagup \\ R'' \end{array}$$

Secondary amine Nitrosamine

For example,

Pyrrolidine + HNO_2 $\xrightarrow[\text{contents}]{\text{Stomach}}$ *N*-nitrosopyrrolidine (NO) + H_2O

Nitrites are widely used to enhance the color of, and to preserve, meats, poultry, and fish. Also, nitrates, which are naturally present in high levels in such foods as celery and beets, can be converted to nitrites by the bacteria in saliva. No matter what their source, nitrites may react in the stomach with amines naturally present in foods, drugs, and cigarette smoke to produce nitrosamines. The study of nitrosamines as a possible environmental cause of human cancer is currently a very active area of research.

16.6 Amides

Amides may be thought of as derivatives of carboxylic acids, in which the hydroxyl (—OH) group on the carboxyl group has been replaced by an amino (—NH_2) group. The general formula for the amides, therefore, can be either $\overset{O}{\overset{\|}{R C}}NH_2$, or $RCONHR'$, or $\overset{R''}{\overset{|}{RCONR'}}$ (Table 16.3). The bond between the carbon and the nitrogen atoms in an amide molecule is known as the **amide linkage.** It is a very stable bond and is found as a repeating linkage in large molecules such as proteins, nylon, and other industrial polymers.

$$\overset{O}{\overset{\|}{CH_3C}}-OH \qquad \overset{O}{\overset{\|}{CH_3C}}-NH_2 \text{ —Amide linkage}$$

Acetic acid Acetamide

Amides with short carbon chains are soluble in water due to the presence of the polar amide group. This solubility decreases as the molecular weight of the amide increases. Amides, like alcohols, have melting and boiling points that are higher than alkanes with comparable

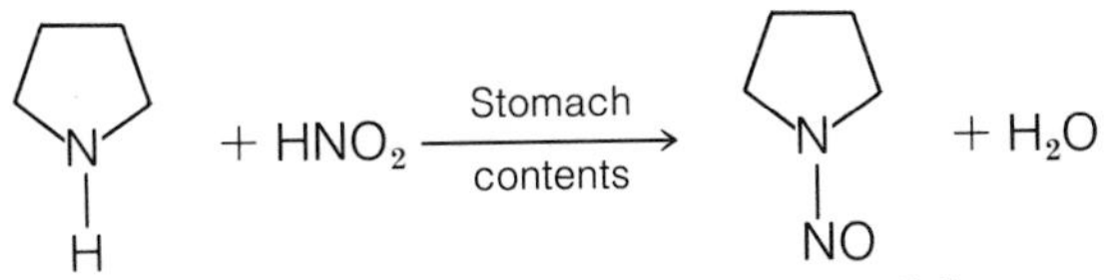

Table 16.3 Some Amides

Name	Structural Formula	Name	Structural Formula
Formamide	$$H-\overset{\overset{\displaystyle O}{\|\|}}{C}-NH_2$$	Acetanilide	$$CH_3\overset{\overset{\displaystyle O}{\|\|}}{C}-\overset{\overset{\displaystyle H}{\|}}{N}-\bigcirc$$
Acetamide	$$CH_3\overset{\overset{\displaystyle O}{\|\|}}{C}-NH_2$$	Benzamide	$$\bigcirc-\overset{\overset{\displaystyle O}{\|\|}}{C}-NH_2$$
Propanamide	$$CH_3CH_2\overset{\overset{\displaystyle O}{\|\|}}{C}-NH_2$$		
Butanamide	$$CH_3CH_2CH_2\overset{\overset{\displaystyle O}{\|\|}}{C}-NH_2$$	Urea	$$H_2N-\overset{\overset{\displaystyle O}{\|\|}}{C}-NH_2$$

molecular weights. These high melting and boiling points result from hydrogen bonding between the polar amide groups on neighboring amide molecules (Figure 16.3).

Amides that have two hydrogens attached to the nitrogen ($R\overset{\overset{\displaystyle O}{\|\|}}{C}NH_2$) are named by dropping the *-ic* or *-oic* ending from the name of the original acid, and adding **-amide.** For example,

$$CH_3CH_2\overset{\overset{\displaystyle O}{\|\|}}{C}-NH_2 \text{ is propanamide, and}$$

$$\bigcirc-\overset{\overset{\displaystyle O}{\|\|}}{C}-NH_2 \text{ is benzamide.}$$

If an amide contains groups attached to the nitrogen, they are named as alkyl groups, but with an *N* in front of the name to indicate they are attached to the nitrogen. For example,

$$CH_3CH_2\overset{\overset{\displaystyle O}{\|\|}}{C}N\underset{\underset{\displaystyle H}{\|}}{C}H_3 \text{ is } N\text{-methylpropanamide, and}$$

$$CH_3CH_2CH_2\overset{\overset{\displaystyle O}{\|\|}}{C}N\underset{\underset{\displaystyle CH_3}{\|}}{C}H_3 \text{ is } N,N\text{-dimethylbutanamide.}$$

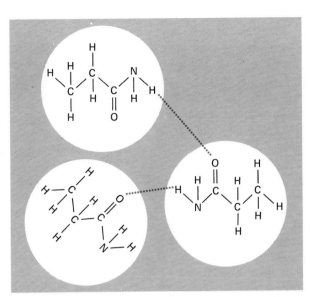

Figure 16.3 Hydrogen bonds can form between the molecules of amides such as propanamide.

Example 16-2 _____

1. Name the following amide:

$$CH_3CH_2CH_2\overset{\displaystyle O}{\overset{\displaystyle \|}{C}}-\underset{\displaystyle \underset{H}{|}}{N}-CH_2CH_3$$

(a) The acid from which this amide is formed has 4 carbons, so it was butanoic acid. Therefore, the last part of the name is butanamide.

(b) The nitrogen has one substituted group: an ethyl group. The name is *N*-ethylbutanamide.

2. Write the structural formula for *N,N*-dimethylacetamide.

(a) The acid from which the amide is formed is acetic acid.

$$CH_3\overset{\displaystyle O}{\overset{\displaystyle \|}{C}}-\underset{\displaystyle |}{N}-$$

(b) The *N,N*-dimethyl indicates that there are two methyl groups attached to the nitrogen.

$$CH_3\overset{\displaystyle O}{\overset{\displaystyle \|}{C}}-\underset{\displaystyle \underset{CH_3}{|}}{N}-CH_3$$

16.7 Some Important Amides

Urea is a waste product of the body's metabolism (see Table 16.3). We have previously seen that compounds containing carbon, hydrogen, and oxygen are broken down by the body to produce waste products of carbon dioxide (CO_2) and water (H_2O). The carbon dioxide is then eliminated through the lungs, and water is eliminated through the breath, sweat, and urine. When nitrogen-containing compounds are broken down by the body, the nitrogen is converted into the compound ammonia, NH_3. Ammonia, however, is extremely toxic to living tissues. Aquatic animals are able to get rid of ammonia quickly into the surrounding water, but land animals must convert this ammonia to a less toxic substance which can be transported through the body. Such animals convert ammonia into urea, a compound that can accumulate in the blood without harm, and which can be eliminated by the kidneys.

An amide with medical importance is xylocaine, a local anesthetic that has largely replaced novocaine for use by dentists.

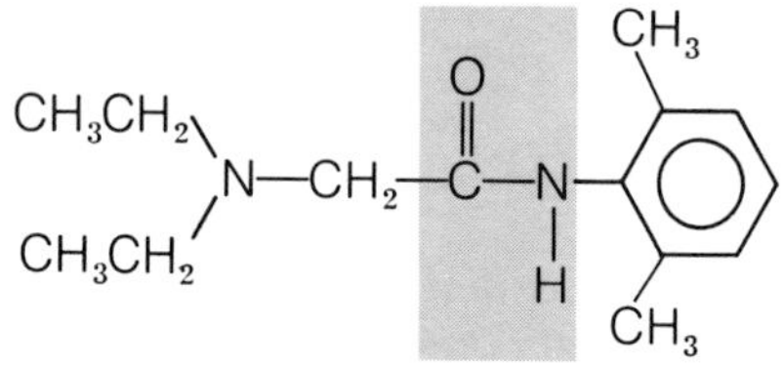

Xylocaine

Another amide of interest is the hallucinogen LSD, lysergic acid diethylamide. LSD seems to disrupt the transmission of nerve impulses in the brain. The structure of LSD has certain features resembling those of serotonin, a chemical produced by the body to control the transmission of nerve impulses. Nerve cells in the brain probably confuse LSD with

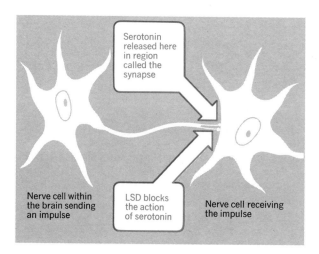

Figure 16.4 The hallucinogen LSD affects brain cells by blocking the action of serotonin.

serotonin, thus preventing the serotonin from carrying out its control function (Figure 16.4). This results in uncontrolled nerve impulses in the brain, creating hallucinations and other behavioral abnormalities.

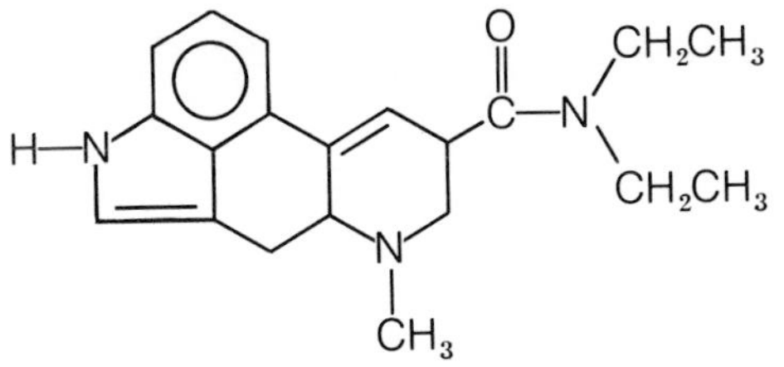

LSD

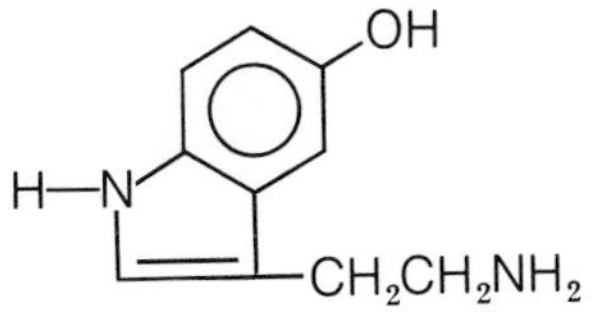

Serotonin

Tetracyclines are complicated molecules containing an amide group together with other functional groups (you might see how many you can identify). The tetracyclines are formed by certain molds, and are widely used for their antibiotic properties. You may be familiar with the names of these compounds as prescription drugs.

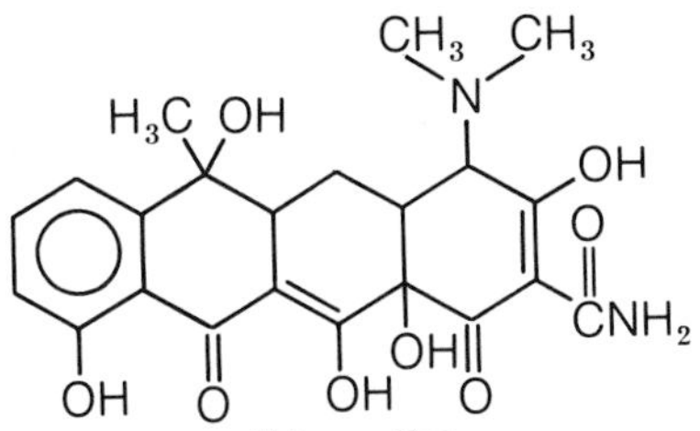

Tetracycline

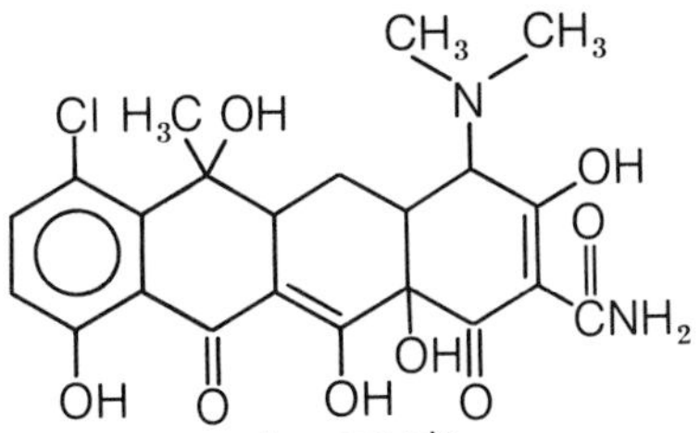

Aureomycin

16.8 Reactions of Amides

Hydrolysis

Compounds that contain amide groups are among the most important in our bodies. For example, proteins are complex compounds containing many amide linkages. Amides hydrolyze (that is, are split apart by water) very slowly. The slowness of this reaction explains the stability of the proteins in our bodies. Amides in which the nitrogen is bonded to two hydrogens will hydrolyze to form acids and ammonia. Other amides will form the corresponding acid and an amine. Under certain conditions, and in the presence of specific enzymes such as those found in the digestive tract, the hydrolysis of amides will occur in a relatively short time.

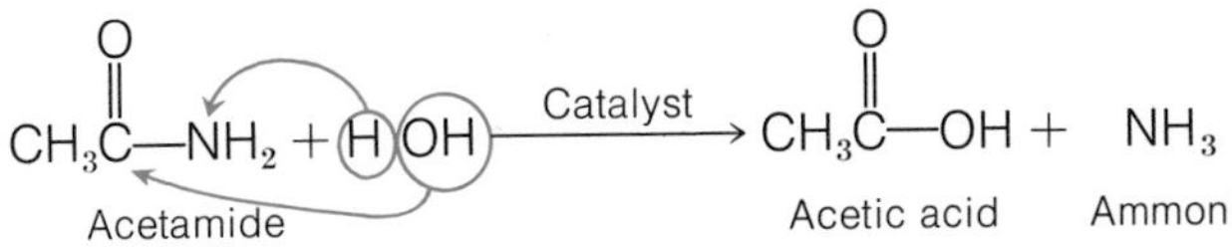

Acetamide — Acetic acid — Ammonia

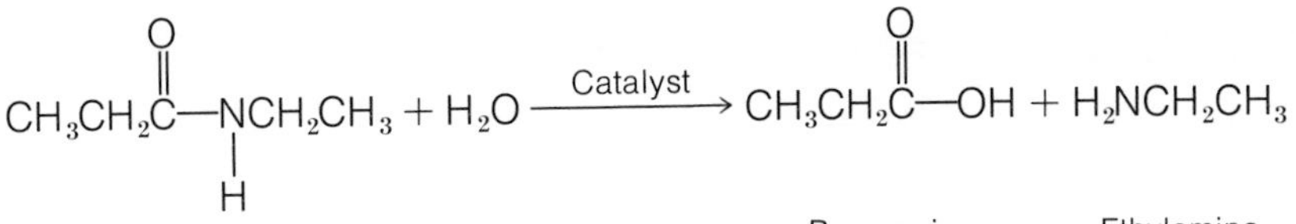

N-Ethylpropanamide — Propanoic acid — Ethylamine

Dehydration

When amides are heated in the presence of a strong dehydrating agent, a molecule of water is removed and a member of the class of compounds called nitriles is formed.

$$\underset{\text{Acetamide}}{CH_3\overset{\overset{\displaystyle O}{\|}}{C}—NH_2} \xrightarrow{P_4O_{10}} \underset{\text{Acetonitrile}}{CH_3C{\equiv}N} + H_2O$$

16.9 Nitriles

Carbon is triply bonded to nitrogen in the extremely toxic compound hydrogen cyanide, H—C≡N:. When inhaled or absorbed through the skin, this compound stops the process by which cells get their energy, thereby killing all cells. However, if the carbon atom triply bonded to the nitrogen is bonded to another carbon atom (instead of to a hydrogen atom), a group of much less toxic compounds called **nitriles** is formed. Industrially, nitriles are polymerized to form many synthetic fabrics such as orlon, acrylan, and dynel, which are used in knitwear and carpets (see Table 14.5).

One very controversial substance containing a nitrile functional group is the compound laetrile. It is extracted from apricot pits, and some doctors claim it is a valuable drug in cancer therapy.

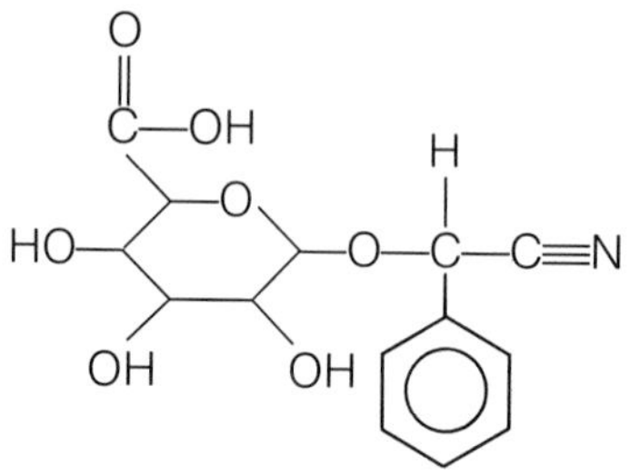

Laetrile

Laetrile is now being carefully studied to see if it is indeed useful for cancer therapy. Laetrile presents a possible health hazard when taken in large doses, however, because enzymes in the digestive tract can hydrolyze the molecule, thereby releasing hydrogen cyanide.

16.10 Imines

Carbon-to-nitrogen double bonds are characteristic of the **imine** functional group, >C=N—H (general formula $R—\overset{\overset{\displaystyle R'}{|}}{C}{=}N—H$). This functional group is found in a molecule of the strong base guanidine. Guanidine forms part of the compound arginine, which is found in very high concentrations in sperm cells and in the proteins of hair and skin. Guanidine groups are also found in molecules of creatine, which (when bonded to a phosphate group) are the phosphocreatine molecules that store energy for our muscles.

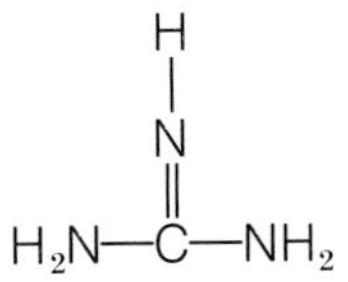

Guanidine

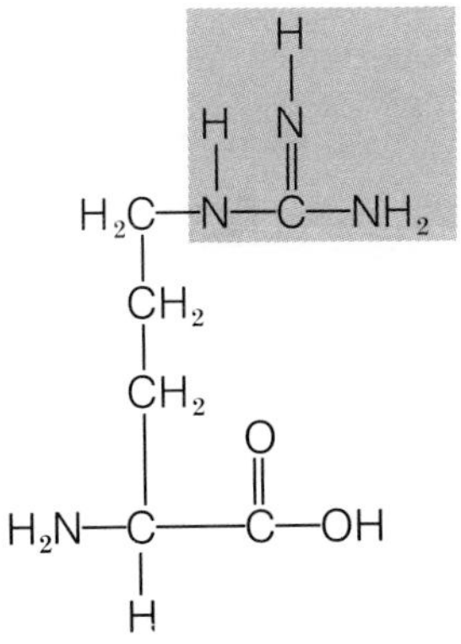

Arginine

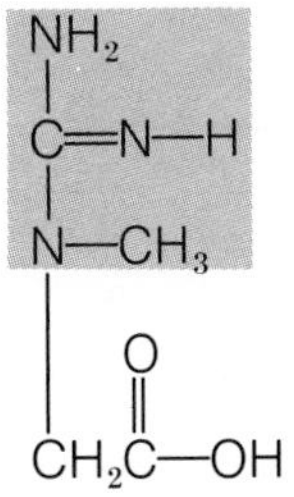

Creatine

16.11 Nitrogen in Rings

Our discussion of compounds containing rings has thus far been limited to molecules whose rings are formed by carbon atoms only. However, the elements nitrogen, oxygen, and sulfur can join with carbon atoms to form a ring. The resulting compounds are called **heterocyclic** compounds, and have two or more types of atoms making up the ring. Heterocyclic rings are found in an enormous number of naturally occurring compounds, and form an entire field of study by themselves. Table 16.4 shows some heterocyclic rings that contain nitrogen. These rings are commonly found in biological systems, and we will study compounds containing such rings in later chapters. Before leaving our study of organic compounds containing nitrogen, however, we will take a brief look at some complex compounds that have important effects on the body.

16.12 Alkaloids

Alkaloids are a large class of nitrogen-containing compounds with complex structures, often including nitrogen in heterocyclic rings. Most of the complicated alkaloids are found in a specific species of plant, and are often part of the plant's defenses. Such alkaloids, in various plant mixtures, have been used as drugs for centuries. Some of the naturally occurring alkaloids have also been synthesized in the laboratory. Moreover, in trying to duplicate the structure of alkaloids, chemists have created compounds with properties far superior to the natural compound when used as a drug. By studying and replacing certain functional groups on these alkaloids, chemists have produced compounds with desirable properties (such as relieving pain or inducing sedation) without the accompanying problems that often result from the use of the natural alkaloid. Such undesirable side effects include dependence (addiction) and tolerance (the need to

Table 16.4 Some Heterocyclic Rings Containing Nitrogen

Name	Structural Formula	Found in the Structure of —
Pyrrolidine		Amino acids
Pyrrole		Chlorophyll, hemoglobin and vitamin B_{12}
Imidazole		An amino acid
Indole		An amino acid
Pyridine		The vitamin niacin
Pyrimidine		Nucleic acids
Purine		Nucleic acids

constantly increase the dosage over time in order to produce the same effect). A complete discussion of the alkaloids and their many uses would be too complicated for this book, but we will briefly mention some interesting examples.

Epinephrine and Norepinephrine

Epinephrine belongs to a class of relatively simple alkaloids containing one benzene ring. We have already discussed the many effects that epinephrine can have on the body. Looking again at the structure of epinephrine we can recognize that it is a secondary amine.

$$HO \overset{OH}{\underset{}{\bigcirc}} CH - CH_2 - N - CH_3$$

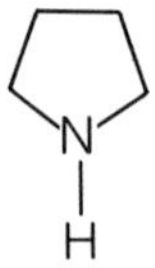

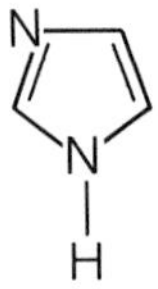

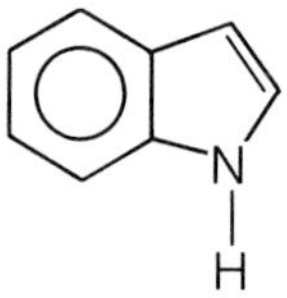

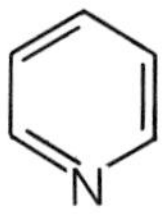

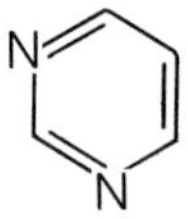

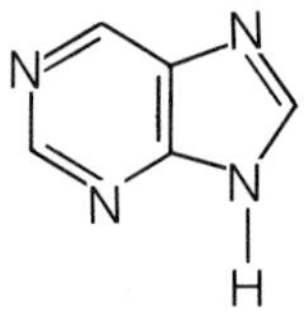

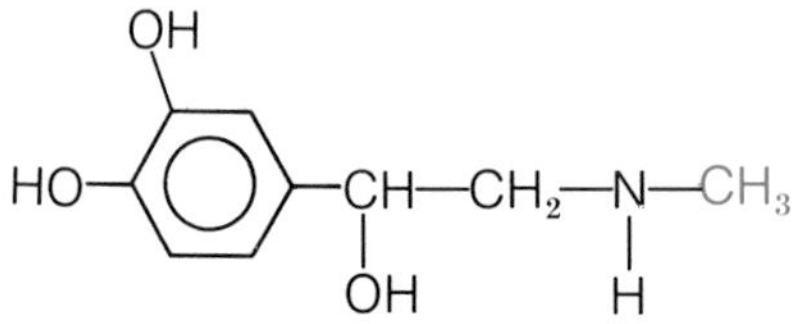

Epinephrine

Another compound with a structure closely related to epinephrine is norepinephrine, also produced by the adrenals. This compound, often called noradrenalin, is a primary amine whose structure differs only slightly from that of epinephrine—it lacks a methyl group on the nitrogen.

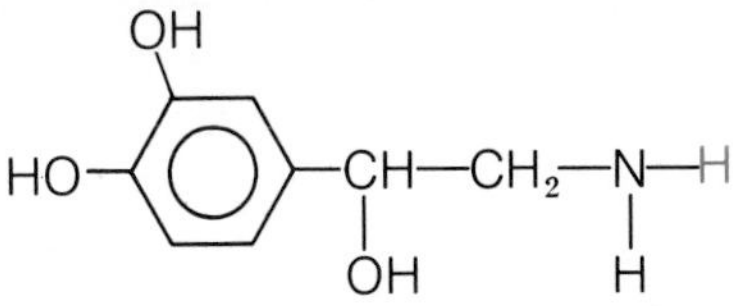

Norepinephrine

Just this small change makes a difference in the action of the two compounds in the body. Norepinephrine, which is also produced by certain nerve endings, acts as a chemical messenger between nerve cells. Its main function in the body is to maintain muscle tone in the blood vessels, and in that way to control blood pressure. For example, reserpine (a drug given to help reduce blood pressure in persons suffering from hypertension) works by greatly reducing the amount of norepinephrine in the nerve endings.

Amphetamines

Benzedrine (also called amphetamine) is a synthetic compound with a structure similar to that of epinephrine. Benzedrine stimulates the cortex of the brain, producing a decreased sense of fatigue and an increased alertness. Together with ephedrine, a natural substance extracted from a tree in the pine family, benzedrine has been prescribed to reduce fatigue, overcome drowziness, and suppress the appetite. Benzedrine has also been used to treat bronchial congestion. With continued use of benzedrine for any of these purposes, however, dependence and tolerance can result. Over the last decade the abuse of this drug has increased to such an extent that it is now under strict regulation and control. Benzedrex, a derivative of benzedrine containing a cyclohexane ring, has the same decongestant effects and is now used in inhalers that relieve nasal congestion. This compound does not cause dependence.

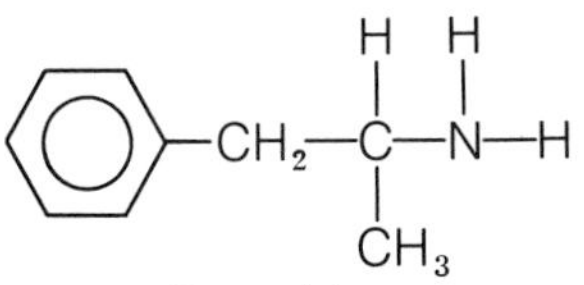

Benzedrine

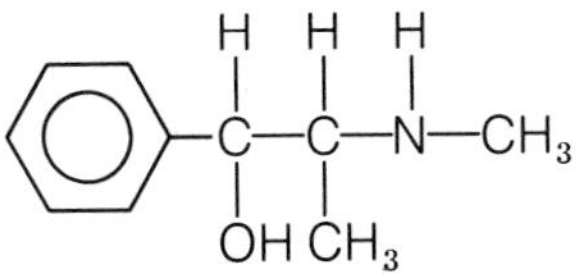

Ephedrine

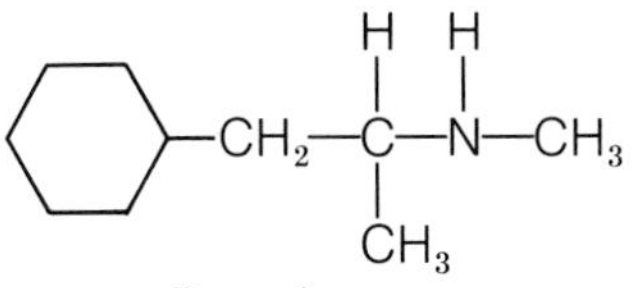

Benzedrex

Nicotine

Some alkaloids contain a nitrogen atom as part of the ring. Nicotine, which is found in the leaves of tobacco, has two nitrogen-containing rings.

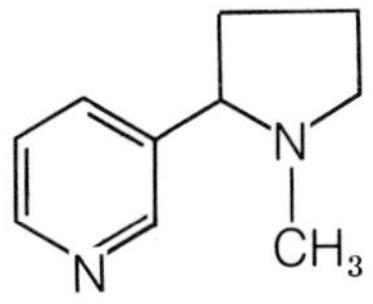

Nicotine

Nicotine in pure form is a rapid acting, extremely toxic drug. There have been reports of gardeners who have died from handling nicotine as an insecticide. In such cases, death from respiratory failure occurred within a few minutes. The action of nicotine in the human body is very complicated. It stimulates the central nervous system (causing irregular heartbeat and blood pressure), induces vomiting and diarrhea, and first stimulates and then inhibits glandular secretions. Nonsmokers can absorb only about 4 mg of nicotine before symptoms of nausea, vomiting, diarrhea, and weakness begin. But a smoker builds up a tolerance to nicotine, and may absorb twice as much without any noticeable effects. The smoke from one cigarette may contain up to 6 mg of nicotine, but only about 0.2 mg is absorbed into the body.

Caffeine

Caffeine, another alkaloid whose structure has two nitrogen-containing rings, is found in coffee beans, tea leaves, and in the seeds of the chocolate tree. In addition to its natural presence in coffee, tea, and chocolate, caffeine is also added to some carbonated beverages and nonprescription medicines (such as some aspirins).

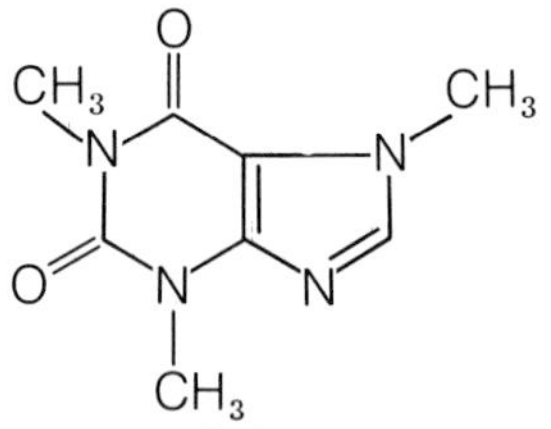

Caffeine

Caffeine is a stimulant to the central nervous system, causing restlessness and mental alertness. Taking caffeine for some people can become a habit—they may become extremely insistent about having their morning coffee—but there is no evidence that caffeine is addictive.

Atropine and Quinine

Some other alkaloids worth mentioning have even more complex nitrogen-containing structures (Table 16.5). Atropine causes dilation of the pupils, and is used in eye surgery. Taken internally it relieves abdominal pain from severe muscle contractions, and it is used as premedication for gas anesthesia because it dries up secretions in the nose and throat. The bark of cinchona trees (found in South America and Java) contains the alkaloid quinine, which is used as an antimalarial drug. An isomer of quinine, called quinidine, is found in a much lower concentration in the bark of these trees, and acts as a heart depressant. It is used in the treatment of irregular heart rhythms.

Cocaine

Cocaine is a local anesthetic, and is a stimulant to the central nervous system. It has effects similar to, but stronger than, the amphetamines. Because it does not produce tolerance, cocaine has been increasingly popular among drug users. In trying to synthesize compounds that contain

Table 16.5 Alkaloids with Nitrogen-Containing Rings

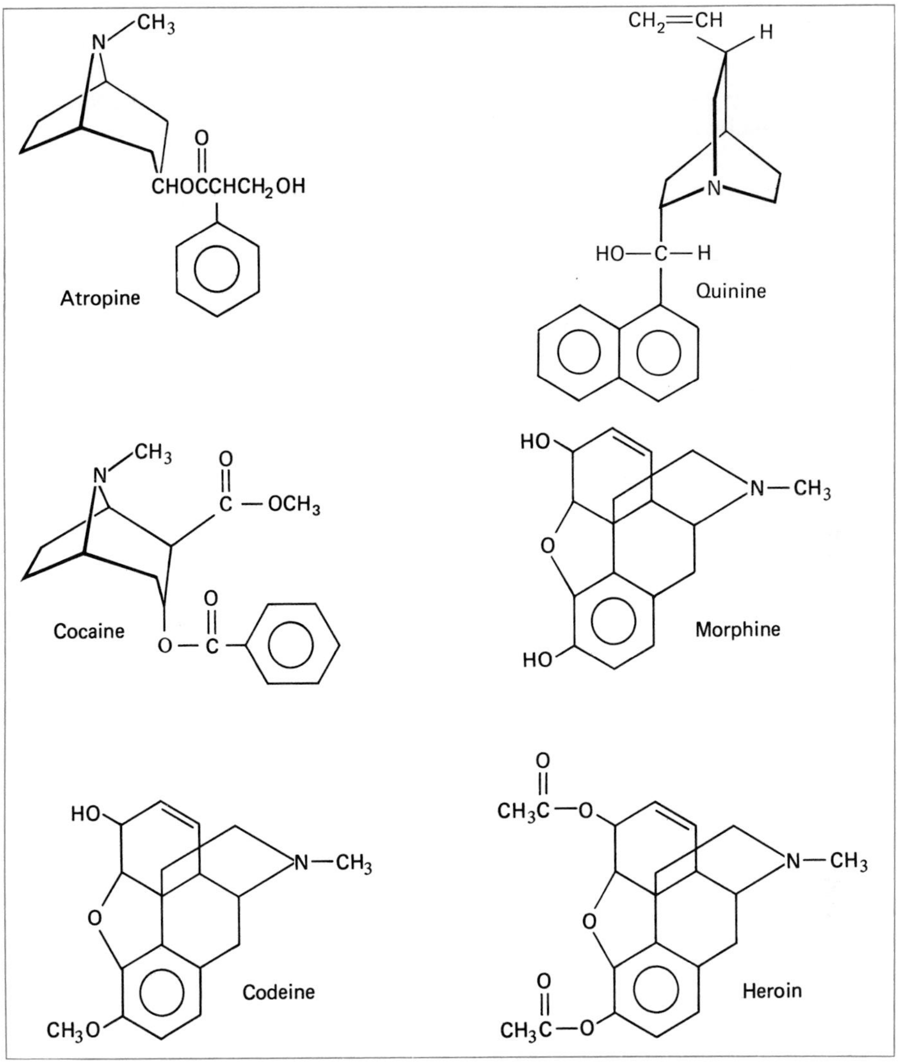

the local anesthetic properties of cocaine (but not its side effects), chemists have developed many useful new anesthetics. One such compound is procaine, which has been widely used in dentistry in the form of its derivative Novocaine.

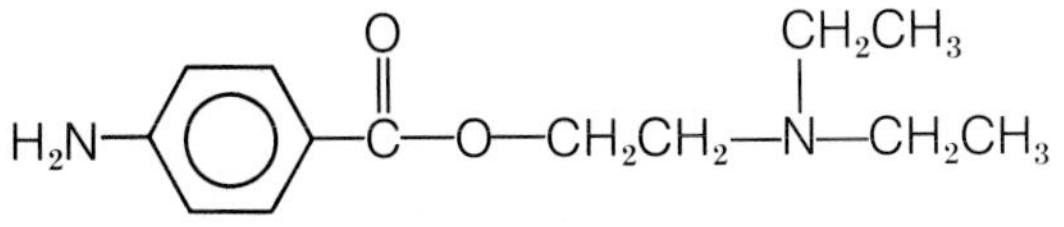

Procaine

Opiates

Morphine and codeine are alkaloids that can be isolated from the opium poppy. Heroin is a synthetic alkaloid which is derived from morphine.

These complicated molecules have a narcotic effect on humans, and have benefited society as well as created great problems through their use and abuse. Morphine reduces pain, causes drowsiness and changes in mood, and produces mental fogginess. It has been used to provide pain relief in radical surgery and in cases of wartime injury. However, repeated use of morphine results in addiction and tolerance. Morphine also causes constriction of the smooth muscles such as those found in the intestines and, therefore, can be used in the treatment of diarrhea. Paregoric, commonly given to children suffering from diarrhea, is a mixture with an opium base. Codeine, although less effective than morphine, is also less likely to cause drug addiction. It is used mainly to relieve coughing.

By studying the way in which morphine acts on nerve cells, researchers recently have discovered a group of substances which exist naturally in the brain and which produce the same effects as the opiates. These compounds, called endorphins, are neurotransmitters (chemical substances that pass information between nerve cells). They seem to be important in the body's regulation of pain and emotions.

Barbiturates

Barbiturates are derivatives of barbituric acid (Table 16.6). These compounds act by depressing the central nervous system, and are thought to inhibit nerve response centers. The different properties of the various barbiturate compounds depend upon the substituted groups on their molecules. Barbiturates can be used as hypnotics, sedatives, anticonvulsants, and anesthetics. Their most common use (and abuse), however, is as a sleep-producing drug; excessive use of these compounds

Table 16.6 Some Barbiturates

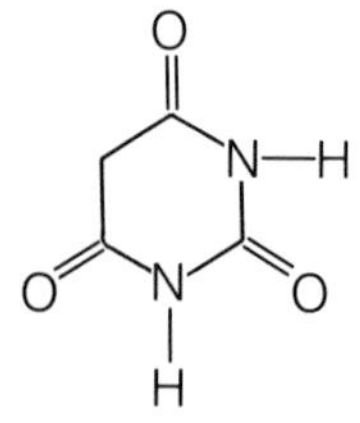

Barbituric acid

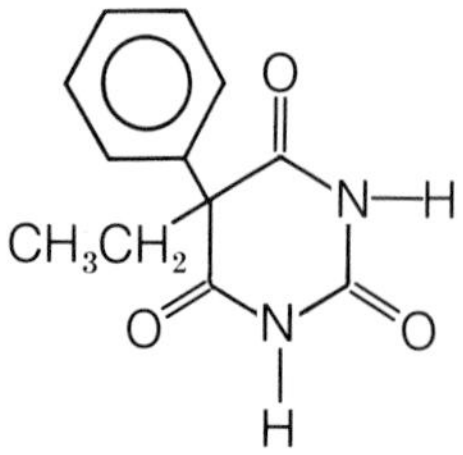

Phenobarbital
(a tranquilizer)

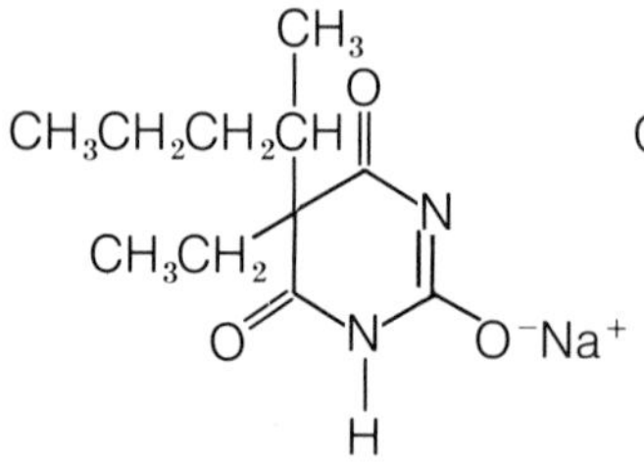

Pentobarbital sodium
(Nembutal-sleeping pills)

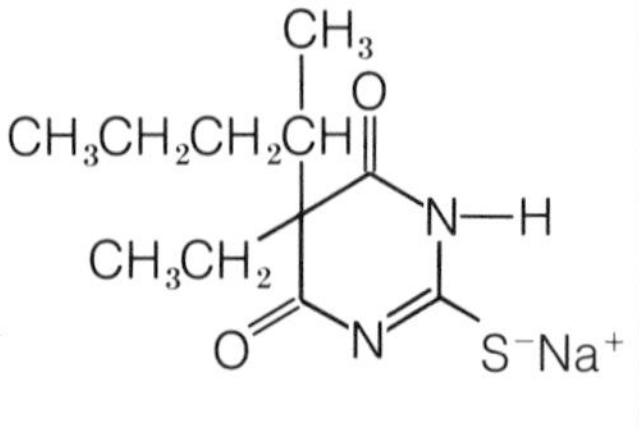

Sodium pentothal
(an anesthetic)

can produce physical dependence. One class of barbituric acid derivatives, the thiobarbiturates, contains a sulfur atom in the molecule. The most important member of this class is sodium pentothal, a fast-acting intravenous anesthetic from which recovery is fairly rapid and often without side effects.

In this chapter we have discussed only a few of the nitrogen-containing compounds found in living systems. Other large classes of nitrogen-containing compounds, such as proteins and nucleic acids, will be the subjects of later chapters.

Chapter Summary

Nitrogen is the fourth most common element in the human body, and is found in many types of compounds. Animals cannot use the nitrogen molecules found in the air, and so must obtain the nitrogen they need through their diets. Amines are weak organic bases that may be classified as primary, secondary, or tertiary according to the groups attached to the nitrogen. Amines will form salts when they react with acids. Amides are derivatives of carboxylic acids, in which the —OH on the carboxyl group has been replaced by —NH$_2$, —NHR, or —NRR'. The bond between the carbon and the nitrogen in these compounds is very stable and is known as an amide linkage. Amides, therefore, undergo hydrolysis very slowly. When dehydrated, amides form nitriles. The nitrile functional group contains a carbon-to-nitrogen triple bond, and the imine functional group contains a carbon-to-nitrogen double bond. Nitrogen atoms can form a ring structure with carbon atoms. Rings containing atoms other than carbon are called heterocyclic rings. Alkaloids are a large class of compounds that often contain nitrogen in ring structures. Most complicated alkaloids occur naturally as part of the defense system of plants, and have been used as drugs for centuries because of their varied physiological effects.

Exercises and Problems

1. Identify each of the following compounds as an amine, amide, imine or nitrile:

$$\text{(a)} \quad CH_3\overset{\overset{\displaystyle H}{|}}{N}-\overset{\overset{\displaystyle O}{||}}{C}CH_2CH_3$$

(b) $CH_3C\equiv N$

(c) ⬡—NH_2

$$\text{(d)} \quad H_2N-\overset{\overset{\displaystyle NH}{||}}{C}-NH_2$$

(e) $CH_3CH_2\overset{\overset{\displaystyle H}{|}}{N}CH_3$

$$\text{(f)} \quad CH_3\overset{\overset{\displaystyle O}{||}}{C}NH_2$$

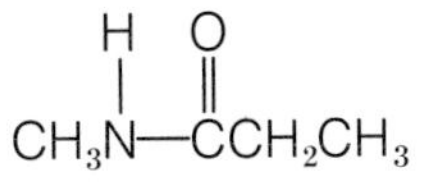

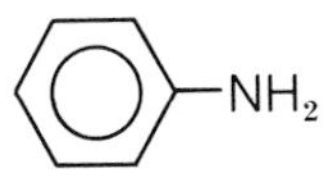

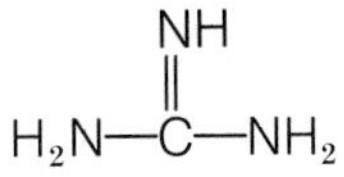

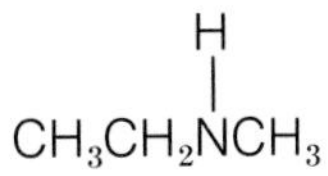

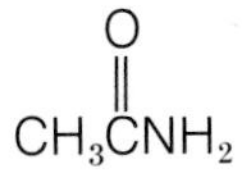

2. Identify each of the following compounds as a primary, secondary, or tertiary amine:

(a)
$$CH_3\text{-}\underset{\underset{CH_3}{|}}{\overset{\overset{CH_3}{|}}{C}}\text{-}NH_2$$

(d) $CH_3\overset{\overset{CH_3}{|}}{C}HNHCH_3$

(b) —N—CH$_2$CH$_3$ with CH$_3$ below N

(e) $CH_3CH_2\overset{\underset{CH_3}{|}}{N}CH_3$

(c) $CH_3CH_2CH_2NH_2$

(f) $CH_3\overset{\overset{CH_3}{|}}{C}HCH_2CH_2NH_2$

3. Name each of the amines in question 2.

4. What substances can cause the foul smell of decaying organic material?

5. Name each of the following amides:

(a) $CH_3CH_2CH_2CH_2\overset{\overset{O}{\|}}{C}NH_2$

(c) —N—$\overset{\overset{O}{\|}}{C}CH_2CH_3$ with H below N

(b) $CH_3\overset{\underset{CH_3}{|}}{N}\overset{\overset{O}{\|}}{C}CH_3$

(d) $CH_3CH_2\overset{\underset{CH_3}{|}}{N}\overset{\overset{O}{\|}}{C}CH_2CH_3$

6. Write the structural formula for each of the following:
 (a) Methylbenzylamine
 (b) Pentanamide
 (c) Triisopropylamine
 (d) *N,N*-Diethylhexanamide
 (e) 2-(*N*-Methylamino)-1-propanol
 (f) *N*-methyl-*N*-phenylpropanamide

7. Identify all the functional groups appearing in the molecule of aureomycin shown on page 349.

8. Draw the structure of two different heterocyclic rings that contain nitrogen.

9. Morphine is a very effective pain reliever. Why must its continued use in a patient be avoided?

10. What changes in functional groups must take place on the morphine molecule to produce heroin?

11. (a) Describe three specific effects of nicotine on the human body.
 (b) Suggest a reason why smokers can tolerate larger concentrations of nicotine before feeling ill effects than can nonsmokers.

12. Describe three different uses of alkaloids in drug therapy.

13. (a) Explain why an amine can act as a base in a chemical reaction.
    (b) Write the equation describing what occurs when propylamine is added to water.
    (c) Which of the following is the strongest base: aniline, dipropylamine, or propylamine?

14. Write the equation for the following reactions:

    (a) Isobutylamine + HCl
    (b) Triethylamine + HCl
    (c) Triethylamine + ethylbromide

15. Write the equation for the hydrolysis of each of the amides in question 5.

chapter 17

The Remaining Twenty Elements

Learning Objectives

By the time you have finished this chapter, you should be able to:

1. Describe three important functions of calcium in the body.

2. Write the reaction for the removal of calcium ions from the blood by the citrate ion.

3. Indicate two ways in which the normal metabolism of calcium can be disrupted.

4. Write the formula of the phosphate ion, and describe three important functions of this ion in our bodies.

5. Describe two functions of magnesium in the living organism.

6. Describe five important functions of the sodium, potassium, and chloride ions in the human body.

7. Write the formula for the sulfhydryl and thioether functional groups.

8. Define *trace element.*

9. Describe the functions of four trace elements.

10. Using specific examples, explain how trace elements can function in both a complementary and antagonistic fashion.

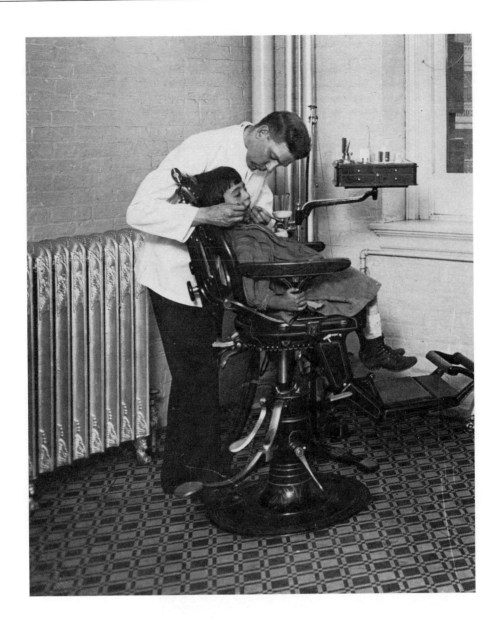

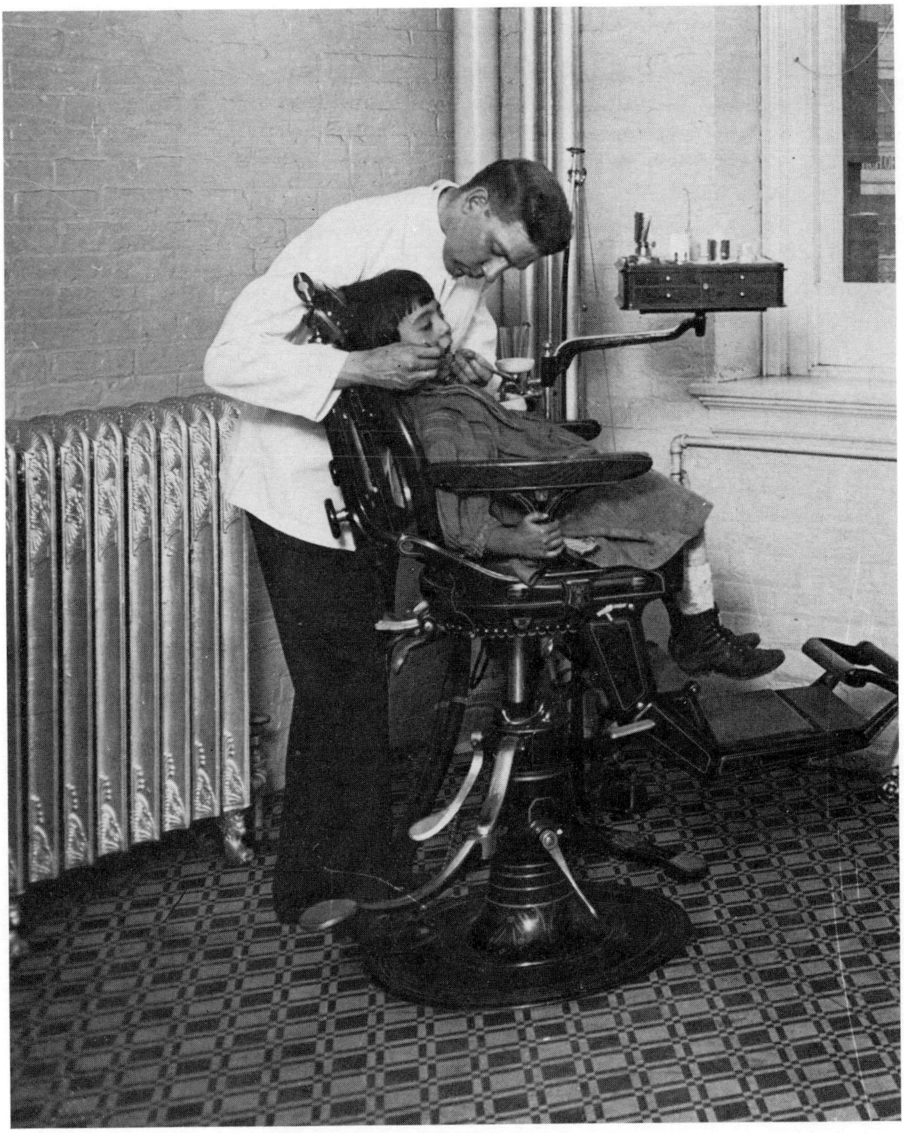

Fred McKay arrived in Colorado Springs, Colorado, in 1901 to begin a practice in dentistry. The young dentist was quite surprised when he discovered that the teeth of people living in that area were badly stained. He was unable to find any mention of such a condition in his dental books, so he named this condition *Colorado Brown Stain.* Sometimes this staining consisted only of small opaque, paper-white areas visible on the surface of normally translucent teeth (a condition called mottling). Other times it was a very noticeable dark brown stain on teeth whose enamel surfaces were often pitted. McKay soon started recording the occurrence and severity of this condition in various parts of the country, and tried to interest other members of the dental profession in joining in a study of Colorado Brown Stain.

By 1916, after a study of 26 communities, McKay had concluded that there was some substance in the drinking water of these communities which was causing the mottled teeth. In addition, he found that this substance had a noticeable effect only during the early stages of permanent tooth development in children. McKay's discovery that Colorado Brown Stain was caused by something in the water prompted communities such as Oakley, Idaho, to switch their source of drinking water. Happily, this change completely eliminated the occurrence of mottled teeth in the children of that town. McKay's investigations continued, and in 1929 he was able to make the important observation that patients suffering from mottled teeth were less likely to have dental caries, commonly known as cavities. In 1931, independent investigations by three different laboratories resulted in the identification of the fluoride ion, F^-, as the substance in drinking water that caused the mottling of children's teeth. Measurements of the fluoride concentration in various water supplies revealed that for half a century the residents of Colorado Springs had been drinking water containing 2 parts per million (2 ppm) of fluoride ion. The original water supply of Oakley, Idaho, was found to contain 6 ppm fluoride (the new water supply for this community contained less than 0.5 ppm fluoride ion). By 1942, it had been established that a level of 1 ppm fluoride in the drinking water would greatly reduce the number of dental caries in a population without producing mottled teeth (Figure 17.1 and Table 17.1). In 1950, the United States Public Health Service officially endorsed the

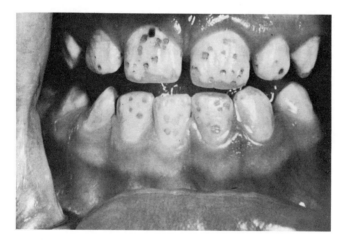

Figure 17.1 These teeth show severe mottling, or fluorosis, caused by the long-term drinking of water containing fluoride in a concentration greater than 10 parts per million. (Courtesy National Institutes of Health)

Table 17.1 Fluoride Levels in Drinking Water

Parts per Million of Fluoride	Effects
Less than 0.5	High incidence of dental caries; no mottling
1	Low incidence of dental caries; little or no mottling
Greater than 2.5	Disfiguring mottling
10 to 20	Severe mottling or fluorosis

fluoridation of public water supplies to a level of 1 ppm fluoride. Many cities and towns quickly acted to fluoridate their water supplies following studies showing that the rate of decayed and missing teeth in children who used such water was reduced by as much as 60%.

Dental caries are caused by bacteria that are found naturally in the mouth. These bacteria use carbohydrates (the subject of the next chapter) as their source of energy, and release carboxylic acids as waste products of their metabolism. Unfortunately, these bacteria also use carbohydrates (particularly sucrose, or table sugar) to produce plaque, a sticky substance that holds the bacteria and the acids they release against the teeth. These acids dissolve minerals in the teeth, causing the enamel to decay.
The number of dental caries that you have depends on three factors: the particular chemical composition of your teeth, the amount of bacterial plaque on the teeth, and both the type and frequency of your carbohydrate consumption. A person's rate of dental caries can be lowered by removing the plaque with flossing and brushing, and by reducing the number of times that sugar is eaten during the day. However, the most effective method of preventing dental caries is by the use of drinking water containing 1 ppm fluoride.

The actual way in which fluoride prevents dental caries is not understood. When children drink fluoridated water, fluoride ions are deposited as a calcium compound in the structure of the tooth enamel while the teeth are forming. This fluoride compound may make the tooth enamel more resistant to the acids produced by mouth bacteria. Scientists also have discovered that fluoride ions in a test tube are very effective in stopping the action of enzymes. Perhaps, therefore, a high concentration of fluoride in the tooth enamel may inhibit the action of bacterial enzymes responsible for converting sugar into the decay-causing acids. Although low levels of fluoride have not been shown to have any adverse effects on the human body, high levels of fluoride intake must be avoided. In cases of high fluoride levels, the body not only deposits fluorides as part of tooth enamel, but also deposits these ions in the bones. This results in enlarged and abnormal bones, a condition called skeletal fluorosis.

Table 17.2 Elements Necessary for Life

Element	Percent of Total Number of Atoms in the Human Body	Number of Grams in a 70-kg Man
Hydrogen	63	6580
Oxygen	25.5	43,550
Carbon	9.5	12,590
Nitrogen	1.4	1815
Calcium	0.31	1700
Phosphorus	0.22	680
Potassium	0.06	250
Sulfur	0.05	100
Chlorine	0.03	115
Sodium	0.03	70
Magnesium	0.01	42
Iron	<0.01	7
Manganese, cobalt, copper, zinc, molybdenum, vanadium, chromium, tin, fluorine, silicon, selenium, iodine	<0.01	<1

The Remaining 20 Elements

If very small concentrations of fluoride can have such varied effects on our bodies, we might ask if there are other elements that have equally important effects in small concentrations. In the last few chapters we discussed carbon, hydrogen, oxygen, and nitrogen. These four elements, we saw, make up 99.3% of all the atoms in the body and form an enormous number of compounds. There are 20 other elements that have thus far been shown in laboratory experiments to be necessary for life (Table 17.2). Although they make up only 0.7% of the atoms in the human body, they perform a wide variety of functions critical to life. Depriving a living organism of any of these elements will result in disease and death.

As shown in Table 17.2, these 20 elements can be divided into two groups. One group contains seven elements, called the **macrominerals,** which are found in greater concentrations in the body than are the remainder. This group of elements contains potassium, magnesium, sodium, calcium, phosphorus, sulfur, and chlorine. The remaining elements are found in such very small amounts that they are called the **trace elements.** All 20 elements are found in living systems either as ions, or covalently

Figure 17.2 These elements must be present in your body for a normal and healthy life.

bonded in organic molecules. Their functions are quite varied, and may depend upon their chemical forms or their locations in the body's tissues and fluids. For example, they maintain the cation/anion balance in the cells and the body fluid. They must be present in particular ratios to permit the contraction of muscles and the transmission of nerve impulses. Their concentrations control the movement of body fluids in the tissues and cells. They are part of the secretions essential to digestion. They activate or form parts of enzymes, and are contained in vitamins, hormones, and in proteins important in carrying materials throughout the body (Figure 17.2).

The Macrominerals

17.1 Calcium

Calcium is the fifth most common element in the body, accounting for 0.31% of the atoms in the body and making up about 2% of an adult's body weight. You've no doubt heard that calcium is important for the formation of strong bones and teeth. Indeed, 99% of the calcium in your body is found in your bones and teeth, in the form of several inorganic salts held in a framework of proteins (Figure 17.3). Saliva contains calcium ions and phosphates that constantly help replace minerals lost from the teeth. This is why cancer patients whose saliva glands have been destroyed during radiation therapy suffer extensive tooth decay. Such decay may be partially prevented by using a mouthwash containing a supersaturated solution of calcium phosphate.

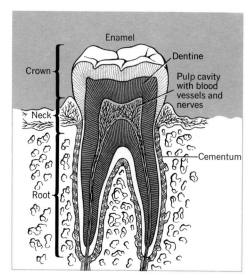

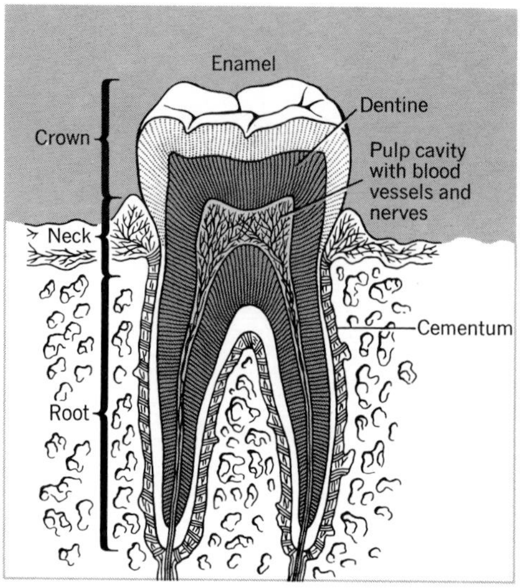

Figure 17.3 The structure of a tooth. Enamel, dentine, and cementum are composed of the following inorganic materials imbedded in a protein framework:
Main Inorganic Material
Apatite $Ca(OH)_2 \cdot 3[Ca_3(PO_4)_2]$
Trace Inorganic Material
Fluorapatite $CaF_2 \cdot 3[Ca_3(PO_4)_2]$
Chlorapatite $CaCl_2 \cdot 3[Ca_3(PO_4)_2]$
Dahllite $CaCO_3 \cdot 3[Ca_3(PO_4)_2]$

Calcium is a group II element; it has two valence electrons and is found in the body as the stable calcium ion, Ca^{2+}. The level of calcium ion in the blood follows a complicated relationship involving two hormones (parathormone and calcitonin) and vitamin D. The level of these substances in the blood controls the amount of calcium absorbed in the intestines, the level of calcium ion in the blood, and the amount of calcium deposited in the bones and teeth.

The calcium ions not found in the skeleton play several other important roles in the body. Calcium ions must be present in the correct concentration to allow the contraction of muscles, and they are especially important in maintaining the rhythmic contraction of the heart muscle. Too low a concentration of calcium can stop muscle contraction completely. Calcium ions affect nerve transmission by their stabilizing effect on the nerve membrane. Too much calcium in the blood results in a deadening of nerve impulses and muscle response, making a person unresponsive to any stimuli. Too little calcium in the blood can result in a high (hyper) irritability of the nerves and muscles. Under such hyperirritability the slightest stimulus, such as a loud noise, cough, or touch, can send a person into convulsive twitching. Such a state is extremely exhausting and will soon result in death.

A specific level of calcium ions in the fluid of the brain is critical in maintaining body temperature. Too high a concentration of calcium will result in the lowering of body temperature. Calcium ions also must be present for the blood to clot. Any condition that removes the calcium ions from the blood will prevent the clotting process from occurring. Leeches, fleas, and other bloodfeeding creatures are able to secrete a substance that reacts with calcium ions in the blood, preventing the blood from clotting while they feed on and digest it. Similarly, citrate ions and oxalate ions will combine with calcium ions in freshly collected blood, decreasing the level of calcium ions left in the blood and, therefore, stopping the clotting reactions. For this reason, sodium citrate is used as an anticoagulant in whole blood to be used for transfusions.

$$2C_6H_5O_7{}^{3-} \; + \; 3Ca^{2+} \; \longrightarrow \; Ca_3(C_6H_5O_7)_2$$

Citrate
ion

Calcium ions, in addition to their role in the clotting process, also activate a variety of enzymes. In fact, the calcium ion may be considered the *coordinator* among the other mineral ions, regulating the flow of these ions in and out of the cell (Figure 17.4).

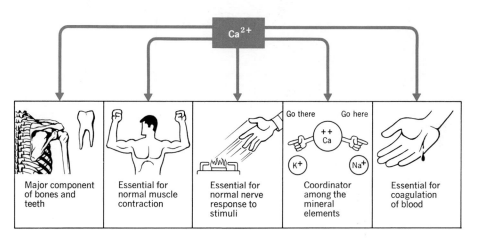

| Major component of bones and teeth | Essential for normal muscle contraction | Essential for normal nerve response to stimuli | Coordinator among the mineral elements | Essential for coagulation of blood |

Figure 17.4 **Functions of calcium in the body.**

Dairy products are the major source of calcium in our diets. An adult who drinks one pint of milk a day is meeting the daily requirement for calcium (1 gram). The complex relationships controlling the level of calcium in the blood can be upset by various factors including low levels of calcium in the diet, or abnormal levels of vitamin D, calcitonin, or parathormone in the blood. Too little vitamin D in the body produces rickets, a disease that causes bones to soften and bend out of shape (Figure 17.5). Although vitamin D is formed in the body when the skin is exposed to sunlight, extra vitamin D has been added to milk supplies in the United States to prevent the occurrence of rickets. However, it has been shown that too much vitamin D can be as harmful as too little; it causes a thickening of bones and a calcification of soft tissues.

Exposure to cadmium ions can lower the level of calcium in our bodies, causing a disruption of normal calcium metabolism. For example, in the late 1950s residents of Japan's Jinzu River basin were affected by a strange malady called the *Hai-Hai* or *Ouch-Ouch* disease. This disease caused severe and painful decalcification of bones, often resulting in multiple fractures. The malady was found to be caused by the presence of cadmium ions, Cd^{2+}, in rice irrigated by polluted water discharged from industries upstream.

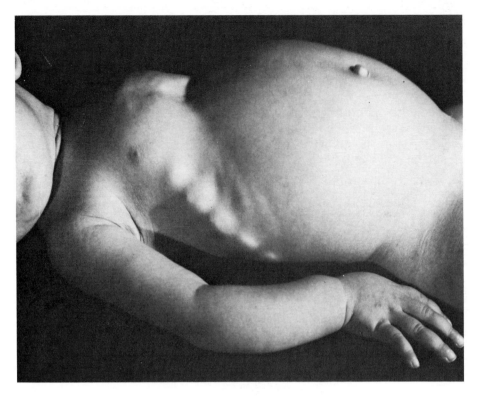

Figure 17.5 This child shows the characteristic deformities of rickets: bones that become soft from a lack of Vitamin D, and bend out of shape. (Courtesy of the Children's Hospital Medical Center, Boston, Mass.)

17.2 Phosphorus

We have seen that calcium plays a major role in the formation of bones and teeth. Phosphorus is an important element in the inorganic calcium salts that are found in these bones and teeth (see Figure 17.3). Ninety percent of the phosphorus in the body is found in the bones and teeth, in the form of the negative phosphate ion, PO_4^{3-}.

$$\begin{array}{c} O \\ \parallel \\ HO{-}P{-}OH \\ \mid \\ OH \end{array}$$

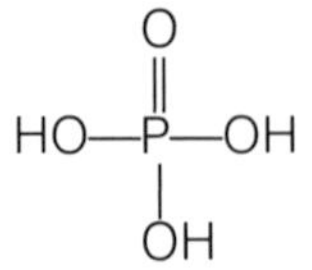

Phosphoric acid, H_3PO_4

$$\begin{array}{c} O \\ \parallel \\ {}^-O{-}P{-}O^- \\ \mid \\ O^- \end{array}$$

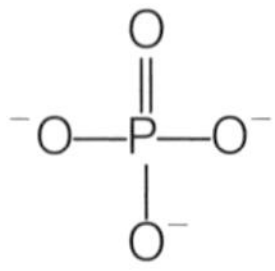

Phosphate ion, PO_4^{3-}

The phosphate ion is formed when phosphoric acid loses three hydrogen ions.

$$H_3PO_4 \xrightarrow{-H^+} H_2PO_4^- \xrightarrow{-H^+} HPO_4^{2-} \xrightarrow{-H^+} PO_4^{3-}$$

| Phosphoric acid | Dihydrogen phosphate ion | Monohydrogen phosphate ion | Phosphate ion |

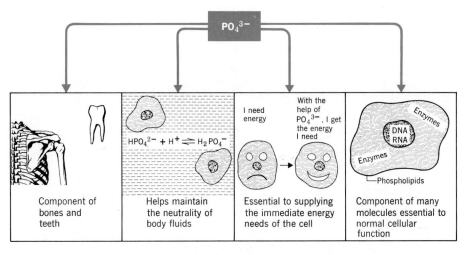

Figure 17.6 Functions of phosphorus in the body.

The dihydrogen and monohydrogen phosphate ions act as buffers, which help to maintain the proper pH of body fluids. Phosphoric acid can also react with an alcohol functional group on an organic compound to form a phosphate ester. Organic phosphate esters are found in phospholipids (which make up cell membranes and nerve tissues), DNA and RNA (which control heredity and protein synthesis), and coenzymes (compounds that work with enzymes in the body). We will be discussing each of these compounds in detail in later chapters. When certain organic phosphate esters undergo hydrolysis, considerable chemical energy is liberated. Such phosphates are known as high-energy phosphates, and are the compounds that supply the immediate energy needs of the cell. These important functions make phosphorus essential to all body tissues (Figure 17.6). Fortunately, phosphorus is so widely distributed in our daily foods that almost everyone is assured of getting at least the 1 gram per day that is required for adults.

17.3 Magnesium

Magnesium ions, Mg^{2+}, make up 0.01% of the atoms in the body. These ions activate many of the enzymes which control the addition and removal of phosphate groups from compounds in the cell (Figure 17.7). Magnesium forms part of the chlorophyll molecule, which traps sunlight in the photosynthesis process and gives plants their green color. Magnesium is found in a wide variety of foods such as green vegetables, nuts, cereals, and seafoods, so we are fairly well assured of enough magnesium in our diets (the recommended daily amount is 400 mg).

If, however, the magnesium level in the body is lowered, a person may suffer emotional irritability and aggressiveness, muscle spasms, and convulsions. Magnesium deficiencies have been observed in chronic alcoholics, in infants having a protein deficiency disease called Kwashiorkor, and in postsurgical patients on restricted diets. On the other

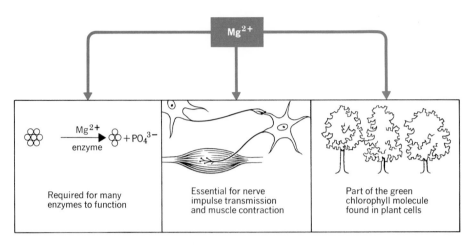

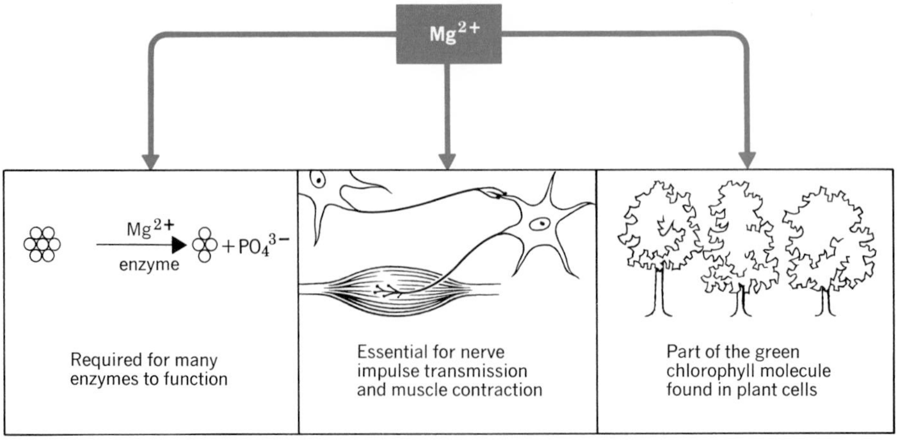

Figure 17.7 Functions of magnesium in living organisms.

hand, too much magnesium decreases muscle and nerve response, and high levels can produce local or general anesthesia and paralysis.

17.4 Potassium, Sodium, and Chlorine

The roles of potassium, sodium, and chlorine in the body are intricately interrelated. Sodium and potassium ions are often present in the form of their chloride salts. The potassium ion is usually found in the intracellular fluid within the cell, whereas the sodium ion is generally found in the extracellular fluids surrounding the cell. The main function of the potassium, sodium, and chloride ions is to control the cation/anion balance in the cells, tissue fluids, and blood. Such cation/anion balance is necessary to maintain the normal flow of fluids and to control the balance between acids and bases in the body. These three ions also play important roles in the transport of oxygen and carbon dioxide in the blood. Sodium and potassium ions are involved in the transport of sugars across cell membranes and in the breakdown of sugars in the cell. In addition, these two ions (together with calcium and magnesium ions) help maintain the proper level of nerve and muscle response. The effects of sodium and potassium are antagonistic to those caused by calcium and magnesium (that is, they have opposite effects). Therefore, the correct balance in the concentrations of these four ions is critical to normal nerve and muscle performance.

Sodium chloride and potassium chloride act to keep large protein molecules in solution, and thus to regulate the proper viscosity or thickness of the blood. The acid in the stomach that begins the digestion of certain foods is hydrochloric acid, HCl, which is derived from the sodium chloride in the blood. Other digestive compounds found in the gastric juices, pancreatic juices, and bile are likewise formed from the sodium and potassium salts in the blood. The response of the retina of the eye to light impulses is another body process which depends upon the correct concentrations of sodium, potassium, and chloride ions (Figure 17.8).

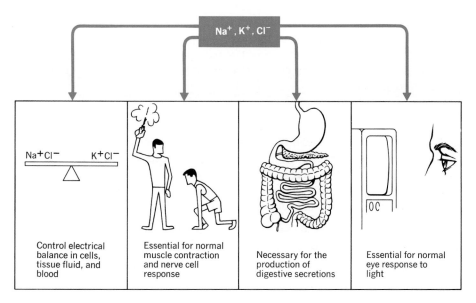

Figure 17.8 Functions of sodium, potassium, and chlorine in the body.

Having learned the many important body functions that depend upon these three ions, it is easy to imagine that an imbalance in the level of any one of them can have serious effects on the body. Experimental animals have been fed diets deficient in these ions, and have experienced slow growth, slow heart rate, muscular atrophy, and sterility. Plant material is high in potassium ions, but high levels of potassium ions in the diet will result in excessive excretion of the sodium ion from the body. This is why herbivorous animals (animals that eat plant material) must be given diets containing a high level of salt or sodium chloride to maintain the proper balance between the sodium and potassium ions in their bodies. Such animals have been known to travel hundreds of miles and to risk their lives in order to reach a salt lick.

You are certainly aware that strenuous exercise, especially in hot weather, results in heavy perspiration. Perspiration is composed mostly of water, but there are also many ions dissolved in this fluid (among which are potassium, sodium, and chloride ions, giving sweat its salty taste). If the concentration of these ions in the body is significantly reduced by heavy perspiration, an imbalance will occur which affects muscle and nerve response. Nausea, vomiting, exhaustion, and muscle cramps can result. For this reason, athletes often guard against such ion loss by taking salt tablets before heavy exercise, or by drinking specially formulated beverages to replenish their lost salts.

Under normal dietary conditions, humans will not experience a deficiency of these ions. A deficiency of potassium can result, however, in people who suffer from prolonged diarrhea or who use diuretics. A major problem with the American diet at the present time, in fact, is not a deficiency of sodium but rather an excess. Although the daily requirement for sodium is about 0.5 grams, we eat 10 to 35 times that much sodium, mostly in the form of salt added to foods. Long-term high levels of sodium may cause an increase in the amount of extracellular fluid (edema), disruption of fat metabolism, changes in gastric secretions, and high blood pressure (hypertension).

17.5 Sulfur

Sulfur is the last of the seven elements in the group of macrominerals. It is found in group VI on the periodic table; it appears just under oxygen and has properties similar to oxygen. It will form two covalent bonds. One functional group in which sulfur is found is the thiol or sulfhydryl group, which closely resembles the hydroxyl group.

$$:\ddot{S}: H \quad —S—H \qquad :\ddot{O}: H \quad —O—H$$
 Sulfhydryl group Hydroxyl group

One class of simple compounds containing the sulfhydryl group is called the mercaptans.

$$CH_3CH_2SH \qquad CH_3CH_2OH$$
 Ethyl mercaptan Ethanol

These compounds have a terrible smell. In fact, it is mercaptans that produce the strong odor given off by skunks. Because natural gas is odorless and toxic, gas companies add mercaptans to it before it is piped to consumers. This allows leaks to be easily detected.

Other compounds containing sulfhydryl groups need not have any odor. Proteins, for example, are polymers of smaller units called amino acids (which we will discuss in detail in Chapter 20). The amino acid called cysteine has a sulfhydryl group, and is very important to the structure and function of protein molecules in which it is found.

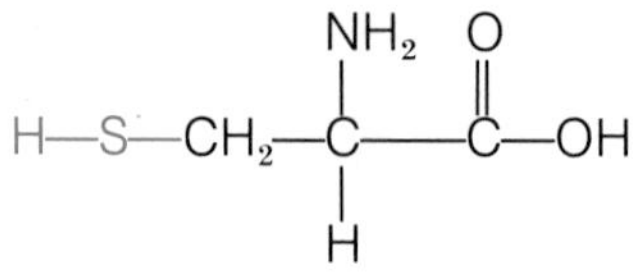

$$H—S—CH_2—\underset{\underset{H}{|}}{\overset{\overset{NH_2}{|}}{C}}—\overset{\overset{O}{\parallel}}{C}—OH$$
 Cysteine

Table 17.3 The Role of Sulfur in the Body

1. Found in proteins and other biological compounds.	
2. Found in several functional groups:	
Thiol or sulfhydryl	—S—H
Disulfide	—S—S—
Thioethers or organic sulfides	—C—S—C—

Sulfur can replace the oxygen atom in an ether linkage to produce a thioether, or organic sulfide, —C—S—C—. Among the compounds containing this functional group are those giving onions and garlic their familiar flavors.

$$H_2C=CH-S-CH=CH_2$$
Divinyl sulfide
(onion flavor)

$$H_2C=CHCH_2-S-CH_2CH=CH_2$$
Diallyl sulfide
(garlic flavor)

Methionine is an amino acid that contains a thioether group. All body proteins, as well as special protein molecules such as enzymes and hormones, contain this amino acid. The body is able to produce cysteine from methionine, making methionine extremely important in sulfur metabolism (Table 17.3).

$$CH_3-S-CH_2CH_2\underset{\underset{H}{|}}{\overset{\overset{NH_2}{|}}{C}}-\overset{\overset{O}{\|}}{C}-OH$$

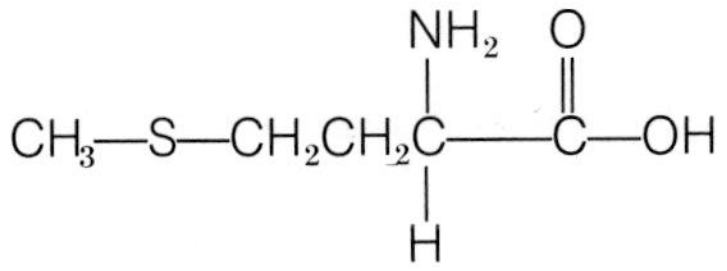

Methionine

We have seen that the roles of the macrominerals in our bodies are varied, complex, and often intricately related. This is why the maintenance of proper concentrations of these elements through good nutrition is so vitally important to assure a healthy body.

The Remaining 13 Elements

17.6 The Trace Elements

The remaining 13 elements which have been shown to be essential to life are often called the **trace elements** because they are found only in trace (extremely small) amounts in the living system. The list of trace elements contains four nonmetals: silicon, fluorine, selenium, and iodine. The others are metals. One is tin, and the remaining eight are transition metals: vanadium, chromium, manganese, iron, cobalt, copper, zinc, and molybdenum (Table 17.4). Although these elements are required in only extremely small amounts, they are critical to the proper functioning of the living system. A total absence of any one of these trace elements can result in death of the organism.

Because these elements are found only in trace amounts in living systems, it has been very difficult for scientists to establish their functions in the body. The chemical action of many of the trace elements is quite complex, and the exact functions of several of them are still largely unknown. Many of these elements, however, form important parts of enzymes. Because enzymes can be used over and over again, they can be effective even when found in only very low concentrations in the cells of the body. The transition metals, in particular, have chemical binding properties that make them especially important in enzyme molecules.

Table 17.4 The Trace Elements

Element	Function
NONMETALS	
Fluorine	Found in bones and teeth, important in the prevention of dental caries
Selenium	Required for prevention of white muscle disease, part of an enzyme that protects against accumulation of peroxides
Iodine	Required for normal thyroid function, found in thyroid hormone
Silicon	Function unknown
METALS	
Chromium	Lowers blood sugar level by increasing the effectiveness of insulin
Manganese	Essential for the function of several enzymes, bone and cartilage growth, and brain and thyroid function
Iron	Found in hemoglobin and many enzymes
Cobalt	Part of vitamin B_{12} molecule
Copper	Part of many enzymes essential for the formation of hemoglobin, blood vessels, bones, tendons, and the myelin sheath
Zinc	Essential for many enzymes, normal liver function, and synthesis of DNA
Molybdenum	Required for the function of several enzymes
Vanadium	Function unknown
Tin	Function unknown

17.7 Iodine

One of the nonmetal trace elements, iodine, has long been known to be essential to life. Seventy to 80% of the iodine in the body is concentrated in the thyroid gland, a small gland in the neck. Iodine is part of a hormone produced by the thyroid gland. This hormone, thyroxin, regulates the body's chemical activity (that is, its metabolism), and is vital to normal growth.

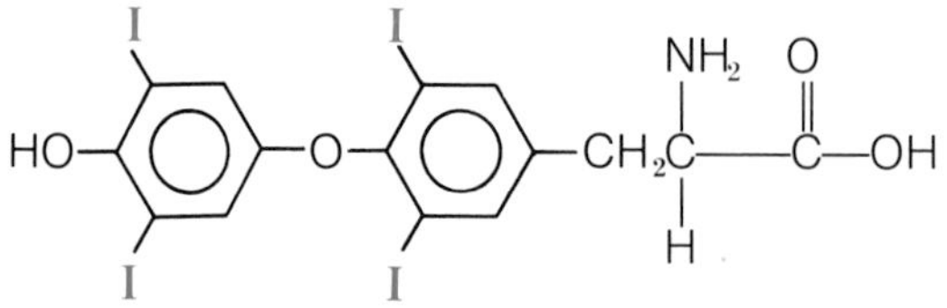

Thyroxin

An insufficient amount of iodine in the diet causes enlargement of the thyroid, a medical condition known as simple goiter (Figure 17.9). Such an

Figure 17.9 These villagers from Paraguay are suffering from endemic goiter. The swollen thyroid gland in their necks results from a lack of the trace element iodine in their diets. (Courtesy World Health Organization. Photo by Paul Almasy)

increase in the size of the thyroid is a compensatory reaction by the body in response to the low level of iodine. The body attempts to increase the production of thyroid hormone by increasing the number of cells in the thyroid, but this attempt cannot succeed as long as the concentration of iodine remains low. Salt-water fish are a rich source of iodine, and many cases of goiter used to be observed in the Midwest where such fish were not commonly available. The amount of goiter in such regions has been greatly reduced by the use of table salt containing the iodide ion, I^-. Although the amount of iodine in your body is only 1/2,500,000th of your entire body weight (an amount of iodine about the size of a pinhead), the absence of this trace amount of iodine would be fatal.

17.8 Iron

Four atoms of iron are found in every molecule of hemoglobin—the molecule in the red cells of our blood which carries oxygen from the lungs to the tissues, and which causes the blood to look red (Figure 17.10). If you were able to extract all of the iron from a healthy body, you would have enough to make only two small nails (about 5 to 7 grams). Yet this amount of iron is critical. Only a small reduction in the level of iron in the blood will result in a condition known as anemia, which causes general body weariness, fatigue, and apathy. Anemias from iron deficiencies result in low hemoglobin levels, and often occur in infants at six months and in women from 30 to 50 years of age.

A healthy adult needs about 18 mg of iron each day. Such dietary iron is found in large amounts in organ meats such as liver, kidney, and heart, as well as in egg yolks, dried vegetables of the pea family, and shellfish. In the body, iron is absorbed in the small intestine in the form of the iron(II) ion, Fe^{2+}. This iron absorption is increased by the presence of vitamin C (ascorbic acid) which reduces the iron(III) ion, Fe^{3+}, in the intestines to the iron(II) ion. Under normal conditions, only 5 to 15% of the iron in the foods we eat is actually absorbed into our bodies. A deficiency of iron in the diet results in listlessness and fatigue, reduced resistance to disease, and

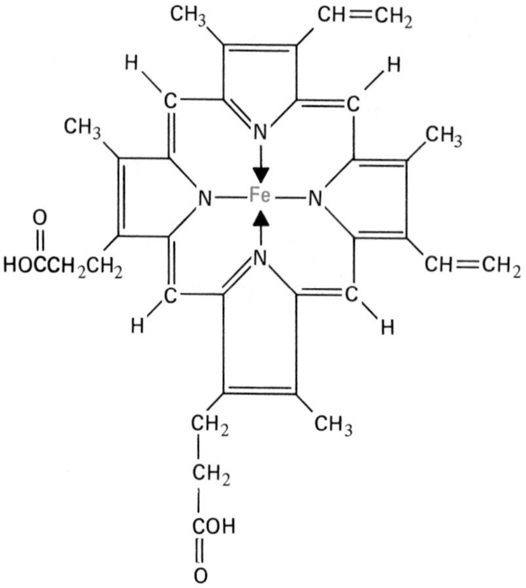

Figure 17.10 Four of these iron-containing heme groups are found in each hemoglobin molecule. Each heme is capable of carrying one oxygen molecule, so one hemoglobin molecule can carry four oxygen molecules.

an increase in heart and respiratory rate. Children with iron deficiencies show a decreased growth rate and abnormal red blood cell growth. High levels of iron, on the other hand, can also be unhealthy. Abnormally high levels will cause cirrhosis of the liver, fibrosis of the pancreas resulting in diabetes mellitus, and congestive heart failure.

The functioning of iron in the body gives a good example of how the concentration of one trace element can be closely dependent upon the concentration of some other trace element. In this case, an enzyme containing copper must be present for a hemoglobin molecule to be formed. Therefore, the concentration of hemoglobin in the body depends not only upon the level of iron, but also upon the concentration of copper. A high concentration of copper will result in a high usage of iron. Such relationships between trace elements can also work in an antagonistic fashion. For example, a high concentration of the trace element molybdenum in the diet will cause a decrease in the absorption of copper by the body, resulting in a decrease in the formation of hemoglobin.

17.9 Copper

Since 1818, copper has been known to be a component of living tissue. The concentration of copper in the human body is very small—less than 1/30th of the amount of copper in a penny—yet this concentration is critical. Copper is necessary for the normal functioning of all living cells. Too much or too little copper will result in malfunctions of the cell.

Copper performs a variety of functions in the body. It is a component of several important enzymes, one of which helps in the formation of blood vessels, tendons, and bones. Copper-deficient animals will exhibit weakness and fragility of blood vessels and bones (Figure 17.11). The formation of the protective sheath around the nerves is dependent upon

Figure 17.11 These three dogs are from the same litter. The dogs at the right and center have been fed diets deficient in copper, and show rough hair and deformed legs not found in the normal dog on the left (who was fed a diet containing copper). (Courtesy Baxter and Van Wyk, Bulletin of the Johns Hopkins Hospital, *93*:1, 1953)

copper, and a deficiency of copper plays a role in a degeneration of the nervous system in which nerve impulses are no longer properly transmitted. Copper also helps protect us from the harmful ultraviolet rays of the sun; this element is part of an enzyme that assists in the formation of melanin, the dark pigment of the skin which is our natural protection from ultraviolet radiation. Our cells would not be able to extract energy from foods without another copper-containing compound. Earlier we mentioned that copper must be present for the formation of hemoglobin. And finally, the ability to taste foods may also depend upon the presence of copper.

Because copper is a part of very important enzymes, low levels of this element can cause serious illness. But high levels of copper can be equally toxic. Because copper is very common in all foods and in drinking water, the body has developed complex ways to regulate the absorption and excretion of copper. The copper level in the body depends upon a balance between copper, molybdenum, and sulfate in the diet. A disorder called Wilson's Disease results from a genetic abnormality in which the body's ability to eliminate copper is impaired. The liver, kidneys, and brain of an individual suffering from this disease will show abnormally high levels of copper, possibly leading to mental illness and death. Wilson's Disease can be treated with a chemical that binds with copper ions and, in that way, detoxifies them.

17.10 Cobalt

A lack of cobalt in the diet results in a disease called pernicious anemia, which produces symptoms of fatigue and general weakness. This disease is not caused by a lack of hemoglobin, but rather results from a lack of erythrocytes (red blood cells that carry the hemoglobin molecules, and thus the oxygen, to the cells). Cobalt is a part of vitamin B_{12}, which is required for the formation of erythrocytes. However, too much vitamin B_{12} in the diet will stimulate the production of too many erythrocytes, producing a condition called polycythemia.

17.11 Zinc

Recent research has shown zinc to be an extremely important trace mineral in development of the fetus and in the nutrition of infants. Zinc is required for the synthesis of DNA, the genetic material in the cell. Therefore, a zinc deficiency in the fetus will result in retarded growth, malformations of the body, and chromosomal abnormalities. Zinc deficiency after birth results in dwarfism, retarded sexual development, loss of hair, and skin lesions (Figure 17.12). Human milk contains zinc at concentrations up to 10 times higher than that found in the blood. In addition, this zinc is in a chemical form that will lead to its increased absorption by the infant. The importance of such zinc was demonstrated by a study in which babies who were fed formula diets fortified with zinc showed increased growth rates and were less prone to vomiting and diarrhea than babies fed on nonfortified diets.

The human body contains about 2.3 g of zinc, which plays a role in over 80 enzymes. For example, it is part of several liver enzymes, one of which is important in the oxidation of alcohol to less toxic substances. A

Figure 17.12 The 16 year-old boy on the left is suffering from dwarfism caused by a deficiency of zinc in his diet. The man on the right is of normal height. (Courtesy Harold H. Sandstead, M.D. Reproduced with permission of *Nutrition Today* Magazine, Annapolis, Md., © March 1968.)

high level of alcohol in the body may cause this enzyme to break down, which will then produce toxic conditions in the liver. Alcoholics with cirrhosis of the liver show a high level of zinc in the urine, indicating the breakdown of this zinc-containing liver enzyme. The normal daily requirement for zinc is 15 mg. Foods rich in zinc include nuts, eggs, beef, and liver.

17.12 Manganese

Manganese is necessary for the functioning of several enzymes. It is found in high concentrations in the mitochondria, areas of the cell where cellular energy is produced. A manganese deficiency will produce abnormalities in the structure of these mitochondria. Manganese is also essential for normal thyroid function and for cartilage and bone growth. Manganese is required for the normal function of the brain and nervous system. In fact, it has been found that as many as one-third of the children who suffer from convulsive disorders such as epilepsy have low levels of manganese in their blood. But, as we have seen with other trace elements, high blood concentrations of manganese can be equally dangerous. Miners who dig for manganese often suffer headaches, psychotic behavior, and drowsiness caused by elevated levels of manganese in the blood.

17.13 Selenium

Selenium, in very small amounts, is an essential nutrient. However, in slightly larger amounts it is quite toxic. In areas of the country where the soil and foliage are rich in selenium, livestock suffer from the "blind staggers"—selenium poisoning that causes impairment of vision, muscle weakness, necrosis of the liver, and death from respiratory failure.

When more than trace amounts of selenium are present in the body, it is substituted for sulfur in many cellular compounds. These selenium compounds have less stability and are more reactive than the corresponding sulfur compounds, and will disrupt the normal functioning of the cell. As an essential nutrient in trace amounts, selenium is necessary to prevent white muscle disease in cattle, sheep, hogs, and poultry. It forms part of the enzyme called glutathione peroxidase, which is the body's main protection against an accumulation of hydrogen peroxide and organic peroxides in the cells. Such organic peroxides are thought to play a role in cancer development. Recent studies have shown that persons living in areas with high dietary levels of selenium are less likely to suffer several types of cancers than individuals living in low selenium areas. However, this beneficial effect of selenium does not seem to occur if the dietary level of zinc is also high.

Elevated concentrations of cobalt can magnify the toxic effects of selenium, producing enlargement of the heart and liver. This may have been the cause of the symptoms known as "beer drinkers' cardiomyopathe," which produced many deaths in several cities in 1965 and 1966. These symptoms resulted when small amounts of cobalt were added to beer to stabilize the foam, multiplying the effect of the selenium that occurred naturally in fairly high amounts in the water used to make the beer.

17.14 Chromium

The level of sugar in the blood is critical to the functioning of body tissues, especially the brain. This blood sugar level stays remarkably constant under normal conditions. Many factors help to control the amount of sugar in the blood; one is the compound called insulin. Insulin is secreted by the pancreas, and works as a control in lowering the blood sugar level. When the pancreas fails to secrete enough insulin, the disease called diabetes mellitus results. Chromium seems to play a role in lowering the blood sugar level in the body by increasing the effectiveness of insulin. This may explain why chromium deficiency produces symptoms similar to diabetes mellitus. Brewer's yeast, whole grain cereals, and liver are good sources of chromium in the diet.

17.15 Molybdenum

Molybdenum participates in the energy transfer reactions in the cell, is necessary for the function of certain intestinal enzymes, and is involved in controlling the amount of copper absorbed by the body. Dietary sources of molybdenum include plants in the pea family, cereals, organ meats, and yeast.

17.16 Silicon, Vanadium, and Tin

The functions of silicon, vanadium, and tin are still under study. Very little is known about their exact roles, but they have been shown to be essential nutrients in the diets of various plants and animals.

Trace element research is increasing our understanding of how these elements influence living organisms, and the types of diseases that result from changes in their concentrations. Doctors are beginning to recognize trace elements as possible aids in treating disease, monitoring stress, and predicting heart disease. Although the interactions among trace elements are quite complex, a thorough understanding of these interactions is becoming more and more essential in our highly industrialized society. The challenge is to maintain the health of the population in a society which is constantly changing the natural levels of these, as well as other, trace elements in the soil, water, food, and air.

Chapter Summary

The remaining 20 elements which are essential to life may be divided into two groups: the macrominerals that are required in the diet in amounts ranging from 500 mg to several grams per day, and the trace elements that are required in much smaller amounts each day. The macrominerals are sodium, potassium, magnesium, calcium, phosphorus, sulfur, and chlorine. The trace elements are silicon, fluorine, selenium, iodine, tin, vanadium, chromium, manganese, iron, cobalt, copper, zinc, and molybdenum. If the normal level of any of these elements in the body is decreased, a deficiency disease will result. These 20 elements play important roles in the control of many bodily processes: the cation/anion balance in cells and body fluids, the contraction of muscles, the transmission of nerve impulses, and the movement of body fluids. They also form a part of digestive secretions, enzymes, hormones, and proteins. The concentrations of these elements in living organisms are closely interrelated, and often can have either complementary or antagonistic effects. The trace elements are essential to life in very small amounts, but many are harmful or even toxic to living organisms in more than trace amounts.

Exercises and Problems

1. In which body tissues is fluoride important?
2. What was the cause of *Colorado Brown Stain?*
3. What concentration of fluoride ion in drinking water has been shown to prevent dental caries without producing mottling?
4. In what ways can fluoride ions be both beneficial and harmful to living tissue?
5. In considering mineral elements necessary for living tissues, calcium and phosphorus are often discussed together. Why? What happens to body growth if there is an inadequate supply of these elements in the diet?
6. What steps would you take to prevent the clotting of a recently drawn blood sample? Write the equation for the reaction that would take place.
7. What three factors control the amount of calcium in the blood and body fluids?
8. Discuss how a trace metal such as cadmium, which is not essential to the body's metabolism, can disrupt the metabolism of an essential element.
9. Write an equation for the formation of the phosphate ion from phosphoric acid.
10. What are three functions of phosphate ions in the body?
11. What effect will high levels of magnesium and calcium have on nerve function? How does this compare with the effect of high levels of sodium and potassium ions?
12. Why might a doctor inject magnesium salts into the blood of a person who is suffering from convulsions?
13. The effects of hydrogen ions on nerve function are similar to those of Ca^{2+} and Mg^{2+}. State whether depression (coma) or convulsions would be a symptom of severe acidosis. Explain your choice.
14. Sodium, potassium, and chlorine are three essential elements whose varied functions in the human body are intricately related. List five essential functions of these elements.
15. Explain, on the cellular level, why high daily dietary concentrations of sodium can cause edema.
16. Why does perspiration taste salty? What are several possible harmful effects of excessive perspiration? Will drinking large amounts of water remedy all of these effects?

17. State which of the following compounds contains a sulfhydryl group, and which contains an organic sulfide.

(a) $CH_3CH_2—S—CH_3$
(b) CH_3CHSH
$\quad\quad\quad\;\,|$
$\quad\quad\quad CH_3$
(c) $CH_3CH_2—S—S—CH_2CH_3$

18. In what large class of compounds in the human body is sulfur found?

19. Suggest a reason for the substitution of selenium for sulfur in compounds in the cell.

20. What is meant by the term *trace element*?

21. What is the function of iodide in the human body? What condition results from too little iodide in the diet?

22. What three trace elements are involved in controlling the rate of synthesis of hemoglobin?

23. Describe the functions of the following trace elements:

(a) copper
(b) cobalt
(c) zinc
(d) manganese
(e) selenium
(f) chromium
(g) molybdenum

24. Using specific examples, explain how:

(a) Trace elements can play complementary roles in living tissues.
(b) Trace elements can play antagonistic roles in living tissues.

section IV
the compounds of life

chapter 18

Carbohydrates

Learning Objectives

By the time you have finished this chapter, you should be able to:

1. Define *monosaccharides, disaccharides,* and *polysaccharides.*

2. Given the structure of a monosaccharide, identify the compound as an aldose or ketose.

3. Draw the linear structure of glucose.

4. List three hexoses and one pentose that play important roles in human metabolism.

5. Name the two polysaccharides making up starch, and describe the difference in their structures.

6. Explain, in terms of its structure, why we cannot digest cellulose.

7. Define *metabolism, catabolism,* and *anabolism.*

8. Write the general reaction for the following processes:

a. Photosynthesis
b. Glycogenesis
c. Glycogenolysis
d. Glycolysis
e. Fermentation

9. Describe in general the digestion and metabolic uses of the starch in a slice of toast.

10. Explain the meaning of *high-energy* bonds and give an example of compounds in which they are found.

11. Describe the two stages of cellular oxidation of glucose, indicating where they occur in the cell and in which step the most energy is released.

12. State two ways in which the body can control carbohydrate metabolism.

13. Define the following terms:

a. Diabetes mellitus
b. Blood sugar level
c. Hypoglycemia
d. Hyperglycemia
e. Renal threshold
f. Glucose tolerance test
g. Insulin shock

Cheryl was an active 10-year-old who enjoyed outdoor activities. She had been looking forward to playing softball during the summer, and was terribly disappointed when a bladder infection kept her from practicing with her team. It was not long after Cheryl's infection had cleared up that her mother began to notice a change in her daughter's behavior. Cheryl became increasingly irritable and tired. She complained that her skin itched and, in fact, had developed several open sores from scratching. She drank large amounts of liquids almost all the time, requiring her to urinate frequently.

One sunny afternoon Cheryl's mother returned home to find her daughter lying on the couch. Cheryl complained that she was extremely thirsty, seemed to have blurred vision, and was having trouble breathing. Her face was flushed, but her skin felt dry and cold. She was breathing deeply and rapidly, and her breath had a strange fruity odor. Cheryl's mother quickly called the doctor, who, upon hearing the symptoms, said he would meet them at the hospital immediately. Meanwhile, Cheryl was becoming less and less responsive, and was moaning with pain. On the way to the hospital she lapsed into a coma.

At the hospital, the doctor found Cheryl's blood pressure to be low, and her temperature to be below normal. An analysis of her urine revealed high concentrations of the sugar glucose, together with ketone compounds. Her blood sugar level was 660 mg/100 ml, and the level of acetone in her blood was 52 mg/100 ml. The doctor diagnosed Cheryl's condition as a diabetic coma, and began treatment by giving her doses of rapid-acting insulin at hourly intervals. The hospital staff monitored Cheryl's blood and urine sugar levels closely. As her blood sugar began to reach a normal level, Cheryl was given doses of sugars to prevent the blood sugar level from falling below normal. She also was given saline intravenously to replace the salts and fluids that she had lost.

As Cheryl was being treated for her diabetes, two floors above her in the hospital a 34-year-old woman named Diane was about to give birth to her first child. Diane's labor had begun early in the eighth month, so there was some concern over the condition of the infant. Unfortunately, the baby was stillborn. To determine the cause of the baby's death, the doctor put Diane through a series of laboratory tests, one of which measured her body's ability to handle a large dose of glucose. From this test and others, Diane was found to have the adult form of the disease diabetes mellitus.

Diabetes mellitus results from a low level or a total lack of the hormone insulin, which is important in controlling the level of the blood sugar glucose. Both Cheryl's juvenile form of diabetes (which accounts for 10% of all diabetes patients) and Diane's adult type can be treated and controlled. However, neither can be cured. Cheryl was placed on a strict low carbohydrate diet, and was taught how to give herself daily doses of insulin by injection. Doctors were able to control Diane's diabetes by strictly regulating her diet.

The blood sugar glucose belongs to a large class of compounds known as carbohydrates. In this chapter we will study the structure, properties, and functions of the carbohydrates. Later, we will return to discuss in greater detail the disease diabetes mellitus.

Carbohydrates

Carbohydrates are a class of compounds which includes polyhydric aldehydes, polyhydric ketones, and large molecules that can be broken down to form polyhydric aldehydes and ketones. These compounds include sugars, glycogen, starches, cellulose, dextrins, and gums. Carbohydrates are found mainly in plants, where they make up about 75% of the solid plant material. They function both as part of the structure that supports the plant and as storehouses for the plant's energy supply. The carbohydrate cellulose is the most important component of the supporting tissue of plants (such as the wood in trees), and the carbohydrate starch is the energy storage molecule (Figure 18.1 and 18.2).

Figure 18.1 This giant sequoia tree is the largest living object in the world. It is as tall as a six-story building, wider at the base than an average city street, and is estimated to be between 2500 and 3000 years old. The support structure of the tree is composed of the carbohydrate cellulose. (Courtesy National Park Service)

18.1 Classification

Carbohydrates are classified according to the size of the molecule. **Monosaccharides** are carbohydrates that cannot be broken into smaller units upon hydrolysis. **Disaccharides** will produce two monosaccharides upon hydrolysis, and **polysaccharides** will produce three or more monosaccharides upon hydrolysis (and can contain as many as 3000 monosaccharide units).

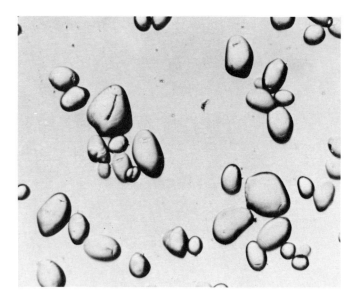

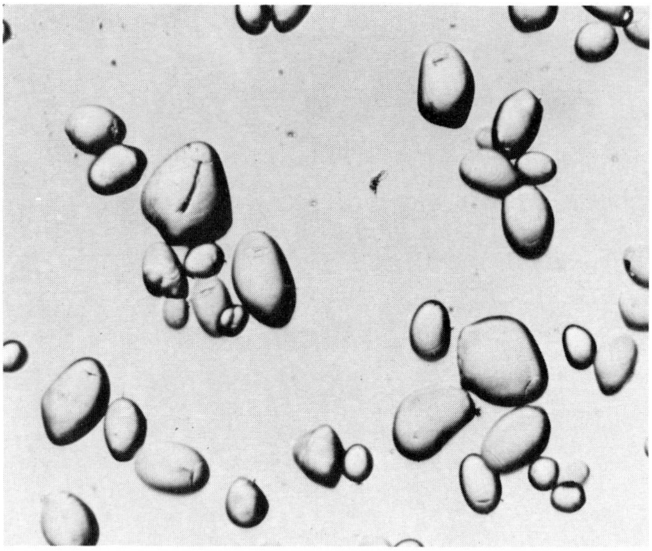

Figure 18.2 Starch granules from a potato. Starch is the energy storage molecule in plants. (Hugh Spencer)

Monosaccharides, also called simple sugars, can be further classified by the number of carbons in the molecule.

3 carbons—triose	5 carbons—pentose
4 carbons—tetrose	6 carbons—hexose, etc.

Monosaccharides may also be classified by the carbonyl functional group found in the molecule.

Aldose—aldehyde functional group
Ketose—ketone functional group

For example, ribose is an aldopentose (a five-carbon sugar molecule containing an aldehyde group) and fructose is a ketohexose (a six-carbon sugar molecule containing a ketone group).

Monosaccharides

Simple monosaccharides are white crystalline solids that are highly soluble in water as a result of their polar hydroxyl groups (which means, therefore, they are not soluble in nonpolar solvents). Most monosaccharides have a sweet taste (for reasons not totally understood). The most common monosaccharides are the hexoses. Table 18.1 shows the structures of some important monosaccharides.

18.2 Glucose

Glucose (also known as blood sugar, grape sugar, and dextrose) is the most common of the hexoses. It is an aldose that is found in the juices of

Table 18.1 Some Important Monosaccharides

PENTOSES			
Aldopentose			Ketopentose

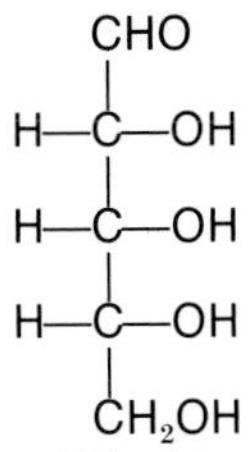

CHO	CHO	CHO	CH₂OH
H—C—OH	HO—C—H	H—C—OH	C=O
H—C—OH	H—C—OH	HO—C—H	H—C—OH
H—C—OH	H—C—OH	H—C—OH	H—C—OH
CH₂OH	CH₂OH	CH₂OH	CH₂OH
D-Ribose	D-Arabinose	D-Xylose	D-Ribulose

HEXOSES			
Aldohexose			Ketohexose

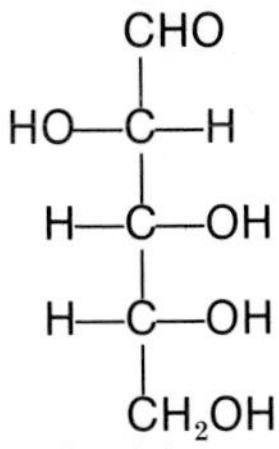

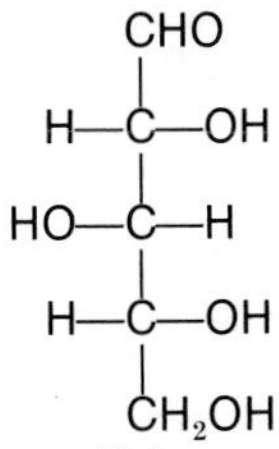

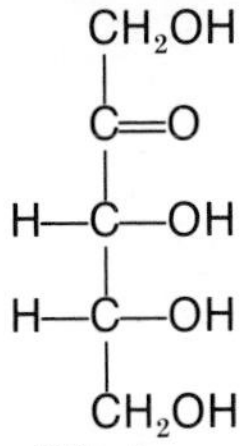

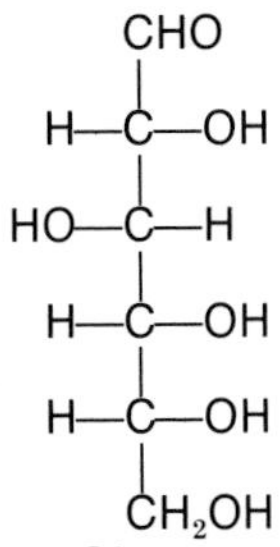

CHO	CHO	CHO	CH₂OH
H—C—OH	HO—C—H	H—C—OH	C=O
HO—C—H	HO—C—H	HO—C—H	HO—C—H
H—C—OH	H—C—OH	HO—C—H	H—C—OH
H—C—OH	H—C—OH	H—C—OH	H—C—OH
CH₂OH	CH₂OH	CH₂OH	CH₂OH
D-Glucose	D-Mannose	D-Galactose	D-Fructose

fruits (especially grape juice), in the saps of plants, and in the blood and tissues of animals. It is the immediate source of energy for energy-requiring cellular reactions such as tissue repair and synthesis, muscle contraction, and nerve transmission. The average adult has 5 to 6 grams of glucose in the blood (about 1 teaspoon). This much glucose will supply the energy needs of the body for only about 15 minutes, so one must continuously replace the glucose in the blood from compounds stored in the liver. The level of glucose in the blood of a normal adult is fairly constant, although it rises after each meal and falls during periods of fasting.

Glucose is a part of many polysaccharides, and can be produced by the hydrolysis of these polysaccharides (it is produced commercially by the hydrolysis of cornstarch). Because glucose is found in most living cells, its chemistry is an important part of the carbohydrate chemistry of the body.

Structure of Glucose
The structure of glucose can be drawn in a straight chain form (Table 18.1). This open chain structure, however, does not help to explain many of the properties of glucose, which actually is found in three forms in water solution. These three forms exist in equilibrium, and are easily converted one into another. The straight chain form of glucose makes up only 0.02% of these molecules. The two other forms are ring compounds that

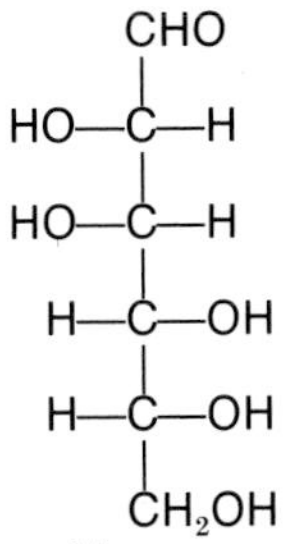

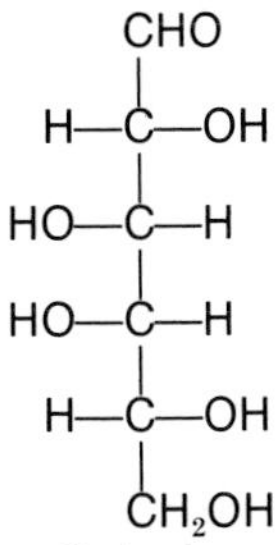

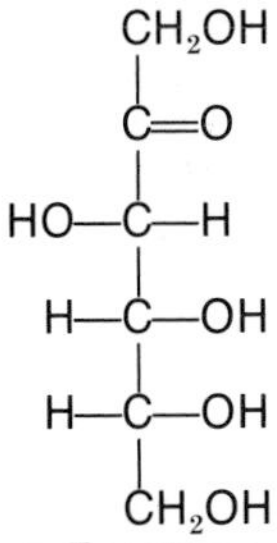

result from the formation of an internal hemiacetal (review Section 15.10). In glucose, the hemiacetal will form between the aldehyde group on carbon number 1 and the alcohol group on carbon number 5.

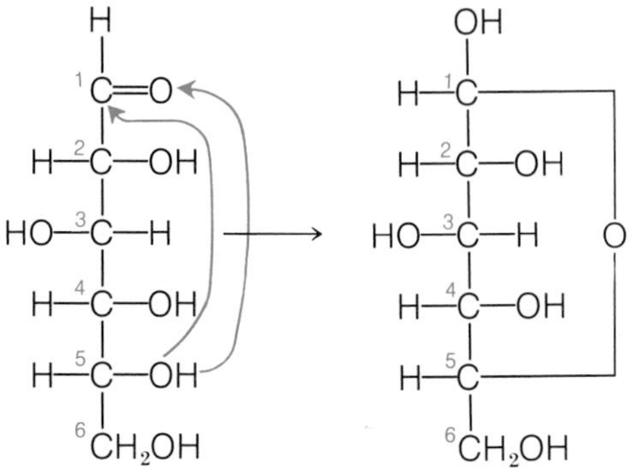

Haworth structural formulas are often used to simplify the drawing of these sugar ring structures. In these formulas the ring formed by the sugar molecule is shown as though we were looking at it from the side (rather than looking down from above the ring). The thickened side of the ring is the one closest to us, and the groups attached to the carbon atoms are then shown either above or below the ring.

Haworth formula

Shorthand forms

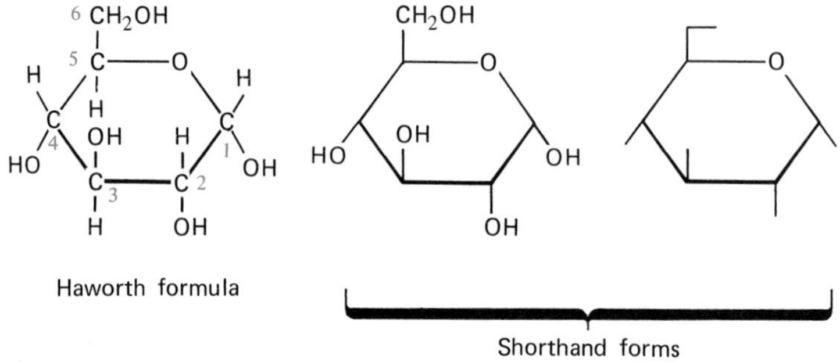

The two ring forms of glucose depend on the placement of the hydrogen and hydroxyl groups on carbon number 1. If the hydroxyl group is below the plane of the ring, this is the alpha (α) form; if it appears above the ring it is the beta (β) form (Figure 18.3).

α – Glucose

Open chain

β – Glucose

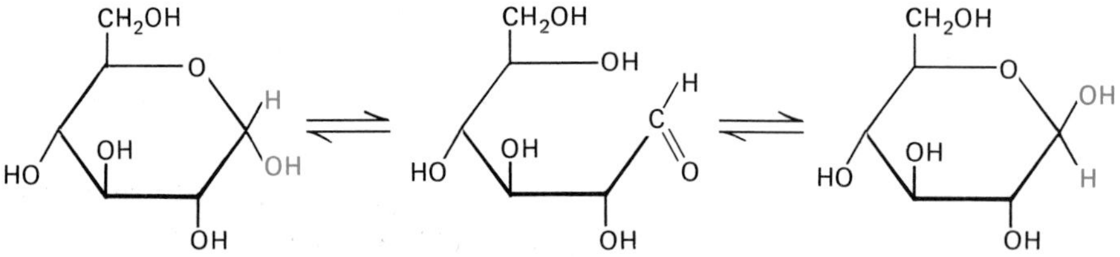

Figure 18.3 Three forms of glucose will exist in equilibrium in water solution: the alpha ring (36%), the open chain (0.02%), and the beta ring (64%).

You may wonder what difference the position of that one hydroxyl group on the ring could possibly make. As we study the metabolism of living organisms, we will see that such small differences can determine whether or not a cell will be able to utilize a molecule. In this case, the difference between starch (a digestible glucose polymer) and cellulose (an indigestible glucose polymer) is in the position of the hydroxyl group on carbon number 1 of the glucose molecule.

Although Haworth formulas are easy to draw, and we will use them throughout the rest of this book, they do not accurately represent the arrangement of groups attached to the ring. You might instead see the ring structures drawn in a "puckered" form, as shown for glucose.

α – Glucose β – Glucose

*18.3 Optical Isomerism

In our study of biochemistry we will find that the position of functional groups on the carbon atom is critical. As we discussed in Section 14.5, a carbon with four different groups attached to it is called a chiral carbon. Compounds containing a chiral carbon will be optically active, and will possess optical isomers. The classification system for optical isomers is based on the three-carbon compound glyceraldehyde.

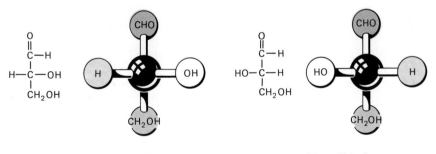

D–Glyceraldehyde L–Glyceraldehyde

The D-family of carbohydrates is represented as having the hydroxyl group located to the right of the chiral carbon farthest from the carbonyl group, and the L-family has the hydroxyl group to the left of the carbon. Most naturally occurring carbohydrates belong to the D-family (Figure 18.4). Of the 16 possible optical isomers in the aldohexoses (8 in the D-family and 8 in the L-family), the three most abundant are D-glucose, D-mannose, and D-galactose.

---

* *This section may be skipped without loss of continuity.*

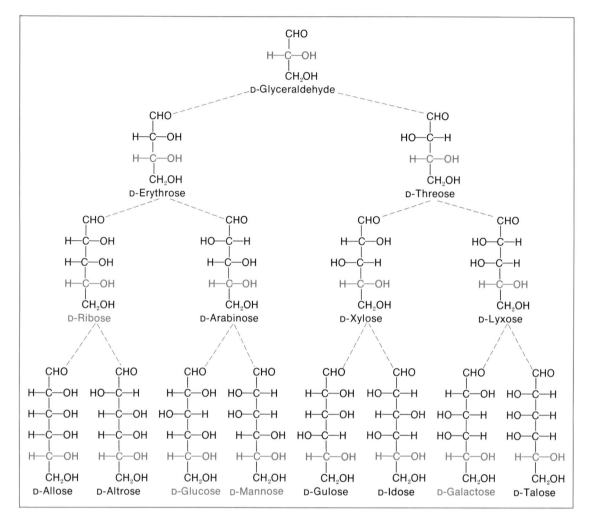

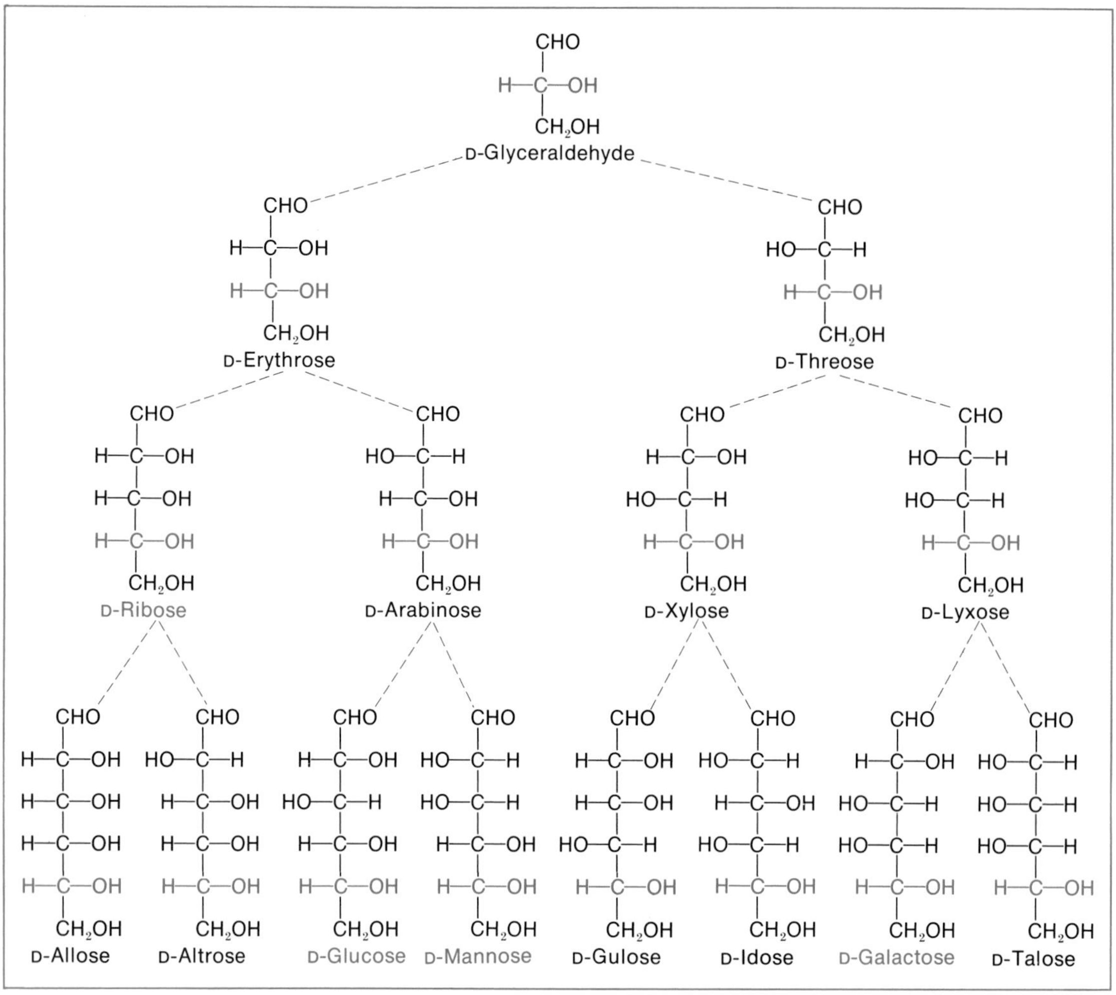

Figure 18.4 The D-family of aldoses.

Again, the small difference in arrangement of atoms between the D and L isomers may seem unimportant to you, but to your body and its cells it is critical. Cells can recognize this difference, and often can use only one of the isomers. For example, yeast can ferment D-glucose to produce alcohol, but not L-glucose. As we will see in Chapter 20, our cells can use only L-amino acids to build proteins. This is because the enzymes that catalyze reactions in cells are asymmetrical compounds themselves, just as your shoes are asymmetrical. To catalyze a reaction, the enzyme must fit the reactant—just as your right shoe can fit only your right foot.

18.4 Fructose

Fructose (also called levulose or fruit sugar) is a ketohexose found in many fruit juices and in honey (Table 18.1). It is the sweetest sugar known, much sweeter than table sugar. Fructose is a part of the disaccharide sucrose (or cane sugar), and is produced by hydrolysis of the polysaccharide inulin. Fructose molecules form internal hemiketals, and exist in ring structures such as the following.

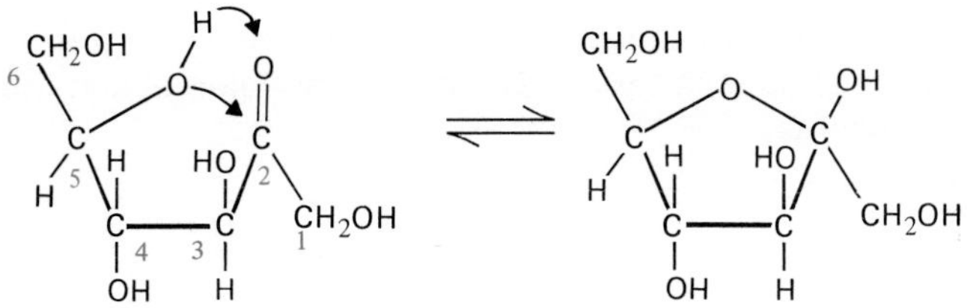

18.5 Galactose

Galactose is not found naturally as a free monosaccharide, but can be formed from the hydrolysis of larger carbohydrates. It is a part of lactose, the sugar found in milk; thus it is found in our diets from birth.

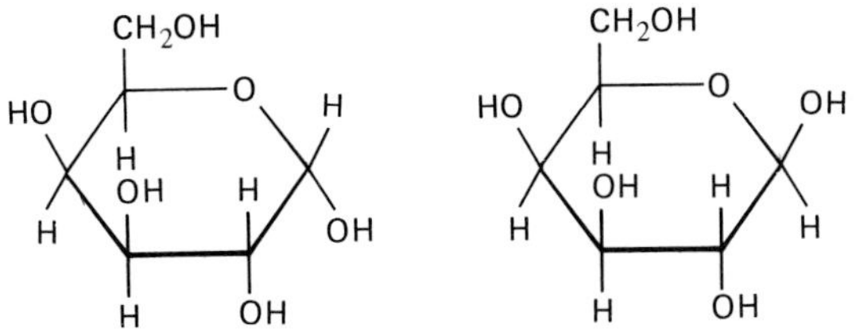

α – Galactose β – Galactose

Galactose is also found in glycolipids, fatlike substances that are components of the brain and nervous system. Agar-agar, a polysaccharide extracted from certain types of seaweed, is a polymer of galactose that humans cannot digest. Nevertheless, it is used as a thickener in sauces and ice creams, as well as in nutrient broths used in microbiology.

18.6 Sugar Acids

Sugar acids are derivatives of carbohydrates in which the aldehyde group (or a hydroxyl group of a monosaccharide) has been oxidized to form a carboxylic acid.

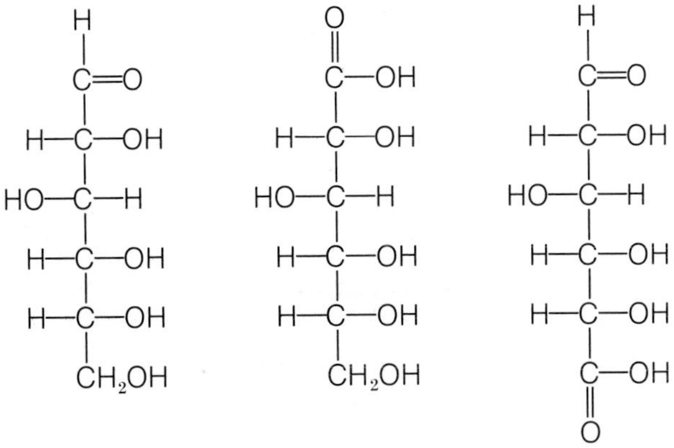

Glucose Gluconic acid Glucuronic acid

Gluconic acid is produced in the breakdown of glucose in some organisms. Glucuronic acid plays a special role in the body. By combining with unwanted toxic materials which have been accidentally swallowed, absorbed as medicine, or produced in normal metabolism, glucuronic acid makes these chemicals less toxic and more water soluble, and thus easier for the body to excrete.

18.7 Pentoses

Some important five-carbon sugars are arabinose, which is formed by hydrolysis of gum arabic and the gum of a cherry tree, and xylose, which is a component of wood, straw, corncobs, and bran. Pentoses that play a role in human metabolism are ribose and deoxyribose, which are parts of nucleic acids (the subject of Chapter 22). Both the alpha and beta forms of ribose exist in solution, but only the beta form is found in nucleic acids and in other metabolically active compounds.

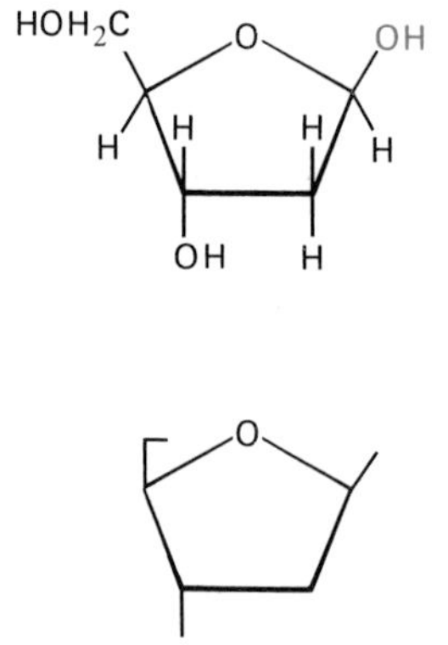

β—Deoxyribose

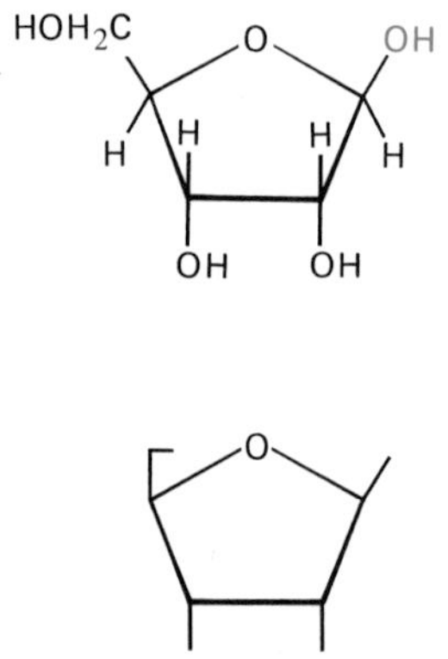

β—Ribose

18.8 Tests for Carbohydrates

Molisch Test

This test is a general test for the presence of a carbohydrate. First, an alcoholic solution of α-napthol is mixed with the unknown solution in a test tube. Then the tube is held at an angle, and cold, concentrated sulfuric acid is poured into the tube. If a red-violet ring forms where the two solutions meet, a carbohydrate is present.

Seliwanoff's Test

This test detects the presence of a polyhydric aldehyde or ketone. The unknown solution is mixed with hot hydrochloric acid and resorcinol. Ketoses will produce a bright red color, whereas aldoses will produce a pink color in the same period of time. This test is frequently used to detect fructose, but sucrose will also give a bright red color because it is hydrolyzed during the test.

Test for Reducing Sugars

Carbohydrates that contain a free, or potentially free, aldehyde or ketone group will reduce alkaline (basic) solutions of mild oxidizing agents such as Cu^{2+} or Ag^{+}. **Benedict's test** is widely used for the detection of reducing sugars. The reagent for this test contains an alkaline solution of copper sulfate, $CuSO_4$. This blue reagent solution is mixed with the unknown solution and is heated. If a reducing sugar is present, the Cu^{2+} ions will be reduced to Cu^{+} ions, and a brick-red precipitate of copper oxide (Cu_2O) will form. The general reaction can be written:

$$\underset{\substack{\text{Blue} \\ (CuSO_4)}}{Cu^{2+}} + \underset{\substack{\text{(Free aldehyde or} \\ \text{ketone)}}}{\text{Reducing sugar}} \xrightarrow[\text{Alkali}]{\text{Heat}} \underset{\text{Brick-red}}{Cu_2O} + \text{Oxidized sugar}$$

The amount of precipitate formed indicates the amount of reducing sugar present. All monosaccharides will give a positive Benedict's test. Clinitest tablets, which are widely used to test for sugar in the urine, are based on the same principle as Benedict's test. In this case a green color indicates very little sugar in the urine, whereas a brick-red color indicates more than 2 grams of reducing sugar per 100 ml of urine.

Note, however, that a false positive Benedict's test for glucose (and an erroneous diagnosis of diabetes mellitus) can occur with patients having a rare condition called pentosuria. These persons have the pentose sugar xylulose in their urine, but suffer no ill effects from it. To guard against this mistake, the Tollens' pentose test can be used to determine if the sugar found in the urine is glucose or xylulose.

Disaccharides

18.9 Maltose

Maltose (malt sugar) is a disaccharide made up of two glucose units. It is produced by the incomplete hydrolysis of starch, glycogen, or dextrins. Maltose that is produced from grains germinated under controlled conditions is called malt, and is used in the manufacture of beer.

Disaccharides are formed by a condensation reaction between two monosaccharides. This reaction involves the formation of an acetal from a hemiacetal and an alcohol.

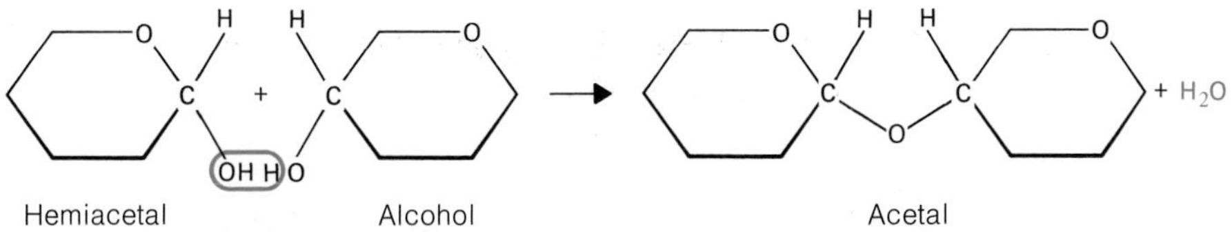

In this condensation reaction, one monosaccharide unit acts as the hemiacetal and the other as the alcohol. The linkage that is formed is called a glycoside linkage (or acetal linkage), and is more stable than the hemiacetal. This linkage will not react with bases; only acids or specific enzymes are able to break the bond.

The glycoside linkage in maltose occurs between carbon 1 of a glucose molecule in the alpha form and carbon 4 on the other glucose. Such a bond is called an α(1-4) linkage. (*Note:* A wavy line connecting the OH on the second glucose is used to indicate that this molecule can be in either the alpha or beta form.)

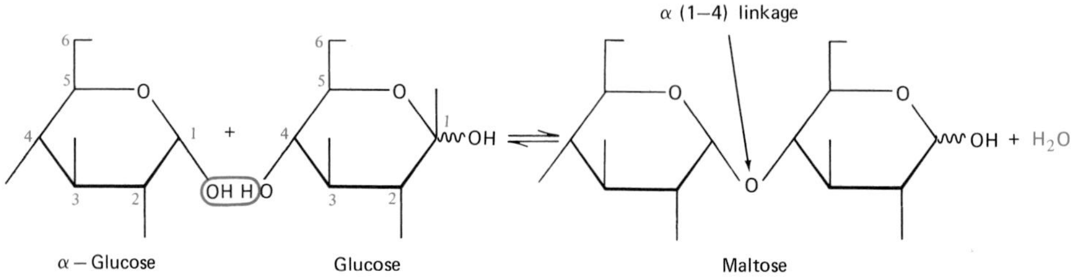

Because the aldehyde group of the second glucose molecule is not involved in the glycoside linkage, maltose can exist in either an alpha or beta form, and is a reducing sugar.

18.10 Lactose

Lactose (milk sugar) is found in the milk of mammals; it is synthesized by the mammary glands from glucose in the blood. Four to 5% of a cow's milk is lactose, whereas human milk contains 6–8%. Lactose itself is a colorless powder that is nearly tasteless. It can, therefore, be used in large amounts in special high calorie diets.

Lactose is formed by a condensation reaction between glucose and galactose. The bond is formed between carbon 1 of galactose in the beta form and carbon 4 of glucose.

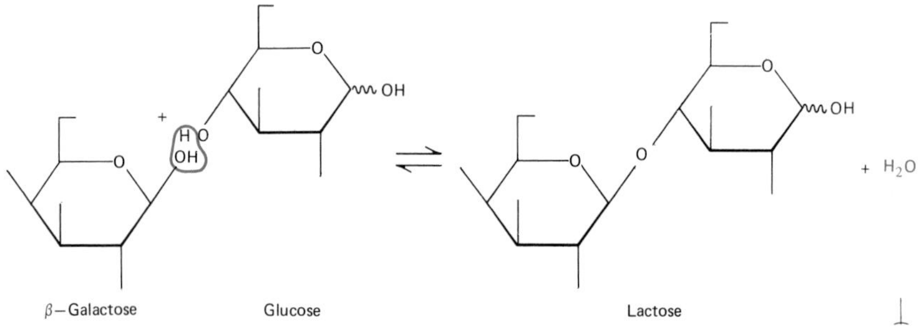

As is the case with maltose, lactose has a potentially free aldehyde group in the glucose unit, and is a reducing sugar.

18.11 Sucrose

Sucrose (also called table sugar, cane sugar, and beet sugar) is found in the juices of fruits and vegetables, and in honey. It is produced commercially from sugar cane or sugar beets, and is the sugar that we use in cooking. We each consume an average of 100 pounds of sucrose a year.

Sucrose is made up of one unit of glucose and one unit of fructose.

The linkage occurs between the aldehyde group of glucose and the ketone group of fructose.

α – Glucose β – Fructose Sucrose

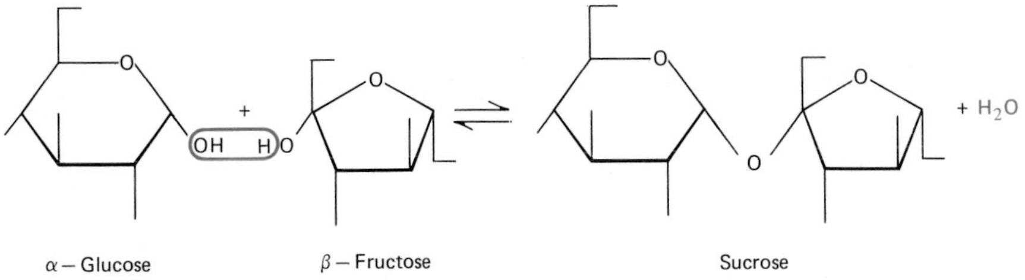

Because both the aldehyde group of glucose and the ketone group of fructose are involved in the linkage, sucrose does not have a potentially free aldehyde or ketone group, and is not a reducing sugar. Sucrose can be hydrolyzed by acids or enzymes found in the intestines and in yeast. Such a hydrolysis of sucrose produces a mixture of fructose and glucose called invert sugar.

Polysaccharides

18.12 Starch

Polysaccharides are polymers containing three or more monosaccharide units. They are used both as a means of storing energy and as part of the structural tissues of the organism. Starch is the storage form of glucose used by plants. It is found in granules in their leaves, roots, and seeds. These granules are insoluble in water; their coating must be broken open for the starch to mix with water. Heat will break open the granules, producing a colloidal suspension whose thickness increases with heating. For this reason, cornstarch is widely used as a thickening agent in cooking.

Natural starches are a mixture of two types of polysaccharides: amylose and amylopectin. Amylose is a linear polysaccharide (molecular weight of about 50,000) whose glucose units are connected by $\alpha(1\text{-}4)$ linkages.

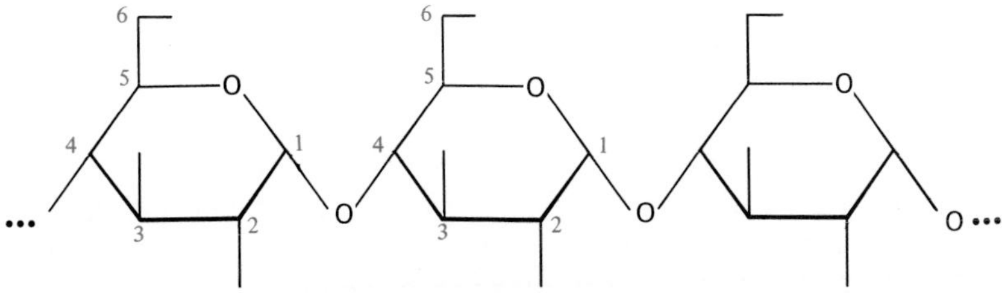

Amylose

Amylopectin (molecular weight of about 300,000) is a highly branched glucose polymer. The nonbranching portion of the molecule consists of glucose units connected by $\alpha(1\text{-}4)$ linkages. The branching occurs every

Figure 18.5 The structure of amylopectin and glycogen. The glucose units are connected by α(1-4) linkages and by α(1-6) linkages.

20 to 24 glucose units, and is a result of α(1-6) linkages between the glucose units (Figure 18.5).

18.13 Dextrins

Dextrins are polysaccharides formed through the partial hydrolysis of starch by acids, enzymes, or dry heat. The golden color of bread crust results from the formation of dextrins. Dextrins get sticky when wet and, therefore, are used as adhesives on stamps and envelopes.

18.14 Glycogen

Glycogen is a heavily branched molecule that is the storage form of glucose in animals (Figure 18.6). It accounts for about 5% of the weight of the liver, and 0.5% of the weight of the muscle in the body. There is enough glucose stored in the form of glycogen in a well-nourished body to supply it with energy for about 18 hours.

The structure of a glycogen molecule is similar to that of amylopectin (Figure 18.5). It consists of straight chains of glucose units connected by α(1-4) linkages. The branching that results from α(1-6) linkages between glucose units in a glycogen molecule is more frequent than in amylopectin, occurring every 8 to 12 glucose units.

The ability of the body to form glycogen from glucose (and glucose from glycogen) is extremely important because glucose is the main source of energy for all cells. When we eat a meal, glucose resulting from the breakdown of carbohydrates enters the bloodstream. If this large amount of glucose were to remain in the blood, the osmotic balance between the blood and the extracellular and intracellular fluids would be completely disrupted. This excess glucose, however, does not circulate in the blood, but instead is converted to glycogen in the liver. The large, branched glycogen molecule is ideally suited for storage because it cannot pass through cell membranes. Later, as glucose is used by the cells, the blood glucose is maintained at its normal level by the breakdown

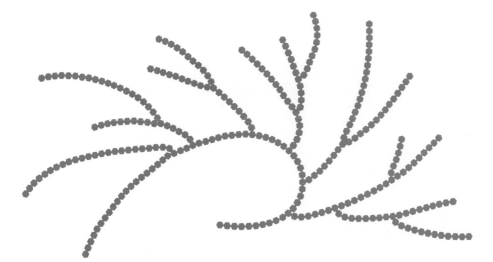

Figure 18.6 The highly branched nature of the amylopectin and glycogen molecules is a result of the α(1-6) linkages that occur about every 20 glucose units. This diagram pictures a larger portion of the molecule than that shown in Figure 18.5. Each ● represents a glucose molecule.

of glycogen in the liver and the resulting release of glucose into the blood. In this way, the blood glucose level is kept relatively constant even though we eat at widely spaced time intervals during the day.

18.15 Cellulose

Cellulose is a glucose polymer (molecular weight from 150,000 to 1,000,000) produced by plants. It makes up the main structural support for plants, whose cells release this compound to form the exterior cell wall. Molecules of cellulose are insoluble in water because of their size and structure. The strength and rigidity that cellulose gives to plants result from hydrogen bonding between the cellulose molecules (Figure 18.7).

The glucose units in cellulose are held together by a glycoside linkage between carbon 1 in the beta position on the first glucose and carbon 4 on the second glucose. In order to indicate that a cellulose molecule is linear, every second glucose unit in the structural diagram is flipped over.

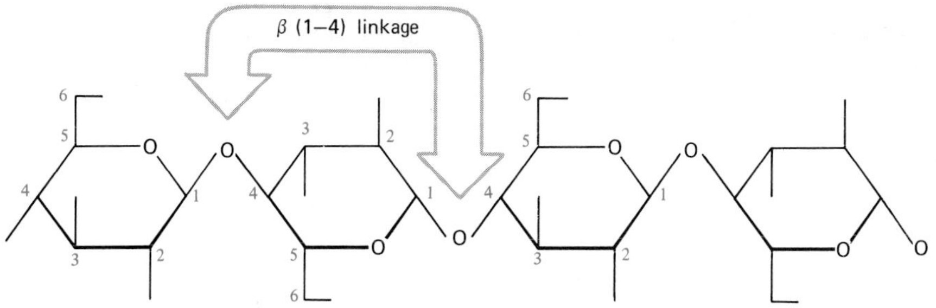

Cellulose

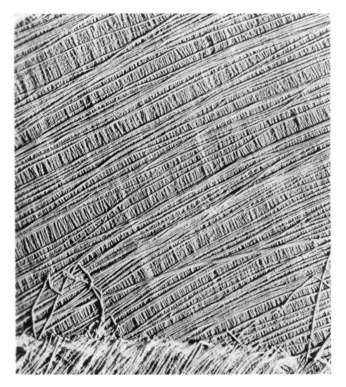

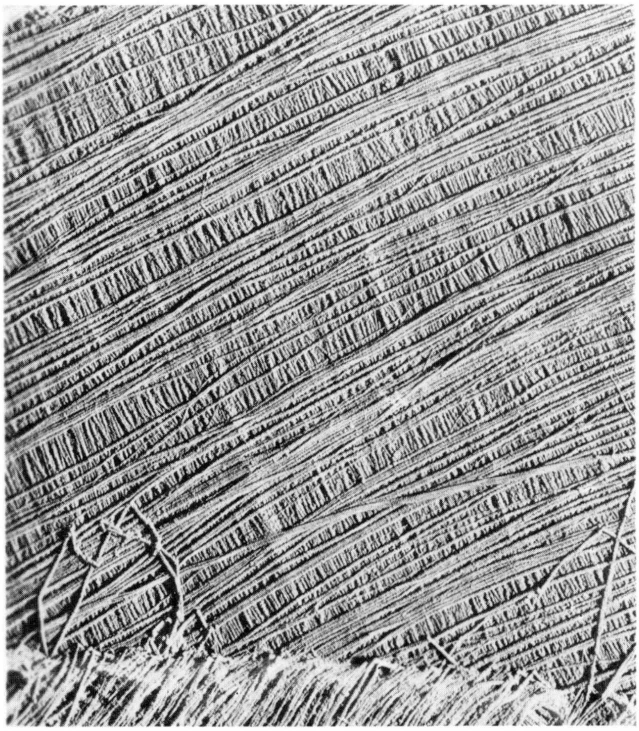

Figure 18.7 This electron micrograph of the cell wall of an alga shows the parallel arrangement of cellulose fibers making up the wall. (Courtesy R. D. Preston)

This linkage in cellulose is called a β(1-4) linkage, and human bodies do not possess the enzymes to break this bond. Therefore, any cellulose we eat passes through the digestive tract undigested, supplying the roughage we need for proper elimination. Some microorganisms, however, can digest cellulose. Grass-feeding animals such as cows have extra stomachs to hold the grass for long periods while these microorganisms break down the cellulose into glucose.

Over 50% of the total organic matter in the living world is cellulose. For example, wood is about 50% cellulose, and cotton is almost pure cellulose. When treated with a wide variety of chemicals, cellulose forms many useful products: celluloid; rayon; guncotton (an explosive); cellulose acetate (used in plastics, food wrapping films, and fingernail polish); methyl cellulose (used in fabric sizing, pastes, and cosmetics); and ethyl cellulose (used in plastic coatings and films).

18.16 Dextran

Dextran (molecular weight over 1,000,000) is a glucose polymer produced by bacteria. Partially hydrolyzed dextrans (molecular weight about 70,000) are used as blood plasma substitutes in the treatment of shock from low blood plasma volume. The dextrans used in this way are gradually eliminated through the urine.

In Chapter 17 we discussed the formation of plaque by bacteria in the mouth. Such plaque contains dextrans that act as glue, holding the plaque against the teeth. If the plaque is not removed, calcium compounds slowly deposit in the plaque, turning it into the hard material called tartar.

18.17 Iodine Test

The iodine test is used to detect small amounts of starch in solution. Starch will produce a dark blue-black color when mixed with the iodine test reagent, a solution of potassium iodide containing iodine. This test can be used to monitor the hydrolysis of starch—the color will slowly change as the starch continues to be broken down into shorter carbon chain products.

Hydrolysis:	Starch	$\xrightarrow{\text{Hydrolysis}}$	Dextrins	$\xrightarrow{\text{Hydrolysis}}$	Maltose
Iodine Test:	*Blue-black*	$\longrightarrow$	*Reddish*	$\longrightarrow$	*Colorless*

18.18 Photosynthesis

Where does the energy stored in these polysaccharides come from? Originally, from the sun. Nuclear reactions occurring on the sun produce energy that radiates out into space. This radiant energy is trapped by plants growing on the earth, and is used to produce carbohydrates and certain amino acids. **Photosynthesis,** the process by which plants capture and use this energy, is quite complex and not totally understood. We do know that the reactions of photosynthesis require the presence of light and molecules of chlorophyll. The overall equation, summarizing the many reactions of this process, shows the eventual formation of glucose from carbon dioxide and water:

$$6CO_2 + 6H_2O \xrightarrow[\text{Chlorophyll}]{\text{Light}} C_6H_{12}O_6 + 6O_2$$

The series of reactions producing these results can be divided into two categories: the light reactions and the dark reactions. The light reactions require the presence of chlorophyll to absorb radiant light energy and to use this energy in the production of oxygen and energy-rich molecules. The dark reactions then use these energy-rich molecules to reduce carbon dioxide to glucose and other organic products.

The reactions of photosynthesis take place in regions in the plant cells called chloroplasts. These chloroplasts contain chlorophyll and all the other enzymes necessary for the light reactions (Figure 18.8).

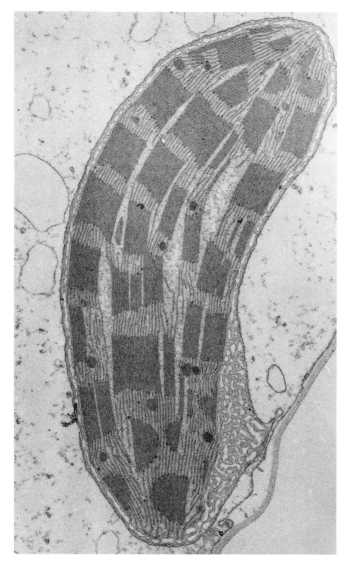

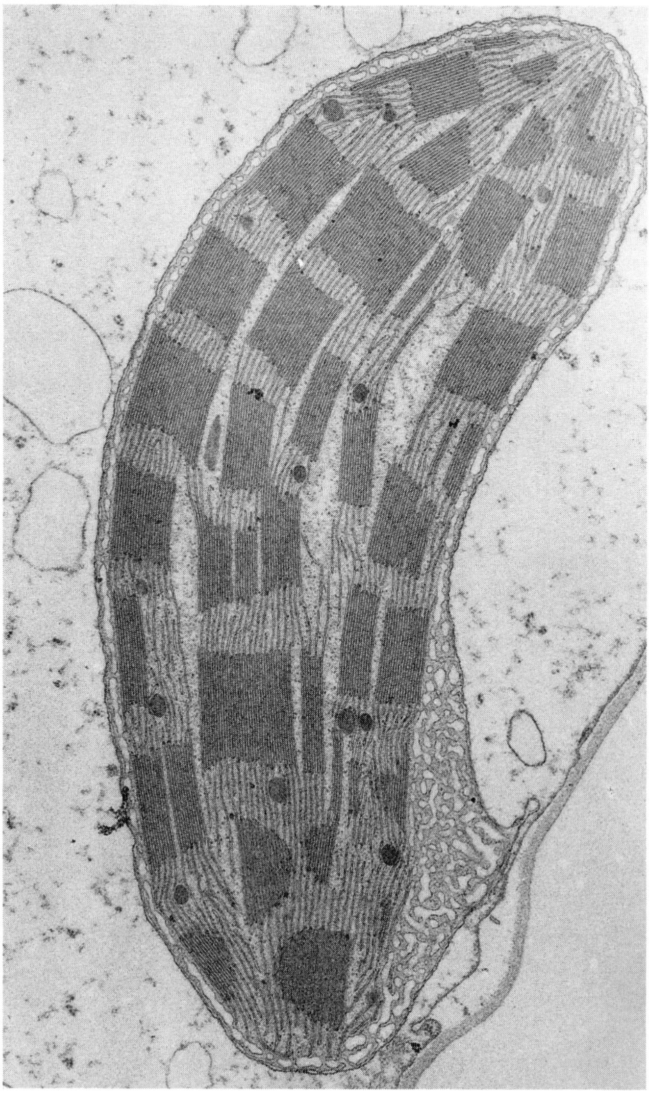

Figure 18.8 An electron micrograph of a chloroplast, the site of the light reactions of photosynthesis. The dense stacks in the chloroplasts are the grana, the main location of the chlorophyll molecules. (Courtesy T. Elliott Weier)

Metabolism of Carbohydrates

18.19 What Is Metabolism?

A superbly trained athlete waits at the starting blocks for the gun, sprints 400 meters at top speed, breaks the tape in record time, and then collapses on the track infield (Figure 18.9). His legs feel like rubber, he painfully gasps for breath, and he feels nauseated. The athlete will require

Figure 18.9 The tremendous exertion required to beat a world record has taken its toll. (Ken Regan/Camera 5)

a rest of $1\frac{1}{2}$ hours to completely recover from the exhaustion of this extreme effort. But where did the energy necessary for such an effort come from in the first place? Exactly how is it produced in the cells of our muscles? Also, what causes the resulting muscle fatigue? And why does it take so long for the muscle cells to recover after strenuous exercise? These are some of the questions we will discuss in this section on carbohydrate metabolism.

Metabolism can be defined as all of the enzyme-catalyzed reactions in the body. These reactions are of two types: **catabolic reactions,** in which molecules are broken down to produce smaller products and cellular energy, and **anabolic reactions,** in which the cell uses energy to produce molecules it needs for growth and repair (these are also called biosynthetic reactions). Anabolic and catabolic reactions occur continuously in the cell, and each involves complicated series of carefully controlled enzyme-catalyzed reactions (Table 18.2).

Table 18.2 Metabolism

METABOLISM	
Catabolic Reactions	Anabolic Reactions
Reactions that break down larger molecules.	Reactions in which larger molecules are produced from smaller components—biosynthetic reactions.
Reactions that produce the energy needed by the cell.	Reactions that require energy to occur.

18.20 Digestion and Absorption of Carbohydrates

Plants are the major source of carbohydrate in our diet, and carbohydrates supply the immediate energy needs of our cells. But exactly how does the starch in, say, a potato chip, get to our muscle cells? **Digestion,** the process by which complex foods are broken down into simple molecules, must take place before this food can be absorbed and used by the body.

The digestion of carbohydrates begins in the mouth, where teeth break large pieces of food into smaller ones. This food is then mixed with saliva, which contains an enzyme that begins the breakdown of starch to maltose. When the food is swallowed it passes into the stomach, where these salivary enzymes are inactivated by the low pH. The stomach further reduces the size of the food particles. As the food then passes into the small intestine, the pancreas and intestinal wall secrete juices which contain enzymes that completely hydrolyze the polysaccharides and disaccharides in the food (Table 18.3). Galactose, glucose, and fructose—the monosaccharides formed by this process—are taken up directly through the intestinal wall into the bloodstream and are carried to the liver. The galactose is converted to glucose by liver enzymes, whereas the fructose may either be converted to glucose or may enter other metabolic reactions. We have seen that glucose may be stored in the liver and muscles as glycogen. The liver normally contains about 100 g of glycogen (about $\frac{1}{2}$ cup), but it may store as much as 400 g.

Table 18.3 Digestion of Carbohydrates

Site of Digestion	Food Digested	End Products	Enzyme Source
Mouth	Starch	Dextrins, maltose	Saliva
Small intestines	Starch	Maltose	Pancreatic juice
	Lactose	Glucose and galactose	Pancreatic juice and intestinal secretions
	Maltose	Glucose	Pancreatic juice and intestinal secretions
	Sucrose	Glucose and fructose	Pancreatic juice and intestinal secretions

18.21 Cellular Energetics

We must now interrupt this potato chip's journey to your muscle cells for a brief discussion of cellular energetics and the compounds that play a part in the anabolic and catabolic reactions in the cell.

The body has many energy needs, one of which is the need for heat to maintain body temperature. This heat is produced by the controlled combustion of compounds such as glucose in the cell. The cell, however,

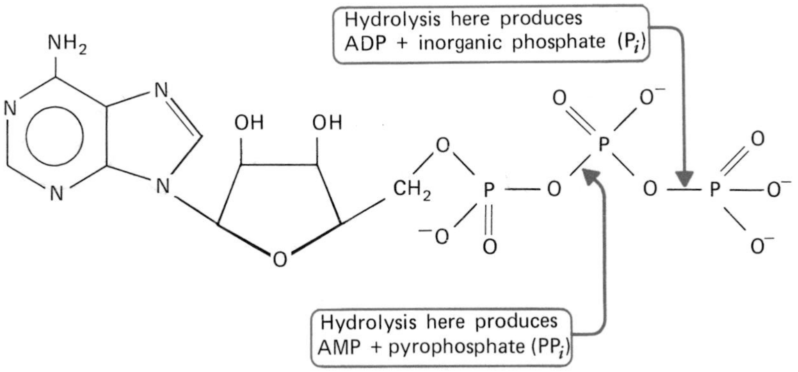

Figure 18.10 The structure of adenosine triphosphate (ATP).

does not operate like a steam engine—converting chemical energy to heat energy and then using the heat energy to do work. Instead, the cell operates much more efficiently, using chemical energy directly to do work such as muscle contraction, transmission of nerve impulses, and biosynthesis. For these tasks the cell requires *high-energy* or *energy-rich* compounds such as adenosine triphosphate, ATP (Figure 18.10). A molecule of ATP (and other molecules similar to ATP) contains two oxygen-to-phosphate bonds that are called *high-energy* phosphate bonds, often represented by a wavy line.

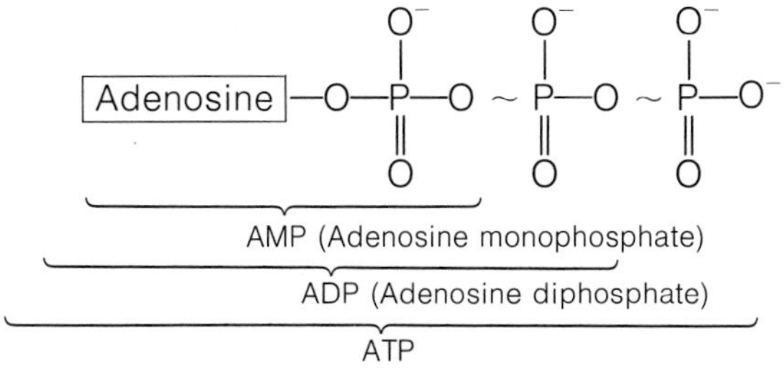

Such phosphate bonds are called high-energy bonds because the hydrolysis of compounds such as ATP is a highly exothermic reaction, and produces about twice as much energy as hydrolysis of compounds containing low-energy phosphate bonds. In energy-releasing reactions in the cell, ATP is hydrolyzed to form ADP and an inorganic phosphate (indicated by P_i).

$$\text{ATP} + H_2O \longrightarrow \text{ADP} + P_i + \text{Energy}$$

If the hydrolysis of ATP is coupled with an endothermic reaction in the cell, such as the contraction of a muscle fiber or the synthesis of a large molecule, the hydrolysis of ATP will supply the energy necessary for the other reaction to occur.

$$\text{Relaxed muscle} + \text{ATP} + H_2O \longrightarrow \text{Contracted muscle} + \text{ADP} + P_i$$

Molecules other than ATP are also needed in the breakdown and synthesis processes of the cell. These molecules belong to a class of

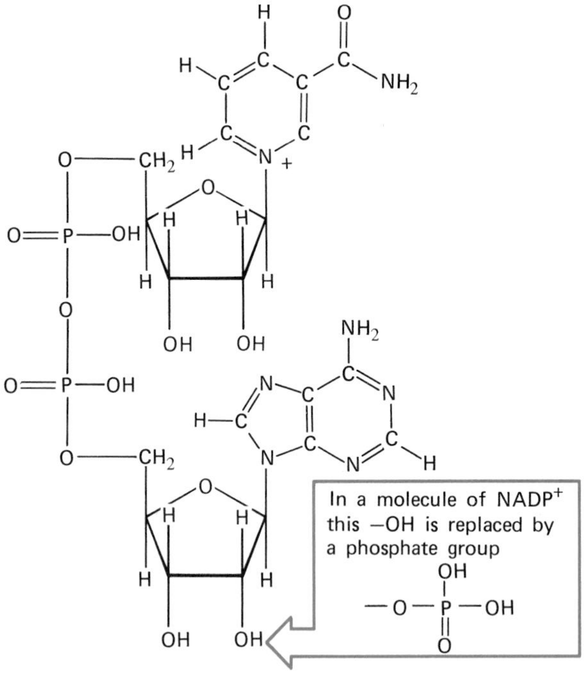

Figure 18.11 The structure of the two coenzymes NAD^+ and $NADP^+$. The difference between the structures of the two molecules is that a hydroxyl group on NAD^+ is replaced by a phosphate group on $NADP^+$. Note that both molecules carry a positive charge.

compounds called coenzymes, which will be discussed in detail in Chapter 21. There are three coenzymes, however, that we will mention here. Two are derivatives of the vitamin niacin, or nicotinic acid, and are called NAD (nicotinamide adenine dinucleotide) and NADP (nicotinamide adenine dinucleotide phosphate) (Figure 18.11). The other coenzyme, a derivative of riboflavin, or vitamin B_2, is FAD (flavin adenine dinucleotide). These three molecules serve as important hydrogen carriers in metabolic reactions. We can indicate how these coenzymes act as hydrogen carriers by writing the following equation.

$$NAD^+ + XH_2 \xrightarrow{\text{Enzyme}_1} NAD{-}H + H^+ + X$$

In this equation NAD^+ represents the positively charged NAD molecule (see Figure 18.11), and XH_2 is a general way of representing some hydrogen-containing reactant molecule. The NADH produced by this reaction can then serve as a hydrogen donor in a reaction requiring hydrogen. In this way, one molecule of NAD^+ can be used over and over again in the cell.

$$NADH + H^+ + Y \xrightarrow{\text{Enzyme}_2} NAD^+ + YH_2$$

18.22 Glycogenesis and Glycogenolysis

When we last saw our potato chip, it had been broken down through a series of steps to form the sugar glucose. If this glucose is part of the excess blood sugar produced right after a meal, we have seen that it will be stored in the liver in the form of glycogen. The process of converting glucose to glycogen is called **glycogenesis.** This process involves the series of reactions shown in Figure 18.12, requiring several enzymes and two high-energy molecules, ATP and UTP (uridine triphosphate).

As the cells of the body use up the glucose in the blood, a process called **glycogenolysis** occurs in the liver. This involves a series of reactions in which glycogen is broken down to form glucose, which can then be released into the bloodstream to maintain the blood sugar level. Notice that glycogenolysis is not the exact reverse of glycogenesis—they have only one step in common (Figure 18.12).

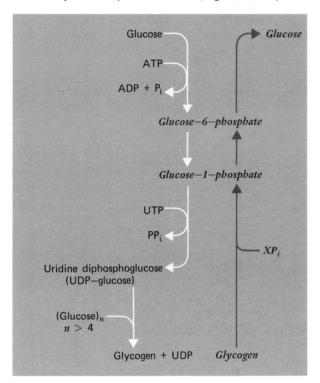

Figure 18.12 The reactions of glycogenesis (the synthesis of glycogen from glucose) and glycogenolysis (the breakdown of glycogen to glucose). Note that these processes have only one reaction in common. (P_i = phosphate and PP_i = pyrophosphate).

18.23 Oxidation of Carbohydrates

Rather than being converted to glycogen in the liver, the glucose from our potato chip might diffuse from the bloodstream into a cell of the body.

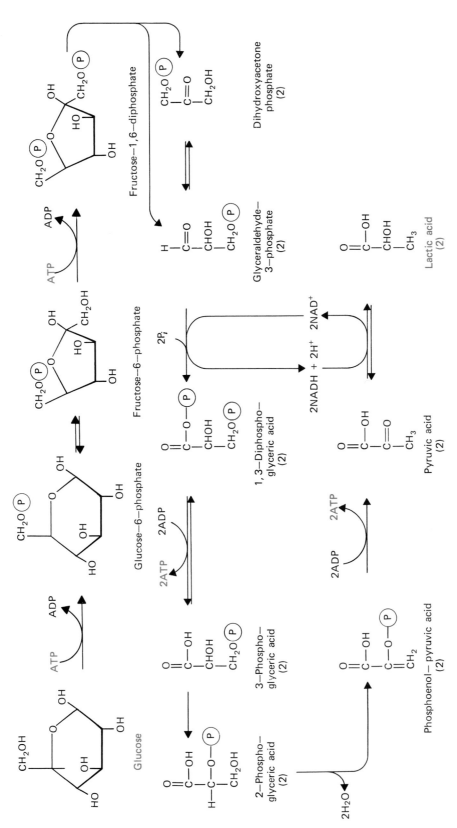

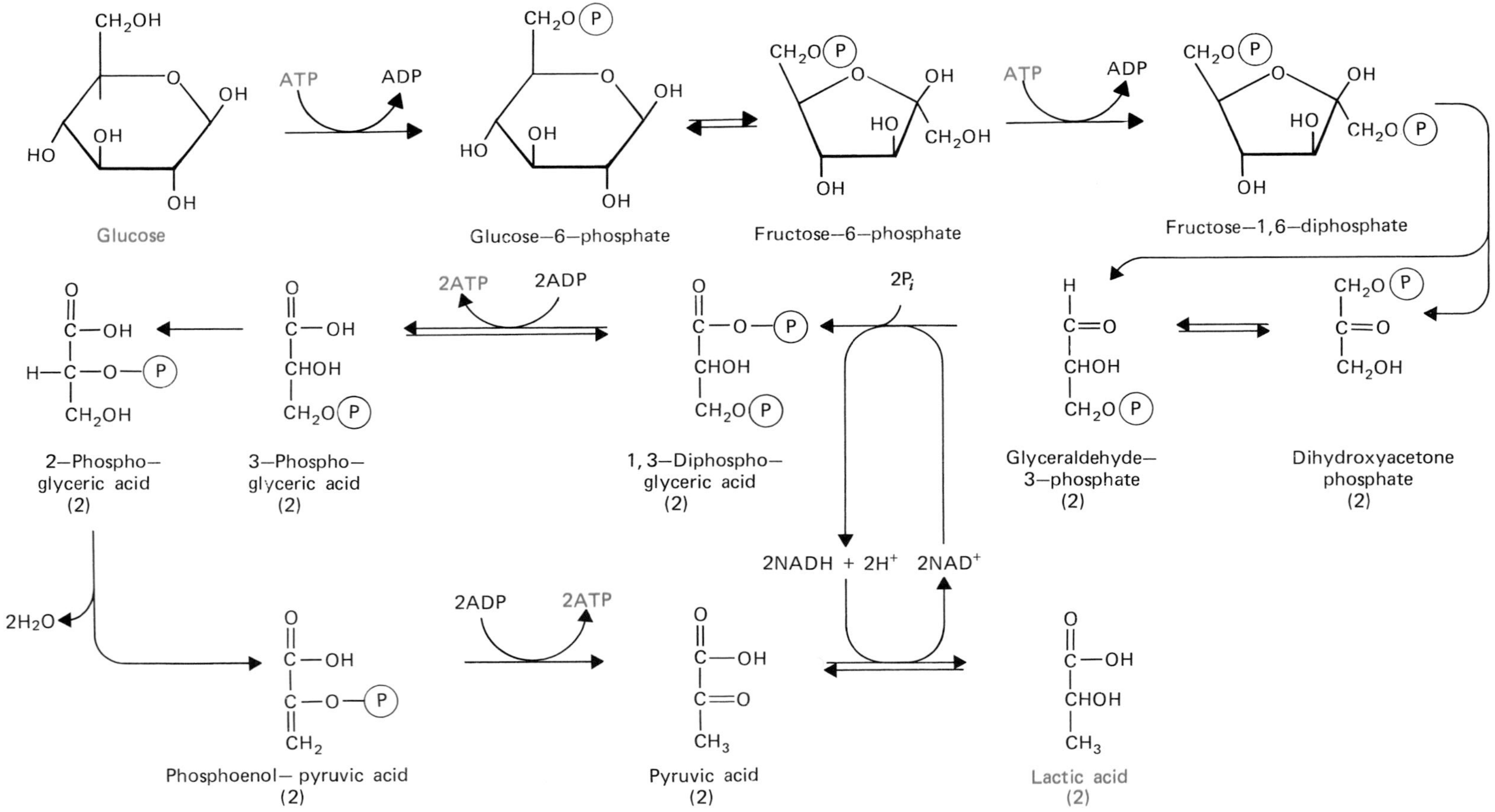

Figure 18.13 The reactions of glycolysis, the anaerobic stage of the oxidation of glucose, in which glucose is converted to lactic acid. Although enzymes are required for each reaction, we have not listed them in this diagram. (Ⓟ = phosphate group, $-PO_3H_2$)

There it can be used in biosynthetic reactions, or it can be broken down to yield useful cellular energy through a process called **cellular respiration.** In cellular respiration, glucose is oxidized to form carbon dioxide and water, and ATP is formed. The cellular processes of respiration are quite efficient; about 44% of the energy released from the oxidation of glucose is trapped in ATP molecules and can be used to do work. For each molecule of glucose that is oxidized, 36 ATP's are produced. But then, what happens to the other 66% of the energy that is released? It is given off as heat to maintain body temperature.

The reactions of cellular respiration are self-regulating, so they will occur only when the cellular supply of ATP is low. Cellular respiration is a very complex process, involving many steps and requiring many enzymes and coenzymes. A specific description of each step must be left to other biochemistry courses, but we can give a general overview of this process.

Anaerobic Stage: Glycolysis

The cellular oxidation of glucose can be divided into two stages. The first, an anaerobic stage, requires no oxygen and occurs in the cytoplasm of the cell. This stage is called **glycolysis,** and involves the breakdown of glucose to form two molecules of lactic acid (Figure 18.13). This process produces two ATP's, and can be summarized by the following overall reaction.

$$\text{Glucose} + 2\text{ADP} + 2\text{P}_i \longrightarrow 2\ \text{Lactic acid} + 2\text{ATP} + 2\text{H}_2\text{O}$$

Aerobic Stage: Citric Acid Cycle and Electron Transport Chain

The second part of glucose oxidation, the aerobic stage, requires oxygen. It is the stage where most of the energy is released in the breakdown of glucose. This second stage also serves as the final stage of oxidation for other compounds used by the cell to supply energy. It involves two series of reactions. The first series is called the **citric acid cycle** [or Krebs cycle, or tricarboxylic acid (TCA) cycle], and involves the final breakdown of the fuel molecule to carbon dioxide. This breakdown consists of a series of oxidation reactions that produce hydrogen atoms. These hydrogen atoms are then used in the second series of reactions, called the **electron transport chain.** This sequence of reactions requires oxygen, and produces ATP and water.

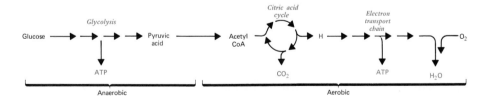

The citric acid cycle and the electron transport chain take place in structures in the cell called the mitochondria. Often also called the "power plants" of the cell, the mitochondria are located near the parts of the cell that require energy, such as the contractile filaments of muscle cells (Figure 18.14).

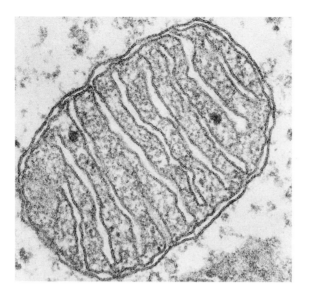

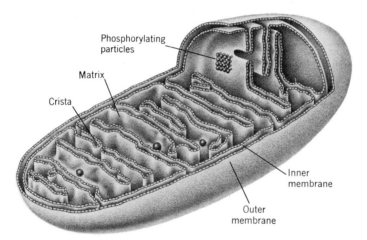

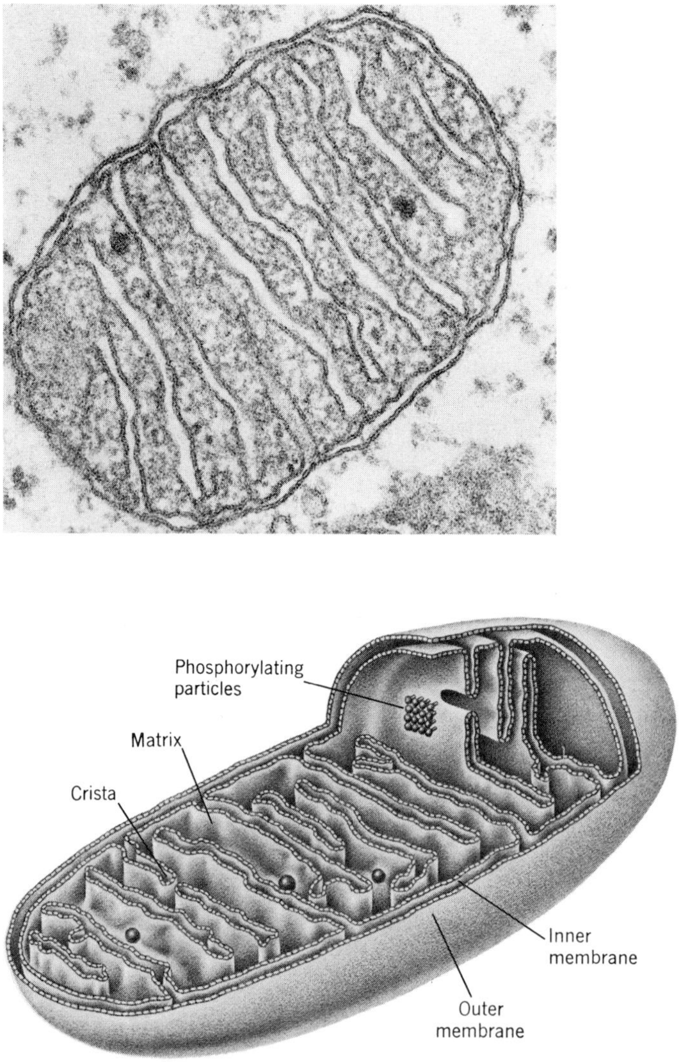

Figure 18.14 A mitochondrion. (a) An electron micrograph of a mitochondrion. (b) A model of a mitochondrion. The inner folds of the mitochondrion, called the cristae, contain the enzymes of the citric acid cycle and the electron transport chain.

18.24 Citric Acid Cycle

If the cell contains enough oxygen, the last step of glycolysis (the conversion of pyruvic acid to lactic acid) does not take place (Figure 18.13). Instead, pyruvic acid is oxidized to acetic acid in the form of acetyl coenzyme A—often referred to as acetyl CoA.

$$CH_3\overset{\overset{\displaystyle O}{\|}}{C}—S—CoA \qquad \text{Acetal CoA}$$

This reaction requires the hydrogen carrier molecule NAD⁺ in addition to coenzyme A, and produces NADH + H⁺ together with one molecule of CO_2.

$$\text{Pyruvic acid} + \text{NAD}^+ + \text{Coenzyme A} \xrightarrow[\text{and coenzymes}]{\text{Other enzymes}}$$

$$\text{Acetyl CoA} + \text{NADH} + \text{H}^+ + \text{CO}_2$$

The acetyl CoA formed in this reaction can enter the citric acid cycle. This series of reactions, shown in Figure 18.15, produces two

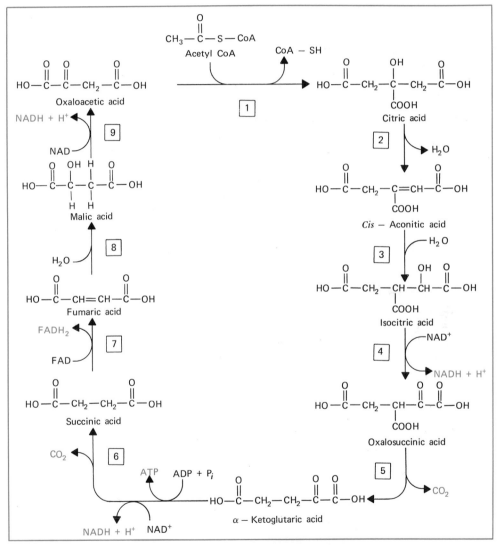

Figure 18.15 The citric acid cycle. Two carbon atoms enter the cycle in the form of acetyl CoA, and are joined to oxaloacetic acid to form citric acid (Step 1). During the series of reactions that then occur, two molecules of carbon dioxide are formed (Steps 5 and 6), and four pairs of hydrogens are produced and carried to the electron transport chain (Steps 4, 6, 7, and 9). At the end of the cycle a molecule of oxaloacetic acid is formed, ready for another turn of the cycle (Step 9).

molecules of carbon dioxide and four pairs of hydrogen atoms. Three pairs of these hydrogen atoms will be carried by NAD^+, and one by FAD. The cell, however, does not have a limitless supply of NAD^+ and FAD, so the citric acid cycle cannot occur independently of the electron transport chain. It is important that you notice the cyclic (or circular) nature of the reactions in the citric acid cycle. The first reaction involves joining acetyl CoA with oxaloacetic acid to form citric acid, and the last reaction again produces the oxaloacetic acid which can then combine with another acetyl CoA.

18.25 The Electron Transport Chain and ATP Formation

The final hydrogen acceptor, or carrier, in aerobic oxidation is the oxygen we breathe. To arrive at this point, the hydrogens carried by NAD^+ and FAD from the citric acid cycle enter a series of reactions called the **electron transport chain.** In this series of reactions, the hydrogens (and later just their electrons) are passed between a series of compounds until they are combined with oxygen to form water. This process produces most of the ATP molecules formed in the oxidation of glucose by the cell. Each transfer is part of an oxidation-reduction reaction in which some energy is released. As the electrons flow through this series of reactions they can do work, just as the electrons flowing through a copper wire can do work. The kind of work done by the electrons in the cell, however, is the production of ATP from ADP and inorganic phosphate. The exact way in which this process works is not well understood, but it is known at what points in the transport chain ATP is produced. From Figure 18.16 we see that each molecule of NADH produces three ATP's, and each molecule of $FADH_2$ produces two ATP's.

Using Figures 18.15 and 18.16 we can summarize the number of ATP's produced by one molecule of acetyl CoA as it passes through the citric acid cycle as follows.

	Citric Acid Cycle	Electron Transport Chain
	Step 4: 1 NADH	3 ATP
	Step 6: 1 NADH	3 ATP
	Step 7: 1 $FADH_2$	2 ATP
	Step 9: 1 NADH	3 ATP
		11 ATP
	1 ATP produced from ADP and P_i in Step 6	1 ATP
	Total	12 ATP

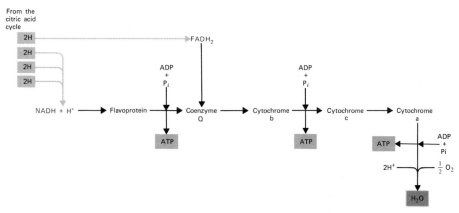

Figure 18.16 The electron transport chain. The hydrogens and their electrons produced in the citric acid cycle are passed between a series of compounds until they are combined with oxygen to form water. It is this series of oxidation-reduction reactions that produces most of the ATP molecules formed in the oxidation of glucose.

18.26 Lactic Acid Cycle

When undergoing only moderate exercise, muscle cells have enough oxygen to carry out the respiration process aerobically. But during strenuous exercise the blood cannot supply oxygen to the muscles fast enough, and the muscle cells must rely upon a back-up system—the production of energy by glycolysis. The lactic acid produced by glycolysis builds up in the muscle cells to the point where it hampers muscle performance, causing muscle fatigue and exhaustion. Such a lactic acid build-up produces mild acidosis, which causes nausea, headache, lack of appetite, and impairment of oxygen transport, resulting in the difficult, painful gulping of air experienced by athletes after extreme physical efforts. Muscle cells are slow to recover from this condition. The lactic acid must be removed either by conversion to pyruvic acid in the muscle cells, or by movement from these cells to the liver. In the liver, about 25% of the lactic acid is oxidized through the citric acid cycle, and the remainder is converted to glycogen. The heavy breathing that occurs after strenuous exercise helps supply the oxygen necessary to oxidize the lactic acid (Figure 18.17). A major difference between a well-conditioned athlete and a nonathlete is that the athlete can supply her muscle cells with the oxygen necessary to maintain aerobic respiration for a longer period than can the nonathlete. This means that her muscle cells will require glycolysis for their supply of ATP only at much higher levels of exertion. All the carbohydrate metabolism that we have discussed is summarized in Figure 18.18.

18.27 Fermentation

Microorganisms such as yeast are able to live quite well without oxygen. Glycolysis occurs in yeast cells in much the same way as it does in animal cells. The process differs in yeast cells only in the last step, where pyruvic acid is converted to ethanol and carbon dioxide rather than to

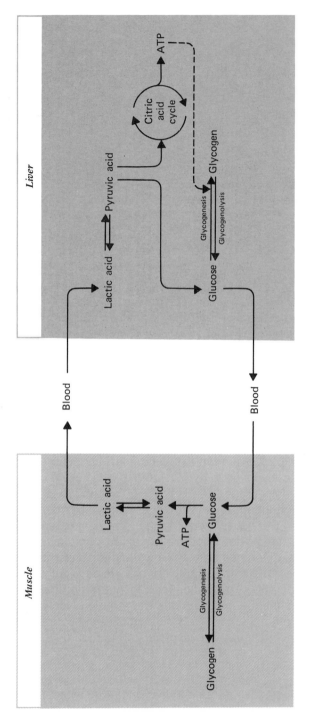

Figure 18.17 The lactic acid cycle. Lactic acid formed in strenuous exercise must be removed either by conversion to pyruvic acid in the muscle cells or by migration from these cells to the liver, where it is converted to glycogen. Some of the lactic acid (about 25%) must be oxidized via the citric acid cycle to provide the energy necessary for this conversion.

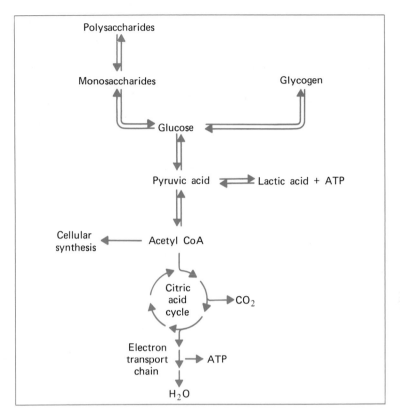

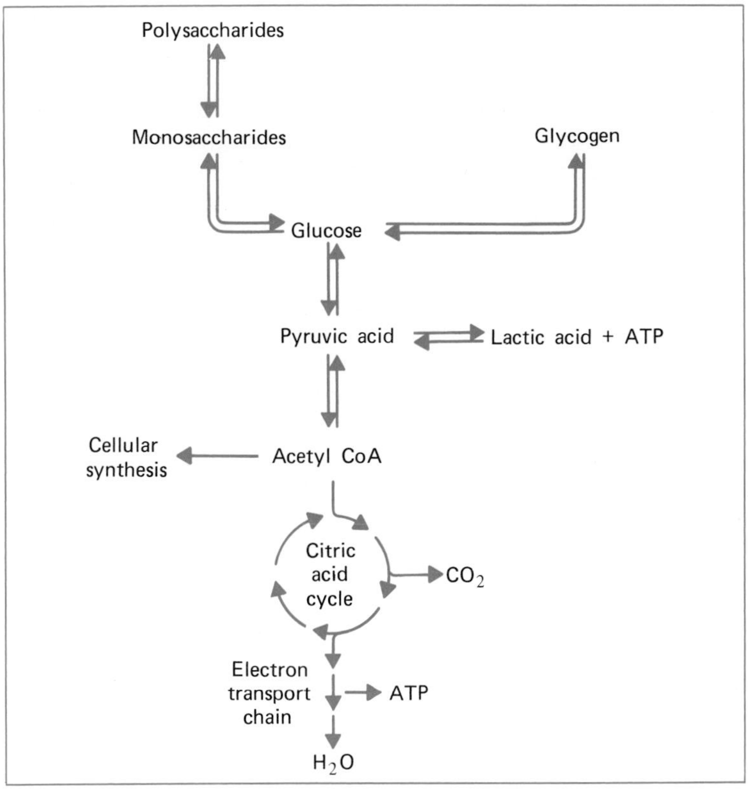

Figure 18.18 A summary of carbohydrate metabolism.

lactic acid. The name given to this form of glycolysis in yeast cells is **fermentation,** and the overall reaction is as follows.

$$\underset{\text{Glucose}}{C_6H_{12}O_6} \xrightarrow[\text{Enzymes}]{\text{Yeast}} \underset{\text{Ethanol}}{2CH_3CH_2OH} + 2CO_2 + \text{Energy}$$

Control of Carbohydrate Metabolism

18.28 Cellular Control

Each cell in our body maintains a sophisticated system for regulating carbohydrate metabolism. Each stage of carbohydrate metabolism has regulatory enzymes—enzymes that are turned on or off by changes in the concentrations of specific substances in the cell. In this way, the overall reactions of metabolism occur at a high rate only when the cell is low in ATP.

18.29 Insulin

Hormones are the body's second level of control of carbohydrate metabolism. Insulin, for example, is a hormone that acts on cell membranes to speed up the passage of glucose from the blood into cells of muscle and fatty tissue. It also increases the production of glycogen and slows the production of glucose in the liver. Insulin is produced by special cells in the pancreas called beta cells, which are sensitive to the level of glucose in the blood. These cells secrete insulin when the glucose level is high. This acts to increase the absorption of glucose by the cells and to increase the conversion of glucose to glycogen in the liver, thus lowering the blood sugar level.

Blood Sugar Level

In a normal adult, the blood sugar level remains fairly constant. Eight to 12 hours after a meal, the level is 60 to 100 mg glucose/100 ml blood. This is called the normal fasting level. The brain, which maintains no storehouse of glucose, depends completely upon the glucose in the blood for its energy requirements. Therefore, the blood glucose level is critical for normal brain function (Figure 18.19).

Hyperglycemia is a condition that results from blood sugar levels which are too high. Mild hyperglycemia occurs after meals. The body reacts to this higher level of glucose, however, by storing it in the form of glycogen or fat, and by oxidizing it. In severe hyperglycemia, the blood sugar level can rise so high that the glucose level exceeds the amount tolerated by the kidneys (called the renal threshold, 160 mg/100 ml blood), and glucose is excreted in the urine. Glycosuria, sugar in the urine, can be detected by using the Benedict's test. Glycosuria can result from conditions such as diabetes mellitus, emotional stress, kidney failure, or the administration of certain drugs.

Hypoglycemia is a condition that results from blood sugar levels which are below normal. Under such conditions the brain becomes starved for glucose. Mild hypoglycemia produces irritability, dizziness, lethargy, grogginess, and fainting. Severe hypoglycemia can produce convulsions, shock, and coma. Such severe hypoglycemia can be produced by overproduction or overinjection of insulin, a condition called hyperinsulinism, in which the blood sugar level drops extremely rapidly and the person goes into convulsions and coma called *insulin shock.* Mild hypoglycemia can be caused by too much carbohydrate in the diet. A meal (especially breakfast) which is rich in carbohydrates can overstimulate the pancreas to produce insulin, causing the blood sugar

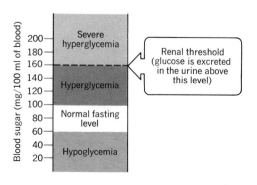

Figure 18.19 Blood sugar levels, and the conditions of hypoglycemia and hyperglycemia.

level to fall below normal. The diet, therefore, can be an important factor in controlling mild hypoglycemia. Breakfast, for example, should be rich in proteins with some fats and carbohydrates. The proteins and fats reduce the speed with which the glucose enters the bloodstream and, in this way, prevent overstimulation of the pancreas.

18.30 Diabetes Mellitus

Diabetes mellitus is a complicated disease whose causes and metabolic reactions are not totally understood. It is a disorder that causes the blood sugar level to rise above normal. In diabetes, the beta cells in the pancreas respond slowly or not at all to the level of glucose in the blood. As we saw at the beginning of this chapter, there are two types of diabetes mellitus. Severe juvenile diabetes, which can begin quite rapidly and whose symptoms are fairly obvious, occurs in individuals who lack the ability to produce insulin. When untreated, the patient will excrete large amounts of glucose in the urine. To supply the body with energy, the untreated diabetic will metabolize large amounts of fats, resulting in the production of high amounts of compounds called ketone bodies (Table 18.4). Acetone can be smelled on the breath of such persons, and acetoacetic acid and β-hydroxybutyric acid will be found in high concentrations in their blood and urine. These conditions can result in severe acidosis, coma, and death. Juvenile diabetes is treated by diet and daily injections of insulin.

Adult diabetes, the second form of this disorder, is much more common than the juvenile form. It is slow in developing, and will occur in individuals after the age of 30 to 40. One out of every five people has a hereditary tendency toward diabetes mellitus, but this type of diabetes often goes undetected. The beta cells of an adult diabetic will respond normally until placed under an unusual or prolonged stress, such as an infection of the pancreas, or obesity. When detected, such diabetes can be treated by diet alone, by diet and insulin injections, or, in special cases, by diet and oral drugs that stimulate the pancreas to produce insulin.

The high concentrations of glucose in the blood of a diabetic result in serious disruption of the normal chemical balance in cells. Tissues that will absorb glucose even when insulin is not present will develop very high concentrations of glucose, approaching that of the blood. Such excess glucose is converted to sorbitol and fructose, compounds that stay in the

Table 18.4 Ketone Bodies

Acetoacetic acid	$CH_3\overset{\displaystyle O}{\overset{\displaystyle \|}{C}}CH_2\overset{\displaystyle O}{\overset{\displaystyle \|}{C}}OH$
3-Hydroxybutanoic acid (β-Hydroxybutyric acid)	$CH_3\overset{\displaystyle OH}{\overset{\displaystyle \|}{C}}HCH_2\overset{\displaystyle O}{\overset{\displaystyle \|}{C}}OH$
Acetone	$CH_3\overset{\displaystyle O}{\overset{\displaystyle \|}{C}}CH_3$

cell and cause electrolyte imbalance, water retention, and other abnormal conditions. For example, accumulation of sorbitol in the lens of the eye can cause cataracts. In the blood vessels, this chemical imbalance can cause atherosclerosis; in the eye it can cause retinopathy, a form of blindness. In fact, diabetes mellitus is one of the major causes of blindness in the United States.

Glucose Tolerance Test

The glucose tolerance test is a useful tool for diagnosing diabetes mellitus. In this test, a patient who has fasted for 8 hours is given 50 to 100 grams of glucose in a solution to drink. His blood sugar level is measured initially, and then during the next several hours. A normal person will reach a maximum blood sugar level of about 160 mg/100 ml during the first hour, and this level will fall to normal (or slightly below normal) by the end of the second hour. A diabetic, however, will begin the test with a higher than normal blood sugar level. The sugar level will peak at the end of the second hour, and will not return to the original level until several hours later (Figure 18.20).

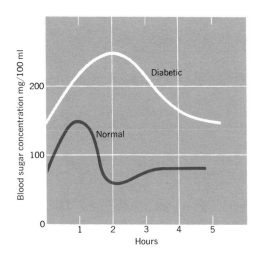

Figure 18.20 The glucose tolerance test.

18.31 Other Hormones

There are other hormones, in addition to insulin, which can affect the blood sugar level. For example, epinephrine will stimulate the release of glucose from the liver, and will increase the blood sugar level to a point that often exceeds the renal threshold. For this reason, glucose will often be found in the urine of individuals under stress. Glucagon is another hormone produced by the pancreas, but its effects are the opposite of those of insulin. Glucagon increases the blood sugar level by increasing the rate of glycogenolysis in the liver. There are also several hormones produced by the pituitary gland that have effects opposite those of insulin. Just these few examples make it obvious that the control of carbohydrate metabolism by the body is an extremely complex process that can be affected by many different factors in our lives.

Chapter Summary

Carbohydrates are a class of compounds which includes the following:

1. Monosaccharides, which are polyhydric aldehydes and polyhydric ketones that cannot be broken down into smaller units upon hydrolysis. Important examples are glucose, fructose, galactose, and ribose.
2. Disaccharides, which produce two monosaccharides upon hydrolysis. Important examples are sucrose (fructose + glucose), maltose (glucose + glucose), and lactose (glucose + galactose).
3. Polysaccharides, which produce three or more monosaccharides upon hydrolysis. Important examples are glycogen (the storage form of glucose in animals), starch (the storage form of glucose in plants), and cellulose (the main structural component of plants).

Monosaccharides exist mainly as five- or six-sided ring structures that result from the formation of an internal hemiacetal or hemiketal. These rings have *cis* and *trans* isomers (α or β forms) that make a great difference in the use of the molecule by an organism. All monosaccharides are reducing sugars and will give a positive Benedict's test.

Disaccharides are formed by a condensation reaction between two monosaccharides. The glycoside linkage between the two monosaccharides is quite stable. Polysaccharides can be linear polymers of glucose, such as amylose [glucose units connected by α(1-4) linkages], or cellulose [glucose units connected by β(1-4) linkages], or can be branched polymers of glucose such as glycogen and amylopectin [glucose units connected by α(1-4) linkages, with branches formed by α(1-6) linkages].

Carbohydrates are produced by green plants in a process called photosynthesis. Carbohydrate metabolism is the term that represents all the enzyme-catalyzed reactions involving the synthesis and breakdown of carbohydrates in the body. Carbohydrate digestion begins in the mouth, and is completed in the small intestine. The glucose (and other monosaccharides) produced during this process enters the blood. It may then be converted by the liver to glycogen in a process called glycogenesis, it may be used by the cell in the synthesis of other compounds, or it may be oxidized by the cell to produce energy through a series of reactions called cellular respiration. Cellular respiration involves an anaerobic stage called glycolysis, in which glucose is broken down to lactic acid, and an aerobic stage, which involves the reactions of the citric acid cycle and the electron transport chain. This second stage produces most of the ATP's that come from the oxidation of glucose. When there is not enough oxygen, muscle cells produce ATP using the process of glycolysis. However, the resulting build-up of lactic acid in the tissues causes muscle fatigue and mild acidosis. Microorganisms such as yeast oxidize glucose under anaerobic conditions, but the end product of this process of fermentation is ethanol, rather than lactic acid.

The control of carbohydrate metabolism is quite complex. Inside the cell, the rate of oxidation of glucose is controlled by the concentration of many substances in the cell, and it occurs at a high rate only when the concentration of ATP is low. The level of glucose in the blood is controlled by several hormones: insulin, epinephrine, and glucagon. Disruption of this control system can cause hyperglycemia, hypoglycemia, or diabetes mellitus. See Figure 18.18 for a summary of the processes of carbohydrate metabolism.

Exercises and Problems

1. Describe four functions of carbohydrates in living organisms.

2. Identify each of the following as (a) an aldose or ketose, and (b) a triose, tetrose, pentose, hexose, or heptose.

(a)
$$
\begin{array}{c}
H \\
| \\
C{=}O \\
| \\
H{-}C{-}OH \\
| \\
H{-}C{-}OH \\
| \\
H{-}C{-}OH \\
| \\
CH_2OH
\end{array}
$$

(b)
$$
\begin{array}{c}
CH_2OH \\
| \\
C{=}O \\
| \\
HO{-}C{-}H \\
| \\
H{-}C{-}OH \\
| \\
H{-}C{-}OH \\
| \\
H{-}C{-}OH \\
| \\
CH_2OH
\end{array}
$$

(c)
$$
\begin{array}{c}
CH_2OH \\
| \\
C{=}O \\
| \\
HO{-}C{-}H \\
| \\
H{-}C{-}OH \\
| \\
H{-}C{-}OH \\
| \\
CH_2OH
\end{array}
$$

(d)
$$
\begin{array}{c}
H \\
| \\
C{=}O \\
| \\
H{-}C{-}OH \\
| \\
CH_2OH
\end{array}
$$

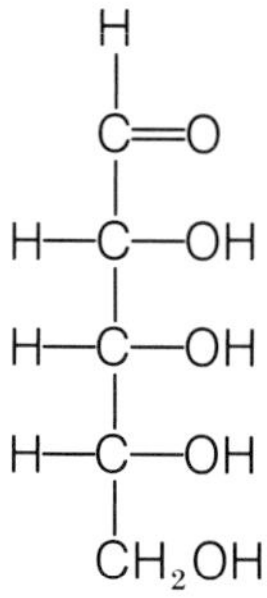

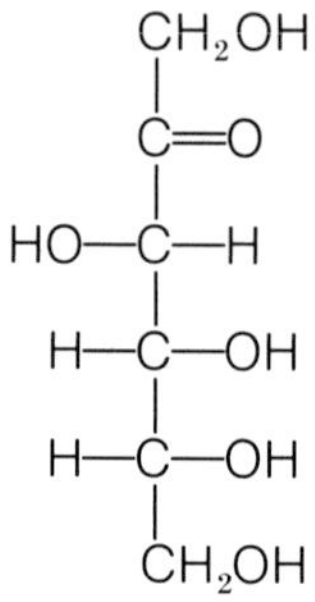

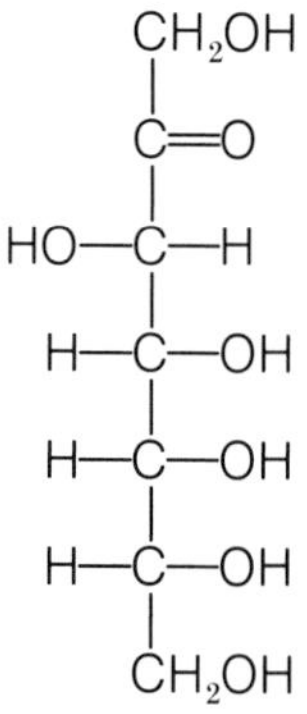

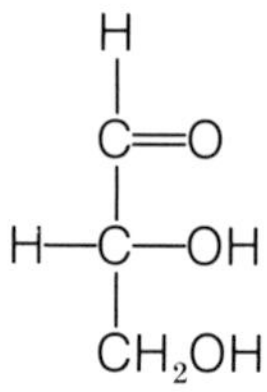

3. Draw (a) the linear structure of glucose, and (b) the ring structure of α-glucose.

4. Which of the following sugars will give a positive Benedict's test? Give the reason for your answer.

   (a) fructose
   (b) ribose
   (c) lactose
   (d) maltose
   (e) sucrose

5. Suppose that a sample of urine from an infant gives a positive test with a Clinitest tablet. Is it correct for the analyst to report glucose in the urine? Why or why not?

6. Why is glucose administered intravenously, whereas sucrose is not?

7. What is the storage form of glucose in animals? How does its structure compare with starch?

8. Describe the effect of a chocolate bar on the blood sugar level of a diabetic and a nondiabetic.

9. Write the equation for the hydrolysis of each of the following compounds, and name each of the resulting products:

   (a) sucrose (b) maltose (c) lactose

10. (a) Name the two polysaccharides making up starch.
    (b) Describe the difference in their structures.
    (c) Design an experiment for the hydrolysis of starch. How would you check the progress of the hydrolysis?

11. Given three unknown solutions labeled A, B, and C, suggest a method for determining which solution contains starch, which contains glucose, and which contains fructose.

12. The disaccharide called melibiose is found in some plant juices.

    (a) What are the two monosaccharides that make up this disaccharide?
    (b) Is melibiose a reducing sugar? Why or why not?

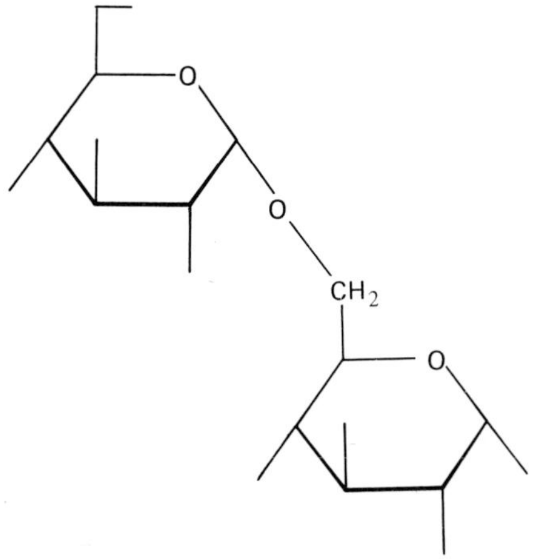

(c) What type of linkage connects the two monosaccharides? What polysaccharide also contains this linkage?

13. Both celery and potato chips are composed of molecules that are polymers of glucose. Explain why celery is a good snack for people on a diet, whereas potato chips are not.

14. The liver protects the body from drugs and foreign materials by inactivating them and joining them with a natural substance (such as glucuronic acid) to aid excretion. For example, the liver metabolizes salicylic acid by forming an ester bond with glucuronic acid. Write a possible equation for this reaction.

15. Write a general reaction for each of the following processes:

    (a) photosynthesis
    (b) glycogenesis
    (c) glycogenolysis
    (d) glycolysis
    (e) fermentation
    (f) citric acid cycle

16. In general terms, describe the possible metabolic reactions that the starch in a piece of toast might undergo from the time you take a bite until the ATP formed from the toast is used in muscle contraction.

17. (a) Draw the structure of AMP, ADP, and ATP.
    (b) What is meant by a *high-energy* phosphate bond?
    (c) Why are molecules such as ATP so important to the survival of a cell?

18. The oxidation of glucose by a cell can be divided into two separate stages: an anaerobic and an aerobic stage.

    (a) What is the name given to the anaerobic stage?
    (b) Write the equation for the overall reaction that occurs in the anaerobic stage.
    (c) Where in the cell does each stage occur?
    (d) Which stage produces more energy?
    (e) Name the two series of reactions that occur in the aerobic stage, and give a general description of the reactions that occur in each.
    (f) Write the equation for the overall reaction that occurs in the aerobic stage.

19. Your alarm clock doesn't go off and you wake up late for a final exam. You jump out of bed, dress quickly, run across campus, and race up three flights of stairs to your classroom. You collapse in your seat gasping for breath, and your legs feel like rubber. Describe the events that have occurred in your muscle cells during this experience. Why do you gasp for breath, and why are your muscles so weak?

20. Your brain requires a constant supply of glucose. Explain how your body maintains a fairly constant blood sugar level even though you eat foods that supply glucose only a few times a day.

21. What is *insulin shock?* How does it occur? Explain why severe hyperinsulinism is more dangerous to a patient than is the lack of insulin over a short period.
22. Suggest a diet for a patient suffering from hypoglycemia.
23. Does glycosuria result only during diabetes mellitus? How would you verify diabetes mellitus in a patient who is suffering from glycosuria?

chapter 19

Lipids

Learning Objectives

By the time you have finished this chapter, you should be able to:

1. Describe the difference between
   a. a simple and compound lipid.
   b. a simple and mixed triglyceride.
   c. a saturated and unsaturated fatty acid.
   d. a saponifiable and nonsaponifiable lipid

2. Explain the meaning of *essential fatty acid.*

3. Describe the process by which butter becomes rancid.

4. Given the iodine number of a lipid, describe the physical properties, and the most likely source of the lipid.

5. Draw the general structure of a triglyceride, and write the equations for its hydrolysis and saponification.

6. Explain how soap is able to remove grease from your hands.

7. Draw the general structure and describe the function of the following compound lipids:
   a. phosphatides
   b. lecithins
   c. cephalins
   d. sphingomyelins
   e. glycolipids

8. Give three examples of nonsaponifiable lipids.

9. Describe the steps of lipid digestion and absorption.

10. Describe fatty acid oxidation and the conditions that will result in ketosis.

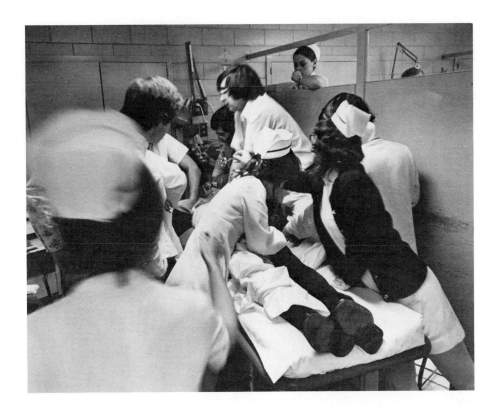

Dan was a man of 34 who, through a great deal of hard work and self-sacrifice, had recently been promoted to branch manager of his bank. What little time he took away from his job was spent playing tennis and hiking with his family. During the past year Dan had been bothered occasionally by a sharp chest pain that occurred during particularly strenuous games of tennis. But this pain was always brief and went away quickly, so Dan never paid much attention to it. One afternoon while Dan was working on plans for a Board of Directors meeting, the pain began again. This time, however, it didn't go away. Instead, it became stronger and soon extended to his left arm as well. Dan found himself sweating very heavily, and became extremely short of breath. His alert secretary recognized the symptoms and called an ambulance, and Dan was rushed to the hospital where he was treated for a heart attack. Luckily for Dan, he reached the hospital in time. In the cardiac care unit he was carefully watched; his condition remained stable for several days, and he then began a slow recovery.

Dan kept asking himself, "Why me? Heart attacks happen only to older men, or to people in poor physical shape." But when he asked his doctor this question, Dan was surprised to learn that coronary (heart) disease is responsible for over half of all deaths reported for people between the ages of 35 and 64. In this particular case, Dan was suffering from

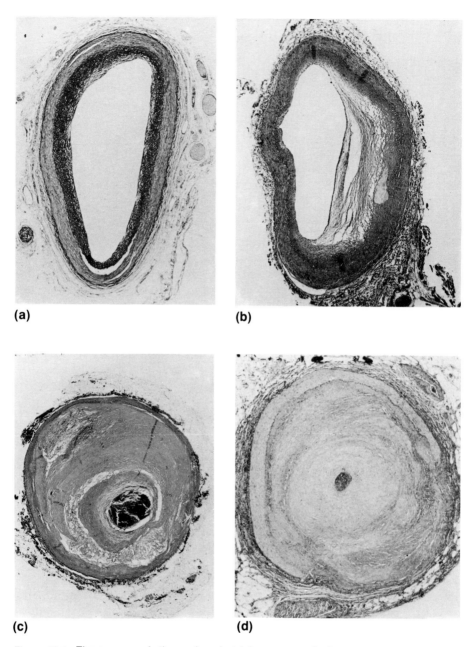

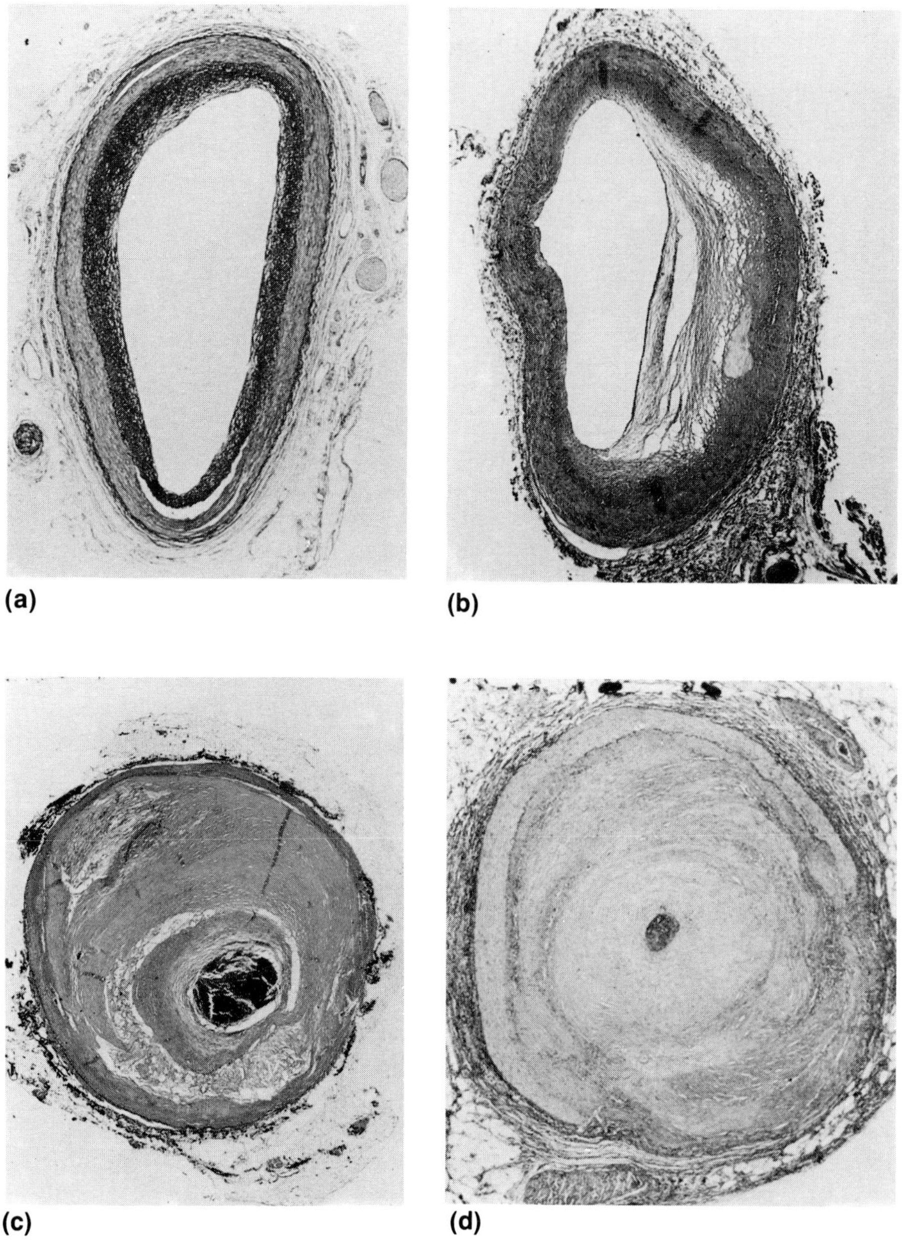

Figure 19.1 The progress of atherosclerosis. (a) A near-normal artery. (b) Plaque forms on the inner lining of the artery. (c) The narrowed channel within the artery is blocked by a blood clot. (d) Atherosclerosis has progressed to the point that this artery is completely blocked. (a and d, Courtesy National Institutes of Health, National Heart & Lung Institute; b and c, Courtesy American Heart Association)

atherosclerosis, the most common form of arteriosclerosis (disease of the arteries). Atherosclerosis is a slow, progressive disease that may begin in childhood, and shows no symptoms for 20 to 50 years. It affects mainly the larger arteries, especially the aorta and the arteries that feed the heart, brain, and kidneys.

Atherosclerosis begins as a yellowish fatty streak that is present in the aortas of most children by the age of three. Such streaks are actually a build-up of cholesterol and various fats, and they appear in most of the other major arteries by the age of 35. The streaks never cause symptoms themselves, but in certain individuals may develop into plaques, the major cause of concern and study in atherosclerosis. Plaques consist mostly of smooth muscle cells from the middle layer of the arterial wall that have multiplied and moved to the inner layer of the wall. Scar tissue and fats (mainly cholesterol) then begin to accumulate on these muscle cells (Figure 19.1). There are two theories suggested to explain why such muscle cells begin to multiply and form plaques. The first is that plaques may form in response to frequent recurring injuries to the arterial wall, which may be caused by high blood pressure, antibodies to the carbon monoxide in cigarette smoke, or high concentrations of blood fats and cholesterol. The second theory is that plaques actually are benign tumors growing from a single, mutated muscle cell.

Plaques may cause serious damage in several ways: They may grow in size and severely restrict the flow of blood to the tissues so that the tissue is damaged or destroyed. Or, a blood clot may form on the rough surface of the plaque and then break off and block essential arteries. Or, the plaque itself may open up and release its contents, which can block blood flow to important areas of the body. And finally, the artery may become "hardened," or lose flexibility as the plaque develops. This can cause a ballooning of the artery, called an aneurysm. Such an aneurysm may then burst, causing hemorrhaging, severe tissue damage, or death.

If atherosclerosis affects the arteries of the heart, as was the case with Dan, a person may suffer the first symptoms of angina pectoris—a short-term, intense chest pain. This happens because the plaques in the arteries leading to the heart have greatly reduced the blood flow, although there is still enough flow to maintain normal activities. However, any activity requiring increased blood flow will result in a starved heart, and an intense pain. This condition can be treated with rest and nitroglycerin (which relaxes the muscles in the coronary arteries), and can occur for long periods of time without harming the heart if the atherosclerosis doesn't progress. However, atherosclerosis can cause heart attacks by completely blocking arteries that supply the heart muscle. Recovery from such heart attacks depends on the amount and the location of damage to the heart muscle.

Lipids

19.1 What Are Lipids?

Cholesterol, the material that plays such an important role in atherosclerosis, belongs to a class of compounds called lipids. **Lipids** are oily or waxy substances that are insoluble in water and that may be

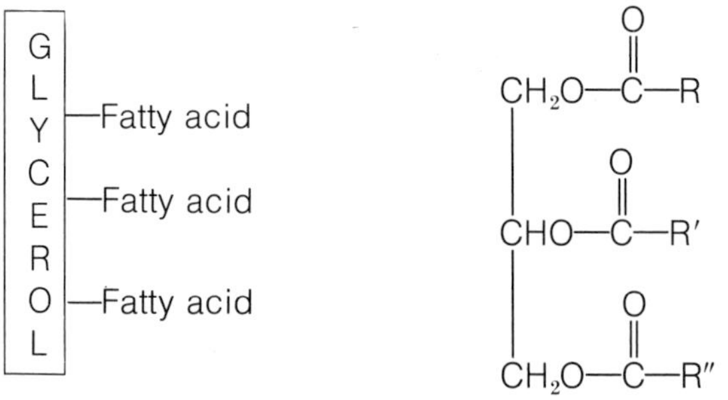

Table 19.1 The Lipids

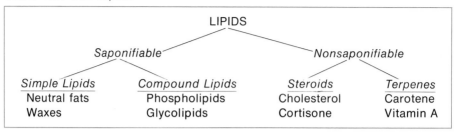

extracted from tissues using nonpolar solvents. Their major functions in living cells are to form part of the structure of membranes and to store energy for the cell.

Lipids are a varied group of compounds that can be categorized in several ways. We can divide the lipids into two major classes: those that can be saponified (hydrolyzed by a base), and those that are nonsaponifiable. The saponifiable lipids can be further subdivided into **simple lipids,** which yield fatty acids and an alcohol upon hydrolysis, and **compound lipids,** which yield fatty acids, alcohol, and some other compounds upon hydrolysis (Table 19.1).

Saponifiable Lipids: Simple Lipids

19.2 Fats and Oils

The simplest and most abundant of the lipid compounds are the neutral lipids, which are also called fats, triglycerides, or triacylglycerols. These compounds are esters of glycerol and three fatty acids, and are the main form of fat storage in plants and in adipose cells (or fat cells) of vertebrates.

$$
\begin{array}{ll}
\text{G} & \\
\text{L} & \\
\text{Y} & \text{—Fatty acid} \\
\text{C} & \\
\text{E} & \text{—Fatty acid} \\
\text{R} & \\
\text{O} & \text{—Fatty acid} \\
\text{L} &
\end{array}
\qquad
\begin{array}{l}
\quad\quad\quad O \\
\quad\quad\quad \| \\
CH_2O\text{—}C\text{—}R \\
\quad\quad\quad O \\
\quad\quad\quad \| \\
CHO\text{—}C\text{—}R' \\
\quad\quad\quad O \\
\quad\quad\quad \| \\
CH_2O\text{—}C\text{—}R''
\end{array}
$$

Fat or Triglyceride

Simple triglycerides contain the same fatty acid in all three positions on the glycerol molecule, whereas **mixed triglycerides** contain two or more different fatty acids. Natural fats are a mixture of simple and mixed triglycerides (Figure 19.2). Triglycerides have very wide commercial use in soaps, paint, varnishes, oilcloth, linoleum, printing inks, ointments, and creams.

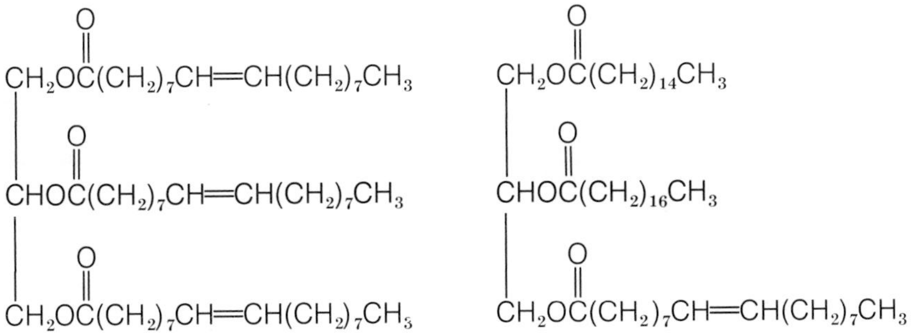

$$CH_2OC(CH_2)_7CH=CH(CH_2)_7CH_3$$

$$\overset{O}{\overset{\|}{CHOC}}(CH_2)_7CH=CH(CH_2)_7CH_3$$

$$CH_2OC(CH_2)_7CH=CH(CH_2)_7CH_3$$

Simple triglyceride
Triolein

$$CH_2OC(CH_2)_{14}CH_3$$

$$\overset{O}{\overset{\|}{CHOC}}(CH_2)_{16}CH_3$$

$$CH_2OC(CH_2)_7CH=CH(CH_2)_7CH_3$$

Mixed triglyceride

Figure 19.2 Simple triglycerides, such as triolein, contain only one type of fatty acid. Mixed triglycerides contain two or more different fatty acids.

19.3 Fatty Acids

Fatty acids are long-chain carboxylic acids that are formed from the hydrolysis of triglycerides. The naturally occurring fatty acids have an even number of carbon atoms and are generally nonbranching. The most common fatty acids have 16 or 18 carbon atoms in their chains. These fatty acids are palmitic, stearic, and oleic acids, with oleic acid making up more than half the total fatty acid content of many fats. **Saturated** fatty acids have only carbon-to-carbon single bonds, are unreactive, and are waxy solids at room temperature. **Unsaturated** fatty acids have one or more carbon-to-carbon double bonds, and are liquids at room temperature (Figure 19.3 and Table 19.2).

 The difference between fats and oils is the number of unsaturated fatty acids that are present. Animal fats, lard, tallow, and butter are mixed fats containing more saturated fatty acids than unsaturated fatty acids. They are waxy, white solids at room temperature. Vegetable oils, olive oil,

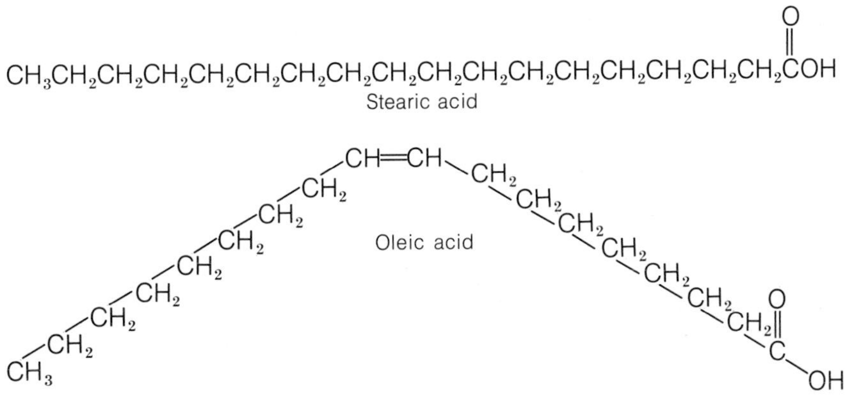

$$CH_3CH_2CH_2CH_2CH_2CH_2CH_2CH_2CH_2CH_2CH_2CH_2CH_2CH_2CH_2CH_2CH_2\overset{O}{\overset{\|}{C}}OH$$
Stearic acid

Oleic acid

Figure 19.3 Stearic and oleic acids are both fatty acids with 18 carbon atoms. Oleic acid is unsaturated; it contains one double bond in the *cis* configuration, which gives the molecule a rigid bend.

Table 19.2 Some Common Fatty Acids

Name (Carbon Atoms)	Formula	Melting Point (°C)
SATURATED		
Butyric (4)	$CH_3(CH_2)_2\overset{\displaystyle O}{\overset{\|}{C}}OH$	− 4.2
Lauric (12)	$CH_3(CH_2)_{10}\overset{\displaystyle O}{\overset{\|}{C}}OH$	44.2
Myristic (14)	$CH_3(CH_2)_{12}\overset{\displaystyle O}{\overset{\|}{C}}OH$	53.9
Palmitic (16)	$CH_3(CH_2)_{14}\overset{\displaystyle O}{\overset{\|}{C}}OH$	63.1
Stearic (18)	$CH_3(CH_2)_{16}\overset{\displaystyle O}{\overset{\|}{C}}OH$	69.6
Arachidic (20)	$CH_3(CH_2)_{18}\overset{\displaystyle O}{\overset{\|}{C}}OH$	76.5
UNSATURATED		
Oleic (18)	$CH_3(CH_2)_7CH{=}CH(CH_2)_7\overset{\displaystyle O}{\overset{\|}{C}}OH$	13.4
Linoleic (18)	$CH_3(CH_2)_4CH{=}CHCH_2CH{=}CH(CH_2)_7\overset{\displaystyle O}{\overset{\|}{C}}OH$	− 5
Linolenic (18)	$CH_3CH_2CH{=}CHCH_2CH{=}CHCH_2CH{=}CH(CH_2)_7\overset{\displaystyle O}{\overset{\|}{C}}OH$	−11
Arachidonic (20)	$CH_3(CH_2)_4CH{=}CHCH_2CH{=}CHCH_2CH{=}CHCH_2CH{=}CH(CH_2)_3\overset{\displaystyle O}{\overset{\|}{C}}OH$	−49.5

corn oil, and cottonseed oil contain a higher concentration of unsaturated fatty acids, and are liquids at room temperature (Table 19.3).

19.4 Essential Fatty Acids

Our bodies can produce saturated fatty acids and unsaturated fatty acids containing one double bond. However, we can't form linoleic, linolenic, or arachidonic acids, which are called the **essential fatty acids.** Infants lacking these fatty acids in their diets will lose weight and develop eczema. The essential fatty acids are used by the body to form prostaglandins, compounds that are found in most mammalian tissues and which have a wide range of physiological effects (see Section 21.14). They are involved in the body's defenses against many sorts of change; in particular, they are powerful inducers of fever and inflammation. Aspirin seems to function as an antipyretic by regulating the production of prostaglandins in the temperature-regulating tissues of the brain.

Table 19.3 Some Common Fats and Oils and Their Fatty Acid Composition*

	Melting Point °C	Percent Composition of the Most Abundant Fatty Acids								Iodine Number
		Saturated				Unsaturated				
ANIMAL FATS		Myristic	Palmitic	Stearic	Arachidic	Palmitoleic	Oleic	Linoleic	Linolenic	
Butter	32	11	29	9	2	5	27	4	–	36
Lard	30	1	28	12	–	3	48	6	–	59
Tallow	N/A	6	27	14	–	–	50	3	–	50
Human fat	15	3	24	8	–	5	47	10	–	68
PLANT OILS										
Corn	−20	1	10	3	–	2	50	34	–	123
Cottonseed	− 1	1	23	1	1	2	23	48	–	106
Linseed	−24	–	6	2	1	–	19	24	47	179
Olive	− 6	–	7	2	–	–	84	5	–	81
Peanut	3	–	8	3	2	–	56	26	–	93
Safflower	N/A	←	7		→	–	19	70	3	145
Soybean	−16	–	10	2	–	–	29	51	6	130

* Values in this table are averages. Extreme variation may occur in the values depending upon the source, treatment, and age of the fat or oil.

19.5 Waxes

Waxes are esters of long-chain fatty acids and long-chain monohydric alcohols (alcohols with one hydroxyl group). For example, beeswax is largely an ester of myricyl alcohol ($C_{30}H_{61}OH$) and palmitic acid. Waxes form protective coatings on skin, fur, feathers, leaves, and fruits. They have properties of water insolubility, flexibility, and nonreactivity that make them excellent coatings. Commercially produced waxes are used in cosmetics, floor waxes, furniture and car polishes, ointments, and creams (Table 19.4).

Table 19.4 Some Common Waxes

Name	Melting Point °C	Source	Uses
Beeswax	61–69	Honeycomb	Candles, polishes
Carnauba	83–86	Carnauba Palm	Floor waxes, polishes
Lanolin	36–43	Wool	Cosmetics, skin ointments
Spermaceti	42–50	Sperm whale	Cosmetics, candles

Chemical Properties of Simple Lipids

19.6 Formation of Acrolein

When heated to high temperatures, fats will hydrolyze. The glycerol that is produced will react to form acrolein, whose vapors are irritating to the nose and eyes. Acrolein is responsible for the unpleasant odor you recognize when oil or fat is burned. Acrolein is also irritating to the digestive tract, and may be responsible for the stomach upset caused occasionally by deep-fried foods.

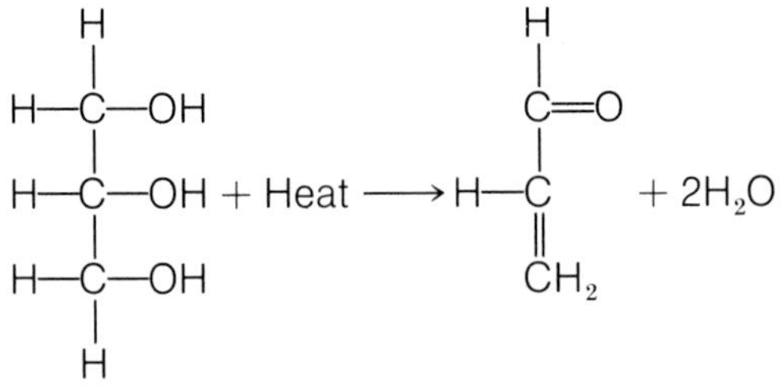

$$
\begin{array}{c}
\underset{\text{Glycerol}}{
\begin{array}{c}
\text{H} \\
| \\
\text{H—C—OH} \\
| \\
\text{H—C—OH} \\
| \\
\text{H—C—OH} \\
| \\
\text{H}
\end{array}}
+ \text{Heat} \longrightarrow
\underset{\text{Acrolein}}{
\begin{array}{c}
\text{H} \\
| \\
\text{C}=\text{O} \\
| \\
\text{H—C} \\
\| \\
\text{CH}_2
\end{array}}
+ 2\text{H}_2\text{O}
\end{array}
$$

19.7 Iodine Number

The unsaturated bonds in a fatty acid will react to add iodine, giving us a useful tool for determining unsaturation. Thus, chemists have defined the iodine number of a simple lipid to be the number of grams of iodine that will react with 100 grams of fat or oil. The higher the iodine number, the more unsaturated the fat. Fats generally have an iodine number below 70, and oils have an iodine number above 70 (Table 19.3).

19.8 Hydrogenation

Oils can be converted to solid fats by hydrogenation, the addition of hydrogen to the double bonds of the molecule.

$$\underset{\text{Linoleic acid}}{CH_3(CH_2)_4CH{=}CHCH_2CH{=}CH(CH_2)_7COOH} + 2H_2 \longrightarrow \underset{\text{Stearic acid}}{CH_3(CH_2)_{16}COOH}$$

Vegetable shortenings are commercially produced by the partial hydrogenation of soybean, corn, or cottonseed oil. (The complete hydrogenation of these oils would produce a hard, brittle product.) These shortenings are usually a mixture of unsaturated oils and hydrogenated oils. Margarine is a mixture of unsaturated oils, hydrogenated oils, flavorings, coloring agents, and vitamins A and D.

19.9 Rancidity

Fats and oils often develop a disagreeable odor and taste, and are then called rancid. There are two causes of rancidity: hydrolysis and oxidation. For example, when left too long at room temperature, butter will become rancid. This occurs because some of the fat in the butter will undergo hydrolysis, accelerated by enzymes produced by microorganisms in the

air. This hydrolysis produces the fatty acid butyric acid, which causes the odor of rancid butter. Also, oxygen in the air can oxidize unsaturated fats or oils to produce short-chain acids or aldehydes with disagreeable odors and tastes. Such oxidation is slowed in manufactured products such as crackers, potato chips, and pastries by adding chemicals called antioxidants.

19.10 Hydrolysis

The hydrolysis of fats can occur in the presence of superheated steam, hot mineral acids, or specific enzymes. Hydrolysis under such conditions produces glycerol and three fatty acids. In general:

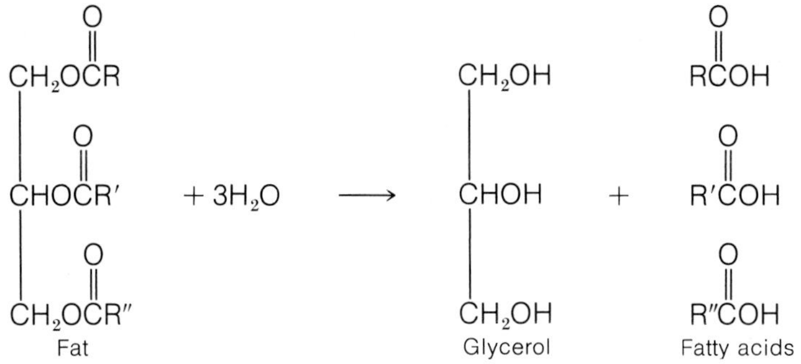

$$
\begin{array}{l}
\text{CH}_2\text{OCR} \\
| \quad\;\; \text{O} \\
\text{CHOCR}' \quad + 3\text{H}_2\text{O} \longrightarrow \\
| \quad\;\; \text{O} \\
\text{CH}_2\text{OCR}'' \\
\text{Fat}
\end{array}
\quad
\begin{array}{l}
\text{CH}_2\text{OH} \\
| \\
\text{CHOH} \quad + \\
| \\
\text{CH}_2\text{OH} \\
\text{Glycerol}
\end{array}
\quad
\begin{array}{l}
\text{RCOH} \\
\;\;\; \text{O} \\
\text{R'COH} \\
\;\;\; \text{O} \\
\text{R''COH} \\
\text{Fatty acids}
\end{array}
$$

For example,

$$
\begin{array}{l}
\text{CH}_2\text{OC(CH}_2)_{16}\text{CH}_3 \\
| \\
\text{CHOC(CH}_2)_7\text{CH}=\text{CH(CH}_2)_7\text{CH}_3 + 3\text{H}_2\text{O} \longrightarrow \\
| \\
\text{CH}_2\text{OC(CH}_2)_{14}\text{CH}_3
\end{array}
$$

$$
\left\{
\begin{array}{l}
\text{CH}_2\text{OH} \\
| \\
\text{CHOH} \\
| \\
\text{CH}_2\text{OH} \\
\text{Glycerol} \\[1em]
+ \text{CH}_3(\text{CH}_2)_{16}\overset{\text{O}}{\overset{\|}{\text{C}}}\text{OH} \\
\text{Stearic acid} \\[1em]
+ \text{CH}_3(\text{CH}_2)_7\text{CH}=\text{CH(CH}_2)_7\overset{\text{O}}{\overset{\|}{\text{C}}}\text{OH} \\
\text{Oleic acid} \\[1em]
+ \text{CH}_3(\text{CH}_2)_{14}\overset{\text{O}}{\overset{\|}{\text{C}}}\text{OH} \\
\text{Palmitic acid}
\end{array}
\right.
$$

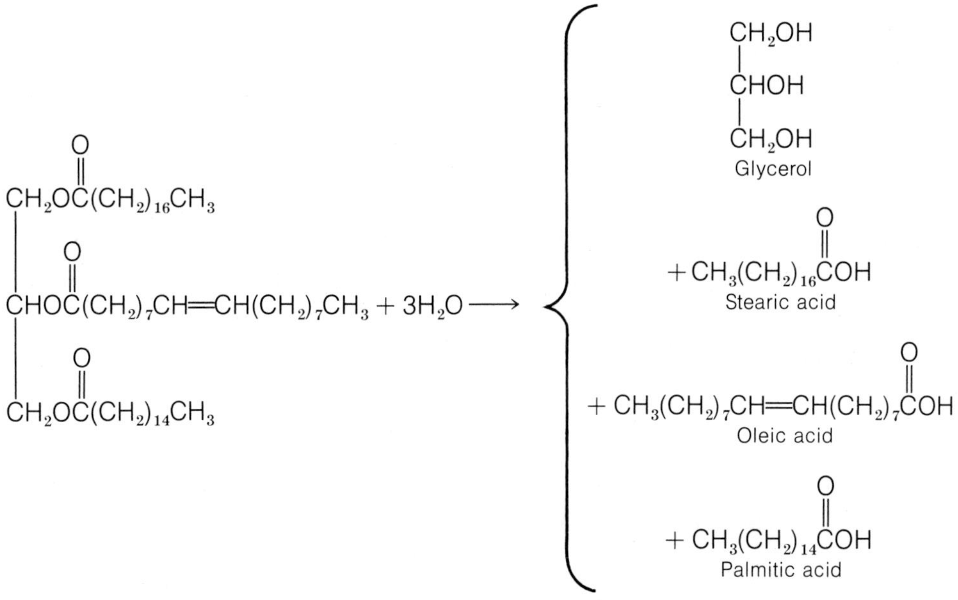

19.11 Saponification

When hydrolysis is carried out in the presence of a strong base such as sodium hydroxide, glycerol and the sodium salts of the fatty acids are produced. In general:

$$\begin{array}{l}
\underset{\text{Fat}}{\begin{array}{l}
\overset{\displaystyle O}{\overset{\|}{CH_2OCR}} \\[4pt]
\overset{\displaystyle O}{\overset{\|}{CHOCR'}} \\[4pt]
\overset{\displaystyle O}{\overset{\|}{CH_2OCR''}}
\end{array}}
\quad + \quad 3NaOH \quad \longrightarrow \quad
\underset{\text{Glycerol}}{\begin{array}{l}
CH_2OH \\[4pt]
CHOH \\[4pt]
CH_2OH
\end{array}}
\quad + \quad
\underset{\substack{\text{Sodium salt}\\(\text{soap})}}{\begin{array}{l}
\overset{\displaystyle O}{\overset{\|}{RCO^-Na^+}} \\[4pt]
\overset{\displaystyle O}{\overset{\|}{R'CO^-Na^+}} \\[4pt]
\overset{\displaystyle O}{\overset{\|}{R''CO^-Na^+}}
\end{array}}
\end{array}$$

For example,

$$\underset{\text{Tristearin}}{\begin{array}{l}
\overset{\displaystyle O}{\overset{\|}{CH_2OC(CH_2)_{16}CH_3}} \\[4pt]
\overset{\displaystyle O}{\overset{\|}{CHOC(CH_2)_{16}CH_3}} \\[4pt]
\overset{\displaystyle O}{\overset{\|}{CH_2OC(CH_2)_{16}CH_3}}
\end{array}}
+ 3NaOH \longrightarrow
\underset{\text{Glycerol}}{\begin{array}{l}
CH_2OH \\[4pt]
CHOH \\[4pt]
CH_2OH
\end{array}}
+ \underset{\text{Sodium stearate}}{3CH_3(CH_2)_{16}\overset{\displaystyle O}{\overset{\|}{C}}O^-Na^+}$$

Such salts are called **soaps.** Sodium salts of fatty acids are used in bar soaps, and potassium salts are used in liquid soaps. A hard soap contains a larger number of saturated fatty acids than a soft soap. Various additives are also used to give commercial soaps their colors and odors. In addition, floating soaps contain air bubbles, and scouring soaps contain abrasives.

Cleansing Action of Soap

Water does a poor job removing grease and oil because water molecules tend to stick together rather than penetrate the nonpolar grease. Soap greatly improves the cleansing power of water. A soap molecule has two portions: a nonpolar "tail" formed by the hydrocarbon chain, and a polar "head" formed by the carboxyl group.

$$\underset{\substack{\text{Nonpolar tail} \\ \\ \text{Soap}}}{CH_3CH_2CH_2CH_2CH_2CH_2CH_2CH_2CH_2CH_2CH_2CH_2\overset{\displaystyle O}{\overset{\|}{C}}} \underset{\substack{\text{Polar} \\ \text{head}}}{-O^-Na^+}$$

The nonpolar tail will readily dissolve in the nonpolar grease, whereas the polar head tends to remain dissolved in the water. In this way the soap

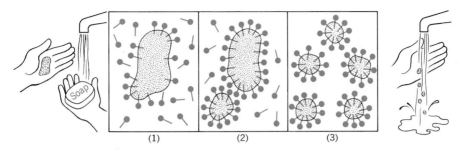

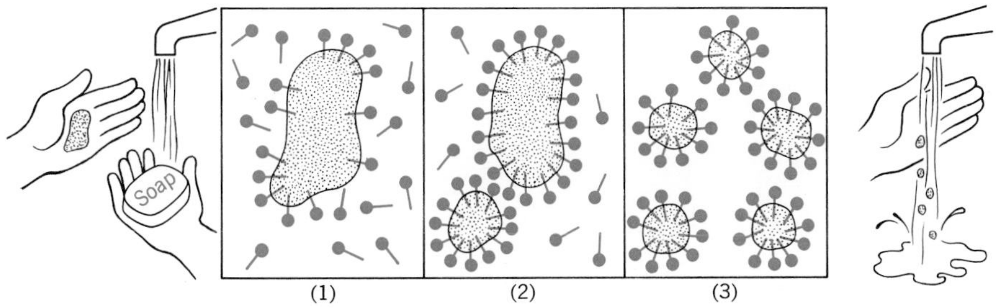

Figure 19.4 The cleansing action of soap. (1) The nonpolar tails of the soap molecules begin to dissolve in the nonpolar grease. (2) As the soap dissolves in the grease, small colloidal grease particles break off and are surrounded by negatively-charged polar heads of the soap molecules. This keeps the grease particles in solution by preventing them from reforming larger droplets. (3) In this manner, the grease can be completely broken up and the colloidal droplets washed away with water.

breaks up the grease into small colloidal droplets; that is, it emulsifies the grease, which can then be washed away (Figure 19.4).

The cleansing power of soap is affected by several factors. For example, hard water contains one or more of the metallic ions Ca^{2+}, Mg^{2+}, Fe^{2+}, or Fe^{3+}, which form insoluble salts with soap. These salts precipitate out (forming soap scum and bathtub ring), leaving less soap in the water to do the cleaning. Water softeners work by replacing such metallic ions with other ions, such as sodium, which do not interfere with the action of soap. Lowering the pH of the water will also decrease the cleansing action of soap by neutralizing the charge on the fatty acid ion.

Detergents, which are mixtures of sodium salts of sulfuric acid esters, have cleaning properties similar to soaps, but have important advantages (Figure 19.5). Their calcium and magnesium salts are water soluble, and they are not affected by pH. However, some early synthetic detergents containing branched chains in their hydrocarbon tails could not be naturally broken down by bacteria in sewage (that is, they were not biodegradable) in the same manner as soaps. Therefore, these products could not be removed from the water by sewage treatment plants. This

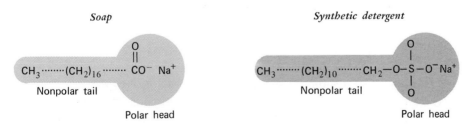

Figure 19.5 Two organic salts that possess cleansing properties. Each has a nonpolar, hydrophobic (water-repelling) tail and a polar, hydrophilic (water-attracting) head.

resulted in rivers and streams being covered with foam and suds, creating an alarming pollution problem. Newer detergents are partially or totally biodegradable, and the suds problem has been eliminated.

Saponifiable Lipids: Compound Lipids

19.12 Glycerol-based Phospholipids: Phosphatides

The **phospholipids** are a class of waxy solids that form part of the structure of cell membranes, and are important in the transport of lipids in the body. They can be divided into two general categories: glycerol-based phospholipids and sphingosine-based phospholipids.

The glycerol-based phospholipids, or **phosphatides,** are derivatives of phosphatidic acid. They contain glycerol, two fatty acids, a phosphate group, and a nitrogen-containing compound that can be choline, ethanolamine, serine, or inositol.

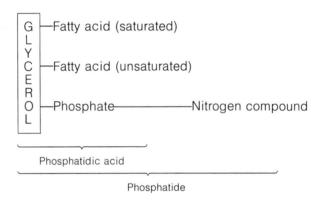

The phosphate group forms a polar head on the phosphatide molecule, and the two fatty acids form two nonpolar tails.

This arrangement gives the phosphatide good emulsifying properties and good membrane-forming properties. Cellular membranes, for example, are composed of proteins and phospholipids. The phospholipids form a double layer with their polar heads on the top and bottom, and their nonpolar tails in the middle. This double layer forms the framework for the membrane, and provides a support structure for the protein (Figure 19.6).

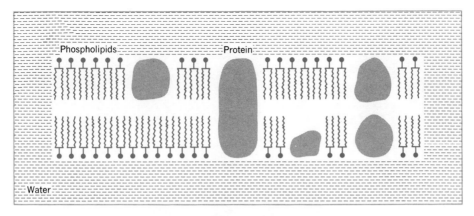

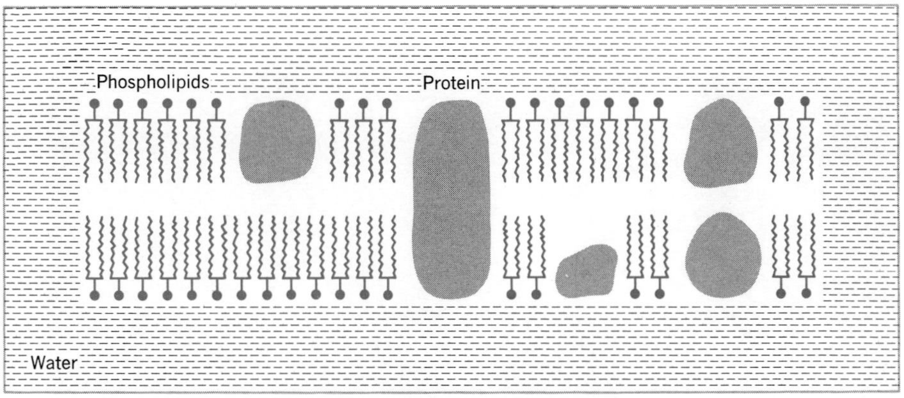

Figure 19.6 Cellular membranes are composed of phospholipids and proteins. The phospholipids are arranged in a double layer (bilayer) with their polar heads to the outside and their nonpolar tails toward the middle. This bilayer serves as an anchor for the protein molecules.

Lecithin

Lecithin is a phosphatide in which the nitrogen compound is choline.

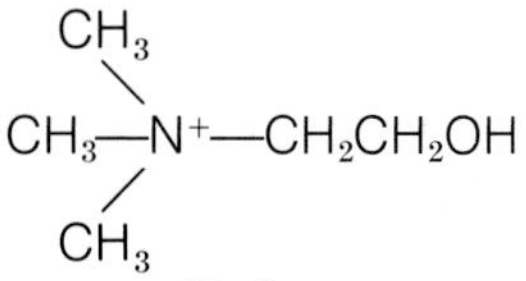

Choline

Lecithin plays an important role in the metabolism of fats in the liver, serves as a source of inorganic phosphate for tissue formation, and is an excellent emulsifying agent. It is important in the transport of fats from one part of the body to another, and is used commercially as an emulsifying agent in such products as chocolate candies, margarine, and medicines. Egg yolks contain a large amount of lecithin, and will emulsify salad oil and vinegar to make mayonnaise. The removal of one fatty acid from lecithin forms lysolecithin, a compound that causes destruction of red blood cells and spasmodic muscle contractions. The venom of poisonous snakes contains enzymes to catalyze the formation of lysolecithin from lecithin.

Premature babies with birthweights of 2 to 3 pounds often suffer from respiratory distress syndrome, or hyaline membrane disease. These small infants have lungs that do not yet function properly, due mostly to a lack of lecithin. In the lungs lecithin acts as a surfactant in the air sacs (or aveoli), reducing the high surface tension of the water in the air sacs and, therefore, preventing them from collapsing. When infants suffering this disease exhale, their air sacs collapse and oxygen cannot pass from the lungs into the bloodstream (Figure 19.7). Babies with severe cases must be

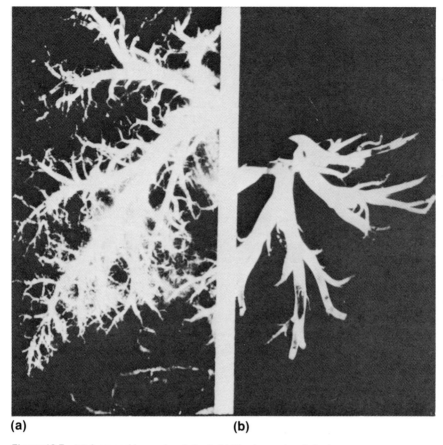

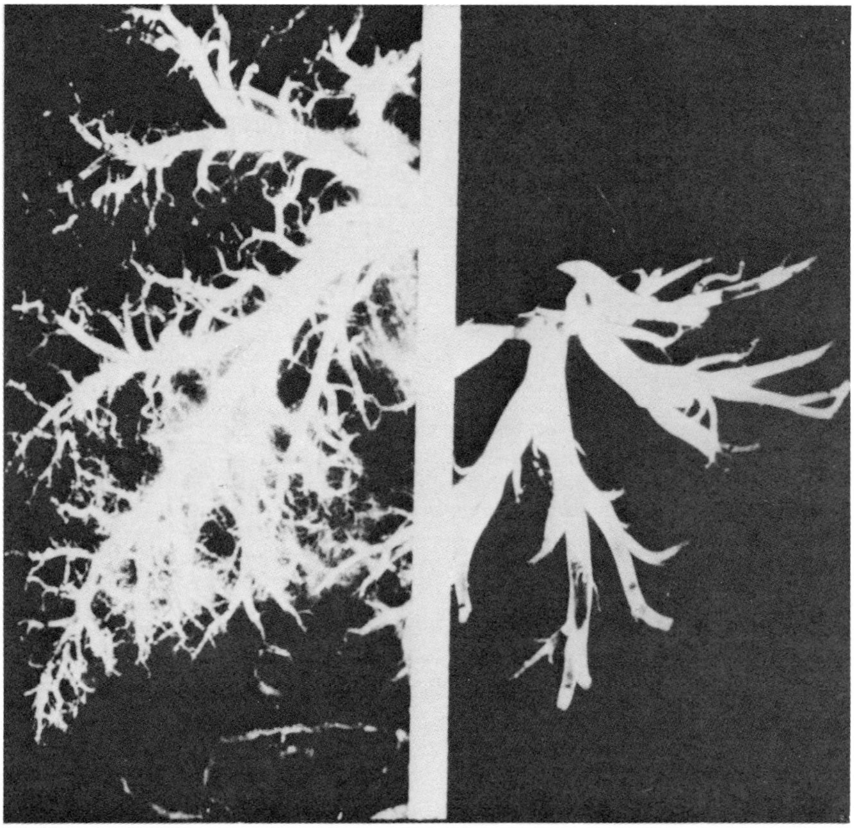

Figure 19.7 (a) A normal lung of an infant. (b) The lung of an infant who died from RDS, respiratory distress syndrome. (University of Virginia Neonatal Intensive Care Unit)

put on a ventilator, which forces high concentrations of oxygen into their lungs. If, after one or two days, these infants do not start producing enough lecithin to be able to breathe on their own, they may suffer blindness from exposure to such high levels of oxygen. After four days without producing their own lecithin, their chances of recovery are very slim. One experimental treatment for this disease is to spray lecithin, dissolved in a substance that will help it stick to the lining of the air sacs, into the air of the ventilator. This helps lower the surface tension in the air sacs, allowing the concentration of oxygen being forced into the lungs to be lowered and reducing the risk of permanent damage.

Cephalins

When the nitrogen compound in a phosphatide is ethanolamine, $H_2NCH_2CH_2OH$, the compound formed is called cephalin. Cephalins are found in blood platelets, and play an important role in the clotting of blood. They also serve as a source of inorganic phosphate for the formation of new tissue.

19.13 Sphingosine-based Phospholipids: Sphingolipids

The alcohol of sphingolipids is not glycerol, but rather is sphingosine. The most common sphingolipid is sphingomyelin.

OH
|
CHCH=CH(CH$_2$)$_{12}$CH$_3$
|
CHNH$_2$
|
CH$_2$OH

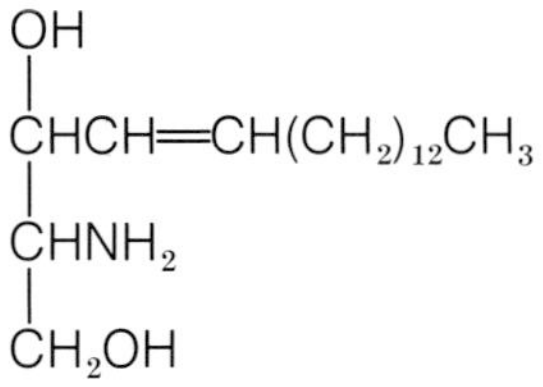

Sphingosine

S
P
H
I
N —Fatty acid
G
O
S
I
N —Phosphate — Choline
E

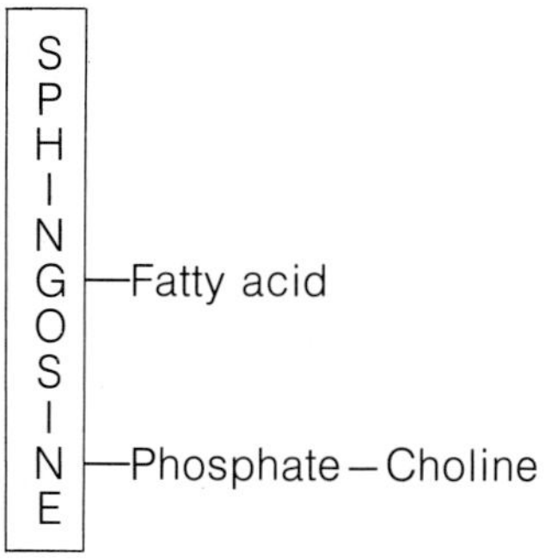

Sphingomyelin

Large amounts of sphingomyelins are found in brain and nervous tissue, and form part of the myelin sheath, the protective coating of nerves. The myelin sheath is very stable, due partly to the interlocking of the long fatty acid chains of the sphingomyelins. Certain diseases such as Neimann-Pick disease and multiple sclerosis result in the production of defective myelin sheaths. Neimann-Pick disease is a hereditary disease in which sphingomyelins build up in the brain, liver, and spleen, resulting in mental retardation and early death. Multiple sclerosis (MS) is a disease in which the body's own immune system attacks and destroys areas of myelin sheath. These areas are then replaced by scar tissue, which interrupts or distorts the flow of nerve impulses, causing paralysis, numbness, loss of coordination and balance, and speech difficulties.

19.14 Glycolipids

The main difference between glycolipids and phospholipids is that a glycolipid contains a sugar group rather than a phosphate group. The sugar group is usually galactose, but may also be glucose. The alcohol is either glycerol or sphingosine.

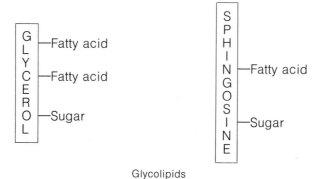

Glycolipids

Table 19.5 The Structure and Function of Some Steroids

Steroid	Structure	Function
Cortisone		One of many hormones produced in the adrenal glands. It is important in controlling carbohydrate metabolism and is used therapeutically to relieve symptoms of inflammation, especially in rheumatoid arthritis.
Vitamin D_2		Irradiation of the steroid hormone ergosterol with ultraviolet light breaks open one of the rings in the steroid nucleus, producing vitamin D_2. This vitamin is essential to prevent rickets, a disease of calcium metabolism.
Digitoxigenin		Extracted from the digitalis plant, this steroid is used in small doses to regulate a diseased heart. In large doses it causes death.
Testosterone		This male sex hormone regulates the development of the male reproductive organs.
Progesterone		This is the female sex hormone that is produced in pregnancy and acts on the uterine lining, preparing it to receive the embryo.

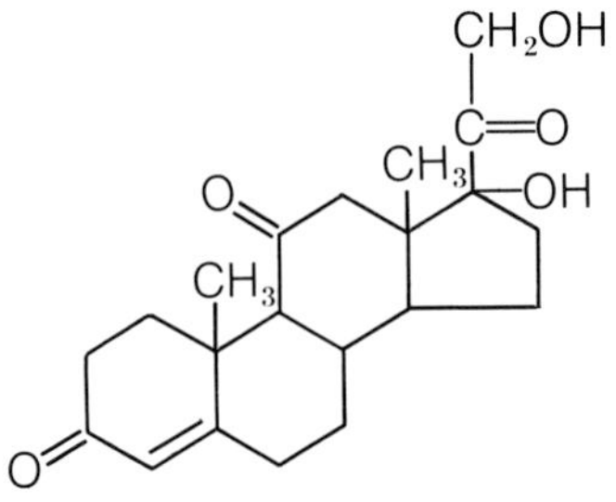

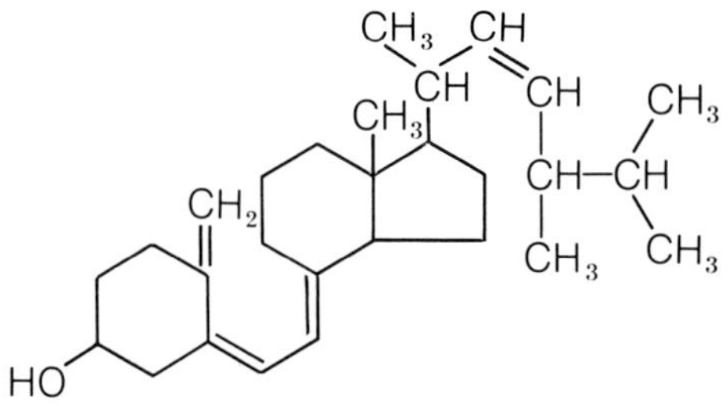

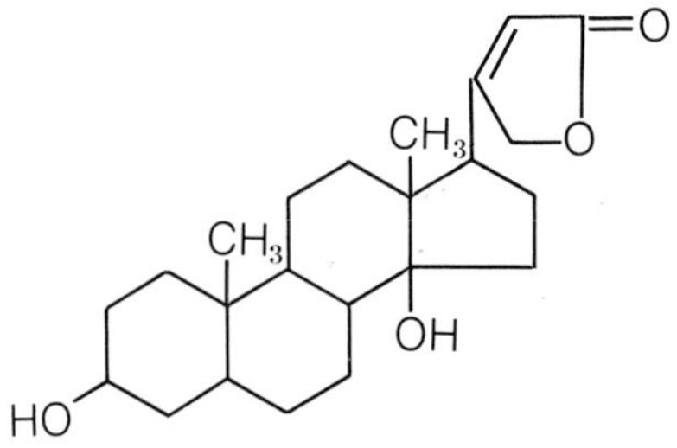

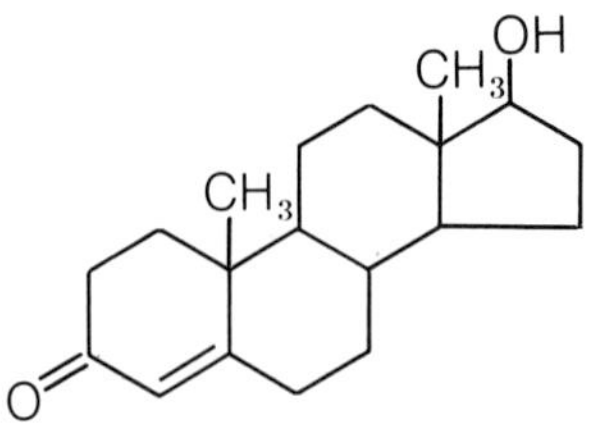

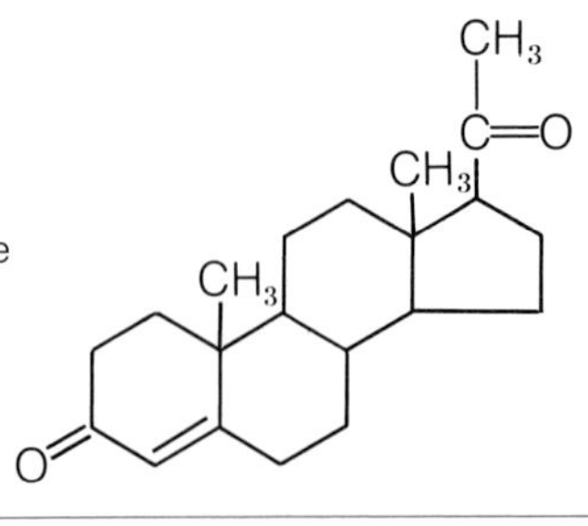

Cerebrosides are glycolipids that contain the base sphingosine. They are found in high concentrations in the brain and nerve cells, especially in the myelin sheath. Several hereditary fat metabolism diseases have been linked to errors in the metabolism of the glycolipids. In Gaucher's disease, the glycolipids contain glucose rather than galactose, and they collect in the spleen and kidneys. In Tay-Sachs disease, the infant lacks an enzyme necessary to break down glycolipids, and they collect in the tissues of the brain and eyes, causing muscular weakness, mental retardation, seizures, blindness, and death by the age of three.

Nonsaponifiable Lipids

19.15 Steroids

Nonsaponifiable lipids are those that are not broken apart by alkaline hydrolysis. Steroids are nonsaponifiable lipids whose structure is based on a complicated four-ring framework consisting of three cyclohexane rings and one cyclopentane ring.

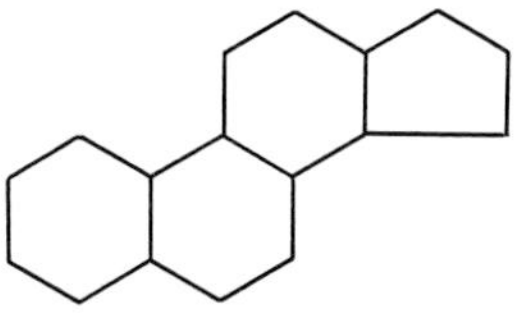

Steroid Nucleus

The steroid nucleus is found in the structure of several vitamins, hormones, drugs, poisons, bile acids, and sterols. The structure and function of some familiar steroids are shown in Table 19.5.

19.16 Cholesterol

Sterols are steroid alcohols. The most common sterol is **cholesterol.**

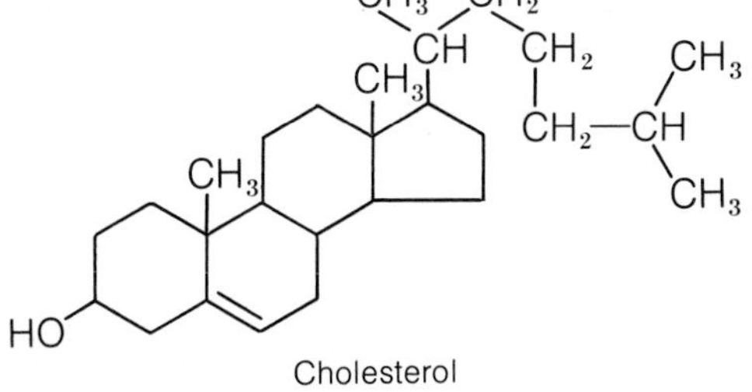

Cholesterol

Cholesterol is a part of all cell membranes, and is the starting material for the synthesis of steroids such as bile acids, sex hormones, and vitamin D. All cells are able to synthesize cholesterol from acetyl CoA, but 90% of the 3 to 5 grams of cholesterol that the body produces each day are produced by the liver. We also ingest some cholesterol each day from foods such as egg yolks and meats; however, high levels of such dietary cholesterol will reduce the liver production.

Cholesterol is carried in the blood in the form of lipoprotein (mainly as low density lipoprotein, or LDL). LDL plays a role in the regulation of cholesterol synthesis in cells other than the liver. These cells have specific places of attachment, or receptor sites, on their membranes for LDL. The interaction of LDL and these sites is the first step in the breakdown of LDL within the cell and the slowdown of cholesterol synthesis by the cell. If this interaction is prevented, which is the case in the genetic disease hypercholesterolemia, large amounts of LDL accumulate in the blood. People suffering from this disease die of atherosclerosis in their early 20s. However, the exact process by which LDL contributes to the development of such atherosclerosis is not well understood.

Lipid Metabolism

19.17 Digestion and Absorption

About 40% of the total caloric consumption of the average American is fat, although this figure will vary greatly with diet. Fats are not acted upon in the digestion process until they reach the small intestine. There they are mixed with bile salts, which emulsify the fats and allow enzymes in the pancreatic and intestinal juices to hydrolyze the fat into glycerol, fatty acids, and mono- and diglycerides.

Bile is a fluid that is continuously manufactured in the liver and stored in the gall bladder. Bile is composed of bile salts, bile pigments, and excess cholesterol. The bile salts, in addition to aiding in the digestion of fats, also help in the absorption of fat-soluble vitamins and the products of fat digestion. Bile pigments, mainly bilirubin, do not play a role in fat digestion. Rather, they help eliminate the waste products from the breakdown of hemoglobin, and are the substance that gives color to the feces. If the bile duct becomes blocked, or if the breakdown of hemoglobin occurs at a very high rate, bilirubin will build up in the blood. This will make the skin appear more yellow, and produce the condition of jaundice. In some people the mucous membranes of the gall bladder will absorb water, concentrating the bile. Under these conditions the cholesterol in bile, which is not very water-soluble, will crystallize out of solution together with bile salts and bile pigments, forming gall stones. These stones can cause infection and pain, and can obstruct the flow of bile, resulting in jaundice.

The glycerol, fatty acids, and mono- and diglycerides that are formed by the hydrolysis of triglycerides cross the intestinal barrier. There they are reformed into glycerides, and are transported to the blood by means of the lymph system. They enter the blood as micro-droplets, and are joined with proteins for transport in the blood. After a meal the blood is rich in fat, and has a milky, opalescent appearance.

19.18 Lipid Storage

Lipids and carbohydrates which are consumed in excess of our energy requirements are stored as triglycerides in the form of adipose tissue under the skin and around major organs. Such storage fat has several functions: It is an energy reserve, a support and a shock absorber for inner organs, and heat insulation for the body. Although the glycogen reserves of our bodies are sufficient to last only a few hours, a 70 kg male has enough stored triglyceride to sustain him for 70 days if there is water available.

The lipids in adipose tissue are in dynamic equilibrium with lipids in the blood. That is, stored fatty acids are constantly being exchanged with food fatty acids. The particular composition of the storage fat differs for different organisms, but it can be altered by controlling the diet. For example, experiments are being carried out to determine if cattle raised on special diets will develop higher concentrations of unsaturated fats in their meat.

19.19 Oxidation of Fatty Acids

Upon oxidation, fats produce much more energy than do carbohydrates; fats give 9 kilocalories/gram, whereas glycogen and starch give 4 kcal/gram. In vertebrates, oxidation of fatty acids provides at least half the energy needed by the liver, kidneys, heart, and the skeletal muscles at rest. In hibernating animals and migrating birds, fat is the only energy source.

The process of fat oxidation starts when fats stored in the adipose tissues are hydrolyzed to fatty acids and glycerol. These compounds are then carried to energy-requiring tissues where they will be oxidized. The glycerol enters the glycolysis pathway, as we discussed in Section 18.23. The fatty acids are oxidized, mainly in the mitochondria of the liver, heart, and skeletal muscles, by means of a repeating series of reactions called the **fatty acid cycle** or **β-oxidation.** This series of reactions is shown in Figure 19.8. We can briefly describe each of the steps in this process as follows:

1. The first step involves joining the fatty acid with coenzyme A (CoA). This requires one ATP, and produces AMP and a diphosphate (or pyrophosphate, denoted PP_i).

2. This step is a dehydrogenation reaction that produces a double bond between the alpha and beta carbon. The hydrogens that are removed in this reaction are attached to the hydrogen carrier molecule FAD, producing $FADH_2$.

3. In this step water is added to the double bond, forming an alcohol group on the beta carbon.

4. The alcohol group on the beta carbon is then oxidized (from which comes the name *beta oxidation*), producing NADH and H^+.

5. In the last step, the bond between the alpha and beta carbon is broken by coenzyme A. This produces one acetyl CoA and a new fatty acid (having two fewer carbon atoms) joined with coenzyme A. This fatty acid will again enter the fatty acid cycle at Step 2.

The cycle will continue to remove two-carbon units from the fatty acid until it has been completely oxidized. Each turn of the cycle produces one

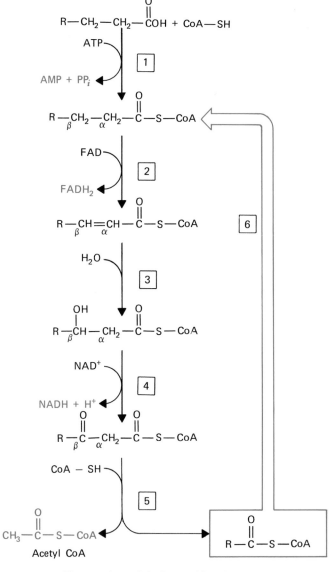

Figure 19.8 The reactions of the fatty acid cycle (see Section 19.19 for an explanation of each step).

FADH$_2$ and one NADH that can enter the electron transport chain, and one molecule of acetyl CoA that can be used in the citric acid cycle for the production of ATP. Overall, about 50% of the energy released in the complete oxidation of a fatty acid is trapped in molecules of ATP.

Example 19-1 _____

How many ATP's would be produced by the complete oxidation of one molecule of palmitic acid?
 (a) The formula of palmitic acid is

$$CH_3CH_2CH_2CH_2CH_2CH_2CH_2CH_2CH_2CH_2CH_2CH_2CH_2CH_2CH_2\overset{\overset{O}{\|}}{C}OH$$

$$\underbrace{\qquad\qquad}_{7}\ \overset{\smile}{6}\ \overset{\smile}{5}\ \overset{\smile}{4}\ \overset{\smile}{3}\ \overset{\smile}{2}\ \overset{\smile}{1}$$

(b) Because each turn of the fatty acid cycle removes a two-carbon unit, palmitic acid will require seven turns of the cycle. (Notice that the seventh turn will produce two molecules of acetyl CoA.)

(c) Therefore, 7 molecules each of FADH$_2$ and NADH which can enter the electron transport chain are formed (see Figure 18.16).

$$7\ FADH_2 \longrightarrow 14\ ATP$$

$$7\ NADH \longrightarrow 21\ ATP$$

(d) Eight molecules of acetyl CoA which can enter the citric acid cycle are produced. From Section 18.25 we have seen that each acetyl CoA produces 12 ATP. Therefore,

$$8\ acetyl\ CoA \longrightarrow 96\ ATP$$

(e) The total number of ATP's formed, therefore, is $14 + 21 + 96 = 131$ ATP. Remember, however, that one ATP was needed at Step 1 for palmitic acid to enter the cycle, so the actual number of new ATP's available is 130.

 The acetyl CoA's produced in the oxidation of fatty acids can be put to many uses in the metabolism of the cell. In addition to entering the citric acid cycle, they can be used in biosynthesis of materials required by the cell (such as cholesterol, amino acids, or other fatty acids), or they can be used in the synthesis of ketone bodies. Figure 19.9 summarizes the key role played by acetyl CoA in the metabolism of all substances in the cell.

19.20 Ketone Bodies

When acetyl CoA is produced in excess by the liver, it is converted into acetoacetic acid, acetone, and β-hydroxybutyric acid—compounds that are

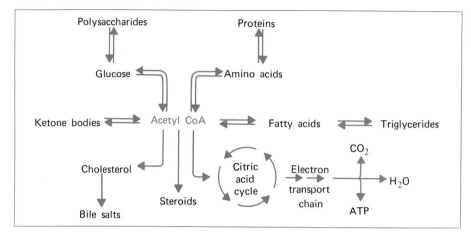

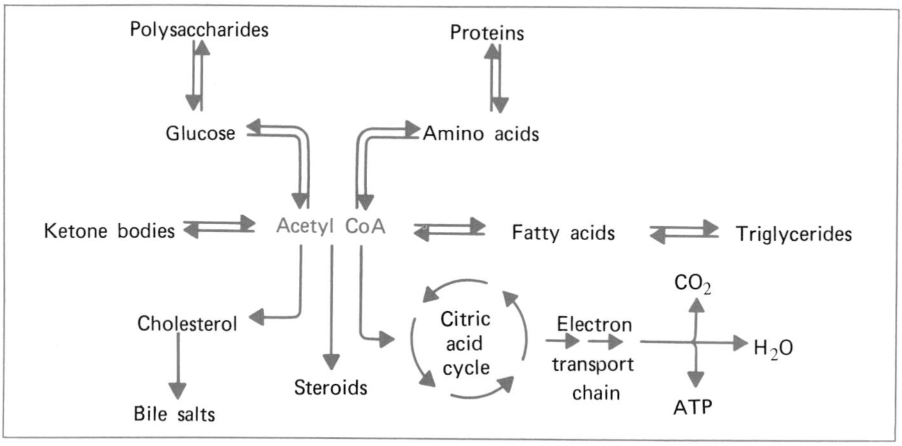

Figure 19.9 Acetyl CoA plays a central role in the anabolic and catabolic reactions of cellular metabolism.

called **ketone bodies** (see Table 18.4). These compounds are then transported to other tissues, where they are oxidized in the citric acid cycle or excreted by the kidneys. Their concentration in the blood is normally very low—about 1 mg/100 ml.

Any disruption of normal metabolism (such as liver damage, diabetes mellitus, starvation, or diets causing a restriction or decrease in glucose metabolism) will increase the level of fat metabolism and, therefore, increase the production of ketone bodies. If the level of ketone bodies in the blood exceeds the amount that can be used by the tissues and excreted by the kidneys, a condition known as **ketosis** occurs. Because two of the ketone bodies are acids, in ketosis the pH of the blood will drop and acidosis will occur. As we have seen, acidosis can lead to nausea, depression of the central nervous system, dehydration, and in extreme cases, coma and death.

Chapter Summary

Lipids are waxy or oily substances that are not soluble in water. Their main functions are to form part of the structure of cell membranes and to store energy for the cell. They are also the starting material for the formation of hormones, vitamins, and bile acids. Triglycerides are esters of glycerol and three fatty acids, and may be either solid (fats) or liquid (oils) at room temperature. Oils have a larger number of unsaturated fatty acids than do fats. The polyunsaturated fatty acids (linoleic, linolenic, and arachidonic acids) are essential fatty acids, and are used in the synthesis of prostaglandins. Waxes are esters of monohydric alcohols and long-chain fatty acids.

Unsaturated fatty acids, or unsaturated fatty acid side chains of tryglycerides, can be hydrogenated to produce saturated fatty acids or saturated fats. Fats and oils can become rancid by hydrolysis of the triglyceride or by oxidation of unsaturated fatty acid side chains. Triglycerides can be hydrolyzed to form three fatty acids and glycerol. If the hydrolysis occurs in the presence of a strong base, glycerol and the salts of the fatty acids, or soaps, are formed. Soaps are good emulsifying agents because they have a nonpolar region of the molecule that will dissolve in the fat or oil, and a polar region of the molecule that will dissolve in the water.

Phospholipids are compound lipids that form the structure of cell membranes and that help to transport other lipids in the body. All the phospholipids contain a phosphate group, fatty acids, an alcohol (glycerol or sphingosine), and a nitrogen-containing compound (choline, ethanolamine, serine, or inositol). Glycolipids are compound lipids containing a sugar group rather than a phosphate group, and are found in high concentrations in brain and nervous tissue. Steroids are nonsaponifiable lipids with a complex ring structure that is found in many vitamins, hormones, drugs, poisons, bile acids, and sterols. Cholesterol is the most common sterol. It is a component of all cells, and is used by the cells as a starting material for the synthesis of many other compounds.

Fat digestion begins in the small intestine, where fats (emulsified by bile salts) are hydrolyzed to glycerol, fatty acids, and mono- and diglycerides. Lipids that are eaten in excess of energy requirements are stored in the fat cells of adipose tissue. Fats are the main energy reserve of the body, producing 9 kilocalories per gram when oxidized. When needed, fats are hydrolyzed to glycerol (which enters the glycolysis pathway) and fatty acids (which are oxidized to acetyl CoA in the fatty acid cycle). The acetyl CoA's that are produced may enter the citric acid cycle or may be used in biosynthesis. If acetyl CoA's are produced in excess of requirements, they are converted into ketone bodies by the liver. In high concentrations, ketone bodies can cause acidosis.

Exercises and Problems

1. State the difference between each of the following:

   (a) Saponifiable lipid and nonsaponifiable lipid
   (b) Simple triglyceride and mixed triglyceride
   (c) Fat and oil
   (d) Saturated fatty acid, unsaturated fatty acid, and polyunsaturated fatty acid
   (e) Wax and triglyceride
   (f) Phospholipid and triglyceride

(g) Lecithin and sphingomyelin
(h) Phospholipid and glycolipid
(i) Hydrolysis and hydrogenation
(j) Hydrolysis and saponification

2. Why does butter turn rancid? How does refrigeration slow this process? What causes the odor of rancid butter?
3. Why would a diet lacking in linoleic acid be bad for a person's health?
4. (a) Describe what is happening, on the molecular level, when you wash salad oil from your hands with soap.
   (b) Suggest a possible reason for the antibacterial action of soap, using the fact that bacterial cell membranes are formed by lipids.
5. In what ways are detergents superior to soaps?
6. Advertising would lead us to believe that cholesterol is bad for us.
   (a) Give reasons to support this statement.
   (b) Then, explain why this statement is not entirely true.
7. Why is it more efficient for our bodies to store excess food calories in the form of body fat than as glycogen?
8. Describe the steps involved in the oxidation of fats by our bodies.
9. Why are diets that severely restrict the intake of carbohydrates potentially dangerous to a person's health?
10. Most general anesthetics are nonpolar compounds. Explain how this property makes them effective in producing anesthesia.
11. Write the general formula for each of the following compounds, and describe the function of each in the human body:
    (a) phosphatide
    (b) lecithin
    (c) cephalin
    (d) sphingomyelin
    (e) glycolipid
12. Write the formula for the following triglycerides:
    (a) tripalmitin
    (b) the mixed triglyceride containing glycerol, arachidic, stearic, and oleic acids
    (c) triolein.
13. Which triglyceride in question 12 would:
    (a) have the highest melting point?
    (b) be most likely to be found in animal fat?
    (c) have the highest iodine number?
14. Write the equation for the complete hydrogenation of the triglycerides in parts (b) and (c) of question 12.

15. The following triglyceride is found in lard.

$$CH_2-O-\overset{\overset{\displaystyle O}{\|}}{C}(CH_2)_{14}CH_3$$

$$CH-O-\overset{\overset{\displaystyle O}{\|}}{C}(CH_2)_{16}CH_3$$

$$CH_2-O-\overset{\overset{\displaystyle O}{\|}}{C}(CH_2)_7CH=CH(CH_2)_7CH_3$$

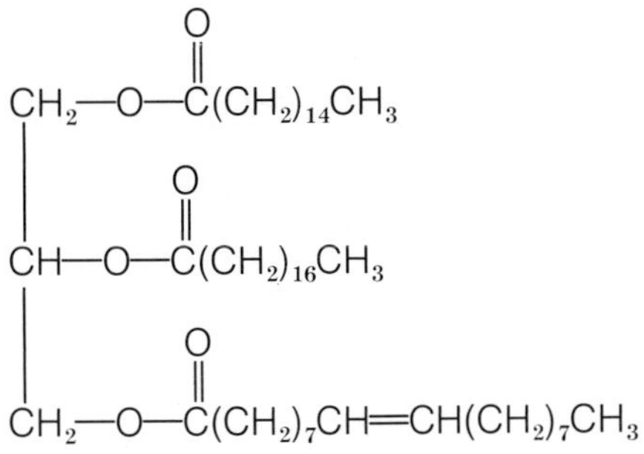

 (a) Write the structure of all the products that would result from the digestion of this triglyceride.
 (b) How does the digestion and absorption of the fats in a strip of bacon differ from the digestion and absorption of the starch in a piece of toast?

16. Before commercially produced soap was available, people made soap from lard and lye that was extracted from wood ashes with a small amount of water. This lye solution contained basic substances such as KOH, Na_2CO_3 and K_2CO_3. Write the equation for the formation of soap from the triglyceride shown in question 15 and KOH.

17. How many ATP's would be produced in the complete oxidation of one molecule of each of the following? Explain your answers.

 (a) lauric acid
 (b) stearic acid
 (c) arachidic acid

chapter 20

Proteins

Learning Objectives

By the time you have finished this chapter, you should be able to:

1. Write the general structure of an amino acid.

2. Describe the difference between
   a. a simple and a conjugated protein
   b. a globular and a fibrous protein

3. Explain how an amino acid can act as a buffer.

4. Define *zwitterion* and *isoelectric point.*

5. Describe the types of bonding found in each of the following protein structures:
   a. primary
   b. secondary
   c. tertiary
   d. quaternary

6. State five methods for denaturing proteins.

7. Describe the steps of protein digestion.

8. Define *essential amino acid* and *adequate protein.*

9. In general terms, identify the reactants and products for each of the following reactions, and describe when these reactions might occur during protein metabolism.
   a. transamination
   b. oxidative deamination
   c. urea cycle

10. Describe four metabolic uses of amino acids.

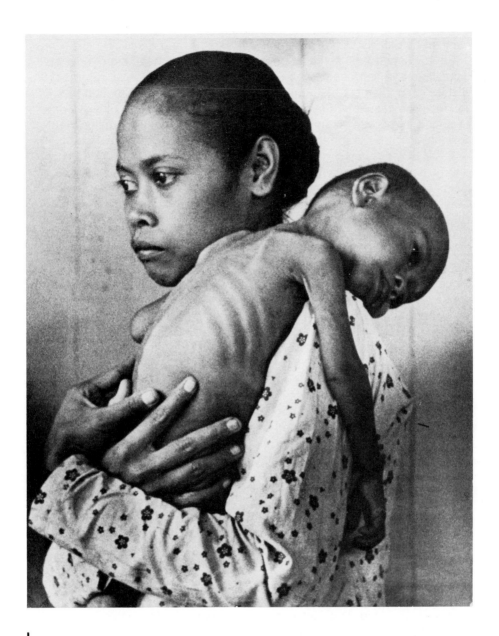

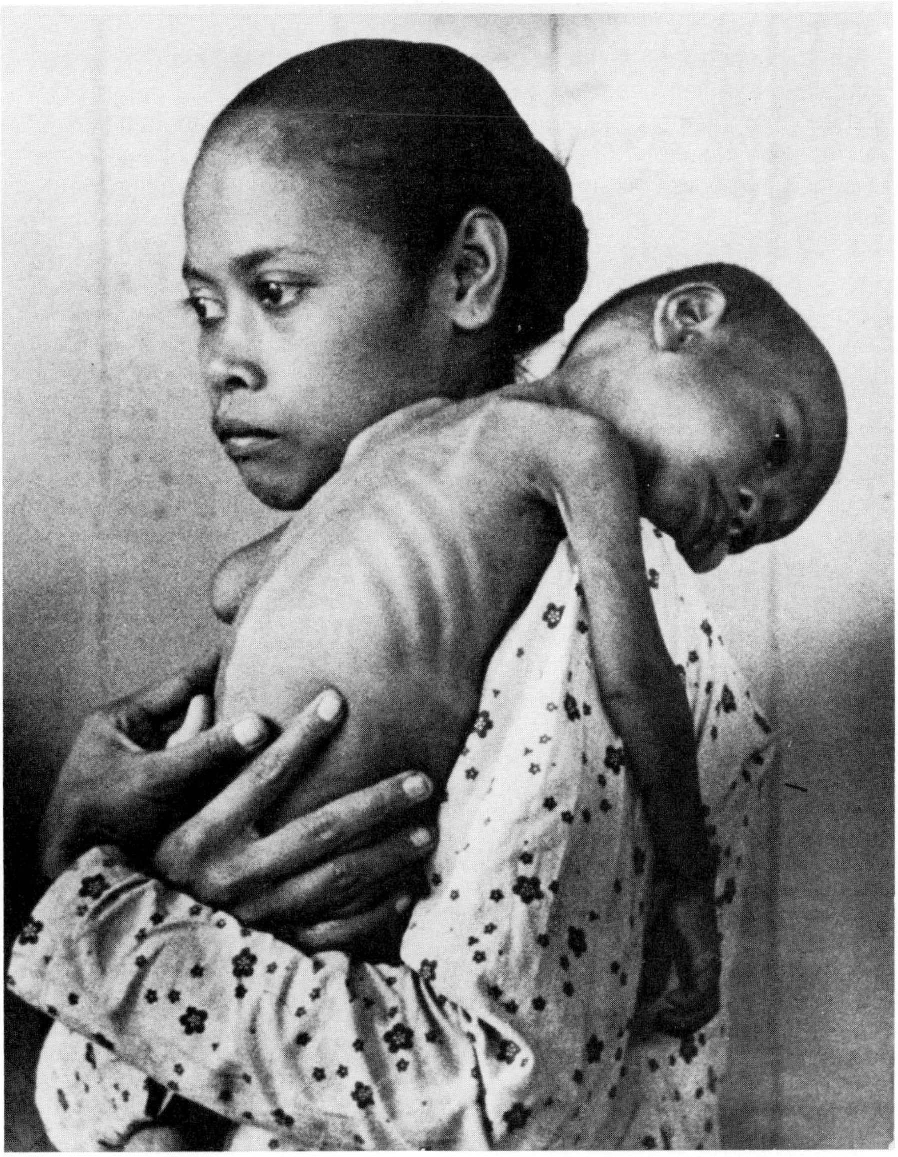

In a small village in western Africa, a young mother nursed her first child. He was a strong, healthy 18-month-old boy who would soon have to be weaned, for his mother was about to deliver her second baby. After the second child was born, the little boy was switched to a diet of mush made from corn. Slowly, day by day, he began to lose energy. He became irritable and apathetic. His growth slowed and he began to lay miserably on the ground, crying and whimpering. After a year his appearance had changed drastically. His face and legs were swollen with fluid, his curly black hair had become soft and brown and was falling out, and his skin was flaky and covered with a rash. His mother had become greatly concerned, but had no way to obtain medical care. She watched helplessly as her son sank into a coma, and died.

In a village to the south, another young boy had been more fortunate. His village had just been given powdered milk and a vegetable-protein mixture to supplement the normal diet of the inhabitants. Although this child had been suffering symptoms similar to the boy in the other village, he began to recover. As he continued on his new diet, the swelling went down, his rashes disappeared, and his energy and appetite returned.

Both of these children were suffering from Kwashiorkor, a disease that affects about 70% of the world's children under six years of age. Kwashiorkor is an African name meaning a disease that affects the first-born when he or she is displaced from the breast by the second child. The first child is then fed a diet of starchy foods that does not contain enough of the proteins essential for normal growth and development. The exact symptoms of Kwashiorkor will vary from country to country depending on the diet, but generally the children show appetite loss, slowed growth, edema (or swelling), disorders of the pigments of the skin and hair, and decreased protection against infection. There is considerable debate as to whether protein deficiency in children also causes mental retardation, but this seems possible because the brain and nervous system grow faster than any other tissues during the first four years of life. At present, more than half of the children affected by this disease will die. The disease, however, can be cured by diets supplemented with milk or vegetable-protein mixtures high in the proteins missing from the child's regular diet. Thus, the widespread occurrence of Kwashiorkor can be decreased only by supplementing the diets of a great portion of the world's people with adequate protein.

Proteins

20.1 Occurrence, Composition, and Function

Proteins are the most complex and varied class of molecules found in the cell. They are found in all living cells, and their biological importance can't be overemphasized. This fact was recognized by the German chemist G. T. Mulder in 1839, when he gave this class of compounds the name *protein*, which means "of prime importance."

All proteins are composed of the elements carbon, nitrogen, oxygen, and hydrogen. Most proteins also contain sulfur, and some have phosphorus and other elements such as iron, zinc, or copper. Proteins are large polymers, and upon hydrolysis will produce monomer units called **amino acids.** Proteins can be divided into two major classes: **simple proteins,** which produce only amino acids upon hydrolysis, and **conjugated proteins,** which produce amino acids and other organic or

Table 20.1 Proteins Have Very Large Molecular Weights Compared to Other Compounds

Compound	Molecular Weight
Inorganic Compounds	
Sodium chloride	58.5
Sulfuric acid	98
Organic Compounds	
Ethanol	48
Urea	60
Carbohydrates	
Glucose	180
Sucrose	342
Cellulose	about 500,000
Lipids	
Cholesterol	384
Triolein	836
Proteins	
Insulin	6,300
Ribonuclease	12,640
Hemoglobin	68,000
γ-Globulin	149,900
Fibrinogen	450,000
Glutanate dehydrogenase	1,000,000
Hemocyanin	9,000,000

inorganic substances upon hydrolysis. These other substances are called **prosthetic groups.**

The molecular weight of most proteins ranges from 12,000 to 1 million or more (Table 20.1). This large size gives protein molecules colloidal properties. For example, they don't pass through differentially permeable membranes. The presence of proteins in urine, therefore, warns doctors of the possibility of damage to the membranes of the kidneys.

Because of their varied natures, proteins can be classified in several different ways. A second classification is based on the physical characteristics of the protein molecule. **Globular proteins** are soluble in water, are quite fragile, and have an active function, such as catalyzing reactions (in the case of enzymes), or transporting other substances (as, for example, hemoglobin). **Fibrous proteins** are insoluble in water, are

Table 20.2 A Classification of Proteins Based on Their Functions in Living Organisms

Class	Example
Enzymes	Pepsin Amylase
Structural proteins	Keratin Collagen
Storage proteins	Ferritin Casein
Transport proteins	Hemoglobin Myoglobin
Hormones	Insulin Parathormone
Contractile proteins	Actin Myosin
Protective proteins	Antibodies Fibrinogen
Toxins	Venoms Botulinus toxin

physically tough, and have a structural or protective function. The keratin in hair and nails, the collagen in tendons, and the contractile fibrils in muscles are examples of such fibrous proteins (Figure 20.1).

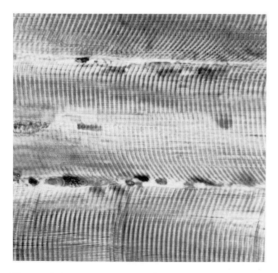

Figure 20.1 The structure of skeletal muscle. The muscle cells contain myofibrils that, when stained, show cross banding or striations, hence the name striated muscle. The myofibrils consist of two types of contractile proteins: actin and myosin. (Courtesy J. Robert McClintic)

The biological importance of proteins results from their wide variety of functions (Table 20.2). Proteins are the body's main dietary source of nitrogen and sulfur. In addition to their catalytic and structural functions, they make up the contractile system of muscles. As antibodies they are the defense system of the body, and as hormones they regulate the body's glandular activity. In the blood they maintain fluid balance, are part of the clotting process, and transport oxygen and lipids. They can act as poisons, such as the venoms in animal bites and stings, or toxins, such as the bacterial toxin producing botulism in improperly produced foods. And some antibiotics that are secretions of bacteria and fungi are protein in nature.

20.2 Amino Acids

The particular function of a given protein is determined by the sequence or order of amino acids in the protein molecule. It is necessary, therefore, for any study of proteins to include a thorough discussion of amino acids. **Amino acids** are carboxylic acids that have an amino group on the alpha carbon—the carbon next to the carboxyl group. The general structure of an amino acid is as follows:

$$\text{H}_2\text{N}-\underset{\underset{\text{R}}{|}}{\overset{\overset{\text{H}}{|}}{\text{C}}}-\overset{\overset{\text{O}}{\|}}{\text{C}}-\text{OH}$$

α-Carbon

Amino group R-group side chain Carboxyl group

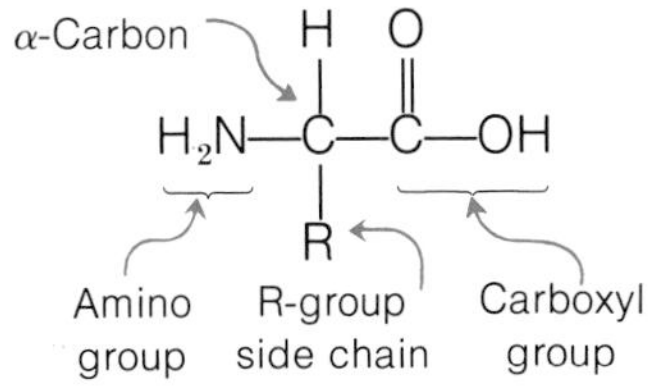

The different R-group side chains on the amino acids make one amino acid different from another. Most naturally occurring proteins are composed of the 20 amino acids shown in Figure 20.2, but there are a few specialized types of proteins that contain other, more rare, amino acids. Still other amino acids not found in proteins exist in a free or combined form; these are, in general, derivatives of the 20 amino acids found in proteins.

*20.3 The L-Family

All amino acids found in proteins, with the exception of glycine, are optically active and belong to the L-family. D-family isomers of amino acids can be found in nature, but never occur in proteins.

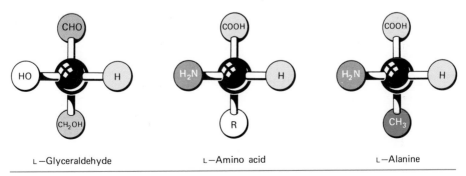

L—Glyceraldehyde L—Amino acid L—Alanine

* This section is optional and may be omitted without loss of continuity.

	Structure of R	Name of Amino Acid	Abbreviation
NONPOLAR R-GROUP	—H	Glycine	Gly
	—CH_3	Alanine	Ala
	—CH—CH_3 CH_3	Valine	Val
	—CH_2—CH—CH_3 CH_3	Leucine	Leu
	—CH—CH_2—CH_3 CH_3	Isoleucine	Ile
	—CH_2—CH_2—S—CH_3	Methionine	Met
	—CH_2⬡	Phenylalanine	Phe
	—CH_2⬡—OH	Tyrosine	Tyr
	—CH_2 (indole ring)	Tryptophan	Trp
	HO—C(=O)—CH—CH_2 HN CH_2 CH_2 (complete structure)	Proline	Pro
POLAR R-GROUP	—CH_2—OH	Serine	Ser
	—CH—OH CH_3	Threonine	Thr
	—CH_2—SH	Cysteine	Cys
	—CH_2—C(=O)—NH_2	Asparagine	Asn
	—CH_2—CH_2—C(=O)—NH_2	Glutamine	Gln
Acidic R-group	—CH_2—C(=O)—OH	Aspartic acid	Asp
	—CH_2—CH_2—C(=O)—OH	Glutamic acid	Glu
Basic R-group	—CH_2—CH_2—CH_2—CH_2—NH_2	Lysine	Lys
	—CH_2—CH_2—CH_2—NH—C(=NH)—NH_2	Arginine	Arg
	—CH_2—C=CH HN N C H	Histidine	His

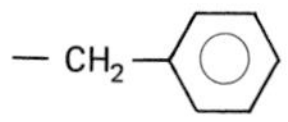

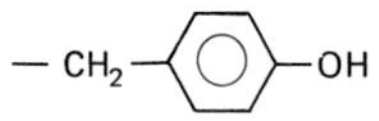

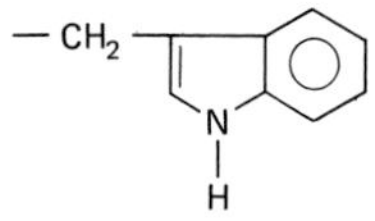

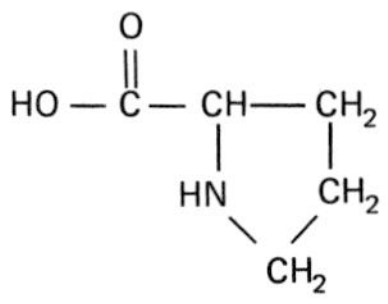

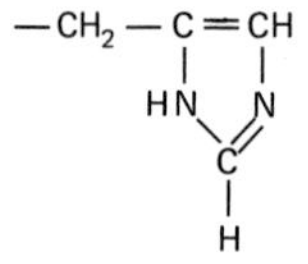

20.4 Acid-Base Properties

Because proteins can, in many cases, be viewed as very large amino acids, a knowledge of the acid-base properties of amino acids can be of great help in understanding some of the properties of proteins. Amino acids contain both an acid group—the carboxyl group—and a basic group—the amino group. In water, amino acids can act as either acids or bases. Molecules having this property are called **amphoteric.**

Amino acids are soluble in water, and have very high melting points. This suggests that they do not exist as uncharged molecules, but rather are found in the form of the highly polar **zwitterion** or dipolar ion.

$$H_3N^+\!-\!\overset{\displaystyle H}{\underset{\displaystyle R}{\overset{|}{\underset{|}{C}}}}\!-\!\overset{\displaystyle O}{\overset{\|}{C}}\!-\!O^-$$

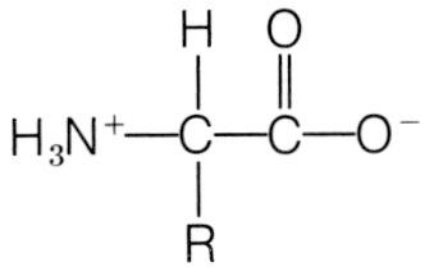

Zwitterion or dipolar ion

The zwitterion is formed when the acidic carboxyl group donates a hydrogen ion to the basic amino group. Although we may write the structure of the amino acid in the uncharged form, keep in mind that it will usually be found as a dipolar ion.

20.5 Isoelectric Point

There is a specific pH at which each amino acid and protein will be electrically neutral, and will not move in an electric field. This pH is called the **isoelectric point** for that molecule. At a pH more basic than the isoelectric point, the amino acid will have a net negative charge and will move toward the positive electrode. At a pH more acidic than the isoelectric point, the amino acid will carry a net positive charge and will move toward the negative electrode. Because some amino acids have an ionizable R-group, each amino acid and protein has a specific isoelectric point (Table 20.3). Proteins can be separated from one another at different pH levels on the basis of their charge by a process called electrophoresis. In this technique, each protein will migrate toward the positive or negative electrode at a different rate depending upon the pH and voltage applied. Paper electrophoresis is a useful tool in analyzing the proteins in human blood serum (Figure 20.3).

At the isoelectric point, the protein will have minimum solubility—the protein molecules can cluster together and are most easily removed from solution. Casein, for example, is the protein found in cow's milk, and has

Figure 20.2 *(Opposite page)* Common amino acids. The general structure for these amino acids is:

$$^+H_3N\!-\!\overset{\displaystyle H}{\underset{\displaystyle R}{\overset{|}{\underset{|}{C}}}}\!-\!\overset{\displaystyle O}{\overset{\|}{C}}\!-\!O^-$$

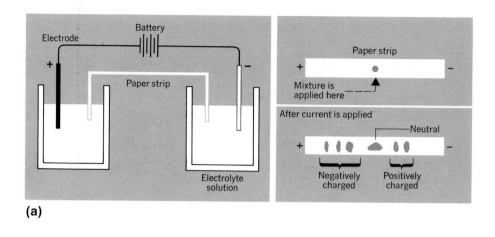

(a)

(b)

Figure 20.3 **Paper electrophoresis. Proteins and amino acids can be separated by their rate of migration in an electric field. (a) Apparatus for paper electrophoresis. (b) Electrophoretic separation of normal hemoglobin (top), and sickle cell hemoglobin (bottom). [(b) Courtesy Department of Pathology, Santa Barbara Cottage Hospital)]**

an isoelectric point of pH 4.7. The normal pH of cow's milk is 6.3. In the production of cheese, however, bacteria produce lactic acid that lowers the pH of the milk. This, then, lowers the solubility of the casein, causing the milk to curdle.

20.6 Buffering Properties

Because amino acids (and proteins) can act as either acids or bases, they are effective buffers in aqueous solution.

$$H_2N-\underset{\underset{R}{|}}{\overset{\overset{H}{|}}{C}}-\overset{\overset{O}{\|}}{C}-O^- \xleftarrow{+OH^-} H_2N-\underset{\underset{R}{|}}{\overset{\overset{H}{|}}{C}}-\overset{\overset{O}{\|}}{C}-OH \xrightarrow{+H^+} H_3N^+-\underset{\underset{R}{|}}{\overset{\overset{H}{|}}{C}}-\overset{\overset{O}{\|}}{C}-OH$$

One of the functions served by the proteins in the blood is to act as buffers, helping to keep the blood pH within its very narrow normal range (pH = 7.35 to 7.45).

Table 20.3 Isoelectric Points of Some Amino Acids and Proteins

Compound	Isoelectric Point
Amino acid	
Glutamic acid	4.0
Alanine	6.0
Lysine	10.5
Protein	
Egg albumin	4.6
Urease	5.0
Hemoglobin	6.8
Myoglobin	7.0
Chymotrypsin	9.5
Lysozyme	11.0

Protein Structure

Proteins are complex molecules that can be classified according to specific primary, secondary, tertiary, and quaternary structural properties.

20.7 Primary Structure: The Amino Acid Sequence and the Peptide Bond

The **primary structure** of a protein is given by the sequence of amino acids in the protein molecule. These amino acids are joined together by covalent bonds called **peptide bonds.** The peptide bond is an amide linkage, formed by joining the carboxyl group of one amino acid to the amino group of a second amino acid through elimination of water (a condensation reaction) (Figure 20.4).

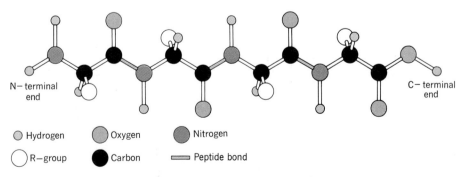

N−terminal end

C−terminal end

○ Hydrogen ◯ Oxygen ◯ Nitrogen

◯ R−group ● Carbon ▭ Peptide bond

Figure 20.4 **Primary structure of proteins. This polypeptide chain contains four amino acids joined by peptide bonds. The end of the molecule having the free amino group is the N-terminal end, and the end of the molecule with the free carboxyl group is the C-terminal end.**

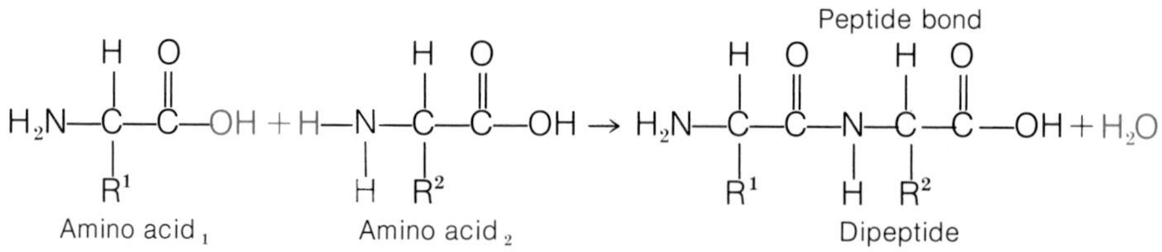

Peptide bond

$$H_2N-\underset{\underset{R^1}{|}}{\overset{\overset{H}{|}}{C}}-\overset{\overset{O}{\|}}{C}-OH + H-\underset{\underset{\underset{R^2}{|}}{H}}{\overset{\overset{H}{|}}{N}}-\overset{\overset{O}{\|}}{C}-OH \rightarrow H_2N-\underset{\underset{R^1}{|}}{\overset{\overset{H}{|}}{C}}-\overset{\overset{O}{\|}}{C}-\underset{\underset{H}{|}}{N}-\underset{\underset{R^2}{|}}{\overset{\overset{H}{|}}{C}}-\overset{\overset{O}{\|}}{C}-OH + H_2O$$

Amino acid $_1$ Amino acid $_2$ Dipeptide

The peptide bond is stable in the face of changes in pH, in solvents, or in salt concentrations. It can be broken only by acid or base hydrolysis, or by specific enzymes. Two amino acids held together by a peptide bond are called a **dipeptide;** three amino acids form a **tripeptide,** and more than three form a **polypeptide.** There is no precise dividing line between polypeptides and proteins. For example, insulin, with 51 amino acids in its primary structure, is a very small protein. Glucagon, with 21 amino acids, however, is considered a large polypeptide. There are many small polypeptides that have important functions in biological systems. Glutathione, which plays a role in oxidation-reduction reactions, is a tripeptide (Figure 20.5). Vasopressin, a hormone produced by the pituitary, is an octapeptide, and the antibiotic bacitracin is a polypeptide with 11 amino acids in its sequence.

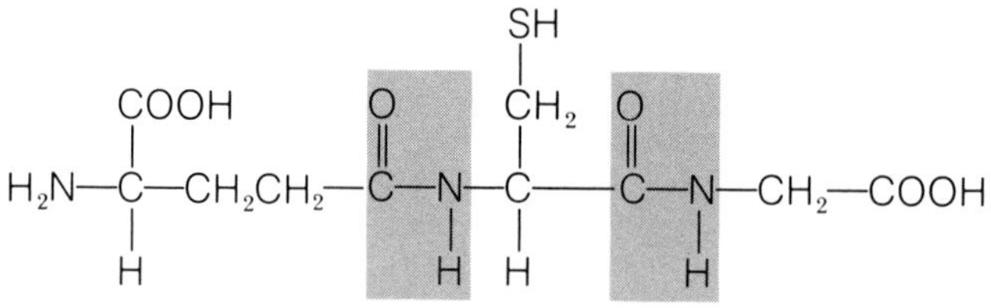

Figure 20.5 The structure of the tripeptide glutathione.

Two different dipeptides can be formed between the amino acids glycine and alanine.

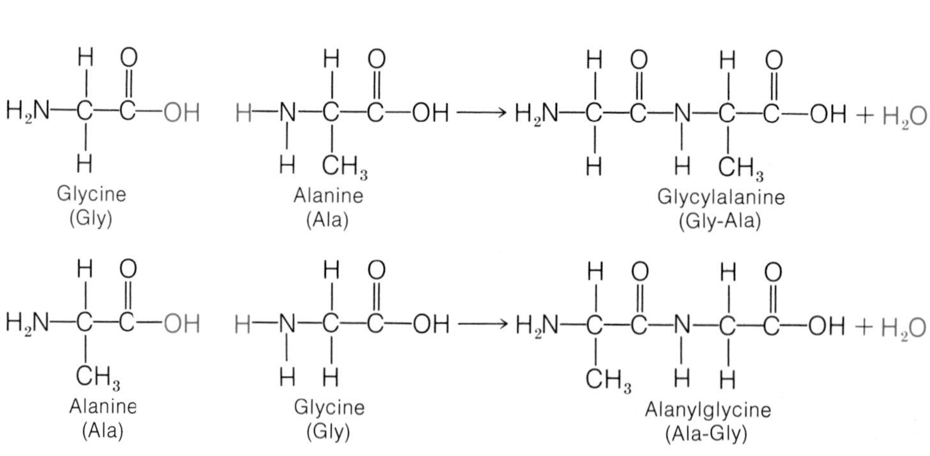

Glycine (Gly) Alanine (Ala) Glycylalanine (Gly-Ala)

Alanine (Ala) Glycine (Gly) Alanylglycine (Ala-Gly)

This gives us a good reason for discussing the standard rules for the naming of peptides and proteins. Because proteins contain very large numbers of amino acids, three-letter abbreviations of the amino acid names are used when writing the amino acid sequence (see Figure 20.2 for the abbreviations). The peptide bond is represented by a dash or dot between the amino acid names. One end of the protein will consist of an amino acid having a free amino group. This is called the N-terminal end of the protein, and the N-terminal amino acid is listed first in the sequence of amino acids. At the other end of the protein there will be an amino acid having a free carboxyl group. This is the C-terminal amino acid, and it is the last amino acid listed in the sequence. The following are two examples of these naming rules:

Lysine–Aspartic acid–Serine–Asparagine–Glutamic acid (I)
Lys–Asp–Ser–Asn–Glu

Valine · Phenylalanine · Alanine · Tryptophan · Leucine (II)
Val · Phe · Ala · Trp · Leu
(N-terminal end) (C-terminal end)

The order of amino acids in a protein will determine its function, and is critical to its biological activity. A change of just one amino acid in the

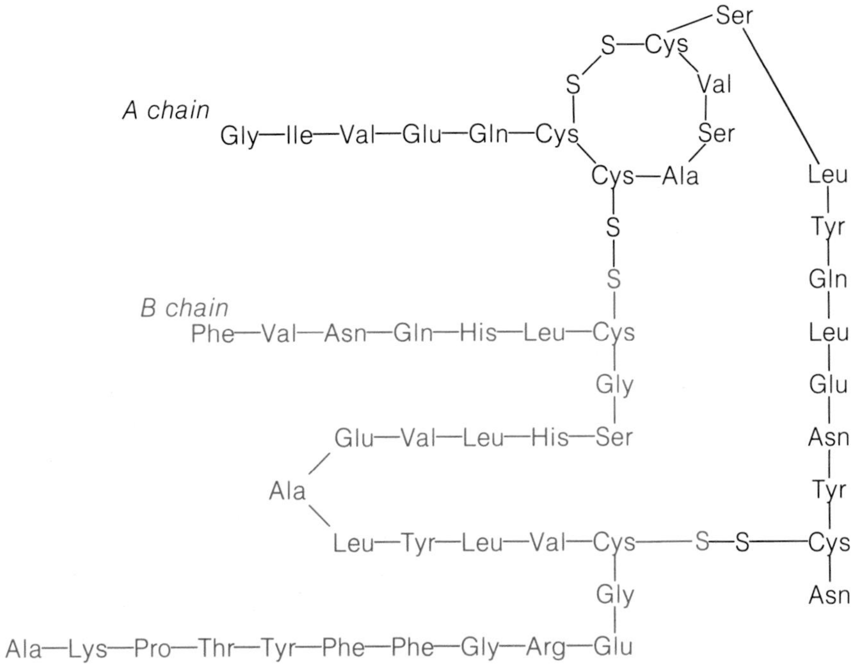

Figure 20.6 The complete amino acid sequence of bovine insulin was first determined by F. Sanger in 1953, for which he was awarded the Nobel prize. The insulin molecule contains 51 amino acids in two polypeptide chains.

sequence can disrupt the entire protein molecule. For example, hemoglobin, the molecule in the blood that carries oxygen, consists of four polypeptide chains having a total of 574 amino acid units. Changing just one specific amino acid in one of the chains results in the defective hemoglobin molecule found in patients with sickle cell anemia.

Adult hemoglobin (Hb-A)	Val–His–Leu–Thr–Pro–Glu–Glu–Lys– . . .
Sickle cell hemoglobin (Hb-S)	Val–His–Leu–Thr–Pro–Val–Glu–Lys– . . .

Recently scientists have found a peptide called levendorphin in the blood of schizophrenics. The structure of this peptide differs from that of the pituitary hormone of normal persons by one amino acid. This change may be the cause of the symptoms of schizophrenia.

Determination of the amino acid sequence in a protein is a complex procedure that was first developed by F. Sanger in 1953. He determined the amino acid sequence of the protein insulin (Figure 20.6).

20.8 Secondary Structure: Noncovalent Bonding

The **secondary structure** of a protein is the specific geometric arrangement of the amino acids in space. These arrangements, resulting from hydrogen bonding, were established by Linus Pauling using X-ray diffraction (Figure 20.7).

Figure 20.7 Hydrogen bonds can form (a) between amino acids on the same polypeptide chain, forming a loop in the molecule, or (b) between amino acids on different polypeptide chains.

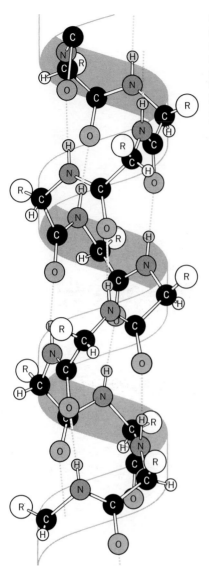

Figure 20.8 Model of a polypeptide chain in an alpha-helix configuration. The amino acids are coiled in a manner resembling a circular staircase, with the loops held together by hydrogen bonding (dotted lines). Note that the R-groups on the amino acids point away from the center of the helix. (Adapted from B. Low and E. T. Edsall, *Currents in Biochemical Research,* Wiley-Interscience, New York, 1956. Used by permission.)

α-Helix

Pauling determined that the polypeptide chains in the protein keratin were curled in an arrangement called an α-helix. In this arrangement, the amino acids form loops in which the hydrogen on the nitrogen atom in the peptide bond is hydrogen bonded to the oxygen attached to the carbon atom of a peptide bond farther down the chain (Figure 20.8). There are 3.6 amino acids in each turn of the α-helix, and the R-groups on these amino acids extend outward from the helix. Hair and wool are made up of several coils of keratin wound around each other and held together by disulfide bridges.

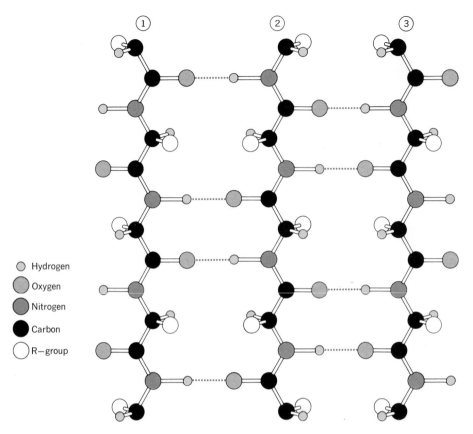

Figure 20.9 Model of polypeptide chains in a pleated sheet structure. In this configuration, several polypeptide chains are held together by hydrogen bonds. The R-groups extend above and below the plane of the sheet.

β-Configuration, or Pleated Sheet

The fibrous protein of silk has a different secondary structure. In silk, several polypeptide chains going in different directions are located next to each other. This gives the protein a zigzag appearance, from which we get the name *pleated sheet* (Figure 20.9). The chains are held together by hydrogen bonds, and the R-groups extend above and below the sheet.

Each protein will have a specific secondary structure depending upon the amino acid sequence. For example, the α-helix is formed when the R-groups are small and uncharged. However, this arrangement will be disrupted by the amino acid proline, which has no amide hydrogen and, therefore, can't form hydrogen bonds. Kinks or bends in the molecule are found at proline positions. In a large protein it is possible to have separate areas of α-helix and pleated sheet arrangements (Figure 20.10). Note that hydrogen bonding is weak, noncovalent bonding, and is easily disrupted by changes in pH, temperature, solvents, or salt concentrations.

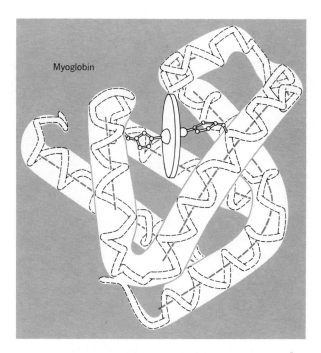

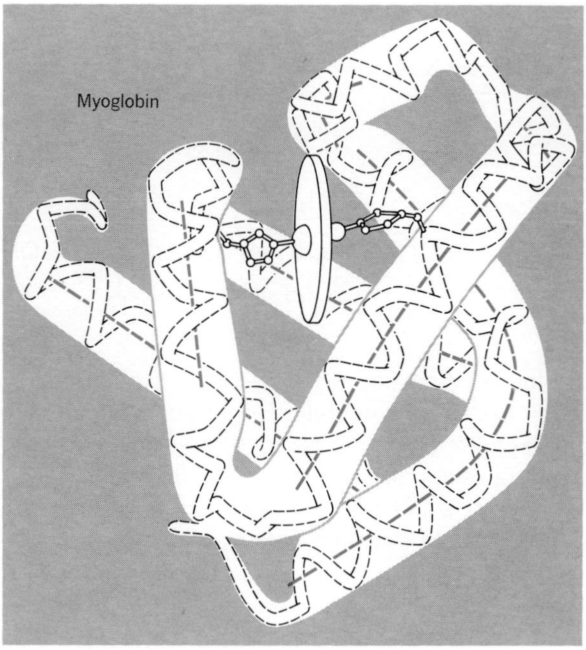

Figure 20.10 The tertiary structure of a molecule of myoglobin consists of eight sections of alpha-helix surrounding a heme group. The main polypeptide chain is shown here without the R-group sidechains. (From R. E. Dickerson and I. Geis, *The Structure and Action of Proteins,* W. A. Benjamin, Inc., Menlo Park. Copyright © 1969 by Dickerson and Geis.)

20.9 Tertiary Structure: Globular Proteins

Tertiary structure is the three-dimensional structure of globular proteins. At normal pH and temperature, each protein will take on a shape that is energetically the most stable given the specific sequence of amino acids and the various types of interactions that may be involved. This shape is called the **native state** or **native configuration** of the protein. In general, globular proteins are very tightly folded into a compact, spherical form. This folding results from interactions between the R-group side chains of amino acids, and may involve hydrogen bonding and the other following interactions:

Disulfide Bridges

Disulfide bridges are interactions that include disulfide linkages between molecules of the amino acid cysteine. Sulfhydryl groups (—S—H) are easily oxidized to form a disulfide (—S—S—). If this reaction occurs between two cysteines, the amino acid cystine is formed.

$$\text{HO}\overset{\overset{\text{O}}{\|}}{\text{C}}-\overset{\overset{\text{NH}_2}{|}}{\underset{\underset{\text{H}}{|}}{\text{C}}}\text{CH}_2-\text{SH} + \text{HS}-\text{CH}_2\overset{\overset{\text{NH}_2}{|}}{\underset{\underset{\text{H}}{|}}{\text{C}}}-\overset{\overset{\text{O}}{\|}}{\text{C}}\text{OH} \xrightarrow{-2\text{H}}$$

Cysteine Cysteine

$$\text{HO}\overset{\overset{\text{O}}{\|}}{\text{C}}-\overset{\overset{\text{NH}_2}{|}}{\underset{\underset{\text{H}}{|}}{\text{C}}}\text{CH}_2-\text{S}-\text{S}-\text{CH}_2\overset{\overset{\text{NH}_2}{|}}{\underset{\underset{\text{H}}{|}}{\text{C}}}-\overset{\overset{\text{O}}{\|}}{\text{C}}\text{OH}$$

Cystine

This disulfide linkage can be formed between two cysteines on different amino acid chains, in this way linking the two chains together to form such tough, strong material as keratin. Or, it can be found between cysteines on the same amino acid chain, creating a loop in the chain. The insulin molecule contains examples of both types of disulfide bridges (Figure 20.6). These bridges are covalent linkages that can be broken by reduction, but which are stable in the face of changes in pH, solvents, or salt concentrations.

The fact that cysteine reacts very easily to form cystine may play an important role in protecting the human body from the effects of free radicals. Free radicals can be produced by ionizing radiation or can be formed from compounds such as caffeine or sodium nitrite, a preservative used in certain meat products. Because of the body's large supply of sulfur-containing compounds such as cysteine, the free radicals are likely to be destroyed by reactions with these compounds rather than in interactions with biologically critical molecules. This theory is supported by the finding that mice injected with cysteine just before exposure to radiation are less likely to develop radiation sickness than mice receiving the same radiation dose without an injection of cysteine.

Salt Bridges

Salt bridges result from ionic interactions between charged carboxyl or amino side chains found on amino acids such as aspartic acid, glutamic acid, lysine, and arginine (see Figure 20.2). These linkages are very easily broken by changes in pH.

Hydrophobic Interactions

Hydrophobic interactions occur between the nonpolar side chains of the amino acids in the protein molecule. Because they are repulsed by the solvent water, these groups tend to be found on the inside of the molecule together with other hydrophobic groups.

20.10 Quaternary Structure: Two or More Polypeptide Chains

The **quaternary structure** of a protein is the manner in which separate polypeptide chains fit together in those proteins containing more than one

Figure 20.11 This model of hemoglobin shows the quaternary arrangement of the four polypeptide chains. The two alpha chains are represented by the light blocks, and the two beta chains by the dark blocks. The iron-containing heme groups that bind oxygen to the molecule are represented by the disks. (Courtesy M. F. Perutz, Medical Research Council, Laboratory of Molecular Biology)

chain. Hydrogen bonding, hydrophobic interactions, and salt bridges may be involved in holding the chains in position. Hemoglobin is just one example of a protein that contains more than one polypeptide chain (Figure 20.11). It consists of four polypeptide chains—two alpha chains and two beta chains—arranged around an iron-containing heme group.

20.11 Denaturation

Various changes in the surroundings of a protein can disrupt the complex secondary, tertiary, or quaternary structure of the molecule. Disruption of the native state of the protein is called **denaturation.** This process involves the uncoiling of the protein molecule into a random state, and will cause the protein to lose its biological activity (Figure 20.12). Denaturation may or may not be permanent; in some cases the protein will return to its

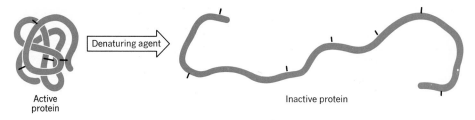

Active
protein

Inactive protein

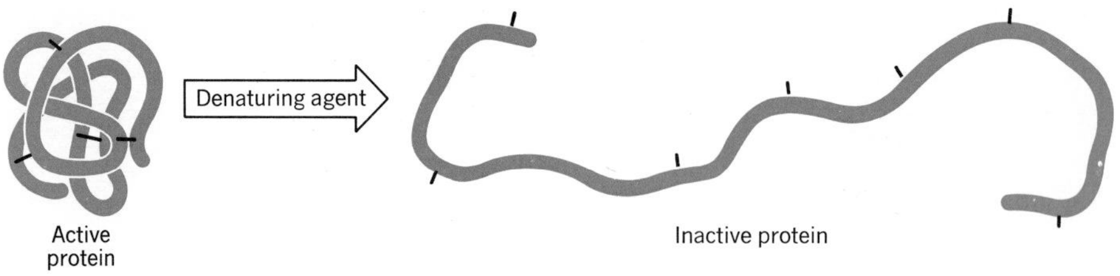

Figure 20.12 Denaturation of a protein disrupts the tertiary structure of the molecule, causing it to uncoil. This results in the loss of protein activity. Denaturation can be either temporary or permanent, depending upon the denaturing agents used.

native state when the denaturing agent is removed. Denaturation may also result in coagulation, with the protein being precipitated from solution. Denaturing agents come in many forms, a few of which are as follows:

pH

Changes in pH have their greatest disruptive effect on hydrogen bonding and salt bridges. For example, the polypeptide polylysine is composed entirely of the amino acid lysine, which has an amino group on its side chain. In acidic pH all the side chains will be positively charged and will repel each other, causing the molecule to uncoil. In basic pH, however, the side chains will be neutral. Therefore, they do not repel, and the molecule will coil into an α-helix.

Heat

Heat causes an increase in thermal vibration of the molecule, disrupting hydrogen bonding and salt bridges. After gentle heating, the protein can usually regain its native state. However, violent heating will result in irreversible denaturation and coagulation of the protein (as, for example, in cooking an egg). Similarly, heat used to sterilize equipment coagulates the protein of microorganisms. Heat can also be used to detect the presence of protein in urine; urine that turns cloudy when heated indicates the presence of protein.

Organic Solvents

Organic solvents such as alcohol or acetone can form hydrogen bonds with protein molecules, which then compete with the hydrogen bonds naturally occurring in the protein. This causes denaturation and coagulation. A 70% alcohol solution is a good disinfectant because it will coagulate the protein in bacteria, thus destroying these organisms.

Heavy Metal Ions

Heavy metal ions such as Pb^{2+}, Hg^{2+}, and Ag^{+} may disrupt the natural salt bridges of the protein by forming salt bridges of their own with the protein molecule. This usually causes coagulation of the protein. The heavy metal ions also bind with sulfhydryl groups, disrupting the disulfide linkages in the protein molecule and denaturing the protein. Such heavy metals, therefore, are toxic to living organisms. As an antidote to heavy metal poisoning, patients are given substances high in protein, such as milk or egg whites. These substances will bond with the metals while they are still in the stomach. The victim must then be made to vomit before the metals are again released through the process of digestion.

Alkaloidal Reagents

Reagents such as tannic or picric acid affect salt bridges and hydrogen bonding, causing proteins to precipitate. Tannic acid is used to precipitate proteins in animal hides; this is the process of tanning used in the manufacture of leather.

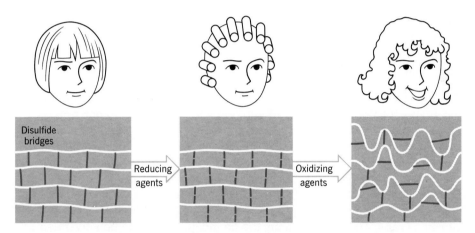

Figure 20.13 Permanent waves work by breaking the disulfide bridges between the cysteines in hair protein molecules, and then reforming the bridges in different locations.

Reducing Agents

Reducing agents disrupt disulfide bridges formed between cysteine molecules. For example, hair permanents work by reducing and disrupting the disulfide bridges in hair. When the hair is curled and oxidizing agents are applied, new disulfide bridges are formed (Figure 20.13).

Radiation

Nonionizing radiation is similar to heat in its effect on proteins. For example, ultraviolet light will burn the skin, causing sunburn.

20.12 Identification of Proteins and Amino Acid Sequences

Many color tests and separation procedures are used in protein research. Two commonly used color tests are the biuret test and ninhydrin test.

Biuret Test

The biuret test detects the presence of two or more peptide linkages. The unknown solution is mixed with sodium hydroxide and a few drops of dilute copper sulfate. If a violet color appears, peptides larger than dipeptides are present. This color test is often used to follow the progress of protein hydrolysis.

Ninhydrin Test

The ninhydrin test is widely used to detect the presence of amino acids, peptides, and proteins. The unknown solution is heated with excess ninhydrin, and a blue color indicates a positive test (except in the case of the amino acid proline, which gives a yellow color).

Protein Metabolism

20.13 Digestion and Absorption

The digestion of proteins begins in the stomach. When food is swallowed, the stomach lining produces hydrochloric acid and inactive forms of protein-digesting enzymes called zymogens, or pre-enzymes. The zymogens travel from the stomach lining into the stomach, where they are then activated by HCl. This process prevents the enzyme from digesting the proteins of the stomach lining as it is traveling through the lining. Protein digestion continues in the small intestine, where enzymes in juices secreted by the pancreas and intestinal wall catalyze the hydrolysis of the proteins (Table 20.4). At this time, the amino acids and some very small polypeptides produced by digestion are absorbed through the intestinal wall, and enter the bloodstream.

20.14 Uses of Amino Acids

Amino acids have many uses in the body. They are utilized by cells to synthesize tissue protein used in the formation of new cells or in the repair of old cells. The largest amounts of proteins are required during periods of rapid growth (such as during infancy, pregnancy, or when a mother is nursing a child), and during periods of extensive repair (such as after surgery, burns, hemorrhage, or infection). Amino acids are also used in the synthesis of other amino acids, enzymes, hormones, antibodies, and nonprotein nitrogen-containing compounds such as nucleic acids or heme groups.

The amino acids not used in such synthesis can be catabolized, or broken down, for energy. When this occurs, the amino group is removed from the amino acid through a process called oxidative deamination, and will enter the urea cycle (which we will discuss shortly). The remainder of the molecule can enter the citric acid cycle to supply needed cellular energy, or can be converted into body fat. This means that you can become fat by eating too much protein, just as you can from eating too much lipid

Table 20.4 The Enzymes of Protein Digestion

Site	Enzyme	Action
Mouth	None	
Stomach	Rennin (infants)	Coagulates milk proteins
	Pepsin	Proteins $\longrightarrow$ polypeptides
Small intestine		
Pancreatic juice	Trypsin, chymotrypsin, and carboxypeptidase	Polypeptides $\longrightarrow$ dipeptides and amino acids
Intestinal juice	Erepsin	Dipeptides $\longrightarrow$ amino acids

or carbohydrate. At the opposite extreme, in cases of severe malnutrition, the body is not supplied with enough lipid and carbohydrate to meet its energy needs. In such a case, tissue proteins will be broken down so that the resulting amino acids can be metabolized to provide energy for the cells.

20.15 Sources of Nitrogen

We mentioned at the beginning of this chapter that proteins are our main dietary source of nitrogen, an element essential to life. The atmosphere is 80% nitrogen but, even though we inhale this molecule with every breath, we have no ability to use elemental nitrogen. We are totally dependent upon microorganisms in the soil or in the roots of leguminous plants to transform nitrogen into forms that plants can use to form amino acids (Figure 20.14).

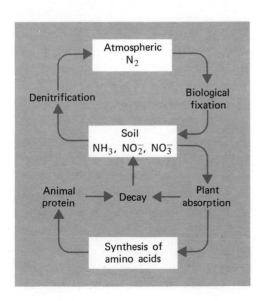

Figure 20.14 **The nitrogen cycle.**

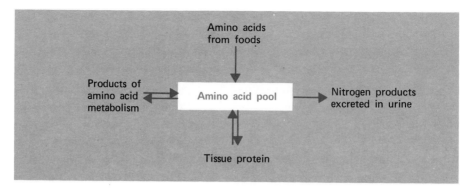

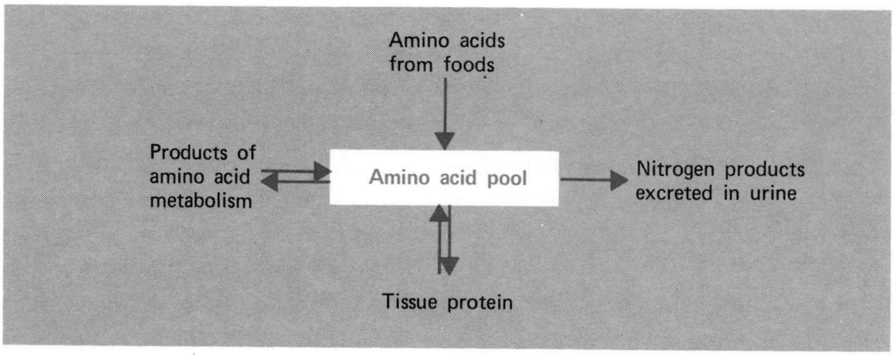

Figure 20.15 The amino acid pool in the human body is constantly changing.

We obtain most of our nitrogen from the plant and animal protein that we eat. There is, however, no storage form of amino acids in the body, as there is for carbohydrates (in the form of glycogen) and lipids (in the form of fat). Instead, there is a pool of amino acids that is constantly changing. Tissue protein is continually being broken down and resynthesized (Figure 20.15). The turnover of amino acids in the liver and blood is relatively rapid—half will be replaced every six days. This turnover is much slower in muscles and supportive tissue—half of the amino acids are replaced every 180 days in muscles, and every 1000 days in the collagen of supportive tissue.

Healthy adults maintain a nitrogen balance within their bodies; they will excrete as much nitrogen as they intake in their diet. Rapidly growing young children will be in a state of positive nitrogen balance. They will not excrete as much nitrogen as they take in, because they need the amino acids to synthesize new tissues. A negative nitrogen balance, in which more nitrogen is excreted than absorbed, occurs in starvation, malnutrition, wasting diseases, and fevers, when amino acids are being used to supply the energy needs of the cell.

In order to maintain a proper nitrogen balance, we need to eat an adequate supply of protein. Although the needs of different people vary considerably, the Recommended Daily Allowance (which exceeds the requirements of most individuals) is 0.8 grams of protein per kilogram of body weight. For the average adult this amounts to 45 to 60 grams of protein, or $\frac{1}{4}$ to $\frac{1}{3}$ pound of protein-rich food, each day. Not only do we need to eat enough protein, however, we must also be sure that it is the right type of protein.

20.16 Essential Amino Acids

There is strong evidence that humans cannot synthesize eight amino acids. These amino acids are called the **essential amino acids,** and must be found in the diet (Table 20.5). Proteins that contain these amino acids are known as **adequate proteins.** Animal protein and milk are adequate proteins,

Table 20.5 The Essential Amino Acids

	Amino Acid	Function in the Human Body
1.	Leucine	Associated with digestive enzymes
2.	Isoleucine	Associated with digestive enzymes
3.	Lysine	Aids in assimilation of other amino acids
4.	Methionine	Associated with fat metabolism
5.	Phenylalanine	Important in utilization of vitamin C and production of thyroxin
6.	Tryptophan	Important in utilization of B vitamins and synthesis of neurotransmitters
7.	Threonine	Important in building tissues and utilization of nutrients
8.	Valine	Important to functioning of the nervous system

but many vegetable proteins are missing one or more of the essential amino acids and, therefore, are inadequate proteins. The protein in soybeans is adequate, but the protein in corn is too low in lysine and tryptophan to support growth in young children. Rice is low in lysine and threonine, and wheat is low in lysine. Deficiency diseases such as Kwashiorkor, therefore, are commonly found in parts of the world where plant protein is the major source of protein.

20.17 Amino Acid Synthesis

Transamination

Amino acids can be synthesized by the body in a process called **transamination.** Transamination involves the transfer of an amino group from one carbon to another. The donor of the amino group is an amino acid, and the receiving molecule is an α-keto acid (an acid containing a ketone functional group on the α-carbon, which is the carbon next to the carboxyl group).

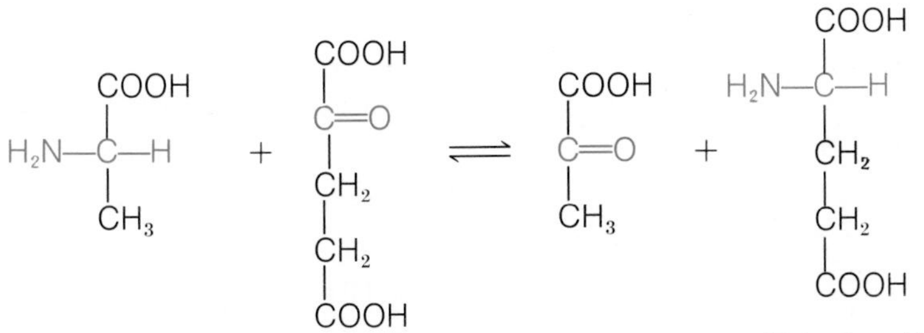

Alanine (Amino acid) | α-Ketoglutaric acid (α-Keto acid) | Pyruvic acid | Glutamic acid (New amino acid)

Specific Syntheses

Amino acids play a role in the synthesis of many metabolic compounds. The following are just a few examples. Tyrosine is used to produce the hormones epinephrine, norepinephrine, and thyroxin, as well as the skin pigment melanin. Tryptophan is used in the synthesis of the nerve transmission chemical serotonin and the coenzymes NAD and NADP. Serine is converted to ethanolamine, which is found in lipid cephalins, and cysteine is used in the synthesis of bile salts.

20.18 Amino Acid Catabolism: Oxidative Deamination

Amino acids that are not used in synthesis can be converted in the liver to ammonia, carbon dioxide, water, and energy. **Oxidative deamination** is a process that involves removing the amino group from an amino acid.

The α-keto acids that are formed in oxidative deamination can be used in several ways. They may be oxidized in the citric acid cycle to produce energy, or they may be converted to other amino acids through transamination. These α-keto acids may also be used in the synthesis of carbohydrates and fats.

20.19 Urea Cycle

The ammonia that is formed in oxidative deamination is toxic to cells; a concentration of 5 mg/100 ml of blood is toxic to humans. The liver disposes of this ammonia by converting it to urea in a cyclic reaction called the **urea cycle,** or the Krebs' ornithine cycle. Ammonia enters the cycle as carbamyl phosphate, and is joined to the amino acid ornithine. In each turn of the cycle, one molecule of urea is produced and a molecule of ornithine is regenerated to begin the next turn (Figure 20.16). The net equation for the reaction is

$$2NH_3 + CO_2 \longrightarrow \underset{\text{Urea}}{H_2N{-}\overset{\overset{\displaystyle O}{\|}}{C}{-}NH_2} + H_2O$$

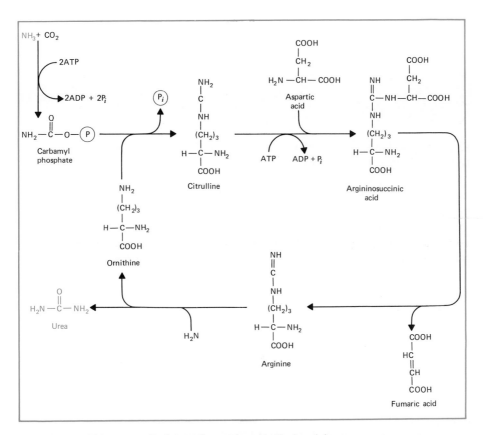

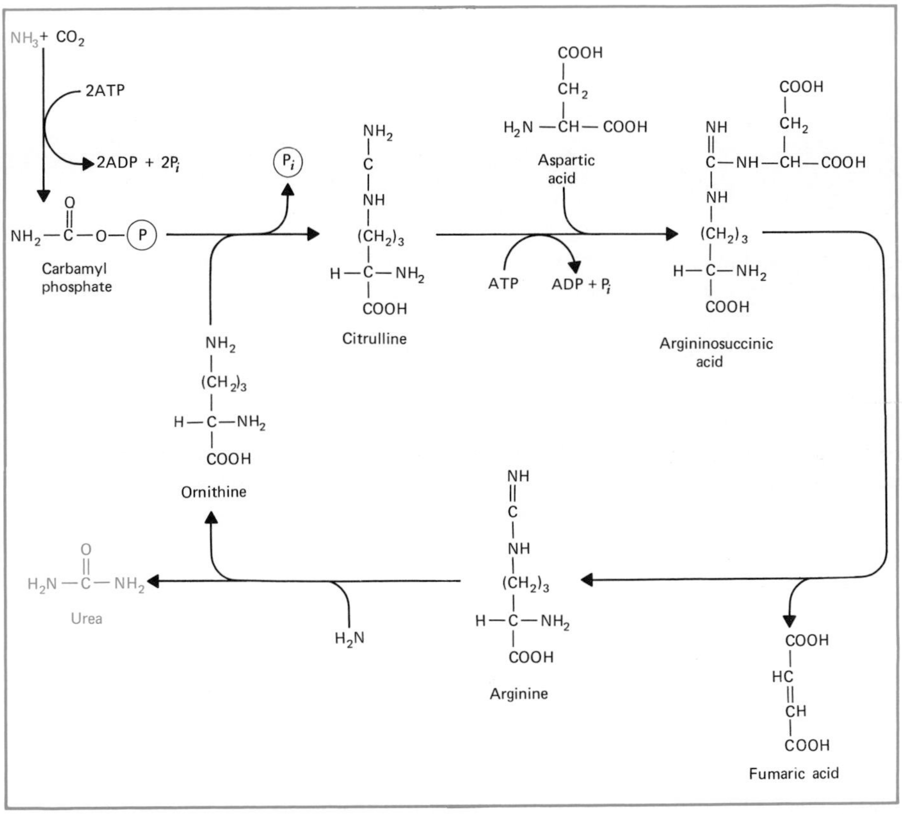

Figure 20.16 The urea cycle. Ammonia, produced in the breakdown of amino acids in the tissues, is converted to the less toxic urea by liver cells in this series of reactions. Ammonia enters the cycle as carbamyl phosphate, and is joined to the amino acid ornithine. In each turn of the cycle, one molecule of urea is produced and a molecule of ornithine is generated to begin the next turh.

The urea can then be safely transported through the body for elimination by the kidneys. Any condition that impairs the elimination of urea by the kidneys can lead to uremia, a build-up of urea and other nitrogen wastes in the blood, which can be fatal. A person suffering from uremia will feel nauseated, irritable, and drowsy. His blood pressure will be elevated and he may be anemic. He may experience hallucinations, and in serious cases may lapse into convulsions and coma. To reverse this condition, either the cause of kidney failure must be removed or the patient must undergo hemodialysis to remove the nitrogen wastes from the blood.

Chapter Summary

Proteins are large polymers of amino acids that have many functions in the cell. They may be simple proteins which contain only amino acids, or complex proteins which contain other nonprotein or inorganic components in addition to amino acids. Enzymes, hormones, antibodies, and transport proteins are all globular proteins. Keratin, muscle fibers, and collagen are examples of fibrous proteins. There are 20 common amino acids found in proteins. These amino acids contain an amino group and a carboxyl group and, as a result, are amphoteric and can act as buffers. Each amino acid and protein will have a characteristic isoelectric point; this is also the pH at which it is the least soluble.

Protein molecules are quite complex. The primary structure of a protein is the sequence of amino acids (held together by peptide bonds) in the polypeptide chain. The secondary structure is the α-helix or pleated sheet arrangement of amino acids that results from hydrogen bonding between amino acids in a protein molecule. Tertiary structure refers to the three-dimensional structure of globular proteins, and results from interactions between the R-group side chains of the amino acids. These interactions include hydrogen bonding, disulfide bridges, salt bridges, and hydrophobic interactions. Some proteins contain more than one polypeptide chain, and the quaternary structure of the protein refers to the way in which these chains fit together. The native state of the protein is the most stable arrangement for that protein. Denaturation is the disruption, permanent or temporary, of the native state of the protein. It can be caused by heat, radiation, changes in pH, or the presence of organic solvents, heavy metal ions, alkaloidal reagents, or reducing agents.

Protein digestion begins in the stomach and continues in the small intestine. The amino acids absorbed in this process can be used in the synthesis of tissue protein, other amino acids, enzymes, hormones, antibodies, and nucleic acids. There are eight amino acids that cannot be synthesized by the human body and, therefore, are essential in the diet. A lack of any one of these can lead to a protein deficiency disease. Amino acids are synthesized in the cell from other amino acids by transamination. Amino acids are catabolized by first removing the amino group in a process called oxidative deamination. The ammonia that is produced is converted to urea through a series of reactions called the urea cycle. The α-keto acids produced then enter the citric acid cycle or other biosynthetic pathways.

The pathways of protein, lipid, and carbohydrate metabolism are intricately related, and can be summarized as follows:

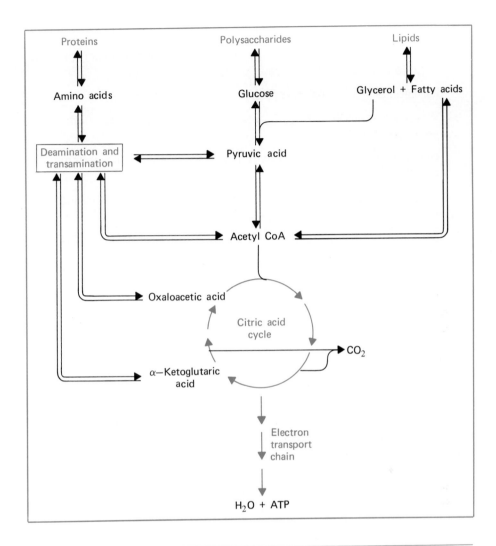

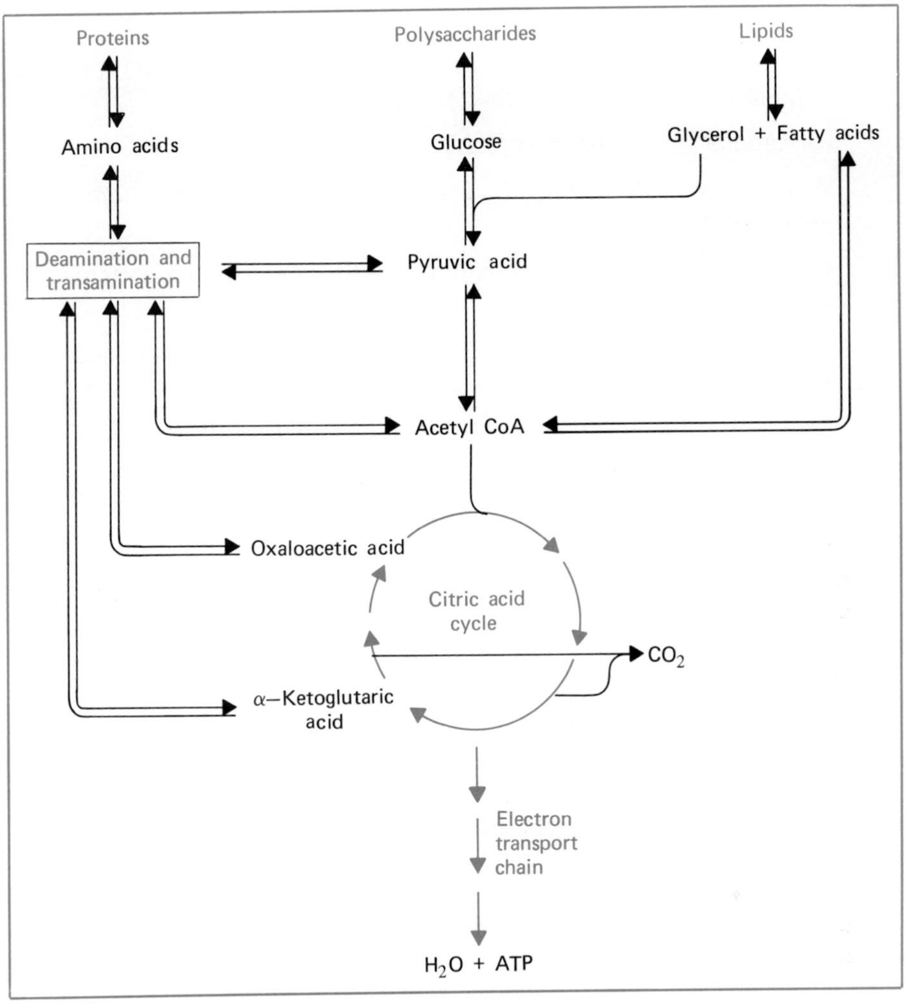

Exercises and Problems

1. Describe the difference between
   (a) a simple protein and a conjugated protein
   (b) a globular protein and a fibrous protein
   (c) the primary structure and the secondary structure of a protein
   (d) a dipeptide and a polypeptide
   (e) α-helix and pleated sheet configuration
   (f) vegetable protein and animal protein
   (g) transamination and oxidative deamination
2. Why does protein in the urine indicate a possible kidney disorder?
3. What explains the difference between the solubility of globular and fibrous proteins?

4. Describe how blood proteins help protect against pH changes.
5. If placed in an electric field, how would each of the following proteins migrate if the pH of the solution were 7?

   (a) Urease (b) Myoglobin (c) Chymotrypsin
6. What fact about the structure of amino acids explains their high melting points?
7. How do the processes of denaturation and digestion of a protein differ?
8. Why are eggs a good antidote for lead poisoning? Why is eating the egg antidote not sufficient to prevent lead poisoning?
9. Explain, on the molecular level, the process of hair straightening.
10. Why can't insulin be given to diabetics in the form of a pill?
11. What are the problems associated with eating a strictly vegetarian diet?
12. (a) Is a diet consisting mainly of rice an adequate diet? Why or why not?
    (b) Suggest several ways in which a diet consisting mainly of corn can be supplemented to provide enough adequate protein.
13. Describe four possible metabolic uses of the amino acids contained in a hamburger.
14. (a) Write the formula for leucine as it would be found in solution at a pH of 7.
    (b) Write the equation for the addition of acid to the solution in (a).
    (c) Write the equation for the addition of base to the solution in (a).
15. Write the structure for the two polypeptides described on page 463.

    (a) Which polypeptide, I or II, would be more soluble in water? Explain why.
    (b) In which polypeptide, I or II, would hydrophobic interactions be more likely to occur? Explain why.
    (c) In which polypeptide, I or II, would salt bridges be more likely to occur? Explain why.
16. Write the structure and identify each of the products of the digestion of the following polypeptide.

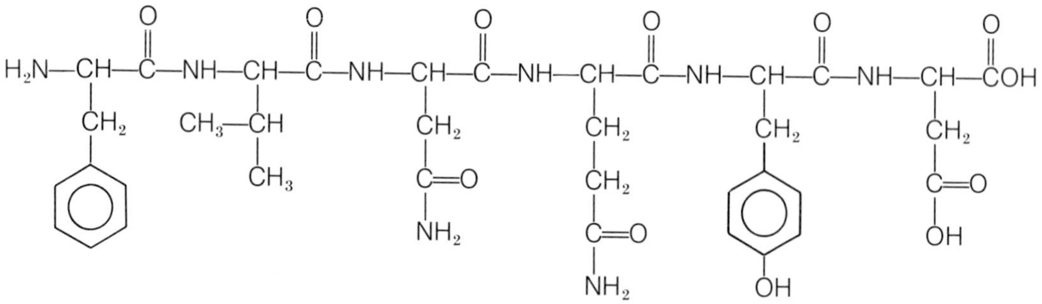

17. Using the three-letter abbreviations for the amino acids, write the amino acid sequence for the polypeptide in question 16.

18. Write the equation for the transamination reaction between threonine and pyruvic acid.

19. Write the equation for the deamination of alanine.

20. Describe in general terms the metabolic fate of the ammonia produced in question 19.

21. A rare inherited disease called *maple syrup urine disease* results when a person lacks the ability to metabolize the α-keto acids that result from the transamination of valine, leucine, and isoleucine.

    (a) Write the structures of the α-keto acids produced in the transamination of valine, leucine, and isoleucine.
    (b) Suggest a possible reason for the name of this disease.

chapter 21

Enzymes, Vitamins, and Hormones

Learning Objectives

By the time you have finished this chapter, you should be able to:

1. Identify the parts of an enzyme molecule.

2. Describe how enzymes work.

3. Indicate how enzymes differ from inorganic catalysts.

4. Explain the *lock-and-key* and *induced-fit* theories of enzyme action.

5. Explain how changes in pH and temperature will affect enzyme activity.

6. Explain why vitamins are necessary for normal cellular function.

7. Explain how multienzyme systems are regulated.

8. State the difference between
   a. reversible and irreversible inhibition.
   b. competitive and noncompetitive inhibition

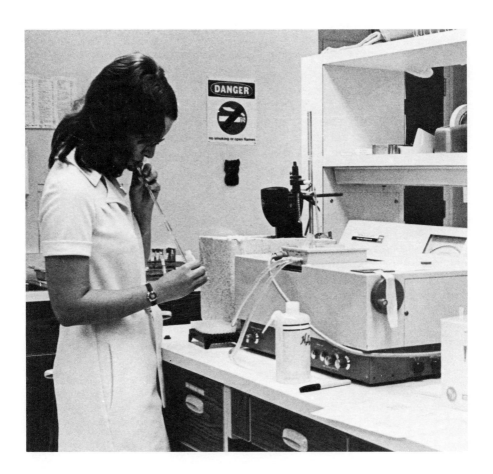

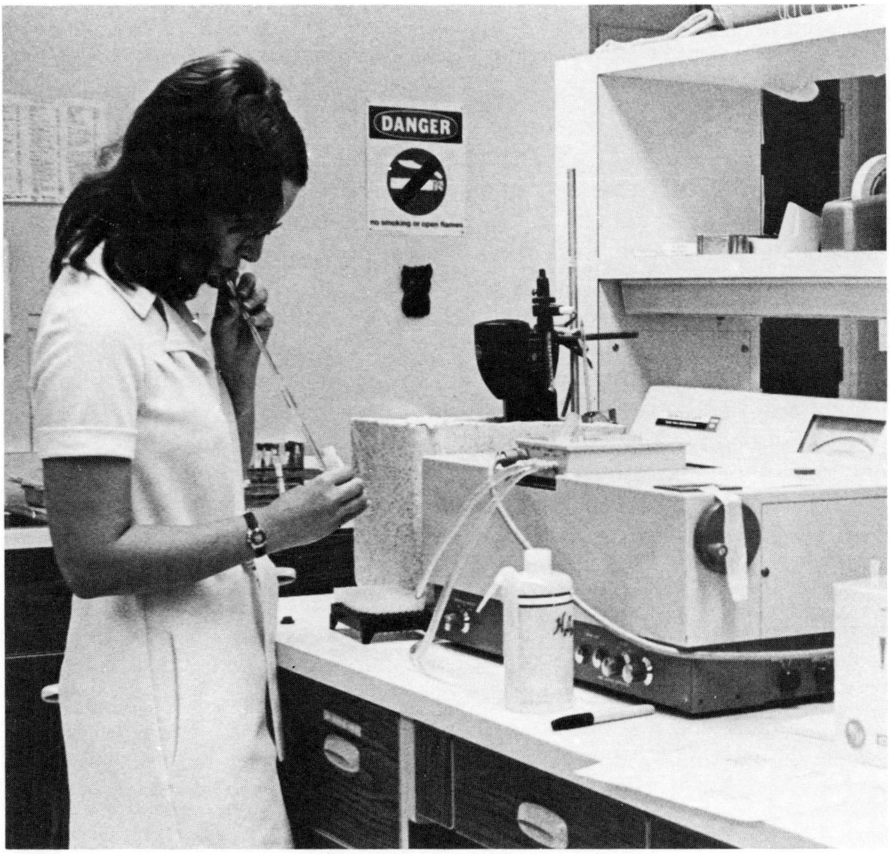

After evening rounds at the hospital, Dr. Feldman was discussing with two medical students some of the patients they had just seen. She was commenting on how diagnostic techniques had progressed in the last 15 years as a result of increased knowledge of human biochemistry. In particular, she mentioned two patients as important illustrations of how enzymes have enabled doctors to design quick and precise diagnostic tests for certain diseases.

The patient in 204A, for example, had been admitted two days earlier with shortness of breath and severe chest pains. Dr. Feldman suspected that he had suffered a heart attack (or myocardial infarction)—in which an artery to the heart muscle is blocked and a portion of the heart muscle dies from lack of oxygen. Until recently, this condition would have been diagnosed using an electrocardiogram and other more general tests. However, it has now been found that certain enzymes are released into the blood by the damaged heart muscle. The concentration of these enzymes in the blood, when measured over a period of two days after symptoms begin, can give a positive diagnosis of myocardial infarction and a direct indication of the extent of damage to the heart muscle. Therefore, on this patient the doctor was able to use a diagnostic technique requiring only analyses of the patient's blood. Samples of blood were analyzed for the concentrations of lactic dehydrogenase and a type of creatine phosphokinase (BM-CPK) which is found only in heart muscle. By evaluating the levels of these enzymes, Dr. Feldman was able to prescribe treatment and to quickly determine the chances for the patient's recovery.

A similar procedure had been used with the patient in 305C. By measuring the blood level of the enzyme glutamic pyruvic transaminase, the doctor had been able to quickly confirm the preliminary diagnosis of infectious hepatitis. Until this test had been developed, the only way to confirm such a diagnosis was by microscopic examination of a sample of the patient's liver tissue. Moreover, because the blood level of this enzyme rises even before any symptoms of the disease appear, Dr. Feldman had arranged to test other members of the patient's family to determine if anyone else had contracted the disease.

Dr. Feldman mentioned two other examples of how learning the functions of specific compounds and their concentrations in body fluids was bringing about new and important diagnostic tests. The enzyme called acid phosphatase is an important indicator of prostate cancer because its concentration in the blood increases when a tumor begins to develop in the prostate gland. A high level of the hormone called human chorionic gonadotropin in the blood of a pregnant woman may signal the presence of twins as early as seven days after conception. The doctor finished her discussion by predicting that the discovery of such enzyme "flags" was just beginning, and that doctors would greatly benefit from research results yet to come.

Enzymes

21.1 What Are Enzymes?

Enzymes are the largest and most highly specialized class of proteins. They function as biological catalysts in the body for the reactions involved in metabolism. Earlier we described catalysts as substances which increase the rate of a chemical reaction by lowering the activation energy of that reaction. The reactions of metabolism would ordinarily occur at extremely slow rates at normal body temperature and pH. Without the enzymes in our digestive tract, for example, it would take us about 50 years to digest a single meal! Enzymes greatly increase this reaction rate, allowing cells to function under normal body conditions. Because a catalyst is not consumed in the reaction, it can be used over and over again. Such compounds, therefore, need to be present in only very small amounts. Enzymes are water-soluble globular proteins that can vary from a molecular weight of 12,000 to over 1 million. The enzyme molecule may consist of a fairly simple single polypeptide chain, or it may be a more complex molecule composed of several polypeptide chains and other nonprotein parts.

21.2 Some Important Definitions

Enzymes may be simple proteins made up entirely of amino acids, or may be conjugated proteins that require a nonprotein group for their biological activity. Before continuing with our discussion of enzymes, however, it will be helpful for us to define some special terms used in the study of these complex molecules (see Figure 21.1).

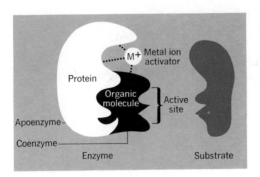

Figure 21.1 Some important terms used in the study of enzymes.

Apoenzyme. Apoenzyme is a term used to describe the protein part of the enzyme molecule.

Cofactor. Cofactors are the additional chemical groups appearing in those enzymes that are conjugated proteins. These cofactors are required for enzyme activity, and may consist of metal ions or complex organic molecules. Some enzymes require both types of cofactors.

Activator. When the cofactor of an enzyme is a metal ion (such as the ions of magnesium, zinc, iron, or manganese), the cofactor is called an activator.

Coenzyme. When the cofactor of an enzyme is a complex organic molecule other than a protein, the cofactor is called a coenzyme.

Prosthetic Group. In the enzyme molecule the cofactor may be weakly attached to the apoenzyme, or it may be tightly bound to the protein. If it is tightly bound, the cofactor is called a prosthetic group.

Zymogen. Zymogen (or pre-enzyme) is the name applied to the inactive form of an enzyme. Enzymes, as we saw in the case of the digestive enzymes, are often secreted in inactive forms as zymogens. They are transported to the place where activity is desired, and are then converted to their active forms.

Substrate. The substrate is the chemical substance or substances upon which the enzyme acts.

Active site. The active site is the specific area of the enzyme to which the substrate attaches during the reaction.

Table 21.1 Classes of Enzymes

Hydrolases: Enzymes that catalyze hydrolysis reactions.	
Example	**Reaction Catalyzed**
Carbohydrases:	Polysaccharides and disaccharides $\xrightarrow{+H_2O}$ Monosaccharides
Esterases:	Ester $\xrightarrow{+H_2O}$ Acid + alcohol
Proteases:	Protein $\xrightarrow{+H_2O}$ Peptides and amino acids
Nucleases:	Nucleic acids $\xrightarrow{+H_2O}$ Pyrimidines + purines + sugars + phosphoric acid

Oxido-Reductases: Enzymes that catalyze oxidation-reduction reactions.	
Example	**Reaction Catalyzed**
Oxidases:	Addition of oxygen to a substrate
Dehydrogenases:	Removal of hydrogen from a substrate

Transferases: Enzymes that catalyze reactions involved in the transfer of functional groups.	
Example	**Reaction Catalyzed**
Transaminases:	Transfer of —NH_2
Transmethylases:	Transfer of —CH_3
Transacylases:	Transfer of $-\overset{\overset{\textstyle O}{\|}}{C}-R$
Transphosphatases: (Kinases)	Transfer of $-O-\overset{\overset{\textstyle O}{\|}}{\underset{\underset{\textstyle OH}{\|}}{P}}-OH$

Lyases: Enzymes that catalyze the addition to double bonds.
Isomerases: Enzymes that catalyze the interconversion of isomers.
Ligases: Enzymes that, in conjunction with ATP, catalyze the formation of new bonds.

21.3 Enzyme Nomenclature and Classification

Because enzymes are the largest class of proteins, a great number of specific enzymes have been isolated and described. At first, the only general rule of nomenclature was to end the enzyme name with the suffix *-in* to indicate a protein. Names such as trypsin, renin, and pepsin are examples of such nomenclature, and are still used today. However, such names give no indication of the substrate involved or the type of reaction taking place. As the number of known enzymes increased, scientists began naming them by adding an *-ase* ending to the name of the substrate. For example, urease catalyzes the hydrolysis of urea; maltase catalyses the hydrolysis of maltose; and phosphatases catalyze the hydrolysis of phosphoric acid esters.

In 1961 the Commission on Enzymes of the International Union of Biochemistry proposed a standard classification of enzymes. They recommended that enzymes be named by the substrate or type of reaction involved, and that they be classified into six major divisions (Table 21.1). Many enzymes, however, still are usually called by their original names, and we will use these more common names in this text.

21.4 Method of Enzyme Action

We have stated that enzymes catalyze reactions in cells by lowering the activation energy. They do this by forming a complex with the substrate (that is, by attaching to the substrate) which then increases the probability that the reaction will occur. We can outline the steps of an enzyme-catalyzed reaction as follows (Figure 21.2):

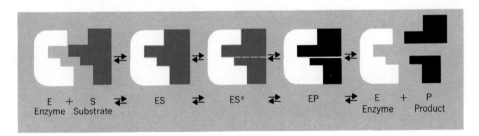

$$E + S \rightleftharpoons ES \rightleftharpoons ES^* \rightleftharpoons EP \rightleftharpoons E + P$$
Enzyme Substrate Enzyme Product

Figure 21.2 **The steps of an enzyme-catalyzed reaction.**

1. The substrate (or substrates) becomes weakly bound to the enzyme surface, forming the enzyme-substrate complex.

$$E + S \rightleftharpoons ES$$

2. The substrate becomes activated; that is, the bonds in the substrate become polarized.

Indicates activated state

$$ES \rightleftharpoons ES^*$$

3. The products of the reaction form on the surface of the enzyme.

$$ES^* \rightleftharpoons EP$$

4. The products are released or "kicked off" the surface of the enzyme, making the enzyme available to catalyze further reactions.

$$EP \rightleftharpoons E + P$$

The rate at which an enzyme catalyzes a reaction will vary with different cellular conditions, and from enzyme to enzyme. Some enzymes, such as carbonic anhydrase, are extremely efficient in catalyzing a reaction. Such efficiency of enzyme action is measured by the **turnover number,** which is the number of substrate molecules transformed per minute by one molecule of enzyme under optimal conditions of temperature and pH (Table 21.2).

Example 21-1

Cholinesterase, the enzyme that catalyzes the breakdown of acetylcholine to acetate and choline at nerve endings, has a turnover number of 1.5×10^6 molecules/minute. How many molecules of acetylcholine can one molecule of cholinesterase hydrolyze in a second?

From the problem:

$$1 \text{ molecule hydrolyzes } \frac{1.5 \times 10^6 \text{ molecules}}{1 \text{ minute}}$$

and

$$1 \text{ minute} = 60 \text{ seconds}$$

Therefore, 1 molecule will hydrolyze

$$\frac{1.5 \times 10^6 \text{ molecules}}{\cancel{1 \text{ minute}}} \times \frac{\cancel{1 \text{ minute}}}{60 \text{ seconds}} = \frac{1.5 \times 10^6 \text{ molecules}}{60 \text{ seconds}}$$

$$= 0.025 \times 10^6 \frac{\text{molecules}}{\text{second}} = 2.5 \times 10^4 \frac{\text{molecules}}{\text{second}}$$

21.5 Specificity

One of the main differences between enzymes and inorganic catalysts is the specificity of enzymes. For example, platinum will catalyze several different types of reactions. However, each enzyme will catalyze only one type of reaction, and in some cases will limit its activity to only one particular type of reactant molecule.

Table 21.2 Turnover Numbers of Some Enzymes

Enzyme	Turnover Number (Molecules of Substrate/Minute)
Carbonic anhydrase	36,000,000
Sucrose invertase	1,000,000
Glutamic dehydrogenase	30,000
Phosphoglucomutase	1,240
Chymotrypsin	100
DNA polymerase	15

Pancreatic lipase, for example, will hydrolyze the ester linkage between glycerol and fatty acids in lipids, but will have no effect on the hydrolysis of proteins or carbohydrates. Kidney phosphatase catalyzes the hydrolysis of esters of phosphoric acid, but at a different rate for each substrate. Urease is even more specialized; it will catalyze only the hydrolysis of urea. An extreme example of the specificity of enzyme action is given by the enzyme aspartase, which will catalyze only the following reversible reaction.

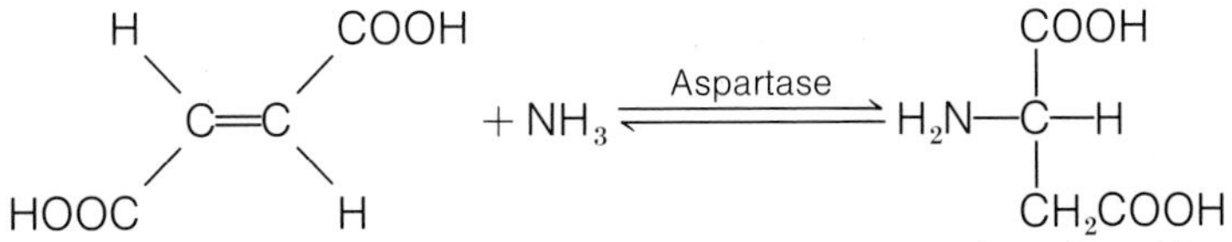

Fumaric acid　　　　L-Aspartic acid

This enzyme will not catalyze the addition of ammonia to any other unsaturated acid—not even maleic acid, which is the *cis*-isomer of fumaric acid.

Maleic acid

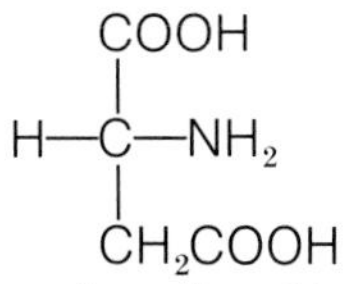

D-Aspartic acid

Moreover, aspartase will not even catalyze the removal of ammonia from D-aspartic acid.

21.6 Lock-and-Key Theory

How can we explain this specificity of enzymes? There is a particular area, called the **active site,** which is found on the surface of the enzyme molecule. This is where the substrate attaches during a reaction. The

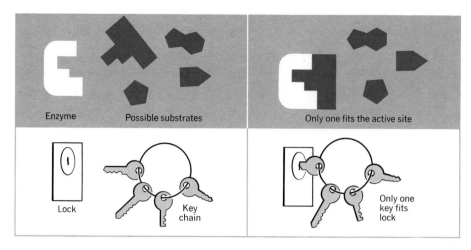

Figure 21.3 Many enzymes are very specific, often limiting their action to only one particular molecule (just as a lock is specific for only one key).

shape, or configuration, of this active site is especially designed for the specific substrate involved. Because the configuration is determined by the amino acid sequence of the enzyme, the native configuration of the entire enzyme molecule must be intact for the active site to have the correct configuration. In such a case, the substrate then fits into the active site of the enzyme in much the same way as a key fits into a lock. The configuration of the lock is specific for only one key; no other keys will turn the lock (Figure 21.3).

Not only is the geometry of the active site important, but so also is the arrangement of charged R-groups around the active site. There is an electrical attraction between the substrate and the enzyme that pulls them together. The reaction products which then occur, however, will have a different distribution of charges, and the repulsion between the products and the active site may cause the products to be "kicked off" the enzyme.

21.7 Induced-Fit Theory

The model of a lock-and-key fit, which helps to explain the action of some enzymes, must be slightly changed for other enzymes. In these cases a better comparison would be a hand slipping into a glove, which then causes, or induces, a fit. Enzyme molecules are flexible, and the active site of some enzymes may not initially match the substrate. However, the substrate itself, as it is drawn to the enzyme, may induce the enzyme to take on a shape that matches the substrate. In such cases the active site is still quite specific to the substrate, just as a left-hand glove will not normally fit a right hand (Figure 21.4).

Enzymes that are secreted as zymogens have their active sites blocked. To activate the enzyme, these sites must be unblocked by the

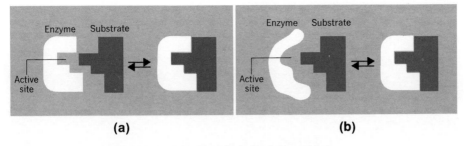

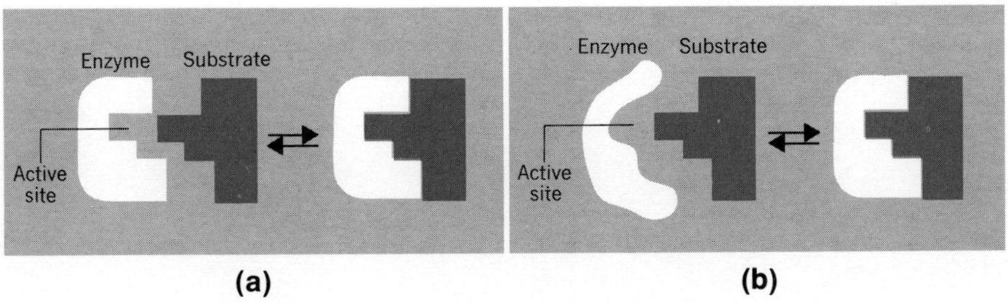

Figure 21.4 Two theories of enzyme action. (a) In the lock-and-key theory, the active site conforms exactly to the substrate molecule. (b) in the induced-fit theory, the substrate induces the active site to take on a shape complementary to the shape of the substrate molecule.

hydrolysis of part of the molecule. Cofactors contribute to the activity of the enzyme either by providing the arrangement of molecules necessary for the active site, or by forming a bridge between the substrate and the enzyme (Figure 21.5).

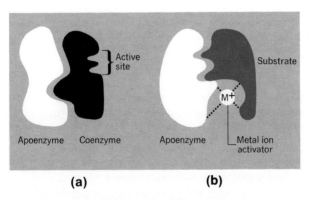

Figure 21.5 Cofactors may contribute to the activity of the enzyme by (a) providing the active site, or (b) forming a bridge between the enzyme and the substrate.

21.8 Factors Affecting Enzyme Activity

We have seen that factors such as pH, temperature, solvents, and salt concentrations can change the structure of a protein. Such factors, therefore, will have an effect on the activity levels of enzymes.

pH

Changes in the pH of the surrounding medium can change the secondary or tertiary structure of an enzyme. This may alter the geometry of the active site or the surrounding charge distribution. Each enzyme has a certain pH at which it is most active. For example, pepsin has an optimum pH of 1.5, whereas trypsin has its maximum activity at pH 8. Increasing or decreasing the pH from these optimal levels will lower the activity of the enzyme (Figure 21.6).

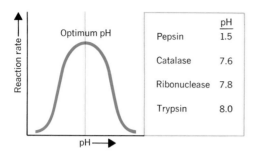

	pH
Pepsin	1.5
Catalase	7.6
Ribonuclease	7.8
Trypsin	8.0

Figure 21.6 Each enzyme has its highest activity at a specific pH.

Temperature

Temperature also affects the rate at which enzyme-catalyzed reactions occur. Most body enzymes have their highest activity at temperatures from 35–45°C. Above this range, the enzyme begins to denature and the reaction rate decreases. When you are running a high fever, for example, you feel ill partly because your high body temperature slows enzyme activity. Above 80°C enzymes will become permanently denatured. Below 35°C the reaction rate slowly decreases until it essentially stops as the enzymes become inactive (Figure 21.7). Enzymes, however, are not denatured at low temperatures, and will resume activity if the temperature is again raised. It is this fact that allows researchers to preserve cell cultures, human tissues for transplants, and organisms such as bacteria for further studies. The preservation of food by refrigeration is based on the

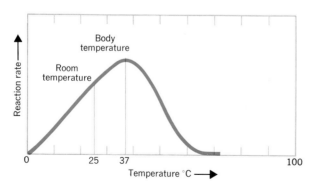

Figure 21.7 Temperature affects enzyme activity. Body enzymes will have a maximum activity at temperatures ranging from 35° to 45° C.

fact that the enzyme-catalyzed reactions which cause spoilage are slowed or stopped at low temperatures.

Vitamins

21.9 Vitamins as Coenzymes

The need for vitamins has been recognized for over 200 years. Long ago, British sailors were given the slang name "limeys" from the lime juice they drank to prevent scurvy. The symptoms of scurvy—swollen gums, painful joints, hemorrhages under the skin, loss of weight, and muscular weakness—are prevented by vitamin C, which was first isolated in 1930.

Exactly what are vitamins? Vitamins are a class of organic nutrients that the body needs, but cannot synthesize. Therefore, vitamins must be present in the diet (in trace amounts) for proper cellular function, growth, and reproduction. Some vitamins or derivatives of vitamins are known to serve as coenzymes in cellular reactions, but the exact cellular function of many vitamins has yet to be determined. When vitamins are lacking in the diet, however, specific deficiency diseases result (Figure 21.8). The U.S. government has set recommended daily allowances (abbreviated, RDA) for vitamins. These figures are based on controlled experiments done over long periods of time, and take into account individual variations in vitamin requirements. Most healthy people who eat a balanced diet will get this required amount of vitamins without taking vitamin supplements. Vitamins may be classified into two major groups, water-soluble and fat-soluble (Table 21.3).

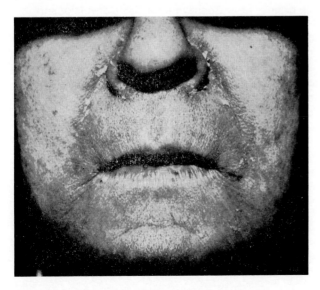

Figure 21.8 The rough, red skin around the nose and chin, and the greasy scaliness of the skin, are caused by a deficiency in vitamin B_6. (Courtesy R. W. Vilter et al., *J. Lab. Med. 42*:355, 1953).

Table 21.3 Vitamins Required in Human Nutrition

Name	Recommended Daily Allowance (RDA)*	Dietary Sources	Function	Symptoms of Deficiency
Water Soluble				
Vitamin C (Ascorbic acid)	60 mg	Citrus fruits, raw green vegetables	Maintenance of normal connective tissue, carbohydrate metabolism	Scurvy, bleeding gums, loosened teeth, swollen joints
Vitamin B_1 (Thiamine)	1.5 mg	Whole grains; liver, brain, heart, kidney; legumes	Formation of enzymes involved in citric acid cycle	Beriberi, heart failure, mental disturbance
Vitamin B_2 (Riboflavin)	1.7 mg	Milk, eggs, liver, yeast, leafy vegetables	Coenzymes in electron transport chain	Fissures of the skin, visual disturbances
Niacin (Nicotinic acid)	20 mg	Yeast, lean meat, liver, whole grains	NAD, NADP; coenzymes in hydrogen transport	Pellagra, skin lesions, diarrhea, dementia
Vitamin B_6 (Pyridoxine)	2 mg	Whole grains, pork, glandular meats, legumes	Coenzyme for amino acid and fatty acid metabolism	Convulsions in infants; skin disorders in adults
Vitamin B_{12} (Cyanocobalamin)	6 micrograms	Liver, kidney, brain. Synthesized by bacteria in gut	Synthesis of nucleoprotein	Pernicious anemia
Pantothenic acid	10 mg	Yeast, liver, kidneys, egg yolk	Forms part of Coenzyme A (CoA)	Neuromotor disorders, digestive disorders, cardiovascular disorders.
Folic acid	0.4 mg	Yeast, organ meats, wheat germ	Synthesis of nucleoprotein	Anemia, inhibition of cell division

Table 21.3 Vitamins Required in Human Nutrition (Continued)

Name	Recommended Daily Allowance (RDA)*	Dietary Sources	Function	Symptoms of Deficiency
Biotin	0.3 mg	Liver, egg white, dried peas and lima beans, synthesized by bacteria in the gut	Protein synthesis, CO_2 fixation, transamination	Skin disorders
Choline (essential factor; not strictly a vitamin)		Egg yolk, wheat germ	Aids in fat oxidation and transport	Inadequate fat absorption, fatty liver
Inositol	Unknown	Found in many fruits and vegetables	Aids in fat metabolism	Fatty liver
Fat Soluble				
Vitamin A (A_1-retinol) (A_2-dehydroretinol)	5000 I.U. (1 I.U. = 0.3 micrograms of retinol)	Green and yellow vegetables and fruits; cod liver oil	Formation of visual pigments; maintenance of normal epithelial structure	Night blindness, skin lesions, eye disease (In excess—vitamin A toxicity, hyperirritability, skin lesions, bone decalcification, increased pressure on the brain)
Vitamin D (D_2-calciferol) (D_3-cholecalciferol)	400 I.U. (1 I.U. = 0.025 micrograms of cholecalciferol)	Fish oils, liver; provitamins in our skin activated by sunlight	Increase Ca^{2+} absorption from gut; important in bone and teeth formation	Rickets (defective bone formation) (In excess—growth retardation in infants above 2000 I.U. per day)
Vitamin E (Tocopherol)	30 I.U. (1. I.U. = 0.81 mg of α-tocopherol)	Green, leafy vegetables	Maintain resistance of red cells to hemolysis	Increased fragility of red blood cells
Vitamin K (K_2-phylloquinone)	5 mg	Produced by bacteria in the intestines	Enables prothrombin synthesis in liver	Failure of coagulation

* For adults and children over 4 years of age from the U.S. Department of Health, Education and Welfare, revised in January 1976.

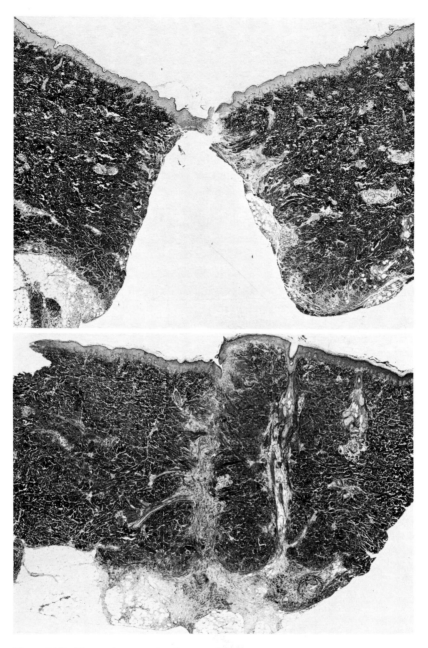

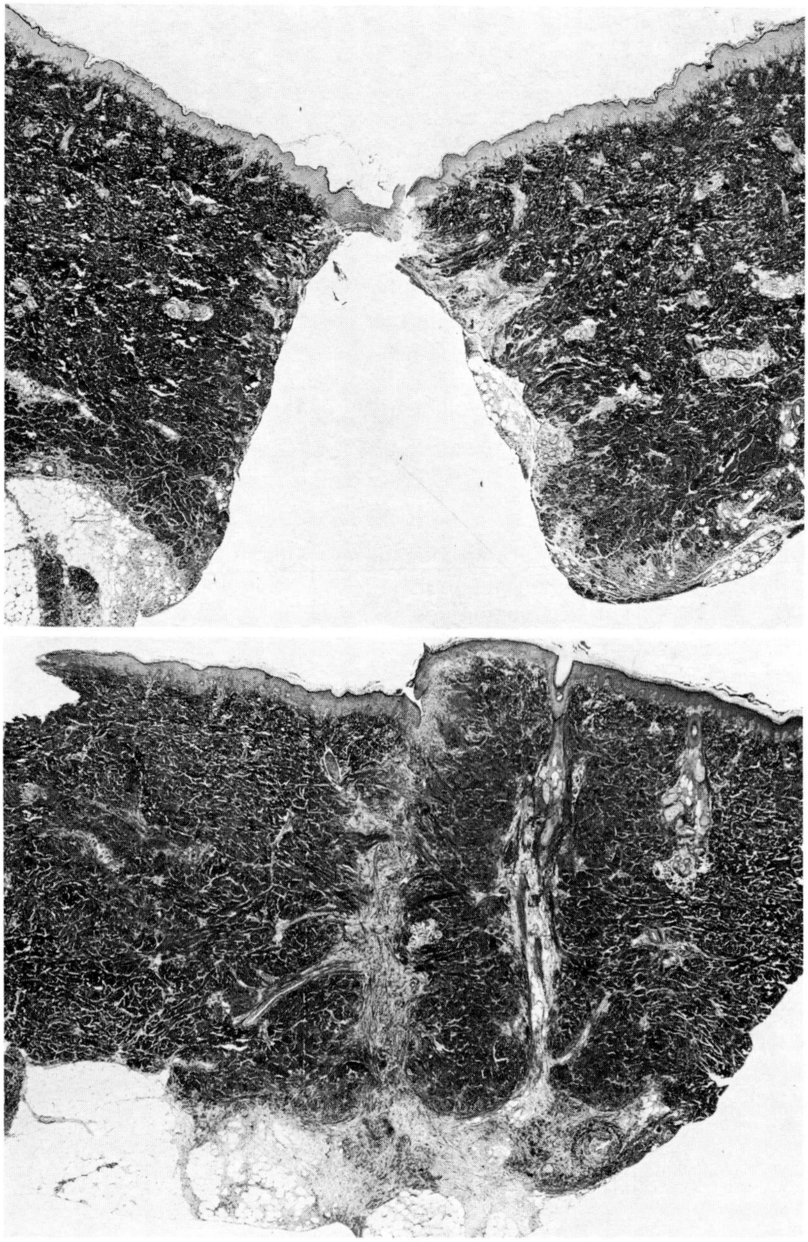

Figure 21.9 These photographs show the first satisfactorily controlled experiment on human scurvy. The top photograph shows a biopsy 10 days after a wound was made in the midback region of a subject who had eaten an ascorbic acid-free diet for six months. The wound shows no healing except in the outer layer of skin cells. The bottom photograph shows a biopsy taken after 10 days of treatment with oral doses of ascorbic acid. The wound is healing, and shows abundant formation of connective tissue. (Courtesy J. H. Crandon and The Upjohn Company).

21.10 Water-Soluble Vitamins

Because the vitamins in this group are water-soluble, they are easily eliminated from the body in the urine. Therefore, they must be present in the diet daily. Fruits and vegetables contain many of these vitamins. Such foods should (if necessary) be cooked quickly in a very small amount of water to prevent both loss of the vitamins into the water and destruction of the vitamin's activity by the heat.

Vitamin C

Vitamin C, or ascorbic acid, is the most unstable of all the vitamins. It is easily destroyed by oxidation, a process which is speeded up by the presence of heat or a base.

Ascorbic acid

Vitamin C has been the subject of much controversy over the last decade. Various people have recommended large doses of vitamin C to prevent colds, heart disease, and cancer. However, much more research is needed to substantiate these claims and to determine all the functions of vitamin C in the body. It is known that vitamin C is necessary for the formation of cells that produce tissues, teeth, and bone. It plays an important role in the metabolism of carbohydrates and proteins, and in the formation of norepinephrine, serotonin, hemoglobin, and collagen. Collagen is the "cement" that holds tissues together, and a lack of it results in the symptoms of scurvy (Figure 21.9). Vitamin C may also aid in the absorption of iron from the intestines and its storage in the liver, as well as assisting in the excretion of adrenal hormones. At present, there are no proven toxic effects from taking large doses of vitamin C.

The B Vitamins

The vitamins in the B family are thiamine, riboflavin, niacin, pyridoxine, cyanocobalamin, pantothenic acid, folic acid, and biotin. They all are identified by the same letter because at the beginning of this century the vitamins in this group were thought to be a single vitamin. Many coenzymes are mono- or diphosphate esters of the B vitamins (Table 21.4). For example, thiamine pyrophosphate is the diphosphate ester of thiamine (vitamin B_1). It is an important coenzyme for several enzyme systems, and is involved in carbohydrate and amino acid metabolism. A lack of thiamine results in beriberi, a disease which affects nerves and causes paralysis of the legs, enlargement of the heart, slowing of the heartbeat, and a reduction in appetite. No single food source is rich enough in vitamin B_1 to provide the complete daily requirement, so it must be obtained from many foods.

The coenzyme flavin adenine dinucleotide, or FAD, is a derivative of

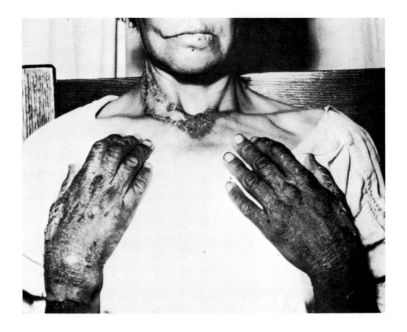

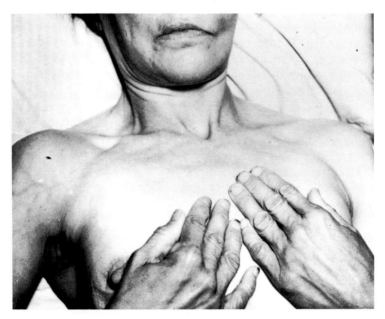

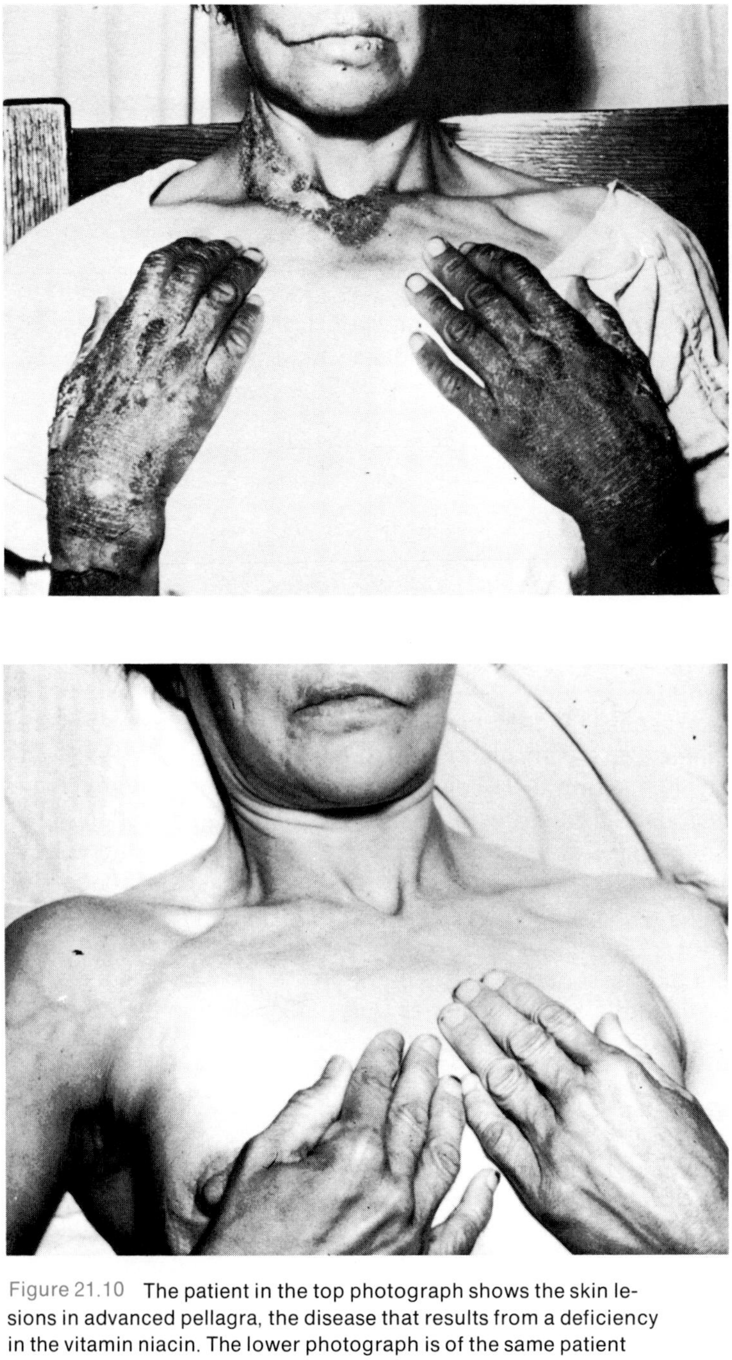

Figure 21.10 The patient in the top photograph shows the skin lesions in advanced pellagra, the disease that results from a deficiency in the vitamin niacin. The lower photograph is of the same patient after niacin therapy. (Courtesy The Upjohn Company)

Table 21.4 Coenzymes Formed From Vitamins in the B Family

Thiamine (B_1)	Thiamine pyrophosphate
Riboflavin (B_2)	Flavin adenine dinucleotide (FAD)
Niacin	Nicotinamide adenine dinucleotide (NAD)
Pyridoxine (B_6)	Pyridoxal phosphate
Cyanocobalamin (B_{12})	Cobamide coenzymes
Pantothenic acid	Coenzyme A (CoA)
Folic acid	Tetrahydrofolate
Biotin	Covalently bonded to carboxylases

riboflavin (vitamin B_2). We have seen that FAD acts as an important hydrogen carrier in the election transport chain (Sections 18.21 and 18.25). Other coenzymes that serve as hydrogen carriers are NAD and NADP (Figure 18.11). These coenzymes are derivatives of the vitamin niacin. A deficiency of niacin results in pellegra, a disease which causes dermatitis, diarrhea, and dementia (Figure 21.10).

21.11 The Fat-Soluble Vitamins

The fat-soluble vitamins (vitamins A, D, E, and K) are fairly stable, are not destroyed by heat, and are not soluble in water. Any excess that is ingested is stored in the liver for release when the body requires additional vitamins (rather than being excreted by the kidney, as with the water-soluble vitamins). As a result, too much of a fat-soluble vitamin can be just as toxic as too little. For example, vitamin A is required for normal growth, for vision in dim light, and for maintaining healthy skin and the inner lining of the body. It is, however, a very toxic compound. As little as 25,000 I.U. each day over a period of 25 days can cause such toxic effects as increases in cerebral spinal fluid pressure, headaches, irritability, patchy loss of hair, and dry skin with ulcerations. In a similar manner, a lack of vitamin D produces rickets, a disease that results from the body's inability to absorb calcium and phosphorus from the intestinal track (see Figure 17.5). But excessive vitamin D (100,000 I.U. for adults and 4,000 I.U. for children) is toxic, and will cause loss of appetite, thickening of bones, and calcification of joints and soft tissues.

Regulation of Enzyme Activity

21.12 Regulatory Enzymes

Enzymes give living systems the ability to act. But if all enzymes were equally active in the cell all the time, the cell would probably "burn itself

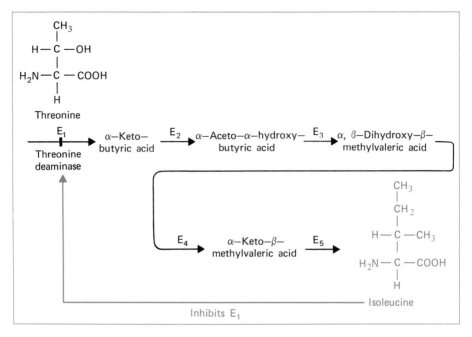

Figure 21.11 In the synthesis of the amino acid isoleucine from the amino acid threonine, the first enzyme (threonine deaminase) is the regulatory enzyme. It is inhibited by the end product of the multienzyme system (isoleucine).

out" and die. Living systems, therefore, not only have the ability to act, but also the ability to control this action. For example, enzymes in the bloodstream catalyze the formation of blood clots when we are bleeding, but they are not active and do not form clots under normal conditions. Enzymes catalyze the contraction of muscle fibers when we walk, but are inactive when we sit or rest. The control mechanisms of the living system involve control of enzyme concentrations and control of enzyme activity.

We have seen that most biological chemical reactions occur in a sequence of reactions that eventually produces a specific metabolic result. Because each reaction in such a sequence is catalyzed by a separate enzyme, these sequences of reactions are called **multienzyme systems.** For example, a multienzyme system is involved in the breakdown of glucose to lactic acid in muscle cells, and other multienzyme systems are involved in the synthesis of different amino acids.

In most multienzyme systems, the enzyme that catalyzes the first reaction of the series is the **regulatory** or **allosteric enzyme.** This enzyme controls the rate of the entire process. Regulatory enzymes are usually complex, high-molecular-weight molecules containing several polypeptide chains and cofactors. These enzymes usually have more than one site for attachment of molecules—one for the active site, and one or more for regulatory molecules. A site for a regulatory molecule is called a **regulatory** or **allosteric** (meaning other space or location) **site.** The **regulatory molecule** itself can either inhibit or increase the activity of the

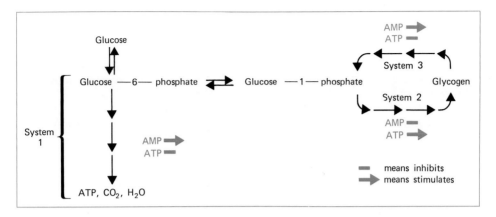

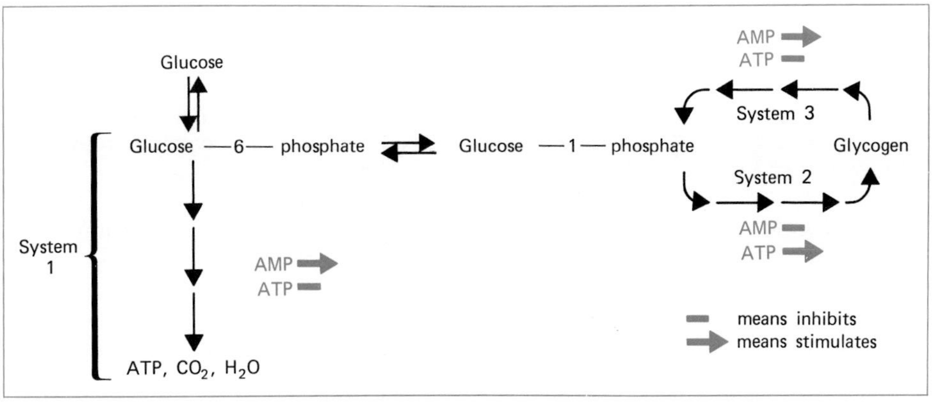

Figure 21.12 ATP and AMP have opposite regulatory effects on the three multienzyme systems controlling the level of ATP in the cell. System 1 involves the oxidation of glucose. System 2 involves glycogenesis, and system 3 glycogenolysis.

enzyme. Such an allosteric enzyme is a flexible molecule, and the regulatory molecule causes a slight change in the shape of the enzyme. This, in turn, changes the shape of the active site, making it either more or less receptive to the substrate.

Many regulatory enzymes are inhibited by the end product of the multienzyme system.

$$A \xrightarrow{E_1} B \xrightarrow{E_2} C \xrightarrow{E_3} D \xrightarrow{E_4} F \xrightarrow{E_5} G$$

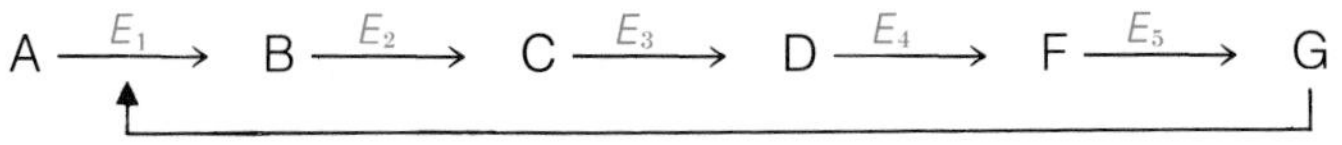

In the above example, E_1 is the regulatory enzyme for the process, and it will be inhibited by high concentrations of product G (Figure 21.11).

All cells need energy in the form of ATP, and there are three multienzyme systems that control the level of ATP in the cell (Figure 21.12). When ingested into the human body, glucose is either broken down in system 1 to form carbon dioxide, water, and ATP, or is converted to glycogen in system 2. System 3 breaks down glycogen to release glucose when demand for this sugar is great. The regulatory enzymes for these systems are affected by cellular levels of ATP and AMP. High levels of AMP will stimulate systems 1 and 3, and inhibit system 2. Similarly, high levels of ATP will inhibit systems 1 and 3, and stimulate system 2.

21.13 Regulatory Genes

As we will see in the next chapter, the function of a gene is to direct the synthesis of a protein molecule. Genes, therefore, also direct the synthesis of cellular enzymes. One way in which a cell can control its metabolic activities is to switch on and off the genes that direct the synthesis of cellular enzymes. In this way the cell can control the concentration of enzymes within itself.

Table 21.5 Hormones of the Principal Endocrine Glands*

Gland or Tissue	Hormone	Major Function of the Hormone
Thyroid	1. Thyroxin 2. Thyrocalcitonin	1. Stimulates rate of oxidative metabolism and regulates general growth and development. 2. Lowers level of calcium in the blood.
Parathyroid	Parathormone	Regulates the levels of calcium and phosphorus in the blood.
Pancreas (Islets of Langerhans)	1. Insulin 2. Glucagon	1. Decreases blood glucose level. 2. Elevates blood glucose level.
Adrenal medulla	Epinephrine	Various "emergency" effects on blood, muscle, temperature.
Adrenal cortex	Cortisone and related hormones	Control carbohydrate, protein, mineral, salt, and water metabolism.
Anterior pituitary	1. Thyrotropic 2. Adenocorticotropic 3. Growth hormone 4. Gonadotropic (two hormones) 5. Prolactin	1. Stimulates thyroid gland functions. 2. Stimulates development and secretion of adrenal cortex. 3. Stimulates body weight and rate of growth of skeleton. 4. Stimulate gonads. 5. Stimulates lactation.
Posterior pituitary	1. Oxytocin 2. Vasopressin	1. Causes contraction of some smooth muscles. 2. Inhibits excretion of water from the body by way of urine.
Ovary (follicle)	Estrogen	Influences development of sex organs and female characteristics.
Ovary (corpus luteum)	Progesterone	Influences menstrual cycle; prepares uterus for pregnancy; maintains pregnancy.
Uterus (placenta)	Estrogen and progesterone	Function in maintenance of pregnancy.
Testis	Androgens (testosterone)	Responsible for development and maintenance of sex organs and secondary male characteristics.
Digestive system	Several gastrointestinal	Integration of digestive processes.

* From G. E. Nelson, G. G. Robinson, and R. A. Boolootian, *Fundamental Concepts of Biology*, Second Edition, copyright © 1970 by John Wiley and Sons, Inc., New York, page 114. Used by permission.

21.14 Regulatory Hormones

The body has control systems in addition to those within individual cells; these systems coordinate the actions between cells in multicellular systems. This higher level of communication between cells, tissues, and organs involves the nervous system and the endocrine system. The endocrine system is a group of glands that produces and secretes chemical messengers called **hormones** into the body fluids, particularly the blood (Figure 21.13 and Table 21.5). Each hormone has target organs or cells that are influenced by its presence. The secretion of a particular hormone by an endocrine gland may be triggered by the nervous system, as in the release of adrenalin triggered by hearing a nearby explosion, or by the concentration of a specific chemical compound, as in the release of insulin triggered by an increase in blood glucose levels.

Hormones are thought to regulate cellular processes in two ways. Steroid hormones such as estrogen and testosterone will enter the cell and bind to a receptor site on a large protein molecule. This steroid-receptor group is then transferred to the nucleus of the cell, where it activates specific genes. On the other hand, peptide hormones such as insulin, human growth hormone, and hormones which regulate gastric and kidney secretions attach to specific receptor sites on the surface of the target cell

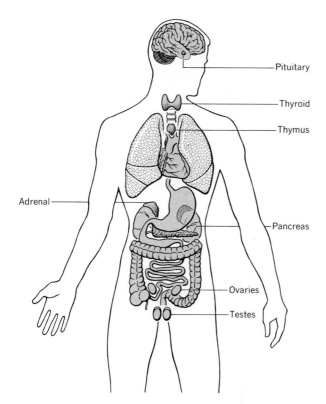

Pituitary

Thyroid

Thymus

Adrenal

Pancreas

Ovaries

Testes

Figure 21.13 The endocrine glands secrete chemical messengers called hormones.

Table 21.6 Some Hormones That Trigger Response in Target Tissues by Increasing the Level of Cyclic AMP in the Tissue

Hormone	Target Tissue	Response
Thyrotropin	Thyroid	Thyroxin secretion
Luteinizing hormone	Ovary	Progesterone secretion
Epinephrine	Muscle	Glycogenolysis
	Fatty Tissue	Fat hydrolysis
	Liver	Glycogenolysis
Parathyroid hormone	Bone	Calcium resorption
	Kidneys	Phosphate excretion

membrane, triggering changes within the cell. The number of receptor sites on a target cell determines the level of response by that cell, and the number may increase or decrease depending upon the concentration of hormones in the blood.

Cyclic AMP

Cyclic AMP is a molecule that acts within the cell as a secondary messenger, or intracellular hormone. Its formation is triggered by the attachment of a hormone molecule to a target cell. The system works as follows: A hormone, the primary messenger, is released into the bloodstream. It travels to a target cell, where it attaches to a specific receptor site on the outside of the cell membrane. This attachment causes a conversion of the enzyme adenyl cyclase from an inactive form to an active form on the inside of the cell membrane. This enzyme then catalyzes the formation of cyclic AMP from ATP (Figure 21.14). The cyclic AMP

Figure 21.14 The formation of cyclic AMP from ATP.

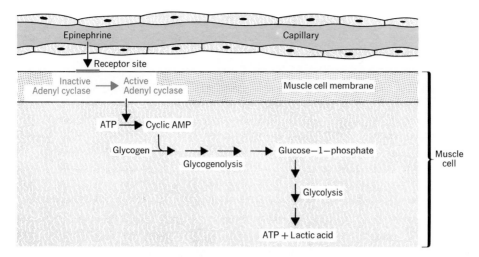

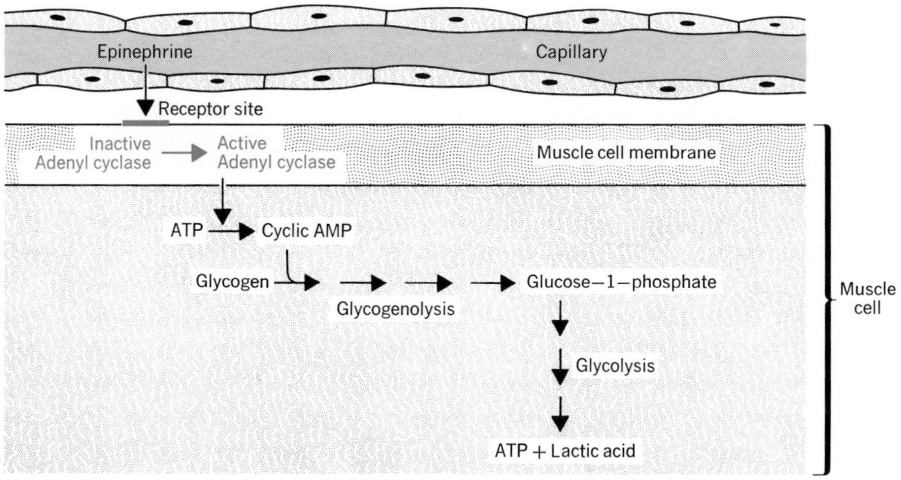

Figure 21.15 Epinephrine stimulates glycogen breakdown and glycolysis in the muscle cell by attaching to a receptor site on the muscle cell membrane. This triggers the activation of the enzyme adenyl cyclase, which catalyzes the formation of cyclic AMP from ATP. The cyclic AMP sets off a series of reactions that result in glycogenolysis and glycolysis, and the production of ATP for use in the contraction of the muscle cell.

spreads through the cell as a secondary messenger, instructing the cell to respond to the hormone in the particular manner characteristic of that cell (Table 21.6 and Figure 21.15). For example, the hormone thyrotropin is secreted by the pituitary gland, and travels to target cells in the thyroid gland. There it triggers the formation of cyclic AMP, to which the thyroid cells respond by secreting more thyroxin.

Prostaglandins

Prostaglandins are a class of compounds that resemble hormones in their effects, but are chemically quite different. They are 20-carbon fatty acids that are synthesized in the cell membrane from unsaturated fatty acids such as arachidonic acid (Figure 21.16). As with cyclic AMP, prostaglandins are present in almost all cells and tissues, and function to carry out the messages that cells receive from hormones. They are among the most potent of all biological agents. They can raise or lower blood pressure, and can regulate gastric secretions. Prostaglandins seem to play a role in both the process that prevents blood platelets from clumping together in healthy blood vessels, and the process that makes them clump together (to form a blood clot) when a vessel is damaged. Fevers and inflammatory reactions are produced by prostaglandins, and aspirin and other anti-inflammatory drugs work by inhibiting prostaglandin synthesis. Prostaglandins also cause uterine contractions, and thus are used to induce labor; they open air passages to the lungs, and thus may be used to relieve asthma. Prostaglandins are under continuing study

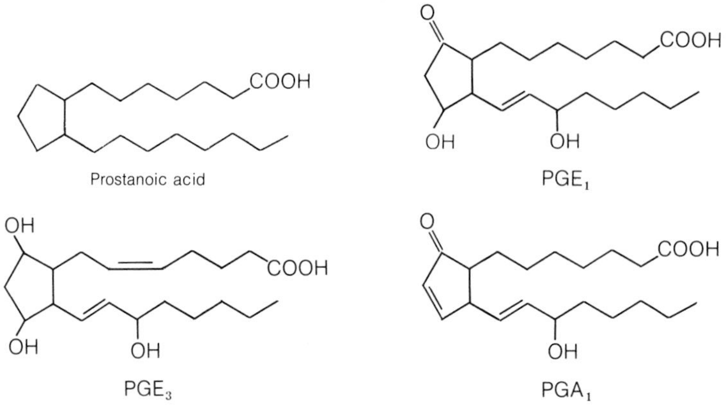

Figure 21.16 Some important prostaglandins. In general the prostaglandins are variants of the basic structure of prostanoic acid. Although their structures are only slightly different, the prostaglandins have widely diverse physiological effects.

to determine their full range of physiological effects and possible therapeutic uses.

Inhibition of Enzyme Activity

21.15 Irreversible Inhibition

The activity of enzymes can be inhibited in several ways. Research into these processes of enzyme inhibition has produced a great deal of knowledge about the specificity of enzymes and the nature of their active sites. The action of many poisons and drugs is due to their ability to inhibit specific enzymes.

Irreversible inhibition of enzyme activity occurs when a functional group or cofactor required for the activity of the enzyme is destroyed or modified. For example, cholinesterase is an enzyme that catalyzes a reaction taking place at the juncture of nerve cells; the enzyme is necessary for normal transmission of nerve impulses. But compounds in nerve gases will combine with the —OH group on a serine molecule which is vital to the active site of the cholinesterase enzyme. In such a case the enzyme then loses its ability to catalyze the reaction, so that animals poisoned by nerve gas become paralyzed (Figure 21.17). Cyanide poisons release the cyanide ion ($C{\equiv}N^-$), which binds very tightly to metal ions, and prevents them from combining with apoenzymes. For example, cyanides inhibit the enzyme cytochrome oxidase (one of the enzymes of the electron transport chain), which requires iron as an activator. Ions of heavy metals such as mercury and lead are extremely toxic because they bind with the —SH functional group on enzymes, permanently denaturing them.

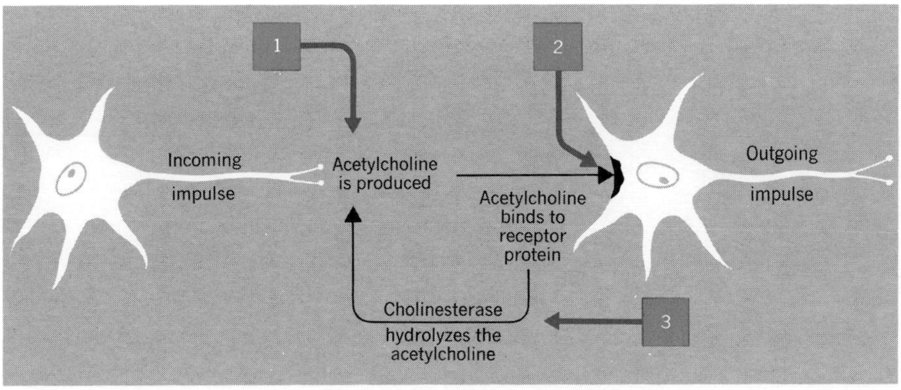

Figure 21.17 The nerve poisons. The formation and breakdown of acetylcholine must occur for normal nerve transmission. Acetylcholine is synthesized at the end of one nerve cell fiber and then migrates to a receptor protein on the next nerve cell, causing the signal to be sent along. Once the signal is sent, the acetylcholine is hydrolyzed by the enzyme cholinesterase, leaving the cell ready for the next signal. Nerve poisons disrupt these processes in the following ways:
(1) Botulism toxin blocks the synthesis of acetylcholine so that nerve signals are not sent, and death is caused by respiratory failure.
(2) Nicotine, curare, atropine, morphine, codeine, cocaine, and local anesthetics such as procaine combine with the receptor protein, blocking its reaction with acetylcholine. Therefore, the second cell does not receive the impulse and it is not sent on.
(3) Anticholinesterases are poisons such as nerve gases, organophosphate insecticides, and some mushroom toxins that inhibit cholinesterase, causing overstimulation of nerve cells by acetylcholine. This results in irregular heart rhythms, convulsions, and death.

Arsenic compounds are poisonous because they contain the arsenate ion (AsO_4^{3-}), which can be mistaken in enzyme synthesis for the phosphate ion (PO_4^{3-}). Enzymes with arsenate ions in their structure will not function properly.

21.16 Reversible Inhibition

Reversible inhibition of enzyme activity takes two forms: competitive and noncompetitive inhibition. **Competitive inhibition** results when a compound having a structure very similar to the substrate competes with the substrate for the active site on the enzyme. When such inhibitors become bound to active sites, there are fewer enzyme molecules available to the substrate. Therefore, enzyme activity will decrease.

$$\{\underset{\text{Enzyme}}{E} + \underset{\text{Inhibitor}}{I} \rightleftharpoons EI\} \text{ competes with } \{\underset{\text{Enzyme}}{E} + \underset{\text{Substrate}}{S} \rightleftharpoons ES\}$$

For example, succinic acid and malonic acid have very similar structures. Succinate dehydrogenase catalyzes the removal of hydrogen from succinic acid, and this reaction is inhibited by malonic acid.

COOH
|
CH$_2$
|
CH$_2$ + E—FAD ⟶
|
COOH

Succinate
dehydrogenase

Succinic
acid

COOH
|
HC
‖
CH + E—FADH$_2$
|
COOH

Fumaric acid

COOH
|
CH$_2$
|
COOH

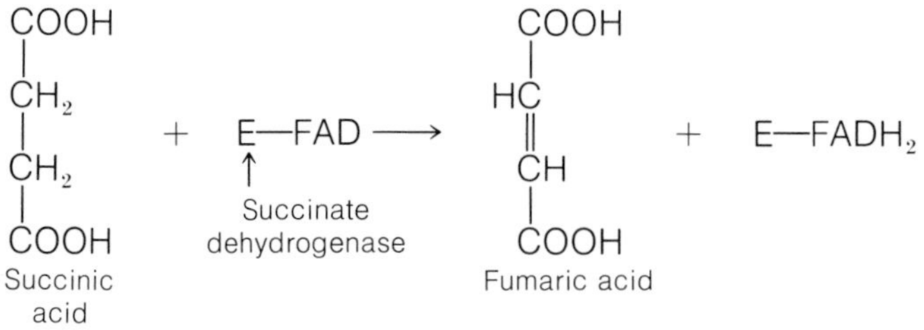

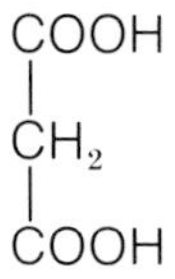

Competitive inhibitor: Malonic acid

Noncompetitive inhibition results when the inhibitor, instead of combining with the active site on the enzyme, combines with some other portion of the enzyme molecule that is essential to its function. A common type of noncompetitive inhibitor combines reversibly with —SH groups that are not located on the active site, but are still essential to the enzyme's activity. For example, iodoacetic acid inhibits the conversion of glucose to lactic acid in muscles by reacting with the —SH groups found on the enzyme glyceraldehyde phosphate dehydrogenase (Figure 21.18).

E—SH + ICH$_2$COOH ⇌ E—S—CH$_2$COOH + HI

Active Iodoacetic Inactive
enzyme acid enzyme

21.17 Antibiotics

Antibiotics are chemicals that are extracted from organisms such as molds, bacteria, and yeasts, and that inhibit growth or destroy other microorganisms. They belong to a large class of chemicals called

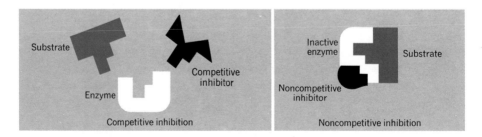

Figure 21.18 In competitive inhibition, the inhibitor and the substrate compete for the active site on the enzyme. In noncompetitive inhibition, the inhibitor combines with some other portion of the enzyme molecule that is essential for the activity of the enzyme.

antimetabolites, which inhibit enzyme function. Antibiotics are used to treat many diseases in humans and animals, and in small amounts are used to speed the growth of poultry and livestock.

Antibiotics can function in many ways to prevent growth or to destroy a disease-causing organism. For example, the sulfa drug sulfanilamide prevents bacterial growth by competitive inhibition. Bacteria use *para*-aminobenzoic acid in the synthesis of the vitamin folic acid.

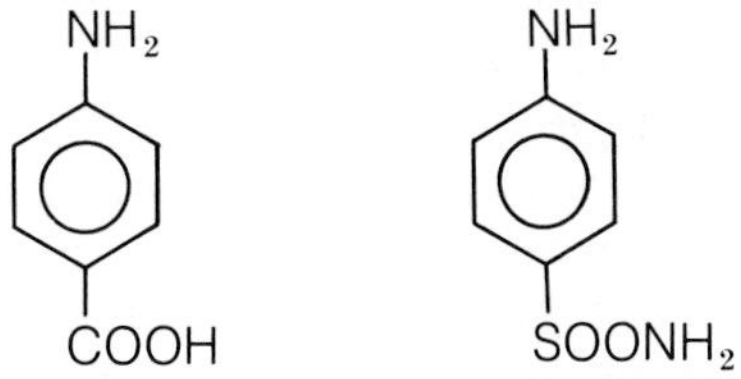

Para-aminobenzoic acid   Sulfanilamide

The structure of sulfanilamide closely resembles that of *para*-aminobenzoic acid, and sulfanilamide molecules compete for the active site on the bacterial enzyme. This, then, inhibits folic acid synthesis. Another way in which antibiotics function is by damaging the structure, or preventing the synthesis, of bacteria cell walls. For example, penicillin G works by inhibiting the reaction which is the final step in the formation of the cell wall, thus causing the bacterial cell to burst. The inhibitory action of novobiocin and bacitracin is very similar to that of penicillin G. Polypeptide antibiotics such as polymyxin and tyrocidine act on the cell membranes of microorganisms, increasing the permeability of the membrane and releasing vital nutrients from the cell. A third way in which antibiotics work is by interfering with protein synthesis within the cell. Chloramphenicol and the tetracyclines alter the nature of the RNA molecules synthesized by the cell, in this way disrupting protein synthesis.

Antimetabolites are also being used in cancer chemotherapy. Methotrexate, an antimetabolite that competes with the vitamin folic acid, has been used very successfully to treat a particular type of uterine tumor. Another antimetabolite, 5-fluorouracil (5-FU), has been used to treat certain forms of intestinal and breast cancers. The fluorine-uracil bond in this compound is very stable and will not enter into a substitution reaction.

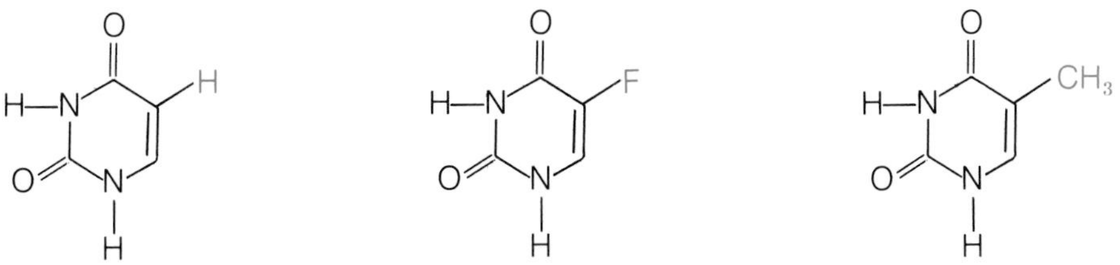

Uracil   5-Fluorouracil   Thymine

5-FU competes with uracil, preventing the formation of thymine (an important part of the DNA molecule, to be discussed in the next chapter). Thus, this compound interferes with DNA synthesis in the cancer cell.

Chapter Summary

Enzymes are proteins that catalyze reactions in living organisms. Enzymes may be simple proteins or conjugated proteins consisting of an apoenzyme and one or more cofactors. A cofactor may be a metal ion (activator) or another organic compound (coenzyme). Enzymes are often secreted in an inactive form as a zymogen or pre-enzyme, and then converted into an active form when needed. When enzymes catalyze reactions, the substrate attaches to the surface of the enzyme in a region called the active site. This activates the substrate and forms products which are then released from the surface of the enzyme. Enzymes catalyze only very specific reactions because the correct substrate fits in the active site just as only one key fits into a specific lock. The activity of an enzyme will be affected by all the factors that can alter the structure of a protein. Each enzyme has an optimum pH at which it is the most active. Metabolic processes are the result of complicated sequences of chemical reactions. Such sequences are controlled by a regulatory or allosteric enzyme—often the enzyme that catalyzes the first step in the sequence. In addition to the active site, this enzyme contains allosteric sites at which a regulatory molecule can attach, thus inhibiting or increasing the activity of the enzyme. Enzyme activity can be inhibited irreversibly when a functional group or cofactor required for the enzyme's activity is destroyed or modified. The activity can be inhibited reversibly when another molecule competes for the active site, or when a portion of the enzyme required for its action is modified reversibly.

Enzyme activity is regulated at many levels, both inside and outside the cell. Hormones, produced by the endocrine system, are one of the body's means of intercellular control. Hormones work in one of two ways: they can attach to a special area on the membrane of a target cell (and thus activate the formation of a substance within the cell, such as cyclic AMP), or they can enter the cell and activate specific genes. Prostaglandins are a group of fatty acids that have effects resembling those of the hormones.

Vitamins are organic substances required in small amounts for normal cellular function, but are not produced by the body. A lack of any vitamin in the diet will cause a specific deficiency disease. Vitamins are often cofactors of important cellular enzymes. They can be classified into two groups: water-soluble vitamins and fat-soluble vitamins.

Exercises and Problems

1. In general terms, describe the similarities and differences between the action of enzymes and inorganic catalysts.
2. What is the difference between an apoenzyme and a coenzyme?

3. (a) Explain how the lock-and-key comparison describes the method of enzyme action.
 (b) How does the induced-fit theory change the lock-and-key theory?

4. Why doesn't pepsin, the enzyme released in the stomach to break down proteins, continue to function in the small intestine?

5. Why is boiling water an effective sterilizing agent?

6. Apples picked before they have completely ripened may be stored at low temperatures and under an atmosphere of nitrogen to keep them fresh for many months. Explain why these storage conditions slow the ripening process.

7. (a) What is an allosteric enzyme?
 (b) Why are allosteric enzymes necessary for the normal operation of the cell?
 (c) What effect does the regulatory molecule have on the allosteric enzyme?

8. Describe three ways that reaction rates are controlled within the human body.

9. What are the differences between vitamins and hormones? (Include both their sources and functions in your discussion.)

10. Prostaglandins are among the most potent biological agents. Describe some of their various physiological effects. In what way do they function like cyclic AMP?

11. What is an antimetabolite?

12. Explain three ways that antibiotics destroy disease-causing organisms.

13. Phosphoglucomutase catalyzes the conversion of glucose-1-phosphate to glucose-6-phosphate.

$$\text{glucose-1-phosphate} \underset{}{\overset{\text{phosphoglucomutase}}{\rightleftharpoons}} \text{glucose-6-phosphate}$$

 (a) In general terms, describe the four steps by which phosphoglucomutase accomplishes this conversion.
 (b) Phosphoglucomutase has a turnover number of 1×10^3. How many molecules of glucose-1-phosphate would be converted in 1 hour by 1 molecule of phosphoglucomutase? How many moles?

14. Under normal conditions a human sperm cell can live 24 to 36 hours. However, several doctors have reported successful artificial insemination using human sperm which have been frozen up to six months. In one case, pregnancy resulted from the use of sperm that had been frozen at $-196.5°C$ for six months. Explain how it is possible for sperm cells to survive for six months and then resume normal activity.

15. Excess vitamin A and D is stored in the body, but excess vitamin C and B_1 is readily excreted.

    (a) What property allows vitamins A and D to be stored in the body, whereas vitamins C and B_1 are excreted?
    (b) Explain the statement that an excess of vitamin D in the diet is potentially as dangerous as a deficiency.

16. Beriberi used to be very common in the Far East where the main food of the diet was polished rice (rice with its outer coating removed). The introduction of enriched rice has mostly eliminated the disease.

    (a) What causes beriberi?
    (b) Why was beriberi occurring among this population?
    (c) With what was the rice enriched?

17. An increasing number of foods are being fortified with vitamins. Breakfast foods such as cereal, bread, margarine, and milk may all be fortified with vitamin D as well as other vitamins.

    (a) Explain why this might pose a health hazard to young children who eat a daily breakfast of cereal with milk, and toast with butter.
    (b) Why should fortification of foods with vitamins be carefully controlled?

18. A 17-year-old girl was admitted to a hospital suffering from headaches, blurred vision, and ringing in her ears. These symptoms may indicate a brain tumor, but none was found to be present. The only drug she was taking was vitamin A prescribed by a dermatologist for her acne. After being questioned by the doctor, the girl admitted taking three 50,000 unit vitamin A pills per day instead of only one as prescribed.

    (a) What was the cause of the girl's symptoms?
    (b) How could her symptoms be alleviated?

19. Patients whose bile is blocked from entering the small intestines have blood that does not clot as readily as normal.

    (a) What vitamin is important in the clotting process?
    (b) Suggest a reason why a lack of bile in the intestines could result in a failure of blood clotting.

20. Hyperthyroidism is a disease that results in increased levels of thyroid hormone in the blood. One of the symptoms of hyperthyroidism is a rapidly beating heart. But the rate of heartbeat is directly controlled by epinephrine, and there is no detectable increase in the concentration of epinephrine in hyperthyroidism. Use Section 21.14 to suggest a way that thyroid hormone might act on the heart muscle cell to cause the increase in heart rate.

21. The insecticide Parathion must be handled with great care because it

is converted by the liver to a molecule that resembles nerve gas in its action. Explain how Parathion acts to poison human beings.

22. Public health officials estimate that more than 200,000 children become ill from lead poisoning each year, many of them from eating lead-based paint chips. What is the chemical action of lead that causes this illness?

23. Both ethanol and methanol are oxidized in the body: ethanol to acetic acid, and methanol to formic acid. It is formic acid that causes the acidosis of methanol poisoning. The first step in the conversion of both ethanol and methanol involves the enzyme alcohol dehydrogenase. One therapy for methanol poisoning is the use of a nearly intoxicating dose of ethanol. Use your knowledge of enzyme inhibition to explain why this therapy is effective.

24. Myasthenia gravis, a rare disease in which the skeletal muscles do not contract properly, is thought to result from a low level of acetylcholine in the muscle cells. Looking at Figure 21.17, would a drug that inhibits step 1, 2, or 3 be most effective in treating this disease? Explain your choice.

chapter 22

Nucleic Acids

Learning Objectives

By the time you have finished this chapter, you should be able to:

1. List the three components of nucleotides.

2. Name the sugars and bases found in DNA and RNA.

3. Describe the structure of a DNA molecule.

4. List the three types of RNA and describe their functions.

5. Describe the genetic code and its relationship to amino acids and polypeptide chains.

6. Describe the steps in protein synthesis.

7. State two ways in which cells can regulate protein synthesis.

8. Define *mutation,* and indicate several ways in which mutations can disrupt the normal functions of an organism.

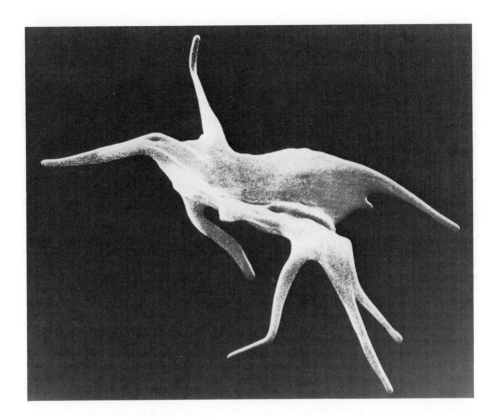

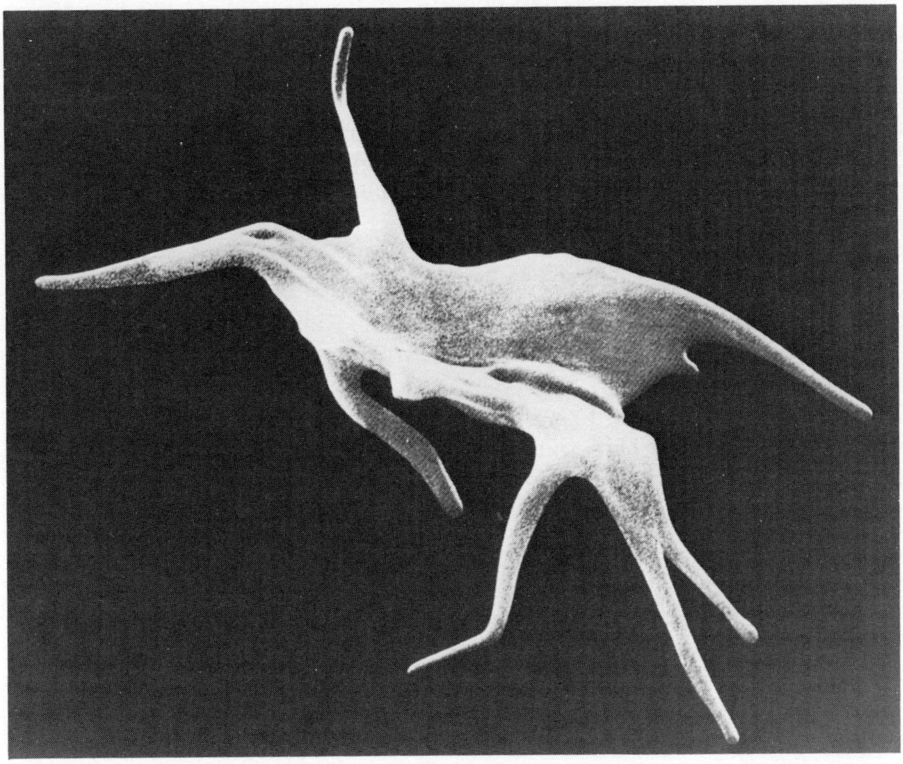

Mrs. Smith watched her son George walk slowly down the street to the school bus stop. She was puzzled and concerned. He'd been complaining on and off for a year about pains in his joints and abdomen. Twice he'd run a low fever and suffered from vomiting, and she had kept him home from school. He was the youngest of her four children, and the only sickly one among them.

Early that afternoon Mrs. Smith received a call from the school nurse, who had just taken George to the hospital—he had collapsed during a vigorous game of kickball, screaming about a pain in his stomach. When Mrs. Smith reached the hospital, she found George under sedation and receiving liquids intravenously. The doctor first thought that George might have appendicitis, but after talking with Mrs. Smith and learning of George's symptoms over the last year, he decided to run several blood tests. As he had suspected, the doctor found that George had sickle cell anemia, and was suffering from a sickle cell crisis. The pain in George's abdomen and the swelling in his joints slowly went away, and a week later he was able to return to school.

A few weeks later George and his mother went to a neighboring town for consultation with doctors at a large clinic specializing in the diagnosis and treatment of sickle cell anemia. There they learned that sickle cell anemia is an inherited disease which affects red blood cells. They were told that long ago a mutation had occurred in the evolution of the African black population, causing an error in the gene for hemoglobin, the molecule in red blood cells that carries oxygen to the body. At present, one out of every two African blacks carries this defective gene, and one out of every 10 American blacks carries this trait. Children who inherit two defective genes from their parents will suffer from sickle cell anemia, and almost all of their hemoglobin will be defective. They will experience periodic sickle cell crises, resulting from the clogging of their capillaries by abnormal red blood cells, and causing severe pain, fever, and swelling of the joints. Such crises can be set off by an infection, cold weather, trauma, strenuous exercise, or emotional stress. Children having sickle cell anemia are anemic, often jaundiced, susceptible to disease, and will usually die in childhood.

People with only one defective gene will carry the sickle cell trait. About one-half of their hemoglobin is defective, but under normal conditions they will show no clinical signs of the disease. However, unusual stress can bring on a crisis in these individuals also.

Sickle cell anemia was first described clinically in 1910, but it was not until 1949 that Linus Pauling demonstrated the molecular basis of the disease. He showed that sickle cell hemoglobin (abbreviated HbS) has different mobility in an electric field than does normal hemoglobin (HbA) (see Figure 20.3). Hemoglobin (molecular weight 64,458) is a molecule consisting of a protein part, called globin, and a nonprotein part, called heme. The globin part consists of four polypeptide chains: two alpha chains containing 141 amino acid units, and two beta chains containing 146 amino acid units. Each chain is wrapped around a heme group containing an Fe^{2+} atom; this is the group that actually carries the oxygen (see Figure 17.10 and Figure 20.11). Therefore, each hemoglobin molecule can carry four molecules of oxygen.

In 1956, Ingram showed that the entire difference between HbS and HbA was in one amino acid on the beta chain. In HbS, a valine molecule is erroneously substituted for a glutamic acid molecule in position 6 on the beta chain.

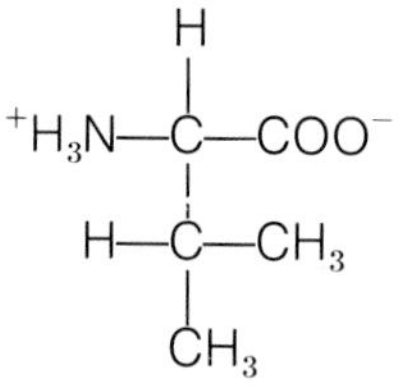

Valine

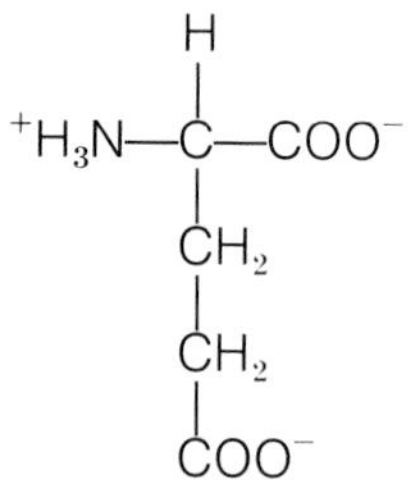

Glutamic acid

At pH 7, the glutamic acid will have a negative charge, whereas the valine will have no charge. This accounts for the difference in mobility of the two hemoglobin molecules in an electric field.

(a)

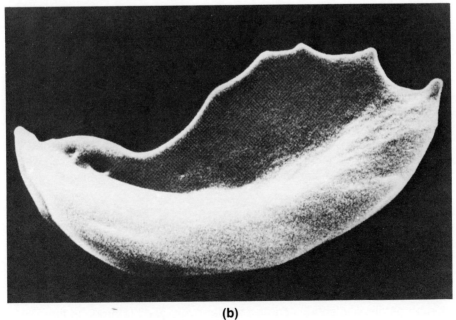

(b)

Figure 22.1 (a) Normal red blood cells. (b) Red blood cells from a person with sickle cell anemia. [(a) Courtesy Francois Morel. From *J. Cell Biol. 48*:91-100, 1971. (b) Courtesy Springer-Verlag, Publishers, from CORPUSCLES by Marcel Bessis.]

At low oxygen concentrations, red blood cells containing HbS will take on abnormal shapes, some resembling sickles (Figure 22.1). This sickling results from the substitution of valine, which contains a nonpolar R-group, for glutamic acid, which contains a polar R-group. At low oxygen concentrations a nonpolar or hydrophobic interaction occurs between HbS molecules. This causes them to group together into long chains or filaments, and results in the strange shapes taken on by the red blood cells. The abnormal shape will correct itself in most red blood cells as the oxygen concentration is increased, but some cells will remain irreversibly sickled.

Such sickling takes place in the capillaries, where the red blood cells release their oxygen. The sickled cells will increase the thickness of the blood and can block the capillaries, depriving tissues of oxygen. Such blockage brings about the clinical symptoms of a sickle cell crisis. The anemia suffered by patients is a result of the decreased life span of their red blood cells.

People suffering from sickle cell anemia, and carriers of the trait, can be identified through blood screening tests. At present, however, there is no effective treatment for the disease. As one approach, drugs are currently being tested that may help prevent the hydrophobic attraction of HbS at low oxygen concentrations.

DNA, Deoxyribonucleic Acid

22.1 Molecular Basis of Heredity

Sickle cell anemia is just one of more than 2000 human diseases known to be caused by disorders in genes. **Genes** are specific segments of molecules called DNA. Each gene contains the information necessary to make one polypeptide chain. To understand how the information carried by genes actually directs the formation of a polypeptide chain, however, we must first become familiar with the molecular nature of the genetic material DNA.

22.2 Nucleotides

DNA (deoxyribonucleic acid) and RNA (ribonucleic acid) belong to a class of polymers called **nucleic acids.** Nucleic acids are polymers of monomer units called **nucleotides.** However, unlike the monomer units we have studied before, nucleotides can be further hydrolyzed to produce three components:

1. A nitrogen-containing base

2. A five-carbon sugar
3. Phosphoric acid.

There are two classes of nitrogen-containing bases found in nucleotides: **pyrimidines** and **purines.** The bases derived from pyrimidine are cytosine (C), thymine (T), and uracil (U). Those derived from purine are adenine (A) and guanine (G). The base uracil is found only in nucleotides of RNA, and the base thymine is found only in nucleotides of DNA.

The Pyrimidines

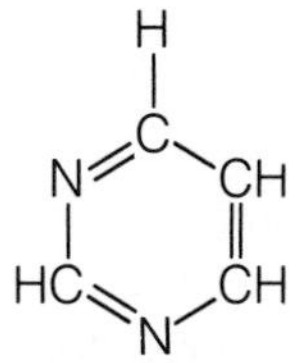

Pyrimidine

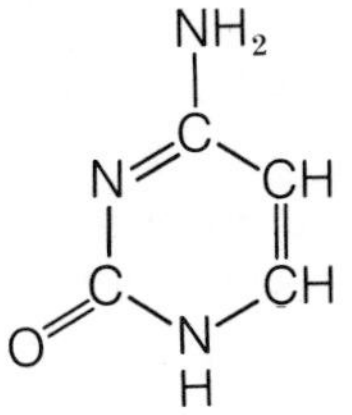

Cytosine (C)

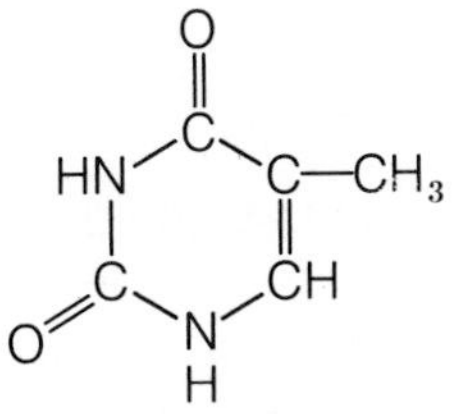

Thymine (T)
(in DNA)

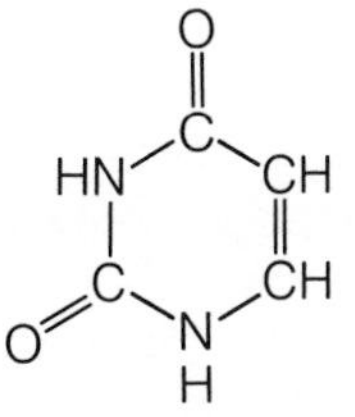

Uracil (U)
(in RNA)

The Purines

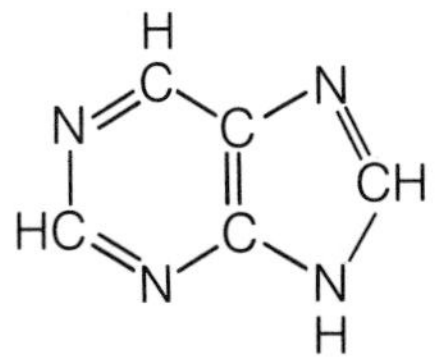

Purine

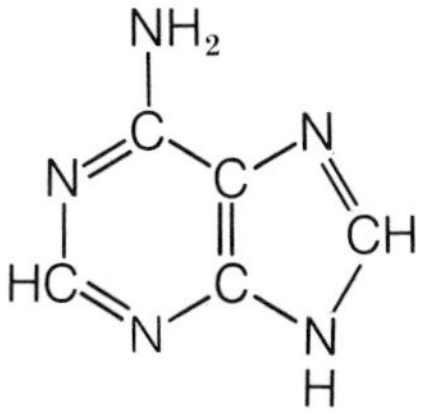

Adenine (A)

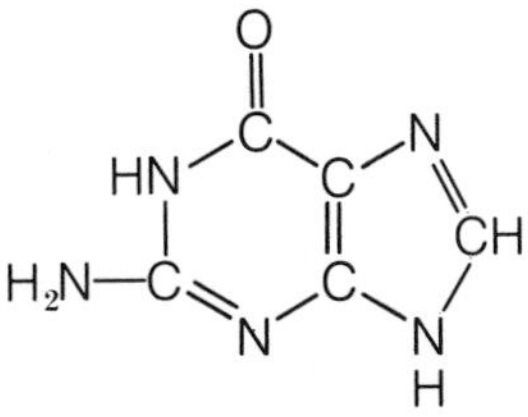

Guanine (G)

The second component of nucleotides is the pentose sugar. RNA contains the sugar ribose, and DNA contains a derivative of ribose, 2-deoxyribose.

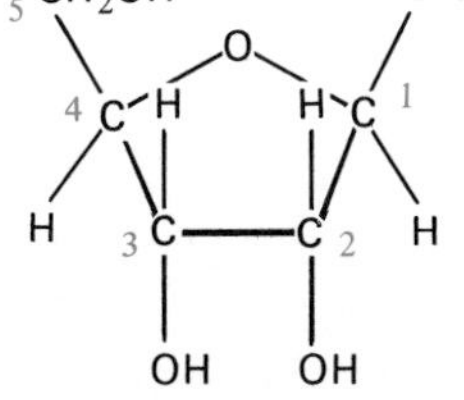

Ribose

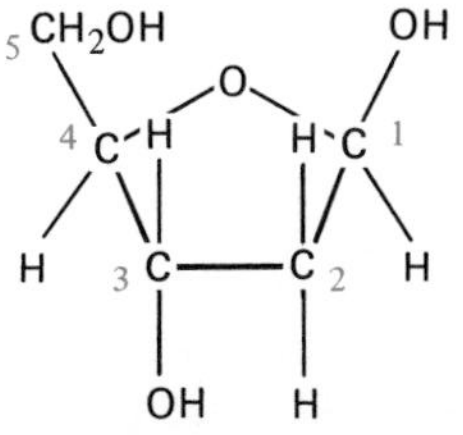

2-Deoxyribose

The three components of the nucleotide are joined together in the following manner.

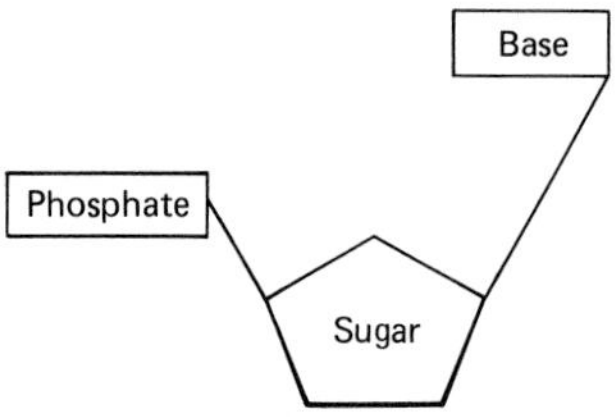

A Nucleotide

Adenylic acid — AMP

This example shows the nucleotide adenylic acid or adenosine monophosphate (AMP), which is made up of adenosine, ribose, and phosphate. Other ribonucleotides will have similar structures, but different bases. The deoxyribonucleotides will contain the sugar deoxyribose instead of ribose.

Free nucleotides are found in large numbers in the cell, and perform many functions. In addition to being the basic structural unit of nucleic acids, they also participate in biosynthetic reactions, serve as coenzymes, and are important in the transport of energy from energy-releasing reactions to energy-requiring reactions. ATP (adenosine triphosphate), GTP (guanosine triphosphate), and UTP (uridine triphosphate) are all energy-carrying mononucleotides.

22.3 The Structure of DNA

Each human cell contains about 2 meters of the nucleic acid DNA, packed into a set of 46 chromosomes having a total length of only about 200 micrometers. This reduction in length is possible because the DNA molecule wraps and folds itself around proteins (called histones) which are tightly bound to the DNA. DNA was first isolated in 1868, but the structure of the molecule was not determined until 1953, when J. D. Watson and F. H. C. Crick proposed a structure that explained the physical and chemical properties of DNA. The structure proposed by Watson and Crick consists of two helical polynucleotide chains coiled around the same axis, forming a double helix (Figure 22.2). The hydrophilic (attracted to water) sugar and phosphate components of the nucleotides are found on the outside of the helix, and the hydrophobic bases are found on the inside.

The nucleotides making up each strand of DNA are connected by ester bonds between the phosphate group and the deoxyribose sugar (Figure 22.3). This forms the "backbone" of each DNA strand, from which the bases extend. The bases of one strand of DNA will pair with bases on the other strand by means of hydrogen bonding. This hydrogen bonding is very specific: Adenine's structure permits it to hydrogen bond only with thymine, and guanine will bond only with cytosine (Figure 22.4). As a

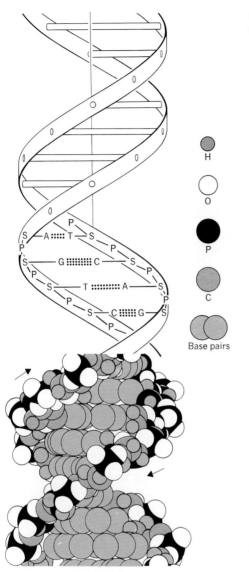

Figure 22.2 The double helix of the DNA molecule.

result, the two strands of DNA are not identical, but rather are complementary—where thymine appears on one strand, adenine will appear on the other.

Example 22-1

What is the sequence of bases on a strand of DNA that would be complementary to a strand having the following sequence of bases?

A A T C G T A G G C A C

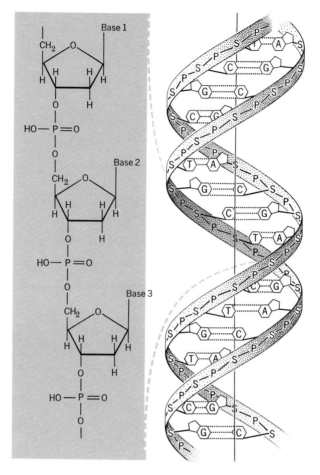

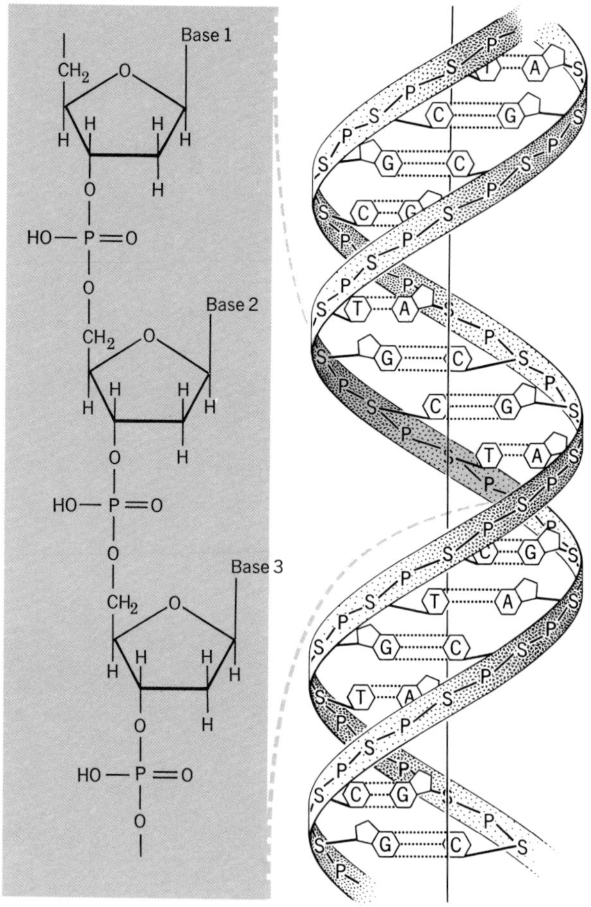

Figure 22.3 The backbone of the DNA molecule consists of the sugar and phosphate groups of the nucleotides. The deoxyribose sugar of one nucleotide is bonded to the phosphate group on the next nucleotide.

Because adenine (A) pairs only with thymine (T), and cytosine (C) only with guanine (G), we can write the sequence of bases as follows:

Original strand	A A T C G T A G G C A C
Complementary strand	T T A G C A T C C G T G

22.4 Replication of DNA

When a cell divides, the DNA molecules must **replicate** (that is, must make exact copies of themselves) so that each daughter cell will have

Figure 22.4 Hydrogen bonding in DNA. The bases on the nucleotides of DNA extend toward the inside of the helix, and the two strands of the helix are held together by hydrogen bonding between the bases. This hydrogen bonding is very specific: thymine will form hydrogen bonds only with adenine, and cytosine only with guanine.

DNA identical to the parent cell. The replication of DNA is catalyzed by the enzyme DNA polymerase. In this process, the two strands of the DNA helix unwind, and each strand serves as a template or pattern for the synthesis of a new strand of DNA (Figure 22.5). Each of the two daughter helixes will contain one original DNA strand and one newly made strand. The genetic information for the cell is contained in the sequence of the bases A, T, C, and G in the DNA molecule. Anything that alters the order of the sequence will cause a change, or mutation, in the genes of the cell.

RNA, Ribonucleic Acid

Molecules of RNA make up 5–10% of the total weight of the cell. The nucleotides of RNA contain ribose instead of deoxyribose, and the base uracil in place of thymine. Unlike DNA, RNA is not double-stranded; it consists of a single strand of nucleic acid. The additional hydroxyl group on the ribose is very important in forming hydrogen bonds that stabilize the tertiary structure of the nonhelical regions of the RNA molecule. There

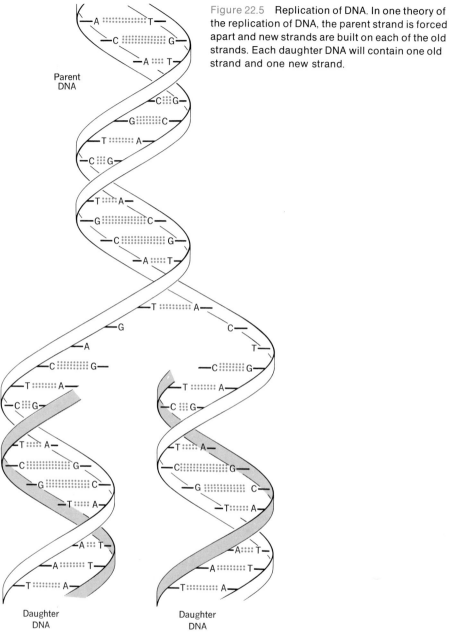

Figure 22.5 Replication of DNA. In one theory of the replication of DNA, the parent strand is forced apart and new strands are built on each of the old strands. Each daughter DNA will contain one old strand and one new strand.

are three distinct types of RNA found in the cell: messenger RNA (*m*RNA), ribosomal RNA (*r*RNA), and transfer RNA (*t*RNA).

22.5 Messenger RNA (*m*RNA)

*m*RNA is synthesized in the nucleus of the cell, and contains the four bases adenine, cytosine, guanine, and uracil. It is synthesized on one strand of a DNA helix, so it will have a sequence of bases complementary to that of the DNA (Figure 22.6).

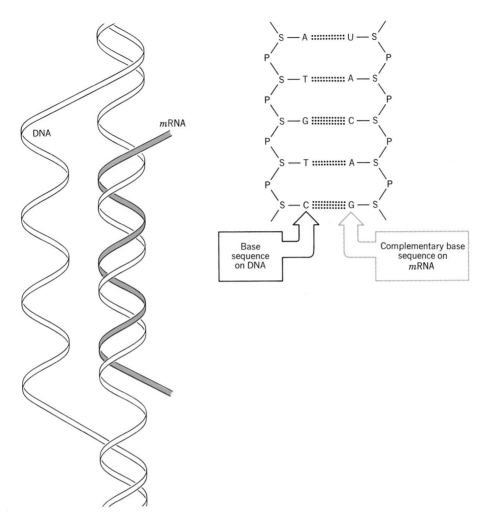

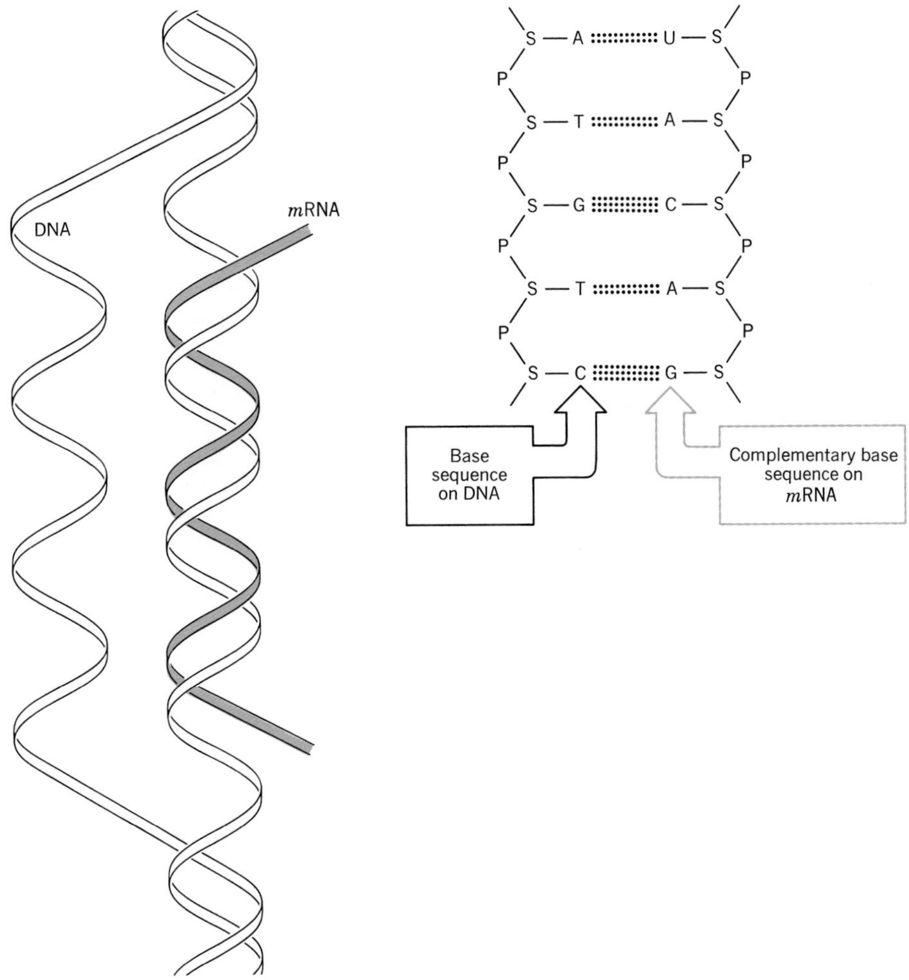

Figure 22.6 The transcription of DNA to *m*RNA occurs in the nucleus of the cell. The double helix of DNA opens up, and the *m*RNA is synthesized on one strand of the DNA. The sequence of bases on *m*RNA is complementary to the sequence on the DNA. However, in making the *m*RNA the cell uses the base uracil instead of thymine.

Example 22-2

What will be the sequence of bases on the *m*RNA molecule that is synthesized on the following strand of DNA?

DNA T A T C T A C C T G G A

Using the same procedure as in Example 22-1, but remembering that *m*RNA contains the base uracil in place of thymine, we have

DNA	T A T C T A C C T G G A
*m*RNA	A U A G A U G G A C C U

*m*RNA is not a stable molecule, and is synthesized by the cell whenever it is needed. After being synthesized, an *m*RNA molecule will migrate to the cytoplasm of the animal cell. There it serves as a template or pattern for the sequencing of amino acids in the synthesis of proteins in the ribosomes.

22.6 Ribosomal RNA (*r*RNA)

*r*RNA is also synthesized in the nucleus of the cell from a DNA template (Figure 22.7). It migrates to the cytoplasm of the cell where, with proteins, it forms the ribosomes. The **ribosomes** are the sites of protein synthesis, and are located on the endoplasmic reticulum (Figure 22.8). The ribosomes are made up of two subunits that combine with *m*RNA to form the "factory" for the production of proteins.

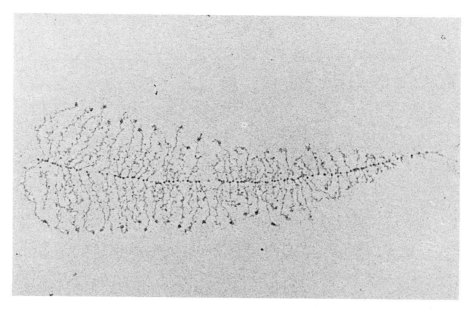

Figure 22.7 Transcription of RNA on a single gene. This electron micrograph shows the formation of fibrils of RNA on the central DNA strand. The shortest fibrils are just beginning to be formed, while the longest fibrils are almost completed. The small dots at the base of each RNA fibril on the DNA strand are molecules of the enzyme RNA polymerase. (Courtesy O. L. Miller, Jr. and Barbara R. Beatty, Biology Division, Oak Ridge National Laboratory).

22.7 Transfer RNA (*t*RNA)

*t*RNA is the smallest of the RNA molecules; it is water-soluble and moves easily within the cell. *t*RNA molecules are synthesized in the nucleus of the cell, and each is specifically designed for a particular amino acid. A *t*RNA molecule becomes *charged* when a specific amino acid is joined to the terminal adenine nucleotide present on each *t*RNA polynucleotide chain (Figure 22.9). The *t*RNA molecule then carries this amino acid to the ribosomes, where the amino acid is used in protein synthesis.

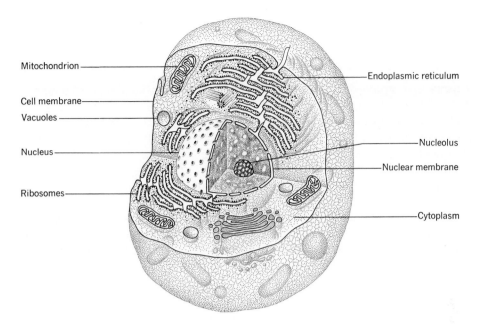

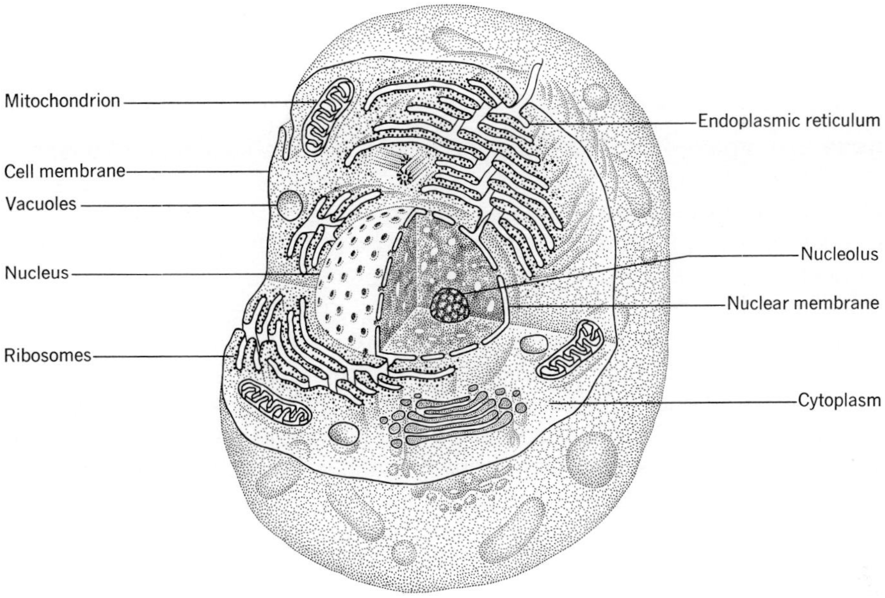

Figure 22.8 The structure of an animal cell.

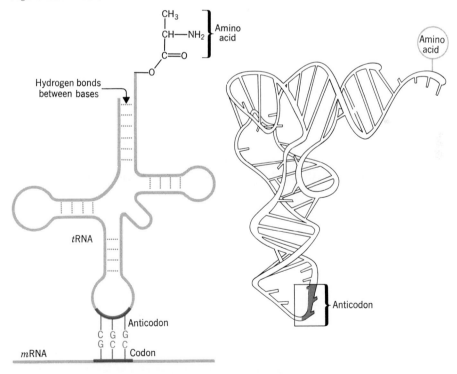

Figure 22.9 The structure of *t*RNA (a) A schematic drawing of a *t*RNA molecule. The three bases in the anticodon region are complementary to a codon on the *m*RNA. (b) The three-dimensional structure of *t*RNA. [Redrawn from G. J. Quigley and A. Rich, *Science* 194, 796-806 (November 19, 1976). Copyright © 1976 by the American Association for the Advancement of Science.]

Protein Synthesis

22.8 The Genetic Code

A **gene** is a region on a DNA molecule comprised of a sequence of bases that will translate into the specific sequence of amino acids making up a protein. The **genetic code** is the general term used to describe the sequence of bases found in the gene. In the early 1960s the "language" of the genetic code was discovered through extensive studies using artificially synthesized *m*RNA molecules. It was discovered that a three-base sequence is necessary to code for each amino acid. This three-base sequence on *m*RNA is called a **codon.** There is more than one codon for most amino acids, as shown in Table 22.1. For example, GGU, GGC, GGA, and GGG are all codons for the amino acid glycine; UUU and UUC are codons for the amino acid phenylalanine. You will notice from the table that the first two bases in the codon for a given amino acid are usually the same. The third base can vary, and is less important in determining the amino acid being specified. The genetic code on the messenger RNA is not punctuated; that is, you read the code one triplet after another, without a break from one end of the *m*RNA molecule to the other.

*m*RNA: G G U C A G U G C U C C . . .

Amino acid: Gly - Gln - Cys - Ser -. . .

Table 22.1 The Codons for the Amino Acids

UUU	Phe	UCU	Ser	UAU	Tyr	UGU	Cys
UUC	Phe	UCC	Ser	UAC	Tyr	UGC	Cys
UUA	Leu	UCA	Ser	UAA	*Stop*	UGA	*Stop*
UUG	Leu	UCG	Ser	UAG	*Stop*	UGG	Trp
CUU	Leu	CCU	Pro	CAU	His	CGU	Arg
CUC	Leu	CCC	Pro	CAC	His	CGC	Arg
CUA	Leu	CCA	Pro	CAA	Gln	CGA	Arg
CUG	Leu	CCG	Pro	CAG	Gln	CGG	Arg
AUU	Ile	ACU	Thr	AAU	Asn	AGU	Ser
AUC	Ile	ACC	Thr	AAC	Asn	AGC	Ser
AUA	Ile	ACA	Thr	AAA	Lys	AGA	Arg
AUG	Met (*Start*)	ACG	Thr	AAG	Lys	AGG	Arg
GUU	Val	GCU	Ala	GAU	Asp	GGU	Gly
GUC	Val	GCC	Ala	GAC	Asp	GGC	Gly
GUA	Val	GCA	Ala	GAA	Glu	GGA	Gly
GUG	Val	GCG	Ala	GAG	Glu	GGG	Gly

The codon AUG codes for methionine and serves as a "start" codon to signal the beginning of the amino acid sequence. Three codons (UAA, UAG, UGA) do not code for any amino acid, but serve as "stop" or terminal codons. They cause the completed protein to be released from the ribosome.

An **anticodon** is the three-base sequence on a transfer RNA molecule that is complementary to the codon on the messenger RNA molecule. Each *t*RNA molecule contains a region carrying an anticodon complementary to the codon on the *m*RNA molecule (Figure 22.9).

22.9 The Steps in Protein Synthesis

The steps involved in the **transcription** of DNA to *m*RNA (that is, the passage of genetic information from DNA to *m*RNA), and then in the **translation** of *m*RNA to protein (that is, the expression of the genetic information in the amino acid sequence of the protein), are identical in all living cells. Many ribosomes can be producing proteins from the same strand of *m*RNA. When viewed under a microscope these groups of ribosomes, called polyribosomes or **polysomes,** appear as a series of dots (Figure 22.10). The following step-by-step description refers to the diagrams in Figures 22.11 and 22.12.

1. *Synthesis of* m*RNA.* *m*RNA is synthesized in the nucleus of the animal cell from a template formed by one strand of the DNA molecule. The *m*RNA then migrates from the nucleus to the cytoplasm.

2. *Attaching amino acids to* t*RNA.* Amino acids are attached to molecules of *t*RNA by enzymes that are specific for each amino acid and each *t*RNA. Each such enzyme has two active sites: one that recognizes the amino acid and one that

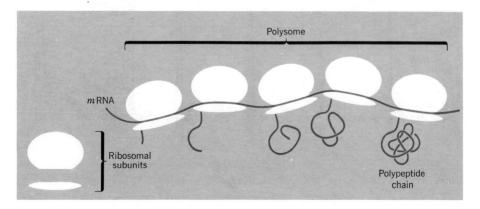

Figure 22.10 Ribosomes consist of two subunits that form a complex with the *m*RNA. A polysome is a cluster of ribosomes all making polypeptide chains on the same strand of *m*RNA.

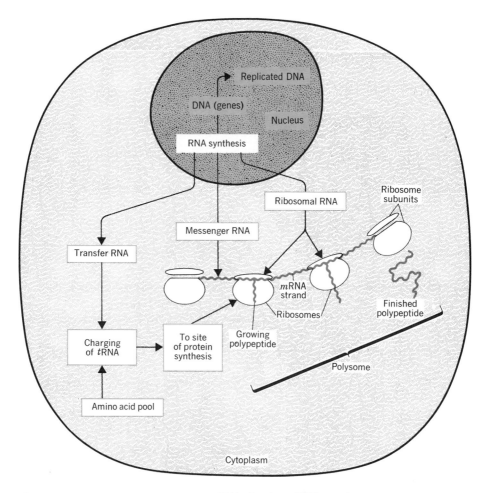

Figure 22.11 The relationship between DNA, the three RNA's, and protein synthesis. (Adapted from J. R. Holum, *Elements of General and Biological Chemistry*. Copyright © 1975, John Wiley and Sons, Inc., New York. Used by permission.)

recognizes the anticodon region of the *t*RNA. The attachment of the amino acid to the *t*RNA (that is, the activation or *charging* of the *t*RNA) is a two-step process requiring the presence of ATP.

3. *Beginning the polypeptide chain.* To begin the synthesis of a protein, the ribosome breaks into its two subunits. The smaller unit attaches to the *m*RNA and to a starting amino acid that is a derivative of *N*-formylmethionine (symbol, fMet). This complex then reforms with the larger subunit of the ribosome. This procedure insures that the synthesis of the protein will not accidentally begin in the middle of the *m*RNA molecule.

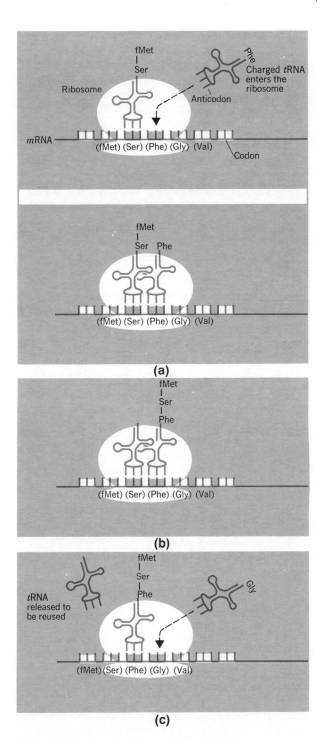

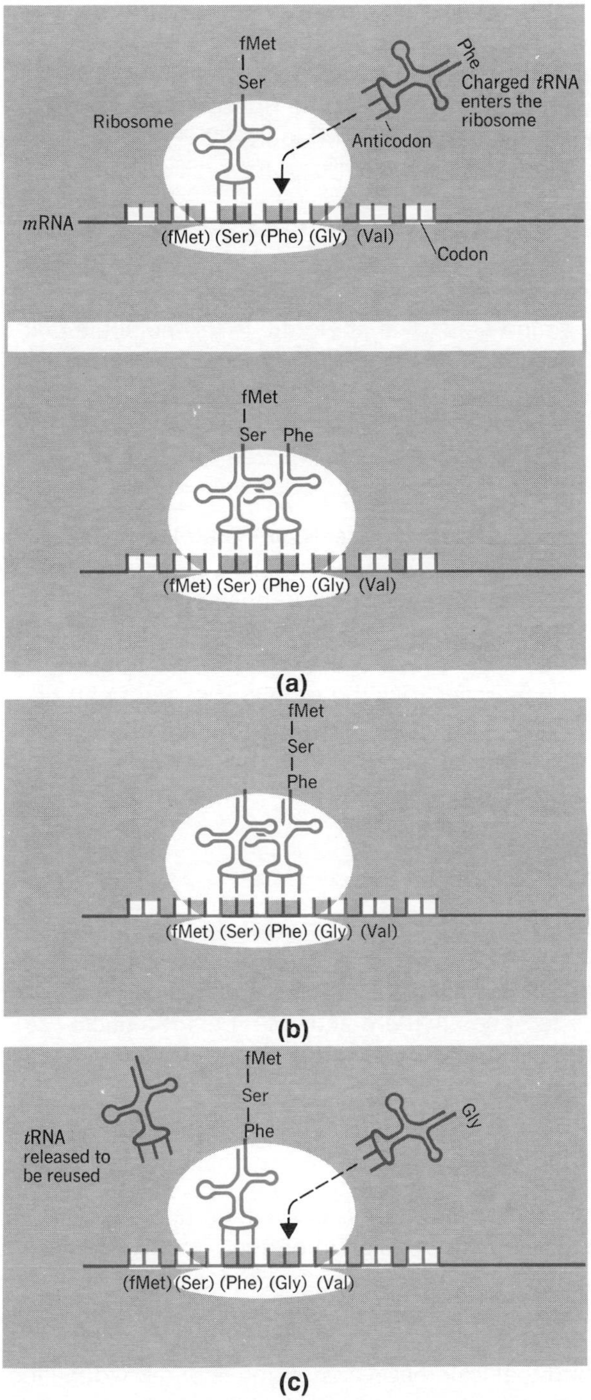

Figure 22.12 Growth of the peptide chain. The individual steps are discussed in Section 22.9.

4. *Adding amino acids to the chain.* The growth or elongation of the polypeptide chain involves a series of repeated steps that take place between *t*RNA and *m*RNA inside the ribosome as it moves along the *m*RNA strand (Figure 22.12):
   a. A charged *t*RNA binds to the next codon on the *m*RNA.
   b. A peptide bond forms between the amino group on the newly attached amino acid and the carboxyl group on the previously bound amino acid, displacing the *t*RNA molecule that had been attached to the previously bound amino acid.
   c. This displaced *t*RNA, now empty, is bumped from the ribosome and is free to become charged once again.

5. *Ending the polypeptide chain.* The elongation of the chain stops when a terminal codon is reached on the *m*RNA molecule. The polypeptide chain separates from the ribosome, and the last *t*RNA drops off the chain. The starting amino acid is now separated from the protein if it is not part of the molecule. As the polypeptide chain has been growing, it has been naturally bending and folding into its native shape.

22.10 Regulation of Protein Synthesis

The control of protein synthesis is important in the differentiation of cells. Each cell in the body contains all the genetic information necessary to produce an entire organism, yet less than 10% of this information will be expressed in any one cell. The genes that are active in a heart muscle cell, for example, are different from those active in a liver cell. As we have seen, cellular processes are regulated by a complex set of biological controls. The *coarse* control mechanism is the synthesis of enzymes by the cell, whereas the *fine* control involves the allosteric or regulatory enzymes.

Very little is known about the exact process of the control of enzyme synthesis in higher animals (that is, the regulation of gene transcription to *m*RNA). What we do know about gene regulation comes mostly from studies on *Escherichia Coli (E. coli),* a single-cell organism found naturally in our digestive tract.

These studies have found that different types of genes can be identified in the control of enzyme synthesis. A region of DNA containing the code for a specific protein is called a **structural gene.** A group of structural genes that codes for all the enzymes catalyzing a multienzyme system is called an **operon;** these genes will be located beside one another along the chromosome. The synthesis of *m*RNA on a structural gene is catalyzed by the enzyme RNA polymerase. However, RNA polymerase can function only if a strip of DNA next to the operon, called the **operator gene,** says it is all right to go ahead. This operator gene is, in turn, controlled by a protein molecule called the **repressor.** When the repressor molecule is bound to the operator gene, no synthesis of *m*RNA can take place. The removal of the repressor molecule from the operator gene requires that a molecule called the **inducer** bind to the repressor. This union changes the shape of the repressor, and it will no longer fit on the operator gene. With the repressor so removed, the operator gene is "turned on," and RNA polymerase is free to catalyze the synthesis of *m*RNA on the operon.

Let's look at the best understood example of this control process: the lac operon of *E. coli* (Figure 22.13). *E. coli* is able to live using the sugar lactose as its only source of carbon. The lactose cannot be used, however, until this disaccharide has been broken down into glucose and galactose by the enzyme β-galactosidase. The structural gene for β-galactosidase does not continuously produce *m*RNA for this enzyme, and there is a repressor molecule on the operator gene. When *E. coli* is placed in a medium containing lactose, the lactose molecule itself acts as the inducer. It attaches to the repressor, causing this molecule to move off the operator gene. This permits the production of the *m*RNA needed for the synthesis of β-galactosidase by the ribosomes. The transcription of *m*RNA will continue until all the lactose has been broken down. At this point the repressor molecule will again be free, and will return to the chromosomes to repress the operator gene. β-Galactosidase production will then stop.

22.11 Theories of Gene Action

In 1958 Crick proposed a theory of gene action which has come to be known as the *central dogma* of molecular biology. This theory states that genetic information can flow in only one direction—from DNA to RNA to protein. However, experimental discoveries since that time have revealed a more complicated pattern of gene action.

In the early 1970s H. M. Temin and D. Baltimore discovered that an enzyme called reverse transcriptase (or RNA-directed DNA polymerase), which is present in RNA tumor viruses, is capable of catalyzing the synthesis of double-stranded DNA from the viral RNA (Figure 22.14). The ability of the RNA virus to synthesize DNA may explain why a virus can disappear after infecting an organism, only to reappear months or years later. The viral-produced DNA may become included in the cellular DNA, and may then be replicated and passed on from cell to cell until it is activated in some unknown way to transform the normal cell to a tumor cell. It is thought that environmental factors such as hormones, chemicals, or radiation may influence the expression of the viral DNA. One theory being actively studied is that human breast cancer and leukemia may involve RNA tumor viruses.

The use of reverse transcriptase in the laboratory has given researchers a technique for determining the sequence of bases on a gene.

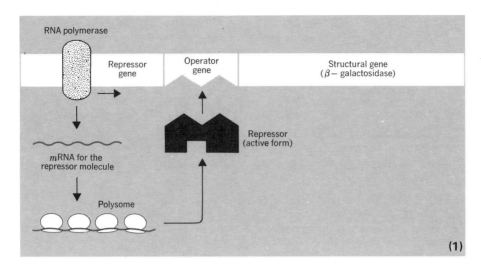

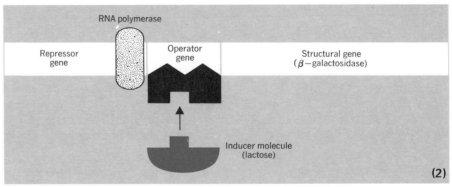

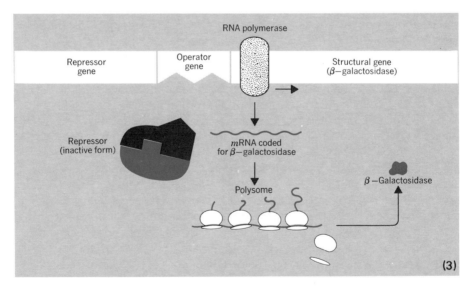

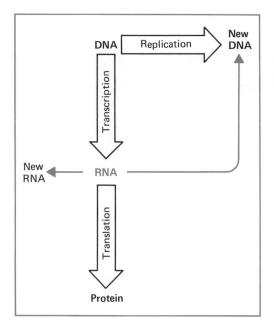

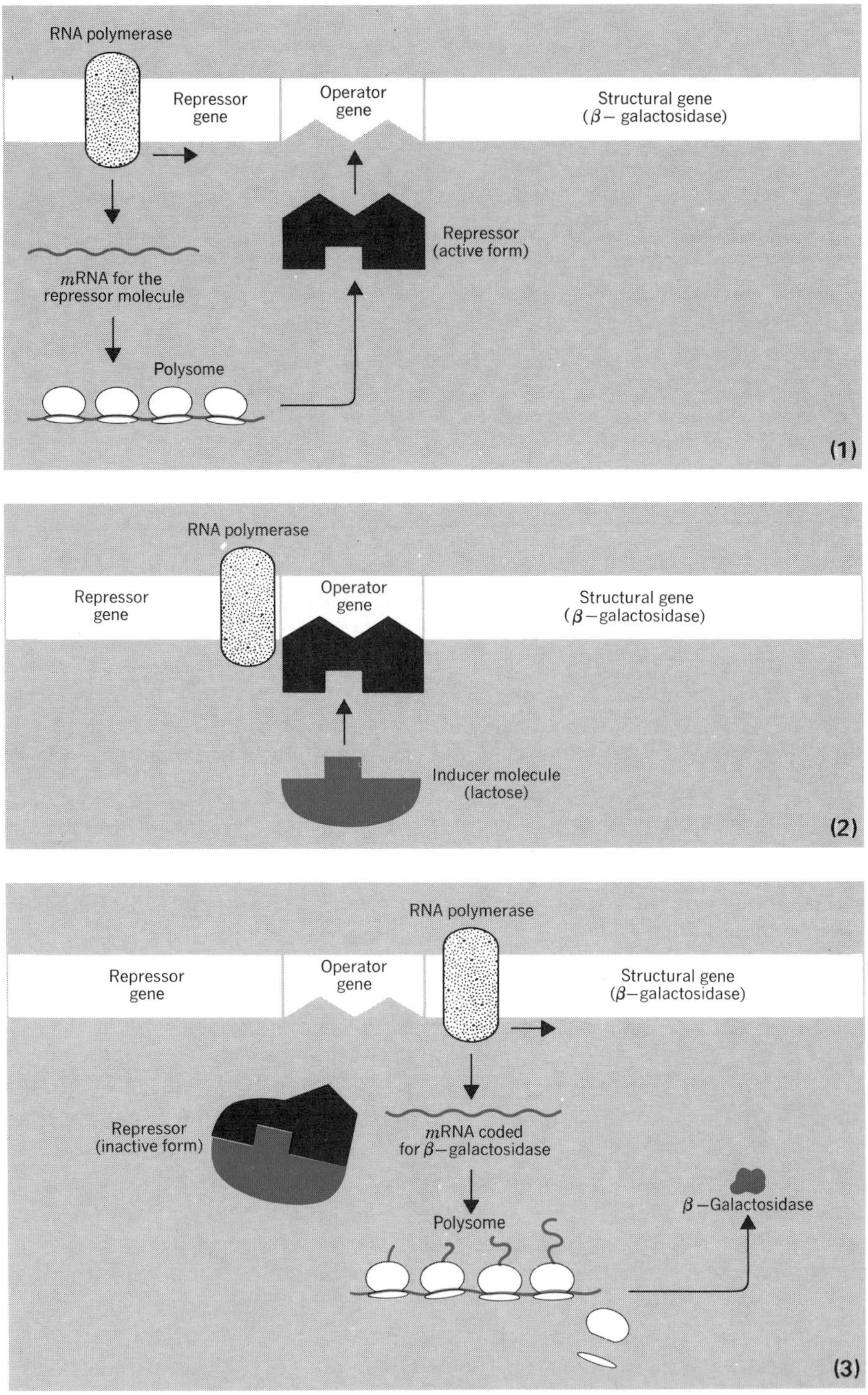

Figure 22.14 The flow of genetic information. The central dogma of molecular biology, which states that genetic information flows from DNA to RNA to protein, has been modified to include evidence that RNA can produce new RNA and, in some cases, DNA. However, it has yet to be shown that the amino acid sequence of a protein can code for a nucleotide sequence in a nucleic acid.

It has led to the discovery that, in some viruses, a gene may code for more than one protein—depending upon the way in which the gene is read to produce *m*RNA. Another recently developed technique, called recombinant DNA, is producing much new information about the exact nucleotide order of genes, how these genes are arranged on chromosomes, and what the function of each gene may be. Recombinant DNA methods may soon be used by pharmaceutical companies to produce human insulin, human growth hormone, and various disease-fighting antibodies. This technique involves splicing a specific segment of DNA (such as the gene that codes for human insulin) into the normal DNA of specially developed strains of *E. coli.* This recombined DNA is then inserted into *E. coli,* and will be reproduced in the organism's normal reproductive cycle (Figure 22.15). Because one *E. coli* can make billions of copies of the transplanted gene in a day, the production of large amounts of the desired DNA (or the

---

Figure 22.13 (*Opposite page*) Control of protein synthesis in the lac operon of *E. coli.* (1) *m*RNA coded for the repressor molecule is synthesized on the repressor gene. The repressor molecule is produced from the *m*RNA template in the ribosomes. (2) When the repressor molecule is attached to the operator gene, the transcription of *m*RNA from the structural gene for β-galactosidase is halted. (3) When the inducer molecule (in this case, lactose) attaches to the repressor molecule, the repressor is no longer able to bind to the operator gene. Transcription of the structural gene is no longer blocked, and *m*RNA coded for β-galactosidase is synthesized. This transcription will terminate when all the lactose has been hydrolyzed, leaving the repressor molecule again free to bind to the operator gene.

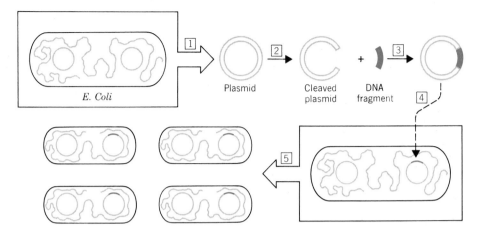

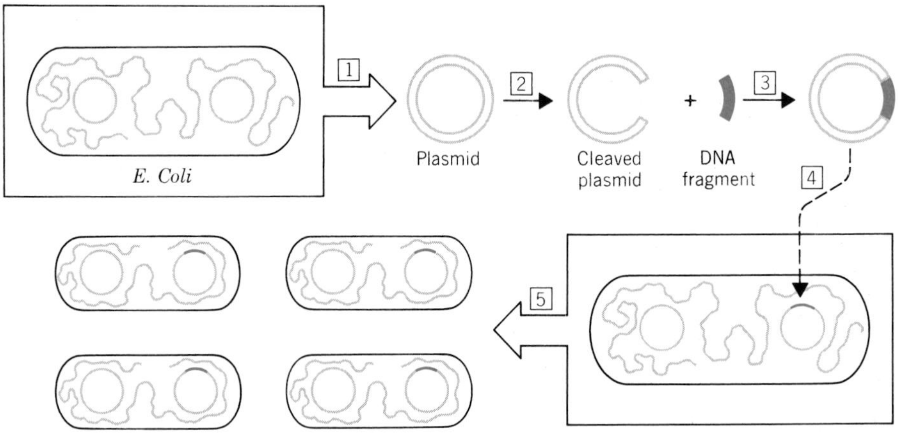

Figure 22.15 The steps in the formation of recombinant DNA. (1) The outer membranes of *E. coli* are dissolved, and small circular pieces of DNA called plasmids are isolated. (2) These plasmids are broken at specific positions by special enzymes that leave single-stranded "sticky" ends on the cleaved plasmids. (3) The cleaved plasmids are mixed with pieces of foreign DNA that are then attached to the plasmids by the enzyme DNA ligase. (4) The recombined plasmid with its foreign DNA is inserted into a new *E. coli* cell. (5) The segment of foreign DNA is reproduced with the rest of the native DNA in the normal reproductive cycle of the *E. coli* cells.

protein it produces) is possible in a relatively short period of time. This very efficient technique, however, may have a risk: an otherwise harmless bacterium could be given some potentially dangerous properties. As a result, special strains of *E. coli* have been developed for use in recombinant DNA research, and special guidelines have been prepared for containment procedures.

22.12 Mutations

A **mutation** is any chemical or physical change in the DNA molecule which results in the synthesis of a protein having an altered amino acid sequence. Some mutations may be beneficial to an organism, but most will be to varying degrees detrimental. There are many ways by which a gene can be changed, or mutated. The sequence of bases in the DNA molecule is critical and very specific. Replacing a base in the sequence, adding a base, or deleting a base will throw off the code and may change the amino acid sequence of the resulting protein. Such changes can occur spontaneously, or may be caused by radiation, chemical agents or, as we have just seen, by viruses.

The altered proteins produced through a mutation might improve the organism's chances of survival by providing new alternative chemical pathways, or they may have no biological activity, resulting in death of the

Table 22.2 Some Diseases of Fat Metabolism That Are Hereditary*

Disease	Symptoms
Fabry's disease	Reddish-purple skin rash, kidney failure, and pain in legs.
Gaucher's disease	Spleen and liver enlargement, mental retardation in infantile form, and erosion of long bones and pelvis.
Generalized gangliosidosis	Mental retardation, liver enlargement, skeletal deformities, and red spot in retina in about 50% of the cases.
Niemann-Pick disease	Mental retardation, liver and spleen enlargement, and red spot in retina in about 30% of the cases.
Tay-Sachs disease	Mental retardation, red spot in retina, blindness, and muscular weakness.

* In these diseases, sphingolipids accumulate in tissues because the enzymes that normally catalyze the cleavage of these lipids are defective.

cell. In other cases, the defective gene may cause abnormalities or disease. At the beginning of this chapter we saw that a mutation in the structural gene for hemoglobin occurred among the African black population, resulting in an error in one amino acid on the beta polypeptide chain. This causes the production of the abnormal hemoglobin found in sickle cell anemia. Among the other diseases that result from mutations in genes are such maladies as cystic fibrosis, albinism, hemophilia, galactosemia, color blindness, and PKU (Table 22.2).

22.13 PKU—A Final Look

We began this book with the story of Billy, a child suffering from the disease phenylketonuria. PKU is a disease resulting from a mutation in the structural gene that codes for the liver enzyme phenylalanine hydroxylase. This defective gene is carried by 2% of the population. When a child inherits the defective gene from both parents, the child will suffer from PKU. In this disease the defective phenylalanine hydroxylase produced in the liver will not catalyze the conversion of phenylalanine to tyrosine (Figure 22.16). This results in low tyrosine concentrations and the build-up in the body of phenylalanine and PKU metabolites—substances produced by the metabolism of the excess phenylalanine.

Melanin, the dark pigment found in hair and skin, is formed from tyrosine, and the low level of tyrosine in untreated PKU children leads to the light hair and skin observed on Billy. The hormones epinephrine, norepinephrine, and thyroxin are also synthesized from tyrosine. A high

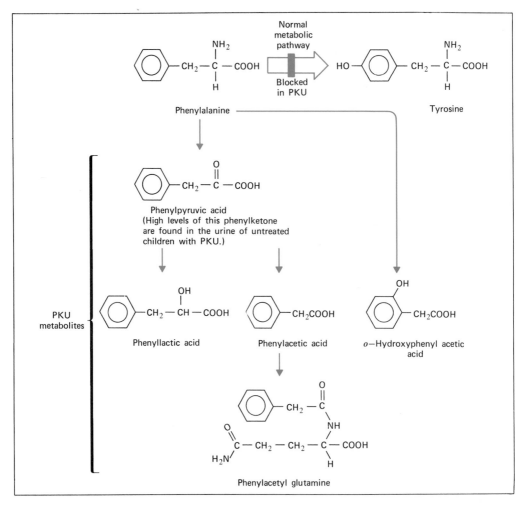

Figure 22.16 **The blockage of a metabolic pathway in PKU results in the production of PKU metabolites as the body tries to metabolize phenylalanine by alternate chemical routes.**

level of phenylalanine and PKU metabolites blocks energy-releasing reactions in the brain of an infant, preventing cells from obtaining the amount of energy necessary for normal functioning. This high level also seems to delay the formation of the myelin sheath, the protective coating around the nerves. In addition, large amounts of phenylalanine in the fluid of the brain prevent the normal uptake of nutrients by the brain cells, so these cells will not have the normal mix of chemicals from which to build their essential and permanent parts. The brain of an infant at birth is only 25% of its mature weight, and it grows rapidly, reaching 89% of its mature weight by six years of age. During these years of rapid growth, it is extremely important that the brain be surrounded by the correct chemical environment. The abnormal chemical environment of the brain of an untreated PKU child results in the formation of defective brain cells, which explains the mental retardation of such children (Figure 22.17).

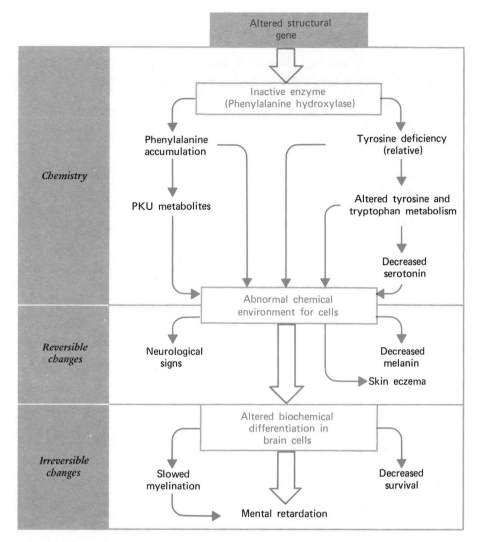

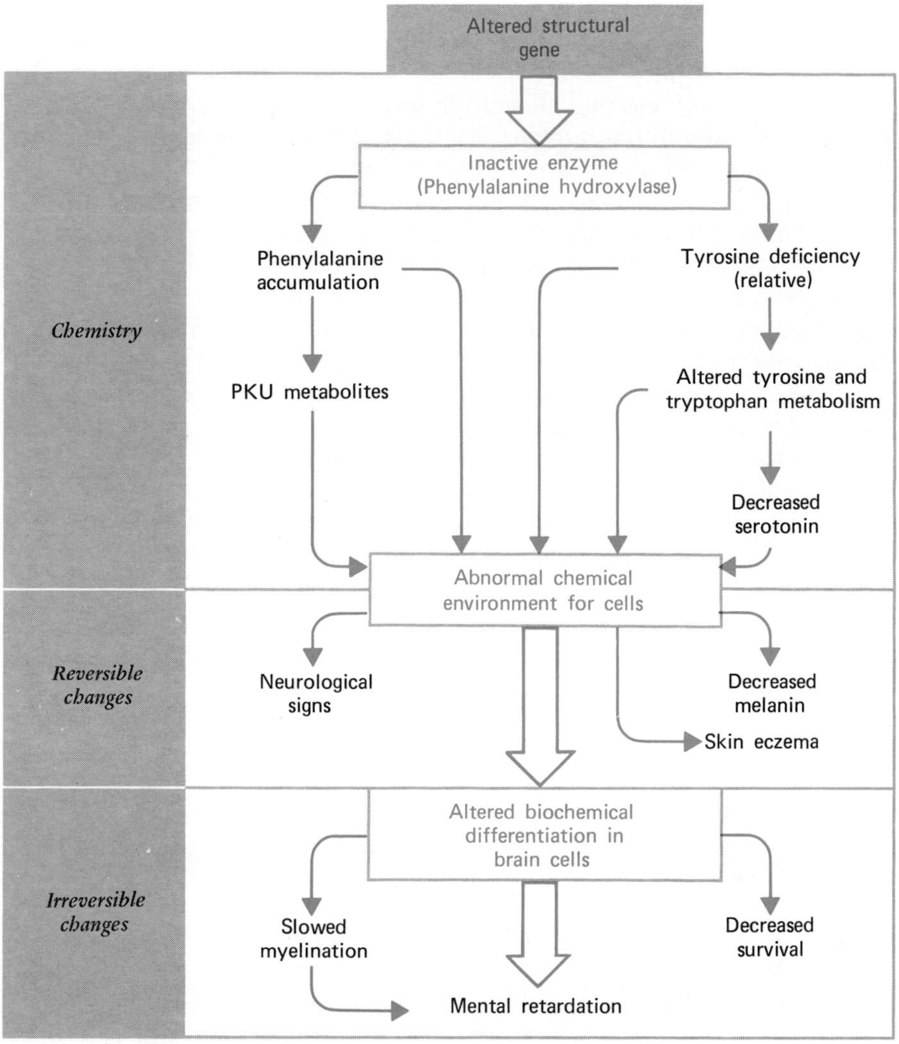

Figure 22.17 The metabolic changes caused by the mutation of PKU. (Adapted from *The Metabolic Basis of Inherited Disease* by Stanbury et al. Copyright © 1972 by McGraw-Hill, Inc. Used with permission of McGraw-Hill Book Company.)

Chapter Summary

A human tissue cell contains 46 chromosomes in its nucleus. Each chromosome is composed of molecules of DNA and proteins. The structure of the nucleic acid DNA is a double helix, consisting of two helical polynucleotide chains coiled around one another. The two chains are held together by hydrogen bonding between the bases of the nucleotides. A nucleotide consists of a sugar molecule, a phosphate group, and a nitrogen base. Genetic information is carried in the sequence of bases on the DNA molecule. A gene is a sequence of such bases which codes for a specific protein molecule. DNA molecules can replicate themselves, allowing genetic information to be passed from a parent cell to daughter cells.

RNA is the second nucleic acid found in cells. Messenger RNA (*m*RNA) is formed in the nucleus of the cell on one DNA strand. It then migrates to the cytoplasm where it joins with ribosomal RNA (*r*RNA) and proteins to form ribosomes, the location of protein synthesis in the cell. The function of transfer RNA (*t*RNA) is to join with specific amino acids in the cytoplasm and bring them to the ribosome. There the anticodon on the *t*RNA is matched with the codon on the *m*RNA, and a protein having a specific amino acid sequence is synthesized. Figure 22.12 summarizes the steps of protein synthesis. The control of gene expression is a complex process; Figure 22.13 summarizes the control mechanism in one microorganism.

A mutation is any change in the order of bases in the DNA molecule which results in the production of a protein having an altered amino acid sequence. A mutation can produce a change that improves an organism's chances of survival, that results in the death of the organism, or that produces a specific genetic disease such as PKU.

Exercises and Problems

1. What are the three components making up a nucleotide?
2. What is the difference between a nucleic acid and a nucleotide?
3. Which nitrogen bases found in DNA are pyrimidines? Which are purines?
4. (a) From the structure of deoxyribose and ribose, explain the meaning of the prefix *deoxy-*.
   (b) What is the important function of the hydroxyl group on carbon 2 of ribose in RNA molecules?
5. How are the two polynucleotide chains held together in the double helix of DNA?
6. "The two strands of DNA are not identical, but rather are complementary." Explain this statement.
7. In general terms describe the replication of a DNA molecule.
8. What are the three main structural differences between a molecule of DNA and a molecule of RNA?
9. What are the three types of RNA found in the cell? Describe the function of each type.
10. How is genetic information carried on the DNA molecule?
11. Describe the process by which the synthesis of β-galactosidase in *E. coli* is stopped when the enzyme is no longer needed.

12. What is the *central dogma* of molecular biology?

13. Write the structure of the nucleotide containing the sugar deoxyribose and the base cytosine.

14. The following is a sequence of bases which might be found on the structural gene that codes for the human pituitary hormone, oxytocin.

    DNA strand: TACACAATGTAAGTTTTGACGGGGGACCCTATC

    (a) What is the sequence of bases on the complementary strand of DNA that would form an α-helix with this strand?
    (b) What would be the sequence of bases found on the *m*RNA molecule synthesized on the original DNA strand?
    (c) Mark off the codons on the *m*RNA strand in (b) and list the bases found on the anticodons of the *t*RNA molecules that would join with the first three codons on the *m*RNA.
    (d) What is the sequence of amino acids in a molecule of oxytocin? (The initiating methionine is not part of the oxytocin molecule.)

15. Sickle cell anemia results from a mutation of a specific region of DNA that codes for the polypeptide chain of a hemoglobin molecule.

    (a) Suggest three possible causes of this mutation.
    (b) The *m*RNA for the β-chain of normal hemoglobin has the base triplet GAA or GAG in the 6th position. What is the change in this base sequence that results in the production of the β-chain of sickle cell hemoglobin?
    (c) In general terms describe the steps involved in the production of one normal β-chain.
    (d) Explain the difference in mobility of normal and sickle cell hemoglobin in an electric field.
    (e) Explain how the substitution of one amino acid in the β-chain of hemoglobin can result in the symptoms of sickle cell anemia.

16. One of the arguments for eliminating the use of fluorocarbons in aerosol sprays is that fluorocarbons might destroy the ozone layer in the upper atmosphere. This layer absorbs ultraviolet radiation coming in from outer space. The destruction of this ozone layer could result in an increase in the number of cases of skin cancer. Explain why such an increase in skin cancer might occur.

appendix 1

Numbers in Exponential Form

Whenever we are working with very large or very small numbers, it is convenient to write them in **exponential form** (also called scientific notation). Numbers written in exponential form are expressed as a number between one and ten, called the **coefficient,** multiplied by 10 raised to some power. The **exponent** is the power to which the number 10 is raised, and is written as a superscript next to the number 10.

$$4.75 \times 10^{3} \leftarrow \text{exponent}$$

If the exponent is a positive number, it indicates how many times the coefficient is to be multiplied by the number 10. For example,

$$10^3 = 1 \times 10^3 = 1 \times 10 \times 10 \times 10 = 1000$$

$$4.5 \times 10^6 = 4.5 \times 10 \times 10 \times 10 \times 10 \times 10 \times 10 = 4{,}500{,}000$$

If the exponent is a negative number, it tells us how many times the coefficient is to be divided by the number 10. For example,

$$10^{-3} = 1 \times 10^{-3} = \frac{1}{10 \times 10 \times 10} = 0.001$$

$$4.5 \times 10^{-6} = \frac{4.5}{10 \times 10 \times 10 \times 10 \times 10 \times 10} = 0.0000045$$

The following table shows what various numbers look like in exponential form.

Number	Exponential Form	Number	Exponential Form
10	1×10^1	45	4.5×10^1
100	1×10^2	356	3.56×10^2
1000	1×10^3	8400	8.4×10^3
10,000	1×10^4	24,500	2.45×10^4
100,000	1×10^5	680,000	6.8×10^5
1,000,000	1×10^6	7,450,000	7.45×10^6
0.1	1×10^{-1}	0.5	5×10^{-1}
0.01	1×10^{-2}	0.037	3.7×10^{-2}
0.001	1×10^{-3}	0.004	4×10^{-3}
0.0001	1×10^{-4}	0.00056	5.6×10^{-4}
0.00001	1×10^{-5}	0.000082	8.2×10^{-5}
0.000001	1×10^{-6}	0.0000091	9.1×10^{-6}
0.0000001	1×10^{-7}	0.0000002	2×10^{-7}

Multiplying Numbers in Exponential Form

To multiply two numbers expressed in exponential form, you

1. *Multiply* the two coefficients.
2. Then, *add* the two exponents to determine the new power of 10 to use in the product.

For example,

(a) $(1 \times 10^4) \times (1 \times 10^6) = 1 \times 10^{(4+6)} = 1 \times 10^{10}$
(b) $(4 \times 10^2) \times (6 \times 10^5) = (4 \times 6) \times 10^{(2+5)} = 24 \times 10^7 = 2.4 \times 10^8$
(c) $(2 \times 10^4) \times (3 \times 10^{-6}) = (2 \times 3) \times 10^{[4+(-6)]} = 6 \times 10^{-2}$

Example (b) illustrates that in exponential form we always rewrite the coefficient so that it represents a number between one and ten.

Dividing Numbers in Exponential Form

To divide two numbers expressed in exponential form, you

1. *Divide* the two coefficients.
2. Then, *subtract* the exponent in the denominator from the exponent in the numerator.

For example,

(a) $\dfrac{1 \times 10^6}{1 \times 10^4} = \dfrac{1}{1} \times 10^{(6-4)} = 1 \times 10^2$

(b) $\dfrac{8 \times 10^7}{2 \times 10^5} = \dfrac{8}{2} \times 10^{(7-5)} = 4 \times 10^2$

(c) $\dfrac{8 \times 10^4}{3 \times 10^{-2}} = \dfrac{8}{3} \times 10^{[4-(-2)]} = 2.67 \times 10^6$

(d) $\dfrac{4 \times 10^{-3}}{8 \times 10^2} = \dfrac{4}{8} \times 10^{(-3-2)} = 0.5 \times 10^{-5} = 5 \times 10^{-6}$

appendix 2

Chemical Calculations

In performing the many calculations needed in any area of science, it is usually very helpful to keep careful track of the units of measure represented by each of the numbers being used. For example, rather than just writing down the number 15, you should write 15 grams, or 15 feet, or 15 gallons, or whatever units of measure the number 15 happens to represent. Doing this will not only prevent confusion in the middle of a calculation, but can often remind you of the steps you must perform to finish solving the problem. For example, suppose you must do a calculation to determine the speed of some object. If you know that speed is commonly measured in such units as miles per hour $\left(\dfrac{\text{miles}}{\text{hour}}\right)$, feet per second $\left(\dfrac{\text{feet}}{\text{second}}\right)$, or meters per second $\left(\dfrac{\text{meters}}{\text{second}}\right)$, this tells you that you will eventually need to divide some measurement of distance (in the appropriate units) by a measurement of time (in the appropriate units). You will find that the calculations required in this textbook are much easier to perform if you make a point of keeping your numbers properly labeled with their units of measure.

The Role of the Number *One*

There are two properties of the number *one* with which you are very familiar, and which are crucial to the performance of chemical calculations. The first property is that any number, when multiplied by one, remains the same. More generally, any quantity remains the same when multiplied by the number one. For example, $36 \times 1 = 36$, 7 apples $\times 1 = 7$ apples, and $92 \dfrac{\text{miles}}{\text{hour}} \times 1 = 92 \dfrac{\text{miles}}{\text{hour}}$.

The second property of the number one is that this number can be written as $\dfrac{2}{2}$, or $\dfrac{156}{156}$, or as the quotient of any number divided by itself. More generally, the number one can be represented by any quantity divided by itself. For example, the number one can be represented by $\dfrac{5 \text{ apples}}{5 \text{ apples}}$, or $\dfrac{156 \text{ camels}}{156 \text{ camels}}$—but cannot be represented by $\dfrac{17 \text{ apples}}{17 \text{ oranges}}$, because the numerator is not the same as the denominator. Taken one step further, the number one can be represented by $\dfrac{12 \text{ inches}}{1 \text{ foot}}$, or by $\dfrac{12 \text{ eggs}}{1 \text{ dozen eggs}}$, or by $\dfrac{1 \text{ inch}}{2.54 \text{ cm}}$, because the quantity in the numerator is equal to the quantity in the denominator, even though they might be expressed in different terms. Such ratios which are equivalent to the number one are called **unit factors,** and are the key to most of the calculations in this textbook.

To illustrate how unit factors are used in a calculation, let's look at a problem you could easily solve: How many inches are there in 2 feet? You would immediately say 24 inches, which you would have calculated by multiplying $2 \times 12 = 24$. But what did you do when you started with a distance which you called "2," and said that it is the same as a distance which you called "24"? What you actually did was to use a unit factor as follows:

$$2 \text{ feet} \times \frac{12 \text{ inches}}{1 \text{ foot}} = \frac{2 \times 12 \; \cancel{\text{(feet)}} \; \text{(inches)}}{1 \; \cancel{\text{(feet)}}} = 24 \text{ inches}$$

You feel confident that the distance you started out with (2 feet) is the same distance that you ended up with (24 inches) because all you really did was to multiply your initial distance by a unit factor—that is, by the number one. When you make a point of keeping close track of the units of measure for each of the numbers in the problem (as we just did), you see that similar units in the numerator and denominator "cancel," leaving you with an answer in the desired units of measure.

Working with the Metric System

Most of the mathematical problems in this textbook can be solved through the use of appropriate unit factors. Let's look at some sample problems using the metric system to see how this is done.

Example ⎯⎯⎯⎯⎯⎯⎯⎯⎯⎯⎯⎯⎯⎯⎯⎯⎯⎯⎯⎯⎯⎯⎯⎯⎯⎯⎯⎯⎯⎯

1. How many centimeters are there in 4 meters?

 We might rewrite this question as follows:

 $$4 \text{ meters} = (?) \text{ centimeters}$$

 Because the problem requires us to change meters into centimeters, we must look for a unit factor that shows the relationship between these two units of measure. Using the equality 1 meter = 100 centimeters, we can write two unit factors that might be useful:

 $$\frac{1 \text{ meter}}{100 \text{ centimeters}} \quad \text{or} \quad \frac{100 \text{ centimeters}}{1 \text{ meter}}$$

 The second unit factor is the correct one to choose because it will allow units of measure to cancel so as to give us an answer in centimeters.

 $$4 \; \cancel{\text{meters}} \times \frac{100 \text{ centimeters}}{1 \; \cancel{\text{meter}}} = \frac{4 \times 100}{1} \text{ centimeters} = 400 \text{ centimeters}$$

2. How many milligrams are there in 0.024 grams?

 We know that 1 gram = 1000 milligrams, so we can write two unit factors showing a relationship between grams and milligrams:

 $$\frac{1 \text{ gram}}{1000 \text{ milligrams}} \quad \text{or} \quad \frac{1000 \text{ milligrams}}{1 \text{ gram}}$$

 Our problem states 0.024 g = (?) mg, so the second unit factor is the correct one to use.

 $$0.024 \; \cancel{\text{g}} \times \frac{1000 \text{ mg}}{1 \; \cancel{\text{g}}} = 0.024 \times 1000 \text{ mg} = 24 \text{ mg}$$

Calculations Involving the Metric System and English System

The following table of conversion factors between the metric and English systems will be helpful in setting up the unit factors for the next set of examples.

Length	Mass	Volume
1 m = 39.37 in	1 kg = 2.2 lb	1 *l* = 1.06 qt
1.6 km = 1 mi	908 kg = 1 ton	3.79 *l* = 1 gal
2.54 cm = 1 in	454 g = 1 lb	(1 ml = 1 cc)

Example

1. The distance between Seattle and San Francisco is 820 miles. How many kilometers is this?

   Because we want to change miles into kilometers, we need a unit factor that shows the relationship between these two units. From the above table, we have a choice of

$$\frac{1.6 \text{ km}}{1 \text{ mi}} \quad \text{or} \quad \frac{1 \text{ mi}}{1.6 \text{ km}}$$

   Our problem states 820 mi = (?) km, so the first unit factor is the correct one to use.

$$820 \cancel{\text{mi}} \times \frac{1.6 \text{ km}}{1 \cancel{\text{mi}}} = 820 \times 1.6 \text{ km} = 1312 \text{ km}$$

2. How many milliliters are there in one pint?

   This problem requires us to use several different unit factors to make the change from pints to milliliters. Our choices are as follows:

$$\frac{2 \text{ pints}}{1 \text{ quart}} \quad \text{or} \quad \frac{1 \text{ quart}}{2 \text{ pints}}$$

$$\frac{1 \text{ liter}}{1.06 \text{ quarts}} \quad \text{or} \quad \frac{1.06 \text{ quarts}}{1 \text{ liter}}$$

$$\frac{1 \text{ liter}}{1000 \text{ ml}} \quad \text{or} \quad \frac{1000 \text{ ml}}{1 \text{ liter}}$$

   We can solve the entire problem in one step if we carefully arrange unit factors so that all the units of measure cancel, leaving us with an answer expressed in milliliters.

$$1 \cancel{\text{pint}} \times \frac{1 \cancel{\text{quart}}}{2 \cancel{\text{pints}}} \times \frac{1 \cancel{\text{liter}}}{1.06 \cancel{\text{quarts}}} \times \frac{1000 \text{ ml}}{1 \cancel{\text{liter}}} = \frac{1000}{2 \times 1.06} \text{ ml} = 472 \text{ ml}$$

appendix 3

IUPAC Nomenclature

The Names of Common Alkyl Groups

Formula	Name	Formula	Name
CH_3—	Methyl	CH_3 | CH_3CH_2CH—	*sec*-Butyl (*s*-Butyl)
CH_3CH_2—	Ethyl	CH_3 | CH_3—C— | CH_3	*tert*-Butyl (*t*-Butyl)
$CH_3CH_2CH_2$—	Propyl (*n*-Propyl)	$CH_3CH_2CH_2CH_2CH_2$—	Pentyl (*n*-Pentyl)
CH_3 | CH_3CH—	Isopropyl	CH_3 | $CH_3CHCH_2CH_2$—	Isopentyl
$CH_3CH_2CH_2CH_2$—	Butyl (*n*-Butyl)	$CH_3CH_2CH_2CH_2CH_2CH_2$—	Hexyl (*n*-Hexyl)
CH_3 | CH_3CHCH_2—	Isobutyl	CH_3 | $CH_3CHCH_2CH_2CH_2$—	Isohexyl

Nomenclature of Benzene Derivatives

Benzene Rings with One Substituted Group

1. When a group is substituted onto a benzene ring, the resulting compound is named by adding the name of the substituted group as a prefix before the name benzene.

Nitrobenzene

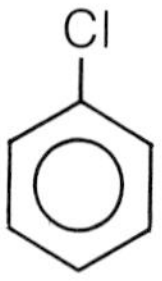

Chlorobenzene

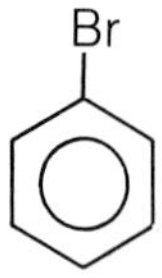

Bromobenzene

2. In the following compounds, however, the benzene ring with the substituted group is given an entirely new name:

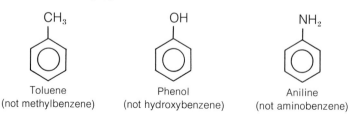

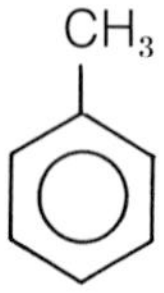

Toluene
(not methylbenzene)

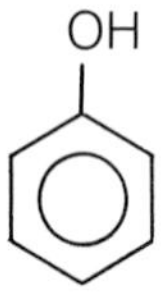

Phenol
(not hydroxybenzene)

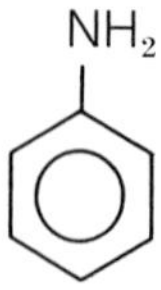

Aniline
(not aminobenzene)

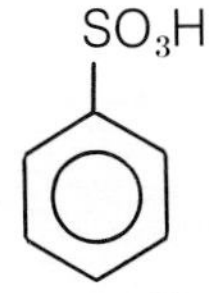

Benzenesulfonic acid

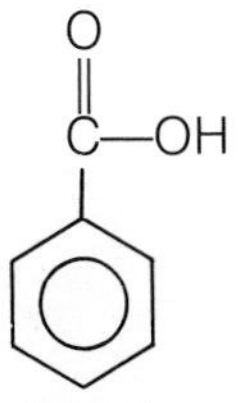

Benzoic acid

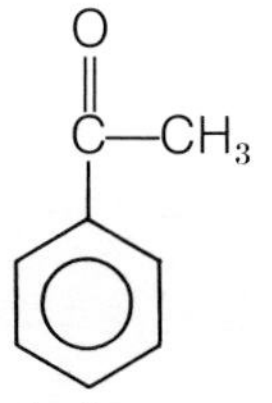

Acetophenone

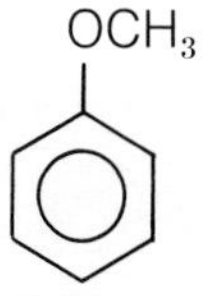

Anisole

Benzene Rings with Two Substituted Groups

The position of the second group on the ring is indicated by the prefixes *ortho (o)*, *meta (m)*, and *para (p)*.

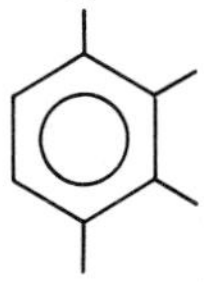

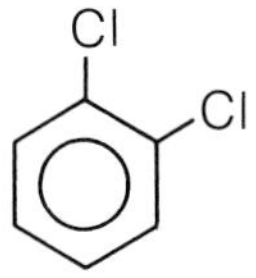

o-Dichlorobenzene

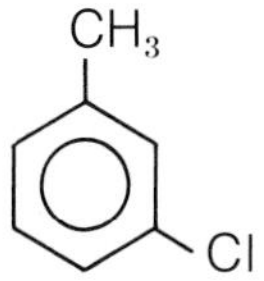

m-Chlorotoluene

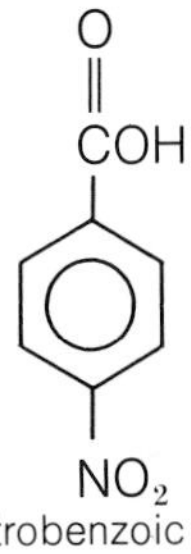

p-Nitrobenzoic acid

Benzene Rings with More Than Two Substituted Groups

If more than two substituted groups are present on the benzene ring, their positions are indicated by numbers. The ring is numbered to give the lowest possible number to the substituted groups.

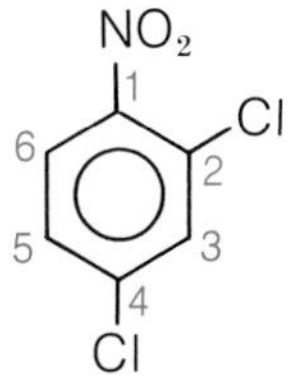

is 2,4-Dichloro-1-nitrobenzene

Benzene Rings as Substituted Groups or R-Groups

When a benzene ring is a substituted group, or R-group, it is called a *phenyl* group.

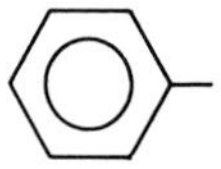

Phenyl group

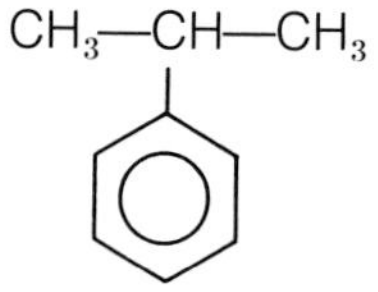

2-Phenylpropane

When the substituted group is derived from toluene, it is called a *benzyl* group.

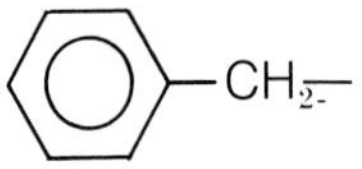

Benzyl group

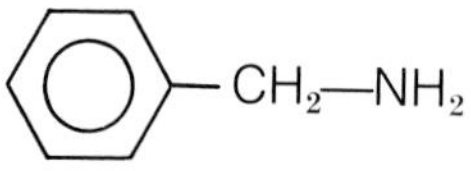

Benzylamine

appendix 4

Answers to Selected Exercises and Problems

Chapter 2

5. (a) Chemical change (c) Physical change
 (b) Physical change (d) Physical change

6. (a) 0.4 kg (i) 30 cm (o) 0.762 m
 (c) 100 mm (k) 0.125 g (q) 2.4 /
 (e) 0.015 / (m) 17.8 ft (s) 7.5 /
 (g) 20 mg (u) 0.231 lb

7. (a) 5 g (e) 5 ml
 (c) 400 mg (g) 15 cm

8. (a) 22° C (f) 423 K
 (b) 194° F

9. (a) (1) 0.79 g/cc (3) 198 g
 (2) 0.79 (4) 1.3 ml

11. 0.8 /, 800 ml $\left(0.8 \text{ qt} \times \dfrac{1 \text{ } l}{1.06 \text{ qt}}\right)$

13. 1000 mg, or 1 g per day $\left(22 \text{ lb} \times \dfrac{1 \text{ kg}}{2.2 \text{ lb}} \times \dfrac{100 \text{ mg}}{1 \text{ kg}}\right)$

15. (a) −196° C (b) −321° F

17. (a) 13.6 g/cc $\left(\dfrac{204 \text{ g}}{15.0 \text{ cc}}\right)$

19. (a) No, patients with diabetes mellitus excrete urine with a specific gravity higher than normal (>1.030).
 (b) 100.2 g $\left(\dfrac{1.002 \text{ g}}{1 \text{ cc}} \times 100.0 \text{ cc}\right)$

Chapter 3

1. (a) steam (b) a car moving at 50 mph (c) a football player

3. (a) 3500 cal (c) 10 cal

4. (a) 175 cal $25 \times (27 - 20)$

 (d) 11,000 cal, or 11 kcal $500 \times (23 - 45)$ Note that a temperature decrease can result when heat is transferred from the water to the reaction that is occurring.

6. (a) Short waves, microwaves, yellow light, ultraviolet light, X rays
 (b) X rays, ultraviolet light, yellow light, microwaves, short waves

8. 11 kcal/cashew
 11 Calories/cashew $[1000 \times (69 - 25) = 44{,}000 \text{ cal/4 cashews}]$

Chapter 4

1. (a) 3800 mm Hg
 (b) 0.25 atm

4. 400 cal $\left(5\ \cancel{g} \times \frac{80\ \text{cal}}{1\ \cancel{g}}\right)$
5. Strongest—sugar Weakest—ammonia
6. (a) Motor oil (b) Honey (c) Water at 10° C
14. 240 ml $(720\ \text{mm Hg} \times 250\ \text{ml} = 750\ \text{mm Hg} \times V_2)$
17. 420° F $\left(\frac{4.8\ \text{atm}}{293\ \text{K}} = \frac{8.0\ \text{atm}}{T_2};\ V_1 = V_2\right)$
19. −8° C $\left(\frac{1.2\ l}{318\ \text{K}} = \frac{1.0\ l}{T_2}\right)$
21. 216 K, −57° C $\left(\frac{400\ \text{ml} \times 800\ \text{mm Hg}}{300\ \text{K}} = \frac{320\ \text{ml} \times 720\ \text{mm Hg}}{T_2}\right)$
24. $P_{N_2} = 684$ mm Hg $(760\ \text{mm Hg} = P_{N_2} + 76\ \text{mm Hg})$

Chapter 5

2.

	Atomic Number	Symbol	Mass Number	Number of p	Number of e^-	Number of n
(a)	3	Li	7	3	3	4
(b)	16	S	32	16	16	16
(c)	20	Ca	40	20	20	20
(d)	35	Br	80	35	35	45
(e)	50	Sn	116	50	50	66
(f)	80	Hg	200	80	80	120

4. (a) 17,35, Cl-35, $^{35}_{17}\text{Cl}$
6. The isotopes of zinc differ only in the number of neutrons in the nucleus.
12. (a) $^{4}_{2}\text{He}$ (c) $^{210}_{84}\text{Po}$ (e) $^{146}_{62}\text{Sm}$
13. (a) $^{4}_{2}\text{He}$
14. (a) $^{90}_{38}\text{Sr} \longrightarrow\ ^{90}_{39}\text{Y} +\ ^{0}_{-1}\text{e}$

    (c) 56 years (d) 6.25 g (e) 0.195 g (f) 7 half-lives
17. $^{222}_{86}\text{Rn} \longrightarrow\ ^{218}_{84}\text{Po} +\ ^{4}_{2}\text{He}$
19. $^{230}_{90}\text{Th} \longrightarrow\ ^{226}_{88}\text{Ra} +\ ^{4}_{2}\text{He} + \gamma$

Chapter 6

5. 9.4×10^{11} disintegrations
7. 100 mice
12. Strontium-90 would be better because it does not give off gamma radiation, which is highly penetrating and could damage other normal tissue.

Chapter 7

2. 4th level — 32 electrons; 6th level — 72 electrons

4.

	Symbol	Atomic Number	Atomic Weight	Electron configuration
(a)	Li	3	6.94	2,1
(b)	N	7	14.01	2,5
(c)	Ne	10	20.18	2,8
(d)	Mg	12	24.31	2,8,2
(e)	Al	13	26.98	2,8,3
(f)	Cl	17	35.45	2,8,7

6. (a) 1 (b) 3 (c) 5 (d) 8 (e) 2 (f) 4 (g) 7 (h) 6

8. (a) Cs· (c) Ċa· (e) ·Äs: (g) ·S̈:

9. (a) 1) Na 2) P (e) 1) Xe 2) Ne
 (c) 1) Rb 2) Li (g) 1) Sr 2) Si

10. (a) Cl

11. (a) Na^+

13. Silicon, phosphorus, sulfur, chlorine

Chapter 8

6. (a) nonpolar (e) polar — O
 (c) polar — N (g) polar — F

7. (a) polar (c) polar (e) polar

9. (a) K_2CO_3 (c) $Ca(NO_3)_2$ (e) $CrCl_3$ (g) $Cu(C_2H_3O_2)_2$

10. (a) Potassium carbonate (e) Chromium chloride
 (c) Calcium nitrate (g) Copper(II) acetate

11. *Set 1:* (a) LiF (c) $MgBr_2$ (e) Fe_2S_3 (g) Al_2O_3
 Set 2: (a) $(NH_4)_2CO_3$ (c) $Mg(HCO_3)_2$ (e) $Ba(NO_3)_2$ (g) $Fe_2(SO_4)_3$
 Set 3: (a) IF_5 (c) N_2O_4 (e) BCl_3 (g) CS_2

12. (a) Ammonium iodide (c) Calcium hydroxide
 (e) Iron(II) chloride (g) Dioxygen difluoride
 (i) Diphosphorus pentoxide (k) Hydrogen iodide
 (m) Lithium sulfide (o) Calcium carbide

13. (a) H:Ï: (c) H (e) :Ö:F̈:
 H:C̈:C̈l: :F̈:
 H

14. (a) H—N—H (c) H—Cl (e) S—F
 | |
 H F

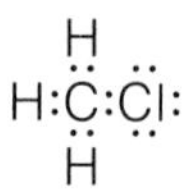

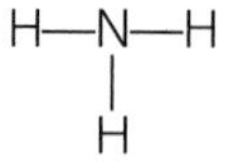

15. (a) N—3, H—12, P—1, O—4

Chapter 9

1. *Group 1:* (a) 40 amu (b) 111 amu (c) 64 amu (d) 164 amu

(e) 233 amu (f) 81 amu (g) 68 amu (h) 18 amu

2. *Group 1:* (a) 6.0 g (b) 278 g (c) 51 g (d) 82 g
(e) 839 g (f) 1.6 g (g) 8.5 g (h) 0.018 g

3. *Group 1:* (a) 3.0×10^{-4} mole (b) 0.300 mole (c) 0.0388 mole
(d) 0.0050 mole (e) 0.0100 mole (f) 0.65 mole
(g) 1.8 moles (h) 3.6 moles

4. (a) 0.6 mole

5. (a) $2Na + 2H_2O \longrightarrow 2NaOH + H_2$
(b) $2KClO_3 \longrightarrow 2KCl + 3O_2$
(c) $MnO_2 + 4HCl \longrightarrow Cl_2 + MnCl_2 + 2H_2O$
(d) $C_3H_8 + 5O_2 \longrightarrow 3CO_2 + 4H_2O$
(e) $4NH_3 + 5O_2 \longrightarrow 4NO + 6H_2O$
(f) $CO_3^{2-} + 2H^+ \longrightarrow CO_2 + H_2O$

7. (a) $H_2 + Br_2 \longrightarrow 2HBr$
(b) $Ca(HCO_3)_2 \longrightarrow CaCO_3 + H_2O + CO_2$
(c) $2AgNO_3 + Cu \longrightarrow Cu(NO_3)_2 + 2Ag$
(d) $3H_2 + N_2 \longrightarrow 2NH_3$
(e) $CH_4 + 4Cl_2 \longrightarrow CCl_4 + 4HCl$

9. 2.56 kg ethyl alcohol

$$(1)\ 5.00\ \cancel{kg} \times \frac{1 \text{ mole } C_6H_{12}O_6}{0.180\ \cancel{kg}} = 27.8 \text{ moles } C_6H_{12}O_6$$

$$(2)\ 27.8\ \cancel{\text{moles}}\ C_6H_{12}O_6 \times \frac{2\ \cancel{\text{moles}}\ C_2H_5OH}{1\ \cancel{\text{mole}}\ C_6H_{12}O_6} \times \frac{0.046 \text{ kg}}{1\ \cancel{\text{mole}}\ C_2H_5OH}$$

11. (a) $2SO_2 + O_2 \longrightarrow 2SO_3$
(b) 800 g SO_2

$$1000\ \cancel{g} \times \frac{1\ \cancel{\text{mole}}\ SO_3}{80\ \cancel{g}} \times \frac{2\ \cancel{\text{moles}}\ SO_2}{2\ \cancel{\text{moles}}\ SO_3} \times \frac{64 \text{ g}}{1\ \cancel{\text{mole}}\ SO_2} = 800 \text{ g } SO_2$$

13. 7.1 tons of lauryl alcohol

$$(1)\ 11\ \cancel{\text{tons}} \times \frac{2000\ \cancel{lb}}{1\ \cancel{\text{ton}}} \times \frac{1 \text{ kg}}{2.2\ \cancel{lb}} = 10{,}000 \text{ kg}$$

$$(2)\ 10{,}000\ \cancel{kg} \times \frac{1\ \cancel{\text{mole}}\ D}{0.288\ \cancel{kg}} \times \frac{1\ \cancel{\text{mole}}\ LA}{1\ \cancel{\text{mole}}\ D} \times \frac{0.186 \text{ kg}}{1\ \cancel{\text{mole}}\ LA} = 6460 \text{ kg}$$

$$(3)\ 6460\ \cancel{kg} \times \frac{2.2\ \cancel{lb}}{1\ \cancel{kg}} \times \frac{1 \text{ ton}}{2000\ \cancel{lb}} = 7.1 \text{ tons}$$

Chapter 10

3. (a) Exothermic (b) Endothermic (c) Endothermic (d) Endothermic

14. (1) Increase the concentration of NO.
(2) Increase the concentration of Cl_2.
(3) Increase the pressure.

17. (a) Exothermic (b) 54 kcal (c) 32 kcal (d) 86 kcal

19. (a) Increase—Increasing the temperature increases the rate of the reverse endothermic reaction more than the forward reaction.

(b) Decrease—Increasing the pressure will drive the reaction to the right toward the smaller number of particles.
(c) Decrease—Decreasing the concentration of Cl_2 will lower the rate of the reverse reaction, causing the system to shift to the right.
(d) Decrease—Increasing the concentration of HCl will increase the rate of the forward reaction, shifting the equilibrium to the right.
(e) No change—A catalyst increases the rate of both the forward and the reverse reactions by an equal amount.

Chapter 11

2. (a) Anion (b) Cation (c) Anion (d) Cation
5. (a) No (b) Yes (c) Yes
7. (a) Side 1
9. (a) Crenation (b) No change (c) Hemolysis

13. $\frac{5 \text{ g dextrose}}{100 \cancel{\text{ml}}} \times 500 \cancel{\text{ml}} = \textit{25 g dextrose}$

$\frac{0.9 \text{ g NaCl}}{100 \cancel{\text{ml}}} \times 500 \cancel{\text{ml}} = \textit{4.5 g NaCl}$

15. (a) Dissolve 0.40 g of NaOH in enough water to make 50 ml of solution.
(c) Dissolve 25 g of KOH in enough water to make 0.50 liter of solution.
(e) Dissolve 1.13 g of NaCl in enough water to make 125 ml of solution.
(g) Dissolve 0.35 g of K_2HPO_4 in enough water to make 10 ml of solution.

17. 0.154M NaCl (1) $0.9\% \text{ NaCl} = \frac{0.9 \text{ g}}{100 \text{ ml}} = \frac{9 \text{ g}}{1000 \text{ ml}}$

(2) $9 \cancel{\text{g}} \times \frac{1 \text{ mole}}{58.5 \cancel{\text{g}}} = 0.154 \text{ mole}$

19. Yes (1) $\frac{7 \text{ g Li}^+}{1 \text{ Eq}} = \frac{0.007 \text{ g}}{1 \text{ mEq}} = \frac{7 \text{ mg}}{1 \text{ mEq}}$

(2) $\frac{1.4 \text{ mg}}{100 \text{ ml}} = \frac{14 \text{ mg}}{1000 \text{ ml}}$

(3) $14 \cancel{\text{mg}} \times \frac{1 \text{ mEq Li}^+}{7 \cancel{\text{mg}}} = 2 \text{ mEq Li}^+$ per liter of blood

21. (a) 200 ml of stock solution plus enough water to make 250 ml of solution
(b) 45 ml of stock solution plus enough water to make 150 ml of solution

23. 1500 ml (1) $\frac{2 \text{ g dextran}}{1 \cancel{\text{kg}}} \times \frac{1 \cancel{\text{kg}}}{2.2 \cancel{\text{lb}}} \times 165 \cancel{\text{lb}} = 150 \text{ g in 24 hours}$

$$(2)\ 150\ \cancel{g} \times \frac{100\ ml}{10\ \cancel{g}} = 1500\ ml$$

Chapter 12

3. (a) $HI + H_2O \rightleftharpoons H_3O^+ + I^-$
 Acid$_1$ Base$_2$ Acid$_2$ Base$_1$

 (c) $CH_3COOH + H_2O \rightleftharpoons H_3O^+ + CH_3COO^-$
 Acid$_1$ Base$_2$ Acid$_2$ Base$_1$

 (e) $O^{2-} + H_2O \rightleftharpoons OH^- + OH^-$
 Base$_1$ Acid$_2$ Base$_2$ Acid$_1$

 (g) $HCl + NH_2OH \rightleftharpoons NH_3OH + Cl^-$
 Acid$_1$ Base$_2$ Acid$_2$ Base$_1$

4. (a) H_3O^+ (c) $H_2PO_4^-$
5. (a) OH^- (c) NH_3
9. $\underbrace{12,\ 9.5,\ 8.4,\ 7.4,}_{\text{Basic}}\ \underbrace{7,}_{\text{Neutral}}\ \underbrace{6.3,\ 5.5,\ 4,\ 1.4}_{\text{Acidic}}$

11. (a) $[H^+] = 1 \times 10^{-1}$ $[OH^-] = \dfrac{1 \times 10^{-14}}{1 \times 10^{-1}} = 1 \times 10^{-13}$

 (b) $[H^+] = 1 \times 10^{-6}$ $[OH^-] = 1 \times 10^{-8}$
 (c) $[H^+] = 1 \times 10^{-12}$ $[OH^-] = 1 \times 10^{-2}$

13. Solution A is more acidic.
 Solution A: $[H^+] = 1 \times 10^{-3}$ Solution B: $[H^+] = 1 \times 10^{-5}$
 The hydrogen ion concentrations differ by a factor of 100.

15. 3 ml (1) $pH = 2 = \dfrac{1 \times 10^{-2}\ mole\ H^+}{1000\ ml}$

 (2) $30\ \cancel{ml} \times \dfrac{1 \times 10^{-2}\ mole\ H^+}{1000\ \cancel{ml}} = 3 \times 10^{-4}\ mole\ H^+$

 (3) $3 \times 10^{-4}\ \cancel{mole} \times \dfrac{1000\ ml}{0.1\ \cancel{mole}\ NaOH} = 3\ ml$

16. (1) (a) 0.0100M HCl $36.5 \times 10^{-3}\ \cancel{g} \times \dfrac{1\ mole\ HCl}{36.5\ \cancel{g}} = 1.00 \times 10^{-3}\ mole$

 $$\dfrac{1.00 \times 10^{-3}\ mole}{100\ ml} = \dfrac{1.00 \times 10^{-2}\ mole}{1000\ ml}$$

 (b) $[H^+] = 0.0100$
 $= 1.00 \times 10^{-2}$
 (c) $pH = 2$

17. (a) Sample 1: acid: 0.052M (1) $52\ \cancel{ml} \times \dfrac{0.100\ mole}{1000\ \cancel{ml}} = 0.0052\ mole$

 (2) $\dfrac{0.0052\ mole}{100\ ml} = \dfrac{0.052\ mole}{1000\ ml}$

Chapter 13

4. (1) $CH_3CH_2CH_2CH_2CH_2CH_3$

 (2) $CH_3CHCH_2CH_2CH_3$
 $|$
 CH_3

 (3) $CH_3CH_2CHCH_2CH_3$
 $|$
 CH_3

 (4) CH_3—CH—CH—CH_3
 $|$ $|$
 CH_3 CH_3

 (5) CH_3
 $|$
 CH_3—C—CH_2—CH_3
 $|$
 CH_3

5. (a) the same
 (e) the same
 (i) the same
 (m) the same

 (c) unrelated
 (g) isomers
 (k) unrelated
 (o) unrelated

Chapter 14

1. (a) alkane, saturated
 (e) alkene, unsaturated
 (i) alkyne, unsaturated
 (m) alkene, unsaturated

 (c) alkane, saturated
 (g) alkane, saturated
 (k) alkane, saturated

2. (a) butane
 (e) 2-methyl-2-butene
 (i) 4,4-dimethyl-2-pentyne
 (m) *trans*-3-hexene

 (c) 2-methyl-3-ethylhexane
 (g) cyclohexane
 (k) 2-ethyl-2-methylhexane

3. (a)

 CH_2—CH_2
 CH_2 CH_2
 CH_2 CH_2
 CH_2

 (c) CH_3
 $|$
 CH_3CHCH=$CHCH_2CH_2CH_3$

 (e) CH_3 CH_3
 $|$ $|$
 CH_3—C=C—CH_3

 (g) CH_2=$CHCH$=$CHCH$=$CHCH_2CH_3$

 (i) CH_3
 $|$
 CH≡C—C——CH—CH_3
 $|$ $|$
 CH_3 CH_3

5. (a) CH_3 CH_2CH_3
 \ /
 C=C
 / \
 H H

 CH_3 H
 \ /
 C=C
 / \
 H CH_2CH_3

7. (a) CH_3CH=CH_2 Alkenes are more reactive than alkanes.
 (b) CH_3C≡CCH_3 Alkynes are more reactive than alkanes.
10. (a) $C_4H_8 + 6O_2 \longrightarrow 4CO_2 + 4H_2O$

(b) $CH_3CH_2CH_2CH_3 + HNO_3 \longrightarrow CH_3CH_2CH_2CH_2NO_2 + H_2O$

(c) $+ Cl_2 \longrightarrow$ $+ HCl$

(d) $CH_3CH{=}CHCH_2CH_3 + H_2 \longrightarrow CH_3CH_2CH_2CH_2CH_3$

Chapter 15

1. (a) carboxylic acid (c) ester (e) ketone
 (g) ether (i) aldehyde (k) ester
2. (a) 2-methylpropanoic acid (c) ethyl butanoate
 (e) 3-hexanone (g) diphenyl ether
 (i) pentanal (k) methyl pentanoate
5. (a) secondary (c) tertiary (e) secondary

10. (a) $RCH_2OH \longrightarrow R\overset{\displaystyle O}{\overset{\|}{C}}H$ (aldehyde) $\longrightarrow R\overset{\displaystyle O}{\overset{\|}{C}}{-}OH$ (carboxylic acid)

(b) $R{-}\overset{\displaystyle OH}{\overset{|}{C}}H{-}R' \longrightarrow R{-}\overset{\displaystyle O}{\overset{\|}{C}}{-}R'$ (ketone)

(c) $R{-}\overset{\displaystyle O}{\overset{\|}{C}}{-}H \longrightarrow R{-}\overset{\displaystyle O}{\overset{\|}{C}}{-}OH$ (carboxylic acid)

(d) $R{-}\overset{\displaystyle O}{\overset{\|}{C}}{-}H + H_2 \longrightarrow R{-}\underset{\displaystyle H}{\overset{\displaystyle OH}{\overset{|}{\underset{|}{C}}}}{-}H$ (alcohol)

(e) $R{-}\overset{\displaystyle O}{\overset{\|}{C}}{-}R' + H_2 \longrightarrow R{-}\underset{\displaystyle H}{\overset{\displaystyle OH}{\overset{|}{\underset{|}{C}}}}{-}R'$ (alcohol)

(f) $R{-}\underset{\displaystyle H}{\overset{\displaystyle H}{\overset{|}{\underset{|}{C}}}}{-}\underset{\displaystyle H}{\overset{\displaystyle OH}{\overset{|}{\underset{|}{C}}}}{-}R' \longrightarrow R{-}CH{=}CH{-}R'$ (alkene) $+ H_2O$

(g) $ROH + HO\overset{\displaystyle O}{\overset{\|}{C}}R' \longrightarrow RO\overset{\displaystyle O}{\overset{\|}{C}}R'$ (ester) $+ H_2O$

(h) $RO\overset{\displaystyle O}{\overset{\|}{C}}R' + H_2O \longrightarrow ROH$ (alcohol) $+ R'\overset{\displaystyle O}{\overset{\|}{C}}OH$ (acid)

(i) $RO\overset{\displaystyle O}{\overset{\|}{C}}R' + NaOH \longrightarrow ROH$ (alcohol) $+ R'\overset{\displaystyle O}{\overset{\|}{C}}ONa$ (acid salt)

11. (a) $CH_3-\overset{\overset{\displaystyle OH}{|}}{\underset{\underset{\displaystyle OCH_3}{|}}{C}}-H$ (c) $CH_3CH_2-\overset{\overset{\displaystyle OH}{|}}{\underset{\underset{\displaystyle OCH_3}{|}}{C}}-CH_3$ (e) $CH_3CH_2-\overset{\overset{\displaystyle OH}{|}}{\underset{\underset{\displaystyle OCH_3}{|}}{C}}-H$

12. (a) $CH_3CH_2\overset{\overset{\displaystyle O}{\|}}{C}H$ (c) $CH_3\overset{\overset{\displaystyle CH_3}{|}}{C}HCH_2\overset{\overset{\displaystyle O}{\|}}{C}H$

(e) $\underset{\underset{\displaystyle CH_2-CH_2}{}}{\overset{\overset{\displaystyle O}{\|}}{\underset{CH_2 \qquad CH_2}{C}}}$ (g) $CH_3CH_2CH_2CH_2\overset{\overset{\displaystyle O}{\|}}{C}H$

13. (a) $CH_3\overset{\overset{\displaystyle O}{\|}}{C}OH$ (c) $CH_3CH_2CH_2CH_2CH_2\overset{\overset{\displaystyle O}{\|}}{C}OH$

14. (a) $CH_3CH_2\overset{\overset{\displaystyle OH}{|}}{C}HCH_3$ (c) ⬡$-CH_2OH$ (e) $CH_3\overset{\overset{\displaystyle OH}{|}}{C}HCH_3$

15. (a) $CH_3CH_2CH=CH_2$

16. (a) $CH_3CH_2CH_2\overset{\overset{\displaystyle O}{\|}}{C}OH + HOCH_3 \longrightarrow CH_3CH_2CH_2\overset{\overset{\displaystyle O}{\|}}{C}OCH_3 + H_2O$

(c) $H\overset{\overset{\displaystyle O}{\|}}{C}OH + HOCH_2CH_2CH_3 \longrightarrow H\overset{\overset{\displaystyle O}{\|}}{C}OCH_2CH_2CH_3 + H_2O$

(e) $\underset{\underset{\displaystyle OH}{}}{\overset{\overset{\displaystyle O}{\|}}{C}OH}$⬡ $+ HOCH_2CH_3 \longrightarrow$ $\underset{\underset{\displaystyle OH}{}}{\overset{\overset{\displaystyle O}{\|}}{C}OCH_2CH_3}$⬡ $+ H_2O$

17. (a) $CH_3CH_2O\overset{\overset{\displaystyle O}{\|}}{C}-$⬡$+ H_2O \longrightarrow CH_3CH_2OH + HO\overset{\overset{\displaystyle O}{\|}}{C}-$⬡

(c) $CH_3(CH_2)_6CH_2O\overset{\overset{\displaystyle O}{\|}}{C}CH_3 + H_2O \longrightarrow CH_3(CH_2)_6CH_2OH + HO\overset{\overset{\displaystyle O}{\|}}{C}CH_3$

18. (a) $CH_3O\overset{\overset{\displaystyle O}{\|}}{C}CH_3 + NaOH \longrightarrow CH_3OH + NaO\overset{\overset{\displaystyle O}{\|}}{C}CH_3$

(c) 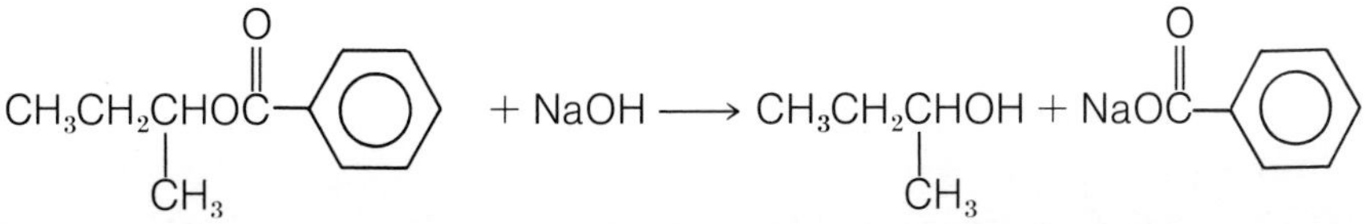 CH_3CH_2CHOC—⬡ + NaOH ⟶ CH_3CH_2CHOH + NaOC—⬡
|
CH_3
|
CH_3

Chapter 16

1. (a) amide (c) amine (e) amine
2. (a) primary (c) primary (e) tertiary
3. (a) *tert*-butylamine (c) propylamine (e) dimethylethylamine
5. (a) pentanamide (c) *N*-phenylpropanamide

6. (a) 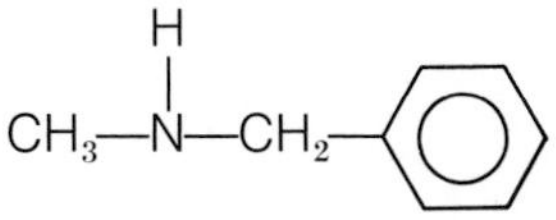 CH_3—N—CH_2—⬡ (with H on N)

(c) CH_3CH—N—$CHCH_3$ (with $CH(CH_3)_2$ on N, CH_3 below each CH)

(e) CH_3—CH—CH_2OH (with $NHCH_3$ above, CH_3 below)

14. (a) $CH_3CHCH_2NH_2$ + HCl ⟶ $CH_3CHCH_2NH_3{}^+Cl^-$ (with CH_3 above each)
(c) $(CH_3CH_2)_3N$ + CH_3CH_2Br ⟶ $(CH_3CH_2)_4N^+Br^-$

15. (a) $CH_3CH_2CH_2CH_2CNH_2$ + H_2O ⟶ $CH_3CH_2CH_2CH_2COH$ + NH_3

(c) 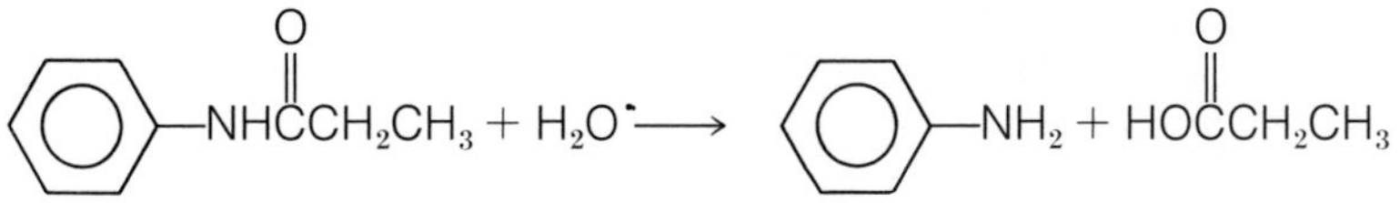 ⬡—$NHCCH_2CH_3$ + H_2O ⟶ ⬡—NH_2 + $HOCCH_2CH_3$

Chapter 17

3. The concentration is 1 ppm.
6. Citrate and oxalate ions can be added to a freshly collected blood sample to prevent it from clotting. These ions react with the calcium ions in the blood, inhibiting the clotting mechanism.

$$2C_6H_5O_7{}^{3-} + 3Ca^{2+} \longrightarrow Ca_3(C_6H_5O_7)_2$$

or

$$C_2O_4{}^{2-} + Ca^{2+} \longrightarrow CaC_2O_4$$

9. $H_3PO_4 \longrightarrow 3H^+ + PO_4{}^{3-}$
17. (a) organic sulfide (b) sulfhydryl

Chapter 18

2. (a) aldose—pentose (b) ketose—heptose (c) ketose—hexose
 (d) aldose—triose

3. (a)

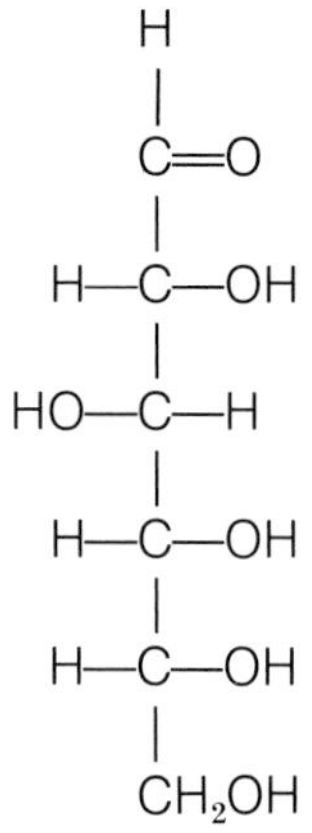

(b)

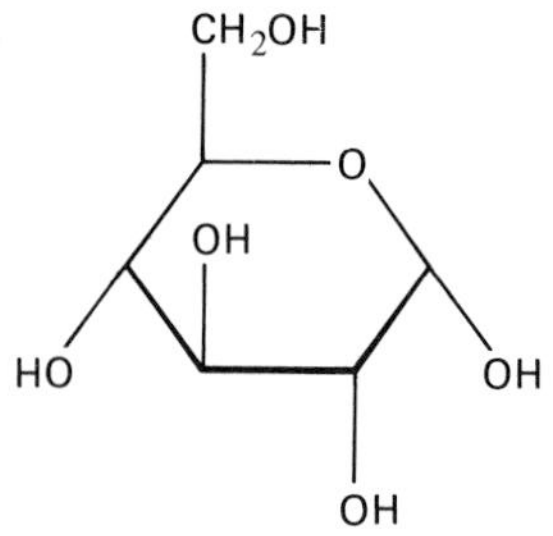

9. (a)

Sucrose

Glucose

Fructose

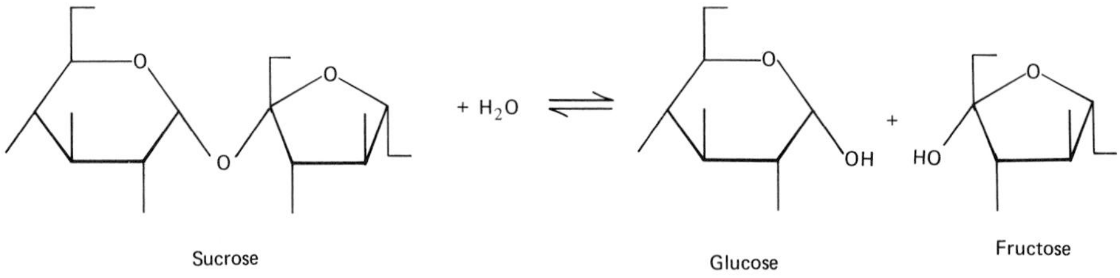

12. (a) Two molecules of glucose
 (b) Yes, it has a free aldehyde group.
 (c) It is a $\alpha(1\text{-}6)$ linkage. This linkage is found in molecules of glycogen and amylopectin.

15. (a) $6CO_2 + 6H_2O \longrightarrow C_6H_{12}O_6 + 6O_2$
 (b) $nC_6H_{12}O_6 \longrightarrow$ Glycogen $+ (n-1)H_2O$
 (c) Glycogen $+ (n-1)H_2O \longrightarrow nC_6H_{12}O_6$
 (d) Glucose $+ 2ADP + 2P_i \longrightarrow$ 2Lactic acid $+ 2ATP + 2H_2O$
 (e) Glucose $+ 2ADP + 2P_i \longrightarrow$ 2Ethanol $+ 2CO_2 + 2ATP + 2H_2O$
 (f) Acetyl CoA $\longrightarrow 2CO_2 + 8H + CoA$

18. (a) Glycolysis
 (b) Glucose $+ 2ADP + 2P_i \longrightarrow$ 2Lactic acid $+ 2ATP + 2H_2O$
 (c) The anaerobic stage occurs in the cytoplasm, and the aerobic stage occurs in the mitochondria.
 (d) Aerobic stage
 (e) The two series of reactions are the citric acid cycle and the electron transport chain. An acetyl CoA enters the citric acid cycle and is oxidized to two molecules of carbon dioxide and four pairs of hydrogens that are bound to hydrogen carriers. These then enter the electron transport chain, which produces water and energy in the form of ATP.
 (f) Pyruvic acid $+$ Oxygen $\longrightarrow$ Carbon dioxide $+$ Water $+$ ATP

Chapter 19

7. A gram of fat contains about twice as many calories as a gram of carbohydrate.

12. (a)

$$CH_2-O-\overset{\overset{\displaystyle O}{\|}}{C}(CH_2)_{14}CH_3$$

$$CH-O-\overset{\overset{\displaystyle O}{\|}}{C}(CH_2)_{14}CH_3$$

$$CH_2-O-\overset{\overset{\displaystyle O}{\|}}{C}(CH_2)_{14}CH_3$$

(b)

$$CH_2-O-\overset{\overset{\displaystyle O}{\|}}{C}(CH_2)_{18}CH_3$$

$$CH-O-\overset{\overset{\displaystyle O}{\|}}{C}(CH_2)_{16}CH_3$$

$$CH_2-O-\overset{\overset{\displaystyle O}{\|}}{C}(CH_2)_7CH=CH(CH_2)_7CH_3$$

(c)

$$CH_2-O-\overset{\overset{\displaystyle O}{\|}}{C}(CH_2)_7CH=CH(CH_2)_7CH_3$$

$$CH-O-\overset{\overset{\displaystyle O}{\|}}{C}(CH_2)_7CH=CH(CH_2)_7CH_3$$

$$CH_2-O-\overset{\overset{\displaystyle O}{\|}}{C}(CH_2)_7CH=CH(CH_2)_7CH_3$$

13. (a) a (b) a (c) c

14.

$$CH_2-O-\overset{\overset{\displaystyle O}{\|}}{C}(CH_2)_{18}CH_3$$

$$CH-O-\overset{\overset{\displaystyle O}{\|}}{C}(CH_2)_{16}CH_3 \quad + H_2 \longrightarrow$$

$$CH_2-O-\overset{\overset{\displaystyle O}{\|}}{C}(CH_2)_7CH=CH(CH_2)_7CH_3$$

$$CH_2-O-\overset{\overset{\displaystyle O}{\|}}{C}(CH_2)_{18}CH_3$$

$$CH-O-\overset{\overset{\displaystyle O}{\|}}{C}(CH_2)_{16}CH_3$$

$$CH_2-O-\overset{\overset{\displaystyle O}{\|}}{C}(CH_2)_{16}CH_3$$

16.

$$CH_2-O-\overset{\overset{\displaystyle O}{\|}}{C}(CH_2)_{14}CH_3$$

$$CH-O-\overset{\overset{\displaystyle O}{\|}}{C}(CH_2)_{16}CH_3 \quad + 3KOH \longrightarrow$$

$$CH_2-O-\overset{\overset{\displaystyle O}{\|}}{C}(CH_2)_7CH=CH(CH_2)_7CH_3$$

$$\begin{cases} CH_2OH \\ CHOH + CH_3(CH_2)_{14}\overset{\overset{\displaystyle O}{\|}}{C}O^-K^+ \\ CH_2OH \\ + CH_3(CH_2)_{16}\overset{\overset{\displaystyle O}{\|}}{C}O^-K^+ \\ + CH_3(CH_2)_7CH=CH(CH_2)_7\overset{\overset{\displaystyle O}{\|}}{C}O^-K^+ \end{cases}$$

17. (a) $\underline{\underset{5}{CH_3CH_2CH_2CH_2}\underset{4}{CH_2CH_2CH_2}\underset{3}{CH_2CH_2}\underset{2}{CH_2CH_2}\underset{1}{\overset{\overset{\displaystyle O}{\|}}{C}OH}}$

$$5 \ FADH_2 \longrightarrow 10 \ ATP$$
$$5 \ NADH \longrightarrow 15 \ ATP$$
$$\underline{6 \ Acetyl \ CoA \longrightarrow 72 \ ATP}$$
$$\overline{97 \ ATP} - 1 \ ATP = 96 \ ATP$$

Chapter 20

5. (a) Urease would migrate toward the positive pole.
 (b) Myoglobin would not migrate.
 (c) Chymotrypsin would migrate toward the negative pole.

14. (a)

(b)

(c)

15. (I)

(II)

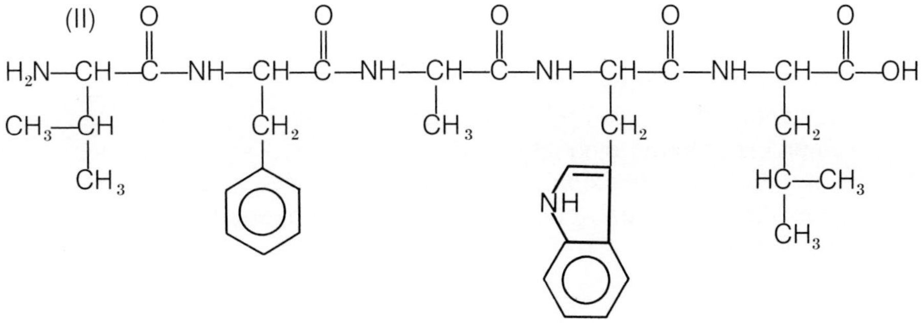

(a) Polypeptide I would be more soluble because its structure contains more polar R-groups.
(b) Hydrophobic interactions would be more likely to occur between the nonpolar R-groups in polypeptide II.
(c) Salt bridges would be more likely to occur between the carboxyl and amino side chains on polypeptide I.

17. Phe-Val-Asn-Gln-Tyr-Asp

18.

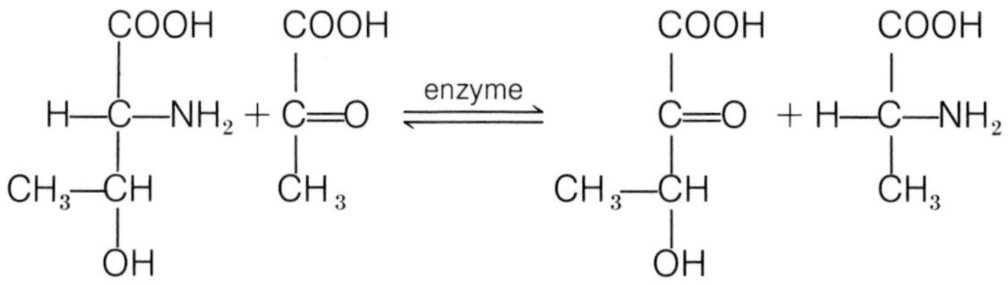

19.

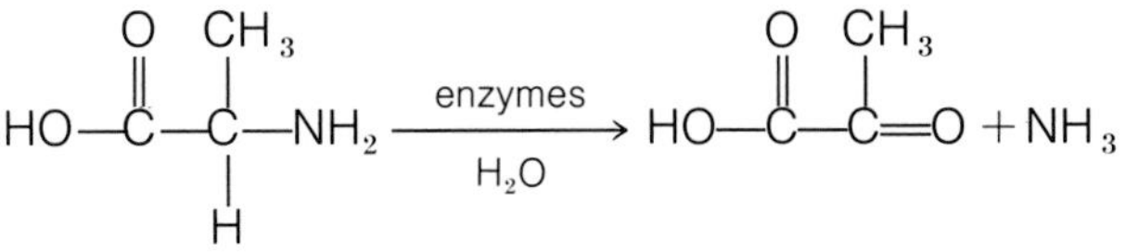

Chapter 21

13. (a) *Step 1.* The glucose-1-phosphate attaches to the active site of the phosphoglucomutase molecule.
Step 2. The glucose-1-phosphate becomes activated.
Step 3. Glucose-6-phosphate forms on the surface of the phosphoglucomutase molecule.
Step 4. The glucose-6-phosphate is released from the active site on the phosphoglucomutase molecule.
(b) 6×10^4 molecules/hour; 1×10^{-19} moles/hour.

16. (a) Beriberi results from a lack of thiamine, vitamin B_1, in the diet.
(b) The thiamine present in the rice was being removed with the outer coating of the rice, and a dietary deficiency was produced by the diet of polished rice.
(c) The rice was enriched with thiamine.

18. (a) Her symptoms were caused by the high level of vitamin A in her tissues.
(b) Her symptoms could be alleviated by stopping the dosage of vitamin A.

20. The thyroid hormone might act directly on the heart muscle to produce cyclic AMP, which then causes the heart to beat faster.
22. Lead ions bind irreversibly to the sulfhydryl groups of enzymes, inactivating them, disrupting normal metabolism, and causing the symptoms of lead poisoning.
24. A drug that inhibits step 3 would prevent the acetylcholine from being hydrolyzed, and thus would raise the concentration of acetylcholine in muscle cells.

Chapter 22

13.

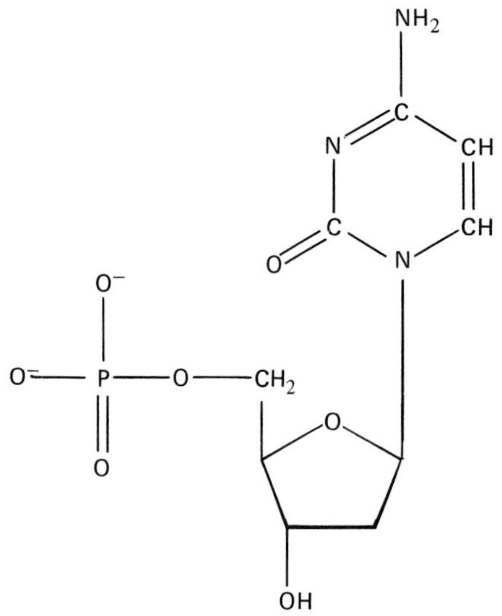

14. (a) ATGTGTTACATTCAAAACTGCCCCCTGGGATAG
(b) AUG UGU UAC AUU CAA AAC UGC CCC CUG GGA UAG
(c) UAC ACA AUG
(d) Cys-Tyr-Ile-Gln-Asn-Cys-Pro-Leu-Gly

Glossary

Absolute zero, 0 K or −273.15°C: The temperature at which all motion within a substance stops.

Acid (Brønsted-Lowry definition), a substance that donates hydrogen ions (protons); a proton-donor.

Acidosis, a condition that occurs when the blood pH falls below 7.3.

Activated complex, an unstable combination of particles that is an intermediate state between reactants and products in a chemical reaction.

Activation energy, the minimum amount of energy with which two particles must collide in order for a chemical reaction to occur.

Activator, a metal ion cofactor for an enzyme.

Active site, the area on an enzyme molecule to which the substrate attaches.

Addition reaction, in organic chemistry, a reaction in which a reagent reacts with a double or triple bond, allowing other substances to be added to the molecule.

Adenosine diphosphate (ADP), a high-energy diphosphate ester that is produced by the hydrolysis of ATP in the cell.

Adenosine monophosphate (AMP), a low-energy monophosphate ester produced by the hydrolysis of ATP and ADP. This nucleotide is also used in the formation of the nucleic acids, DNA and RNA.

Adenosine triphosphate (ATP), a very high-energy triphosphate ester that provides the energy required for the reactions of metabolism.

Adequate protein, a protein that contains all the essential amino acids required by humans.

Adipose tissue, tissue that contains fat storage cells.

ADP, (see Adenosine diphosphate)

Aerobic, requiring oxygen.

Alcohol, an organic compound containing a hydroxyl group (−OH) attached to a saturated carbon atom.

Aldehyde, an organic compound containing a terminal carbonyl group

$$\left(\begin{array}{c} O \\ \| \\ -C-H \end{array}\right).$$

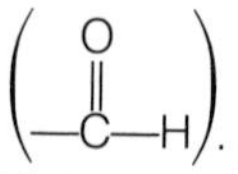

Aldose, general term for a monosaccharide containing an aldehyde group.

Alkali metals, the elements in group I on the periodic table.

Alkaline, basic (not acidic); pH > 7.

Alkaline earth metals, the elements in group II on the periodic table.

Alkaloids, a large class of complex nitrogen-containing compounds, many of which are produced by plants as part of their defense system and which often have strong physiological effects on humans.

Alkalosis, a condition that occurs when the blood pH rises above 7.5.

Alkane, any hydrocarbon containing only single bonds.

Alkene, any hydrocarbon containing one or more carbon-to-carbon double bonds.

Alkyl group, a substituted group that is derived from an alkane.

Alkyne, any hydrocarbon that contains one or more carbon-to-carbon triple bonds.

Allosteric enzyme, a regulatory enzyme that controls the rate of a series of reactions.

Allosteric site, the site on the allosteric enzyme (other than the active site) to which the regulatory molecule attaches to inhibit or increase the activity of the enzyme.

Alpha radiation, ionizing radiation consisting of streams of high-energy helium nuclei (symbol: $_2^4$He).

Amide, an organic compound that contains the amide group

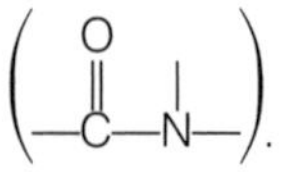

Amine, an organic compound containing a nitrogen attached to one, two, or three carbons.

Amino acid, a monomer unit of a protein, containing both an amino

group and a carboxylic acid group.

$$\text{(general formula: } H_2N-\overset{\overset{\displaystyle COOH}{|}}{\underset{\underset{\displaystyle R}{|}}{C}}-H \text{)}$$

Amino group, the group of atoms $-NH_2$.

Amorphous, arranged in a disordered fashion, with no specific shape.

AMP, (see Adenosine monophosphate)

Amphoteric, having both acidic and basic properties.

Amylopectin, the highly branched polymer of α-glucose found in starch.

Amylose, the linear polymer of α-glucose found in starch.

Anabolic reactions, reactions in which the cell uses energy to produce molecules needed for growth and repair of the cell.

Anaerobic, not requiring oxygen.

Analgesic, a drug that reduces pain without causing the loss of consciousness.

Anemia, a group of diseases that results in a low number of red blood cells. Examples are hemolytic anemia, iron deficiency anemia, and pernicious anemia.

Aneurysm, a ballooning of an artery caused by atherosclerosis.

Angina pectoris, a sharp, intense pain in the chest caused by reduced blood flow to the heart.

Anion, a negatively charged ion.

Antibiotic, a drug that is extracted from microorganisms and that acts as an antimetabolite.

Anticodon, the three-base sequence on *t*RNA that is complementary to the codon on *m*RNA.

Antimetabolite, any compound that inhibits enzyme activity.

Antioxidant, a substance added to food to prevent the oxidation of unsaturated fats or oils present in the food.

Antipyretic, a drug that reduces or prevents fever.

Apoenzyme, the protein portion of an enzyme molecule.

Aqueous solution, a solution containing water as the solvent.

Aromatic hydrocarbons, the class of hydrocarbons containing benzene and its derivatives.

Arteriosclerosis, the disease of the arteries commonly known as hardening of the arteries.

Atherosclerosis, the most common form of arteriosclerosis, in which the inner layer of the arterial wall becomes thickened by lipid deposits.

Atmosphere (atm), the unit of pressure equal to the force per unit area that will support a column of mercury 760 mm high.

Atom, the smallest unit of an element having the properties of that element.

Atomic mass unit (amu), an arbitrary unit of measure established to allow comparison of the relative masses of the elements.

Atomic number, for each element, the number of protons in the nucleus of any atom of that element.

Atomic weight, the weighted average of the masses of the naturally occurring isotopes of an element, expressed in atomic mass units.

ATP, (see Adenosine triphosphate)

Avogadro's number, the number of particles in one mole: equals 6.02×10^{23}.

Background radiation, ionizing radiation that comes from natural sources.

Basal metabolism rate (BMR), the minimum amount of energy required daily by the body to maintain the basic processes of life.

Base (Brønsted-Lowry definition), a substance that accepts hydrogen ions (protons): a proton-acceptor.

Bends, a painful disorder caused by the formation of nitrogen bubbles in the tissues of deep-sea divers who are brought to the ocean's surface too quickly.

Benedict's test, the test widely used for the detection of a reducing sugar in the urine.

Beriberi, deficiency disease resulting from a lack of thiamine, vitamin B_1.

Beta oxidation, (see Fatty acid cycle)

Beta radiation, ionizing radiation consisting of streams of high-energy electrons (symbol: $_{-1}^{0}e$).

Bile, a fluid (containing bile salts, bile pigments, and cholesterol) that is produced in the liver, stored in the gall bladder, and released in the small intestine.

Binary compound, a compound containing only two elements.

Blood sugar, term often used for glucose.

Blood sugar level, the concentration of glucose in the blood (usually expressed in mg per 100 ml of blood).

Boiling point, the temperature at which a substance boils (i.e., changes state from liquid to gas). The normal boiling point of a substance is the temperature at which a substance boils when atmospheric pressure equals 760 mm Hg.

Boyle's law, this law states that the volume of a gas is inversely proportional to the pressure when the temperature remains constant.

Brachytherapy, a procedure in which "seeds" containing radioisotopes are inserted by means of a needle into the area of tissue that requires treatment.

Breeder reactor, a type of nuclear reactor in which more nuclear fuel is produced than is used up in the process of producing energy.

Brownian movement, the erratic, random movement of particles in a colloid.

Buffer, any substance that, when added to a solution, protects against sudden changes in the pH of that solution.

Burner reactor, a nuclear reactor in which the nuclear fuel is used up as energy is produced.

Calcitonin, a hormone produced by the thyroid gland which lowers the level of calcium in the blood.

Calorie (cal), the amount of energy necessary to raise the temperature of one gram of water exactly one degree Celsius.

Calorimeter, the instrument used to determine the caloric content of a substance.

Carbohydrate, any of a class of compounds containing polyhydric aldehydes, polyhydric ketones, or their polymers, whose function is energy storage or structural support in living organisms.

Carbonyl group, the group containing a carbon double-bonded to an oxygen

$$\left(-\overset{\displaystyle O}{\underset{\displaystyle \|}{C}}- \right).$$

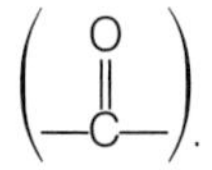

Carboxylic acid, an organic acid containing the carboxyl group

$$\left(-\overset{\displaystyle O}{\underset{\displaystyle \|}{C}}-OH \right).$$

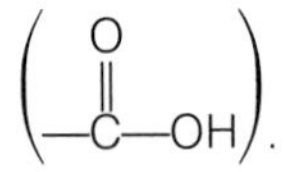

Carcinogen, any chemical that can cause cancer in animals.

Catabolic reaction, a cellular reaction in which large molecules are broken down to produce smaller molecules and cellular energy.

Catalyst, a substance that increases the rate of a chemical reaction without being consumed in the reaction.

Cation, a positively charged ion.

Cellular respiration, the series of reactions by which glucose is oxidized to form CO_2, H_2O, and ATP.

Cellulose, a linear polymer of β-glucose that is the main structural molecule in plants.

Celsius (°C), the temperature scale in the metric system with 100 degrees between the freezing point of water (set at 0°C) and the boiling point of water (100°C).

Charles' law, this law states that the volume of a gas is inversely proportional to the temperature (in degrees Kelvin) if the pressure remains constant.

Chemical change, a change that involves a basic change in the nature of the substances involved.

Chemical formula, a shorthand way of representing the composition of a substance by using the chemical symbols of the elements and subscripts to indicate the number of atoms of each element in a reacting unit of the substance.

Chemical symbol, a one- or two-letter abbreviation representing one atom of an element.

Chiral carbon, a carbon that has four different groups attached to it and that, therefore, is asymmetric.

Chlorophyll, the green pigment found in plant cells that is involved in the light reactions of photosynthesis.

Chloroplast, a site of the light reactions of photosynthesis in a plant cell.

Cholesterol, a sterol that is produced by all cells and is found in cell membranes. It is used to synthesize bile salts and some hormones.

Cirrhosis, a chronic disease of the liver caused by nutritional deficiency, poisons, or previous infections.

***Cis-trans* isomerism,** (see Geometric isomerism)

Citric acid cycle, a cyclic series of reactions (occurring in the mitochondria of cells) that convert acetyl CoA into two molecules of CO_2, one ATP, and eight hydrogens that enter into the electron transport chain.

Codon, the three-base sequence on *m*RNA that codes for a specific amino acid.

Coenzyme, a cofactor that is a complex organic molecule other than a protein.

Cofactor, a metal ion or organic molecule required for an enzyme to function properly.

Colloid, a homogeneous mixture containing relatively large particles (from 1 to 1000 millimicrons) that do not settle out.

Competitive inhibition, inhibition of an enzyme that occurs when another molecule competes with the substrate for the active site.

Compound, a substance formed in a chemical change and composed of two or more elements that combine in a definite proportion by weight.

Compound lipid, a saponifiable lipid that when hydrolyzed yields fatty acids, an alcohol, and some other compound.

Condensation reaction, in organic chemistry, a reaction in which a water molecule is removed from two reactant molecules, thereby forming one product molecule.

Conjugate acid, the substance formed when a base accepts a hydrogen ion.

Conjugate base, the substance formed when an acid donates a hydrogen ion.

Conjugated double bonds, an arrangement of double and single bonds alternating between carbon atoms in a hydrocarbon molecule.

Coordinate covalent bond, a covalent bond in which one atom donates both of the electrons shared in the bond.

Covalence number, the number of electrons that an atom of an element will share in a covalent bond.

Covalent bond, the type of bond formed when electrons are shared between two atomic nuclei.

Crenation, the shrinking of red blood cells when they are placed in a hypertonic solution.

Critical mass, the minimum amount of a fissionable isotope that must be present for a nuclear chain reaction to occur.

Crystalline, arranged in a highly ordered pattern.

Crystalloid, substance containing particles that are small in size (less than 1 millimicron), and that form a true solution when placed in water.

Curie, the unit of measure that describes the activity of a radioactive source. ($1 \text{ curie} = 3.7 \times 10^{10}$ disintegrations/sec)

Cyclic AMP, a chemical messenger within cells, whose formation is triggered when a hormone attaches to a receptor site on the cell membrane.

Dalton's law, this law states that the total pressure of a mixture of gases is equal to the sum of the partial pressures of each gas in the mixture.

Decay series, a series of radioactive decays or disintegrations, by which an unstable nucleus becomes a stable nucleus.

Dehydration, *(a)* a medical condition resulting from excessive water loss. *(b)* In organic chemistry, a reaction involving the removal of a molecule of water from another molecule.

Dehydrogenation, in organic chemistry, a reaction in which two hydrogen atoms are removed from a molecule.

Denaturation, a reversible or irreversible disruption of the normal arrangement of atoms (the native state) of a protein.

Density, the mass of one unit of volume of a substance (commonly expressed in grams per cubic centimeter).

Deoxyribonucleic acid (DNA), the nucleic acid that is a polymer of deoxyribonucleotides, and whose base sequence carries genetic information.

Dextrin, a short-chain polymer of glucose produced by the partial hydrolysis of starch.

Diabetes insipidus, a disease in which the body no longer produces the hormone vasopressin that controls the amount of water reabsorbed by the kidneys, resulting in the excretion of large quantities of urine.

Diabetes mellitus, a disease resulting from low levels or a total lack of the hormone insulin; commonly known as "diabetes."

Dialysis, the movement of ions and small molecules (but not colloidal particles) through a membrane.

Diatomic, containing two atoms.

Digestion, the process by which food is broken down into simple molecules that can be absorbed through the lining of the intestinal tract.

Dipeptide, a molecule that can be hydrolyzed to form two amino acids.

Dipolar ion, an ion, such as an amino acid, that has a positive and a negative region.

Disaccharide, a compound composed of two monosaccharides.

Disintegration series (see Decay series)

Disulfide, any organic compound containing the group (—S—S—).

DNA (see Deoxyribonucleic acid)

Double bond, a covalent bond in which two pairs of electrons (four electrons) are shared by two atomic nuclei.

Double helix, the structure of a DNA molecule—two helical polynucleotide chains coiled around the same axis.

Edema, a swelling of the tissues caused by an increase in the amount of water in the extracellular fluid.

Electrolyte, any substance that, in water solution, conducts electricity.

Electromagnetic spectrum, the entire range of radiant energy, of which visible light is only a small part.

Electron, a subatomic particle that is found in certain regions (called orbitals) around the nucleus. It has a mass of 1/1837 amu and a negative (−1) charge.

Electron affinity, the amount of energy released when an electron is added to a neutral atom.

Electron configuration, the most stable arrangement of electrons in probability regions (orbitals) around the nucleus of an atom.

Electronegativity, a measure of the ability of an atom to attract toward itself the electrons that it shares in a covalent bond.

Electron transport chain, the series of oxidation-reduction reactions that is linked to the citric acid cycle and that produces the majority of ATP's in the oxidation of molecules within the cell.

Electrovalence, the kind of charge (+ or −) and the amount of charge found on ions of an element in an ionic bond.

Element, a pure substance that cannot be broken down by ordinary chemical processes.

Emphysema, a disease in which the lung tissue is so badly damaged that adequate levels of oxygen cannot be maintained in the blood resulting in very labored breathing.

Endocrine gland, any of a specialized group of glands in the body that produce hormones and regulate their secretion into the blood.

Endothermic reaction, a reaction in which energy in the form of heat is required to keep the reaction occurring.

Energy, the capacity to do work.

Energy level, an energy region around the nucleus occupied by electrons.

Entropy, a measure of the randomness or disorder in a system.

Enzyme, a protein molecule that functions as a biological catalyst.

Epinephrine, an adrenal hormone more commonly known as adrenalin.

Equilibrium, a dynamic state in which the rate of the forward reaction equals the rate of the reverse reaction.

Equivalence point, the pH at which all the hydrogen (or hydroxide) ions in a solution have been neutralized.

Equivalent (Eq), one mole of charge (either + or −).

Erythrocyte, a red blood cell.

Essential, a term referring to eight amino acids and three fatty acids that cannot be synthesized by the body, and so must be present in the diet for normal growth and development.

Ester, any organic compound formed in a condensation reaction between an alcohol and an organic acid, and containing the functional group

$$\left(\begin{array}{c} \overset{\displaystyle O}{\underset{\displaystyle |}{\overset{\displaystyle \|}{-C}-O-C-}} \\ | \end{array}\right)$$

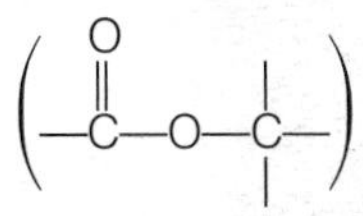

Ether, any organic compound containing an oxygen bonded to two carbons $\left(\begin{array}{c} | \quad | \\ -C-O-C- \\ | \quad | \end{array}\right)$.

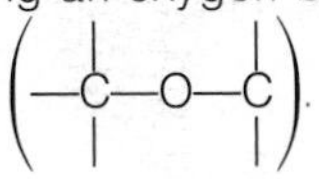

Evaporation, the conversion of a substance in the liquid state to the gaseous state.

Excited atom, an atom having one or more electrons in an energy level higher than normal.

Exothermic reaction, a reaction in which energy in the form of heat is produced.

Extracellular fluids, any fluids found in the body tissues, but not contained inside the cells.

Fahrenheit (°F), temperature scale in the English system with 180 degrees between the freezing point of water (set at 32°F) and the boiling point of water (212°F).

Family (see Group).

Fat, a triglyceride that is a solid at room temperature and that contains mainly saturated fatty acids.

Fatty acid, an organic compound containing one carboxylic acid group and usually having a long carbon chain. Fatty acids are usually produced by the hydrolysis of fats and oils.

Fatty acid cycle, a repeating series of reactions in which a fatty acid is oxidized in two-carbon units.

Fermentation, the process by which yeast cells convert glucose to ethanol, carbon dioxide, and energy.

Fertile isotope, an isotope that can be converted into a fissionable isotope.

Fibrosis, the abnormal formation of fibrous tissue.

First law of thermodynamics, this law states that energy can neither be created nor destroyed, but only changed in form.

Fission, the process by which a large unstable nucleus, when bombarded by neutrons, breaks apart to form two smaller nuclei, several neutrons, and a tremendous amount of energy.

Fissionable isotope, an isotope whose nucleus will break apart when bombarded by neutrons.

Fluorescence, a type of luminescence in which a substance stops giving off light as soon as the external source of energy is removed.

Fluorosis, an enlargement of bones and abnormal bone growth caused by a high concentration of fluoride ions in the diet.

Force, a push or pull on an object that causes the object to start moving or to change its speed or direction once it is moving.
Formula weight, the sum of the atomic weights of all the atoms appearing in the chemical formula of a substance.
Free radical, a highly reactive uncharged particle.
Functional group, a group of atoms that gives characteristic chemical properties to all molecules containing that group.
Fusion, the process by which several small nuclei combine to form a larger, more stable nucleus and tremendous amounts of energy.

Gamma radiation, a naturally occurring high-energy electromagnetic radiation, similar to X rays, with high penetrating power.
Gene, a sequence of bases on a DNA molecule that codes for a specific protein.
Genetic code, the sequence of bases on a DNA molecule.
Geometric isomerism, isomerism that results from the different geometric arrangements of atoms around a double bond or around a ring.
Glucose tolerance test, a series of tests for blood sugar level taken after the ingestion of a high dose of glucose; often used to diagnose diabetes mellitus.
Glycogen, the highly branched polymer of α-glucose that is the glucose storage molecule in animals.
Glycogenesis, the series of reactions by which glucose molecules are joined together to form glycogen.
Glycogenolysis, the series of reactions by which glycogen molecules are hydrolyzed or broken down to form glucose molecules.
Glycolipid, a compound lipid that contains fatty acids, an alcohol, and a sugar group that is either galactose or glucose.
Glycolysis, the series of reactions by which glucose is converted to two molecules of lactic acid and two molecules of ATP; occurs in the cellular cytoplasm.
Glycosuria, a medical condition in which sugar molecules are found in the urine.
Goiter, a swelling of the thyroid gland caused by a deficiency of iodine in the diet.
Graham's law, this law states that lighter gases will diffuse more rapidly than heavier gases.
Gram, a unit of mass; 1 gram = 0.035 oz.
Gram-equivalent weight, the weight of a substance, in grams, that contains one equivalent.
Ground state, term applied to an atom having all of its electrons in the lowest possible energy level.
Group, a vertical column of elements on the periodic table.

Half-life, the length of time required for one-half of the atoms of a radioactive element in a given sample to undergo radioactive decay.
Halogen, any of the elements in group VII: fluorine, chlorine, bromine, iodine, or astatine.
Heat of fusion, the amount of energy (in calories) required to change one gram of a substance (at the melting point) from a solid to a liquid.
Heat of reaction, the amount of heat energy (in calories/mole) absorbed or released in a chemical reaction.
Heat of vaporization, the amount of energy (in calories) required to change one gram of a substance (at the boiling point) from a liquid to a gas.
α-Helix (alpha-helix), a secondary structure of a protein in which the polypeptide chain is coiled into a helix held together by hydrogen bonds.
Hemodialysis, the process of removing wastes from the blood by dialysis.
Hemoglobin, the oxygen-carrying molecule in red blood cells. It consists of four polypeptide chains

and four nonprotein heme groups, each of which can carry one molecule of oxygen.

Hemolysis, the rupturing of red blood cells that results when the cells are placed in a hypotonic solution (or from other causes).

Henry's law, this law states that the higher the pressure, the greater the solubility of a gas in a liquid.

Heterocyclic compound, a compound having a ring structure that contains two or more different types of atoms making up the ring.

Heterogeneous, nonuniform.

High-energy phosphate bond, the phosphorus-to-oxygen bond, found in molecules such as ATP and ADP, which, upon hydrolysis, releases large amounts of energy.

Homogeneous, uniform throughout.

Hormone, any of the chemical messengers produced by the endocrine glands.

Hydrated, surrounded by water molecules.

Hydration reaction, in organic chemistry, the addition of a water molecule to an unsaturated bond.

Hydrocarbon, an organic compound containing only atoms of carbon and hydrogen.

Hydrogenation, in organic chemistry, the addition of a hydrogen molecule to an unsaturated bond.

Hydrogen bond, a weak force of attraction between a partially positive hydrogen and a partially negative atom such as oxygen, fluorine, or nitrogen on another molecule or on another region of the same molecule.

Hydrogen ion, a proton. This term is also often used in acid-base chemistry to mean the hydronium ion (H_3O^+).

Hydrohalogenation, the addition of a hydrohalogen (such as HCl or HBr) to an unsaturated bond.

Hydrolysis, the addition of a water molecule to a reactant, thereby breaking the reactant into two product molecules.

Hydrometer, an instrument used to measure the specific gravity of a liquid.

Hydronium ion, H_3O^+, the ion formed when a hydrogen ion (a proton) joins to a water molecule.

Hydrophilic, water-attracting; term given to substances or groups of atoms that are generally very polar or ionic.

Hydrophobic, water-repelling; term given to substances or groups of atoms that are nonpolar.

Hyper, prefix used to indicate "higher than normal." For example, hypertension, hypercholesteremia, hypertonic, etc.

Hyperglycemia, a condition resulting from higher than normal blood glucose levels.

Hyperthyroid, a condition in which the thyroid gland produces higher than normal amounts of the hormone thyroxin.

Hypertension, high blood pressure.

Hypertonic solution, a solution with a higher solute concentration than the standard solution.

Hypo, prefix used to indicate "lower than normal." For example, hypoglycemia, hypotonic, hypothyroid, etc.

Hypoglycemia, a condition resulting from lower than normal blood glucose levels.

Hypothermia, a lowering of the body's interior temperature.

Hypotonic solution, a solution with a lower solute concentration than the standard solution.

Imine, an organic compound containing a carbon doubly bonded to a nitrogen $\left(-\overset{|}{C}{=}N{-}H\right)$.

Indicator, a chemical dye that changes color at a specific hydrogen ion concentration.

Induced-fit theory, theory of enzyme action stating that the active site of some enzymes is induced by the substrate to fit the shape of the substrate molecule.

Inducer, in protein synthesis, a

molecule that binds to the repressor molecule, causing it to detach from the operator gene and thus allowing the synthesis of *m*RNA.

Inhibition, the prevention of, or interference with, the action of an enzyme, thus lowering its activity. May be reversible or irreversible.

Insulin, the hormone, produced by beta cells in the pancreas, that controls blood glucose levels by increasing the absorption of glucose from the blood and the rate of glycogenesis.

Insulin shock, convulsion and coma resulting from an overproduction or overinjection of insulin, which causes the blood glucose level to decrease very fast.

Intracellular fluid, fluid found within cells.

Intravenous, administered by means of a vein.

Inverse square law, this law states that the intensity of radiation on a given surface area decreases by the square of the distance from the source.

Iodine number, indicates the amount of unsaturation in a compound: the higher the iodine number, the more unsaturated the compound.

Ion, a positively or negatively charged particle.

Ionic bond, the attraction between ions formed when one or more electrons are transferred from one atom to another.

Ionic compound, a compound consisting of an orderly arrangement of oppositely charged ions, which are combined in a ratio such that the compound is electrically neutral.

Ionization energy, the amount of energy that must be added to an atom to remove one electron from its outermost energy level.

Ionizing radiation, radiation, such as alpha, beta, or gamma radiation, that can produce unstable and highly reactive ions in living tissue.

Ion product constant of water (K_w), $K_w = [\text{H}^+][\text{OH}^-] = 1 \times 10^{-14}$.

Isoelectric point, the pH at which an amino acid or protein is electrically neutral and will not migrate in an electric field.

Isomers, compounds having the same molecular formula, but different structures.

Isotonic solution, a solution with a solute concentration equal to the standard solution.

Isotopes, atoms of the same element that differ in the number of neutrons in their nuclei.

Jaundice, condition caused by a high level of bilirubin in the blood, which results from a blockage of the bile duct or malfunction of the liver.

Kelvin (K), the temperature scale in the SI system, with 100 degrees between the freezing point of water (273.15 K) and the boiling point of water (373.15 K). 0 K corresponds to absolute zero, $-273.15°C$.

Keratosis, skin condition characterized by thick warty growths.

Ketone, an organic compound containing a carbonyl group bonded to two other carbons $\left(-\overset{|}{\underset{|}{C}}-\overset{O}{\overset{\|}{C}}-\overset{|}{\underset{|}{C}}-\right)$.

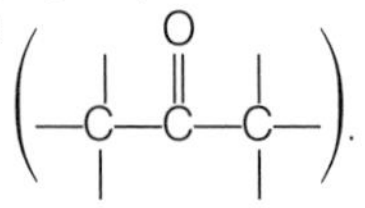

Ketone bodies, compounds produced from the metabolism of excess acetyl CoA (which, in turn, is produced when large amounts of fats are oxidized to supply cellular energy).

Ketose, general term for a monosaccharide containing a ketone group.

Kilocalorie (Kcal), the amount of heat energy necessary to raise the temperature of 1000 grams of water one degree Celsius.

Kinetic energy, energy of motion.

Kinetic-molecular theory, the theory that explains the behavior of a gas in terms of the motion of its particles.

Kwashiorkor, a disease that results

from a lack of essential amino acids in the diet.

Law of definite proportions, this law states that a compound is composed of specific elements in a definite proportion by weight.

$LD_{50/30}$, abbreviation for the dose of radiation sufficient to kill 50 percent of the exposed population within 30 days.

Le Châtelier's principle, this principle states that when a stress is applied to an equilibrium system, the system will change in a way to remove the stress.

LET (linear energy transfer), the amount of energy transferred to a tissue, per unit of path length traveled by ionizing radiation.

Lipid, any of a large class of nonpolar, organic compounds that have oily or waxy properties.

Liter (litre), a unit of volume equal to 1000 cubic centimeters; 1 liter = 1.06 quarts.

Lock-and-key theory, the theory of enzyme action stating that only a specific substrate molecular structure will fit the active site of an enzyme, just as only a specific key will fit in a lock.

Luminescence, the release of energy as visible light by an excited atom.

Macromineral, one of seven elements (K, Mg, Na, Ca, P, S, Cl) that are required in small amounts for normal cell growth and development.

Malignant, term describing cells that are growing and dividing in an uncontrolled fashion.

Mammography, a technique, employing regular X-ray equipment but using a printing process similar to a photocopy, used for the early detection of breast cancer.

Mass, a measure of the resistance of an object to a change in speed or direction.

Mass number, the sum of the number of protons and neutrons in the nucleus of an atom.

Matter, anything that has mass and occupies space.

Melanin, a brown skin pigment produced by the body to protect against the effects of ultraviolet radiation.

Melting point, the temperature at which a solid breaks down to form a liquid.

Messenger RNA (*m*RNA), the RNA that carries the genetic code for a specific protein. *m*RNA is synthesized in the nucleus, and then migrates to the cytoplasm where it attaches to ribosomes and serves as the template for protein synthesis.

Metabolism, all of the enzyme-catalyzed reactions in the body.

Metal, an element that is shiny, dense, and easily worked, that has a high melting point, and that conducts electricity.

Metalloid, an element that acts like a metal in some ways, and like a nonmetal in other ways. These elements are found between the metals and the nonmetals on the periodic chart.

Metastable, in an energy state higher than normal.

Metastasis, the spread of cancer cells from a tumor to other parts of the body.

Meter (metre), unit of length in the metric system. 1 meter = 1.09 yards.

Metric system, a system of measure based on the decimal system.

Milliequivalent (mEq), unit of measure used to express the concentration of ions in the blood; 1000 mEq = 1 Eq.

Millimeters of mercury (mm Hg), unit of pressure equal to 1/760 atmosphere.

Mitochondria, structures in the cell where the citric acid cycle and the electron transport chain occur.

Mixture, two or more substances combined in any proportion.

Molarity (*M*), unit of solution concentration defined as the number of moles of solute per liter of solution.

Mole, the amount of a substance that has the same number of particles as there are atoms in 12 grams of carbon-12 (6.02×10^{23} atoms).

Molecule, an electrically neutral unit formed when two or more atoms are joined together by covalent bonds.

Monomer, a single unit that joins with many other identical units to form a polymer.

Monosaccharide, a carbohydrate that cannot be broken into smaller units by hydrolysis.

Mutagen, any chemical or physical agent that is capable of producing a mutation.

Mutant, containing altered DNA.

Mutation, a chemical or physical change in a DNA molecule that results in the synthesis of a protein with an altered amino acid sequence.

Myelin sheath, the protective coating surrounding nerves.

Native state (Native configuration), the shape of a protein that is energetically the most stable.

Net-ionic equation, a chemical equation showing only the ions that take part in the reaction.

Neutral, term applied to a solution that has neither acidic nor basic properties, or to a particle that has no net electric charge.

Neutralization, the process by which an acidic or basic solution is converted to a neutral solution.

Neutron, a subatomic nuclear particle with a mass of 1 amu and no charge.

Nitrile, an organic compound containing a carbon triply bonded to a nitrogen and singly bonded to another carbon

$$\left(\mathrm{-\overset{\displaystyle |}{\underset{\displaystyle |}{C}}-C\equiv N} \right)$$

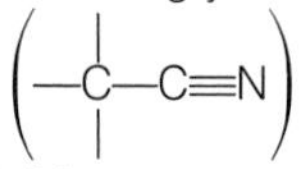

Noble gas, any of the elements in group 0 (helium, neon, argon. krypton, xenon, or radon), all of which have great chemical stability.

Nonelectrolyte, a substance that does not conduct electricity when placed in solution.

Nonmetal, an element that is brittle, has low density and a low melting point, and does not conduct electricity.

Nonpolar, term applied to covalent bonds and covalent molecules when the centers of positive and negative charge coincide.

Nonsaponifiable lipid, any lipid that cannot be hydrolyzed by an aqueous solution of base.

Nuclear transmutation, a reaction in which a high-speed nuclear particle collides with a nucleus to produce a different nucleus.

Nucleic acid, a polymer of nucleotides: either DNA (deoxyribonucleic acid) or RNA (ribonucleic acid).

Nucleotide, the monomer unit of nucleic acids, whose structure contains a five-carbon sugar, a nitrogen-containing base, and a phosphate group.

Nucleus, the dense center of an atom, containing protons and neutrons.

Octet rule, the tendency of elements in groups I to VII to form bonds that result in eight valence electrons in the outer energy level of each atom (except in the first energy level, where the tendency is toward two electrons).

Oil, a triglyceride, extracted from vegetable seeds or fruits, that is a liquid at room temperature and that contains mainly unsaturated fatty acids.

Operator gene, a gene that controls the synthesis of *m*RNA on a specific segment of DNA.

Operon, a group of genes that codes for the enzymes in a multienzyme system.

Orbital, a region around the nucleus of an atom in which there is a high probability of finding one or two electrons. The four kinds of orbitals are called *s*-, *p*-, *d*-, and *f*-orbitals.

Organic chemistry, the study of carbon compounds.

Osmol, the amount of osmotic pressure created by 1 mole of any type of particle in 1 liter of solution.

Osmolarity, a unit of concentration that describes the total number of particles in solution; expressed in osmols/liter.

Osmosis, the movement of water molecules through a differentially permeable membrane from a region of lower solute concentration to a region of higher solute concentration.

Osmotic pressure, the amount of pressure that would have to be applied to a solution to prevent osmosis if the solution were separated from pure water by a differentially permeable membrane.

Oxidation, the loss of one or more electrons by an atom, ion, or molecule. In organic chemistry, the loss of hydrogen or the gain of oxygen by an organic molecule or ion.

Oxidative deamination, the removal of an amino group from an amino acid, producing an α-keto acid and ammonia.

Oxidizing agent, a substance that causes the oxidation of a reactant molecule.

Partial pressure, the pressure exerted by a specified gas in a mixture of gases.

Parts per billion (ppb), unit of concentration equivalent to 1 microgram of solute per liter of solution.

Parts per million (ppm), unit of concentration equivalent to 1 milligram of solute per liter of solution.

Pellegra, a deficiency disease resulting from a lack of the vitamin niacin.

Peptide bond, an amide linkage formed by a condensation reaction between two amino acids.

Period, a horizontal row on the periodic table.

Periodicity, the repeating nature of chemical properties of the elements when they are arranged in order of atomic number.

pH, unit of measure of the hydrogen ion concentration of an aqueous solution; $[H^+] = 1 \times 10^{-pH}$.

Phenylalanine, an amino acid that accumulates in the body of a child with PKU.

Phenylketonuria, PKU, an inherited disease in which an enzyme responsible for the conversion of phenylalanine to tyrosine is defective. If untreated, PKU results in permanent mental retardation.

Phospholipid, a compound lipid whose structure contains an alcohol, fatty acids, and a phosphate group. Phospholipids form the structure of cell membranes, and are important for the transport of lipids in the blood.

Phosphorescence, a type of luminescence in which the substance continues to give off light for a short period of time after the external source of energy is removed.

Photosynthesis, the process by which green plants use sunlight as the source of energy to produce glucose and oxygen from water and carbon dioxide.

Physical change, a transformation during which a substance changes form, but keeps its chemical identity.

Plaque, in dentistry, the sticky substance produced by bacteria in the mouth that adheres to the teeth. In cardiology, a deposit of smooth muscle cells, fats, and scar tissue on the interior of an arterial wall.

Pleated sheet (β-configuration), a secondary structure of a protein, in which polypeptide chains lie next to one another, held together by hydrogen bonds and having the R-groups extending above and below the sheet.

Polar, term applied to covalent bonds and covalent molecules when the center of positive charge and the center of negative charge do not coincide, thus forming an electric dipole.

Polyatomic ion, an electrically charged group of covalently bonded atoms that stays together as a unit in most chemical reactions.

Polycythemia, the excessive formation of red blood cells; called *polycythemia vera* when the increase is caused by a tumor.

Polyester, a polymer of ester molecules, used to make fibers for fabrics.

Polyhydric, containing more than one hydroxyl group.

Polymer, a very large molecule made up of repeating units called monomers.

Polymerization, a chemical reaction in which single molecules called monomers react with each other to form large molecules called polymers.

Polypeptide, a polymer composed of amino acids connected by peptide bonds. A protein may be composed of one or more polypeptide chains.

Polysaccharide, a polymer of three or more monosaccharide molecules.

Polysome, a group of ribosomes all synthesizing protein on the same molecule of *m*RNA.

Polyunsaturated, term describing a triglyceride whose molecules have two or more double bonds.

Potential energy, energy of position.

Pressure, a force exerted per unit of area.

Primary structure, the sequence of amino acids (connected by peptide bonds) in the polypeptide chain of a protein.

Product, a substance that results from a chemical reaction.

Prostaglandin, any of a class of 20-carbon fatty acids that are derived from prostanoic acid and that have a wide variety of potent physiological effects.

Prosthetic group, a cofactor that is tightly bound to an apoenzyme.

Protein, a polymer of amino acids. Proteins perform a wide variety of critical functions in the cell.

Proton, a subatomic nuclear particle having a mass of 1 amu and a positive (+1) charge.

Ptomaine, any amine produced in the natural decay of living organisms.

Pulmonary, having to do with the lungs.

Purines, heterocyclic amines whose derivatives (adenine and guanine) are essential parts of DNA and RNA molecules.

Pyrimidines, heterocyclic amines whose derivatives (cytosine, thymine, and uracil) are essential parts of DNA and RNA molecules.

Quaternary structure, the overall structure of a protein that contains more than one polypeptide chain.

Rad, unit of radiation dosage used to describe the amount of energy absorbed by the irradiated tissue.

Radioactive decay, the process by which an unstable nucleus gives off nuclear particles and/or gamma radiation to become more stable.

Radioactive tracer, a chemical that contains radioactive atoms, but that has the same chemical nature and behavior as naturally occurring compounds; used to follow metabolic pathways.

Radioactivity, the giving off, or emission, of radiation from certain isotopes.

Radiopharmaceutical therapy, the administration of radioisotopes in chemical forms designed to be concentrated in certain regions of the body or in cancerous tissue.

Rancid, term applied to foods containing fats and oils that have undergone hydrolysis or oxidation, forming substances that give the food a bad smell or taste.

Reactant, a starting substance in a chemical reaction.

Recommended daily allowance (RDA), the amount of a nutrient needed in the diet to meet the daily requirements of a healthy individual (as established by the Food and Nutrition Board of the National Academy of Science).

Redox reaction, abbreviation for *reduction-oxidation* reaction, a reaction in which electrons are transferred from one reactant to another.

Reducing agent, a substance that causes the reduction of a reactant molecule.

Reducing sugar, any carbohydrate that

can act as a reducing agent and produce a positive Benedict's test.

Reduction, the gaining of electrons by a reagent. In organic chemistry, reduction occurs when an organic molecule or ion gains hydrogen atoms or loses oxygen atoms.

Rem, the unit of absorbed dose of radiation that will produce the same biological effect as one rad of therapeutic X rays.

Renal threshold, the concentration of glucose in the blood above which glucose begins to appear in the urine.

Replicate, to make an exact copy; in cell division the DNA molecules replicate, producing two daughter cells with identical DNA.

Representative element, any element in groups I-VII and group 0 on the periodic table.

Repressor, in protein synthesis, the protein molecule that binds to the operator gene preventing the synthesis of mRNA.

Retinopathy, a form of blindness caused by a chemical imbalance that results from excess glucose in the blood.

Ribonucleic acid (RNA), a group of nucleic acids, polymers of ribonucleotides, synthesized on the DNA strand and having different cellular functions.

Ribosomal RNA (rRNA), the RNA which, with proteins, forms granules called ribosomes in the cytoplasm.

Ribosomes, granules in the cytoplasm made up of two subunits formed by rRNA and protein that combine with mRNA to synthesize polypeptides.

Rickets, a disease that results from too little vitamin D in the diet, and that causes bones to soften and bend out of shape.

RNA (see Ribonucleic acid).

Salt bridge, a force of attraction that occurs between the charged R-groups

$$\left(\begin{array}{c} \overset{O}{\underset{\|}{}} \\ -CH_2CO^- \text{ and } -(CH_2)_4NH_3^+ \end{array}\right)$$

on the polypeptide chains of a

protein; similar to the attraction between ions in an ionic crystal.

Saponifiable lipid, a lipid that can be hydrolyzed in an aqueous basic solution.

Saponification, the hydrolysis of an ester in an aqueous solution of strong base.

Saturated compound, any hydrocarbon or its derivative that contains only carbon-to-carbon single bonds.

Saturated solution, a solution that contains as many solute particles as can dissolve in the solvent at that temperature.

Scurvy, a deficiency disease caused by a lack of vitamin C, ascorbic acid.

Secondary structure, the shape of a protein molecule (or several protein molecules) that results from bonding forces other than the peptide bond — for example, hydrogen bonding. Two examples of secondary structure are the α-helix and the pleated sheet.

Second law of thermodynamics, this law states that the entropy or disorder of the universe is increasing.

Sickle cell anemia, anemia caused by an inherited defect in the hemoglobin molecule.

Simple lipid, a saponifiable lipid that, when hydrolyzed, yields fatty acids and an alcohol.

SI units, abbreviation for the International System of Units, a system of weights and measures that is the successor to the metric system. Reference units are the kilogram (mass), the meter (length), the degree Kelvin (temperature), the second (time), the ampere (electric current), and the candela (luminous intensity). Various other units are formed by multiplying or dividing a reference unit by 10.

Single bond, a chemical bond in which one pair of electrons (two electrons) are shared by two atomic nuclei.

Soap, a salt of a fatty acid, produced by the saponification of triglycerides.

Solubility, the amount of a solute that will dissolve in a fixed volume or weight of a solvent.

Solute, the substance being dissolved in a solution.

Solution, a homogeneous mixture of two or more substances.

Solvent, the substance in a solution in which the solute is being dissolved.

Specific gravity, a comparison of the mass of a liquid with the mass of the same volume of pure water.

Specific heat, the amount of heat energy required to raise the temperature of one gram of a substance from 15 to 16°C.

Sphingolipid, a phospholipid containing the alcohol sphingosine. Sphingolipids are found in high concentrations in brain and nerve tissue.

Standard atmospheric pressure, the pressure that will support a column of mercury 760 mm high at a temperature of 0°C.

Starch, a mixture of amylose and amylopectin; the energy storage molecule in plants.

Steroid, any of a large class of nonsaponifiable lipids, all of which contain a complicated four-ring framework.

STP, standard temperature and pressure: 0°C and 1 atm.

Structural formula, a diagram that shows the arrangement of atoms in the molecule.

Structural gene, a gene that codes for a specific protein.

Structural isomers, isomers that differ in the sequence of atoms in their molecules.

Substitution reaction, a chemical reaction in which an atom or group of atoms is substituted for some atom on a reactant molecule.

Substrate, a reactant in an enzyme-catalyzed reaction which attaches to the surface of the enzyme.

Sulfhydryl group, the functional group —S—H.

Supersaturated, term applied to a solution that has more solute particles dissolved in it than it will hold at equilibrium at that temperature.

Surface tension, the resistance of the particles on the surface of a liquid to the expansion of that liquid.

Surfactant (surface active agent), a substance that acts to reduce the surface tension of water.

Tartar, material that forms on teeth when calcium compounds are deposited in unremoved plaque.

Teletherapy, the use of high-intensity radiation, such as gamma rays or X rays, to destroy cancerous tissue.

Temperature, a measure of the average kinetic energy of the particles of a substance.

Tertiary structure, the shape of a protein caused by the folding of the secondary structure, resulting from interactions between the R-groups such as disulfide bridges, salt bridges, and hydrophobic interactions.

Thioether group (organic sulfide), the functional group $-\overset{|}{\underset{|}{C}}-S-\overset{|}{\underset{|}{C}}-$

Thyroid, the endocrine gland, located in the neck, that produces the hormones thyroxin and calcitonin.

Thyroxin, the iodine-containing hormone produced by the thyroid.

Titration, a laboratory procedure for measuring the unknown concentration of an acidic or basic solution.

Torr, a unit of pressure equal to 1 mm Hg.

Trace element, an element required in minute amounts for normal cell growth and development.

Transamination, the transfer of an amino group from an amino acid to an α-keto acid, thus producing a new amino acid.

Transcription, the transfer of genetic information from DNA to *m*RNA; the synthesis of *m*RNA on a segment of DNA.

Transfer RNA (*t*RNA), RNA's that each carry a specific amino acid to the ribosomes and place the amino acid, by pairing the bases in the anticodon region of *t*RNA with the codon of *m*RNA, in the proper sequence for the formation of the polypeptide chain.

Translation, the expression of genetic information in the amino acid sequence of a protein; the synthesis of protein molecules on one *m*RNA molecule.

Triacylglycerol, a triglyceride.

Triglyceride, an ester of glycerol and three fatty acids; general term for fats and oils.

Triple bond, a chemical bond in which three pairs of electrons (6 electrons) are shared by two atomic nuclei.

Tyndall effect, the scattering of light by the particles in a colloid.

Turnover number, the number of substrate molecules transformed per minute by one molecule of enzyme under optimal conditions.

Tyrosine, an amino acid that is lacking in an untreated PKU child.

Unsaturated compound, any hydrocarbon or its derivative that contains one or more double or triple bonds.

Unsaturated solution, a solution in which more solute can be dissolved at that temperature.

Urea cycle (Krebs' ornithine cycle), the cyclic series of reactions in the body by which urea is produced from ammonia and carbon dioxide.

Valence electron, an electron in the outermost energy level of an atom.

Vassopressin, an antidiuretic hormone; that is, a substance that controls the release of water into the urine by the kidneys.

Viscosity, a measure of how easily a liquid flows.

Vitamin, an organic nutrient that the body cannot synthesize, but which is necessary for normal body function. Vitamins can be classified as water soluble or fat soluble.

Vitamin deficiency disease, a disease that results only from the lack of a specific vitamin in the diet.

Wax, an ester of a long-carbon-chain fatty acid and a long-carbon-chain alcohol.

Weight, a measure of the attraction of gravity on an object.

Weight/volume percent, a unit of concentration expressed as the number of grams of solute per 100 milliliters of solution.

X Ray, high-energy ionizing radiation, similar to gamma rays, produced in X-ray tubes.

Zwitterion, a dipolar ion; one that has a positively charged area and a negatively charged area. For example, an amino acid.

Zymogen (preenzyme), name given to the inactive form of an enzyme. Usually a portion of the protein molecule must be removed for the enzyme to become active.

Chapter Opening Photo Credits

Chapter 1	Fredric D. Bodin/Stock, Boston
Chapter 2	Max & Kit Hunn/Frederic Lewis
Chapter 3	Fritz Henle/Photo Researchers
Chapter 4	Stefan Bloomfield
Chapter 5	U.S. Air Force
Chapter 6	Courtesy Sheldon Baum, M.D., Director of Nuclear Medicine, The Milton S. Hershey Medical Center
Chapter 7	Argonne National Laboratory
Chapter 8	United Press International
Chapter 9	Keith R. Porter
Chapter 10	Wide World Photos
Chapter 11	Ann McQueen/Stock, Boston
Chapter 12	Ron Nelson
Chapter 13	From Alexander I. Oparin, *Chemical Origin of Life,* translated by Ann Synge, 1964. Courtesy Charles C. Thomas, Publisher.
Chapter 14	Photo by Al Kamuda, The Detroit Free Press
Chapter 15	Hugh Rogers/Monkmeyer
Chapter 16	Mimi Forsyth/Monkmeyer
Chapter 17	Culver Pictures
Chapter 18	Michal Heron/Monkmeyer
Chapter 19	Dan Budnick/Woodfin Camp
Chapter 20	World Health Organization
Chapter 21	Phyllis Lefohn
Chapter 22	From Marcel Bessis, *Corpuscles: Atlas of Red Blood Cell Shapes,* 1973. Courtesy Springer-Verlag, New York, Inc.

Index*

The page numbers in italics indicate pages containing tables.

COMMON CONVERSION FACTORS

LENGTH

1 inch (in) = 2.54 centimeters (cm)	$\left(\frac{2.54\ \text{cm}}{1\ \text{in}}\right)$	$\left(\frac{1\ \text{in}}{2.54\ \text{cm}}\right)$
1 yard (yd) = 0.91 meter (m)	$\left(\frac{0.91\ \text{m}}{1\ \text{yd}}\right)$	$\left(\frac{1\ \text{yd}}{0.91\ \text{m}}\right)$
1 mile (mi) = 1.6 kilometers (km)	$\left(\frac{1.6\ \text{km}}{1\ \text{mi}}\right)$	$\left(\frac{1\ \text{mi}}{1.6\ \text{km}}\right)$

MASS

1 ounce (oz) = 28.4 grams (g)	$\left(\frac{28.4\ \text{g}}{1\ \text{oz}}\right)$	$\left(\frac{1\ \text{oz}}{28.4\ \text{g}}\right)$
1 pound (lb) = 454 grams (g)	$\left(\frac{454\ \text{g}}{1\ \text{lb}}\right)$	$\left(\frac{1\ \text{lb}}{454\ \text{g}}\right)$
2.2 pounds (lb) = 1 kilogram (kg)	$\left(\frac{1\ \text{kg}}{2.2\ \text{lb}}\right)$	$\left(\frac{2.2\ \text{lb}}{1\ \text{kg}}\right)$

VOLUME

1 pint (pt) = 0.47 liter *(l)*	$\left(\frac{0.47\ l}{1\ \text{pt}}\right)$	$\left(\frac{1\ \text{pt}}{0.47\ l}\right)$
1.06 quart (qt) = 1 liter *(l)*	$\left(\frac{1\ l}{1.06\ \text{qt}}\right)$	$\left(\frac{1.06\ \text{qt}}{1\ l}\right)$
1 gallon (gal) = 3.79 liters *(l)*	$\left(\frac{3.79\ l}{1\ \text{gal}}\right)$	$\left(\frac{1\ \text{gal}}{3.79\ l}\right)$
1 milliliter (ml) = 1 cubic centimeter (cm^3)	$\left(\frac{1\ \text{cm}^3}{1\ \text{ml}}\right)$	$\left(\frac{1\ \text{ml}}{1\ \text{cm}^3}\right)$
1 liter *(l)* = 1000 milliliters (ml)	$\left(\frac{1000\ \text{ml}}{1\ l}\right)$	$\left(\frac{1\ l}{1000\ \text{ml}}\right)$

PRESSURE

1 (atm) = 760 mm Hg	$\left(\frac{760\ \text{mm Hg}}{1\ \text{atm}}\right)$	$\left(\frac{1\ \text{atm}}{760\ \text{mm Hg}}\right)$
1 mm Hg = 1 torr	$\left(\frac{1\ \text{torr}}{1\ \text{mm Hg}}\right)$	$\left(\frac{1\ \text{mm Hg}}{1\ \text{torr}}\right)$